Energy, Environment, and Climate

Energy, Environment, and Climate

Richard Wolfson

Middlebury College

W. W. Norton & Company
New York · London

W. W. Norton & Company has been independent since its founding in 1923, when William Warder Norton and Mary D. Herter Norton first published lectures delivered at the People's Institute, the adult education division of New York City's Cooper Union. The Nortons soon expanded their program beyond the Institute, publishing books by celebrated academics from America and abroad. By mid-century, the two major pillars of Norton's publishing program—trade books and college texts—were firmly established. In the 1950s, the Norton family transferred control of the company to its employees, and today—with a staff of four hundred and a comparable number of trade, college, and professional titles published each year—W. W. Norton & Company stands as the largest and oldest publishing house owned wholly by its employees.

This title is printed on permanent paper containing 20 percent post-consumer waste recycled fiber.

Manufacturing by: R.R. Donnelly & Sons Company
Book design by: Alice MacKenzie
Editor: Leo Wiegman
Project editor: Lory A. Frenkel
Production manager: Eric Pier-Hocking
Editorial assistants: Lisa Rand, Jennifer Cantelmi

Library of Congress Cataloging-in-Publication Data

Wolfson, Richard.
    Energy, environment, and climate / Richard Wolfson. — 1st ed.
        p. cm.
    Includes bibliographical references and index.
    ISBN 978-0-393-92763-4 (pbk.)
    1. Climatic changes. 2. Energy consumption—Climatic factors. 3. Energy consumption—Environmental aspects. 4. Power resources—Climatic factors. I. Title.
    QC981.8.C5W645 2008
    551.6—dc22
                                        2007018215

W.W. Norton & Company, Inc., 500 Fifth Avenue, New York, N.Y. 10110
        www.wwnorton.com

W.W. Norton & Company, Ltd., Castle House, 75/76 Wells Street, London W1T 3QT

1 2 3 4 5 6 7 8 9 0

BRIEF CONTENTS

CONTENTS

PREFACE

Many behaviors distinguish the human species from our fellow inhabitants of Planet Earth. Of these, our use of energy in amounts far exceeding what our own bodies can produce affects the environment in an unprecedented way. Centuries ago, pollution from coal burning was already a serious urban problem. Despite regulatory and technological progress in pollution control, diminished air and water quality continue to be major consequences of our ever-growing energy consumption. Further environmental degradation results as we scour the planet for fuels that contain the stored energy of which we demand an unending supply. Our energy-intensive society also enables other environmentally damaging developments such as sprawl, large-scale mechanized agriculture, and massive deforestation. At the same time, energy brings us higher standards of living and allows our planet to sustain a larger population.

In recent decades, a new and truly global impact of humankind's energy consumption has overshadowed the long-standing and still significant consequences associated with traditional pollution, resource extraction, and energy-enabled development. That impact is global climate change, brought about largely by the emissions from fossil fuel combustion. Climate change is a problem that knows no national or even continental boundaries. It will affect us all—although not all equally. It won't have the civilization-ending impact of an all-out nuclear war or a major asteroid hit, but climate change will greatly stress an already overcrowded, divided, and combative world.

Achieving a healthier planet with a stable, supportive climate means either using less energy or using energy in ways that minimize adverse environmental impacts. Here we have choices: To use less energy, we can either deprive ourselves of energy's benefits or we can use energy more intelligently, getting the same benefits from less energy. To minimize environmental and especially climate impacts, we can shift from fossil fuels to energy sources that don't produce as much pollution or climate-changing emissions. Or we can learn to capture the emissions from fossil fuels and sequester them away from Earth's surface environment.

Earth's energy resources are limited to relatively few naturally occurring stores of energy—the fuels—and to energy flows such as running water, sunlight, wind, geothermal heat, and tides. A realistic grasp of our energy prospects demands that we understand these energy resources. We need to know, first and foremost, if a given resource or combination of resources is sufficient to meet humankind's energy demand. For fuels, we need a good estimate of the remaining resource and a time frame over which we can expect supplies to last. We need to understand the technologies that deliver useful energy from fuels and flows, to assess their environmental impacts, and to recognize that none is without adverse effects. And we need to be realistic about the near-term and long-term prospects for different energy sources in the economic context.

The oil shortages of the 1970s spawned a serious exploration of energy alternatives. Governments and industries sponsored research programs, while tax credits encouraged the installation of alternative energy systems. Vehicle mileage and other measures of energy efficiency increased significantly. At the same time, colleges and universities developed specialized courses in energy issues and the relationship between energy and environment. These courses emerged in traditional departments such as physics, chemistry, and engineering; in interdisciplinary programs dealing with technology and society; and in the burgeoning new programs in environmental studies and environmental science that have sprung up with the emergence of a widespread environmental conscience in the last decades of the twentieth century. Textbooks written for such courses addressed the science and policy issues surrounding energy and the environment.

Energy, Environment, and Climate also focuses on energy and its impact on the environment. Unlike its predecessors, it's built from the ground up on the premise that climate change is the dominant energy-related environmental issue of the twenty-first century. More traditional concerns such as pollution and energy resources remain important, and they, too, are covered here. But a full five chapters—about one-third of the book—are devoted to climate and the energy-climate link.

Energy, Environment, and Climate begins with a survey of Earth's history and the origin of the planet's energy resources. A quantitative look at past and present patterns of human energy consumption follows, including a discussion of the link between energy, economic development, and human well-being. Chapters 3 and 4 provide an introduction to the science of energy, including the all-important role of the second law of thermodynamics. Chapters 5 through 11 describe specific energy sources and their resource bases, the role each plays in today's global energy system, their associated technologies and prospects for future technological development, and their environmental impacts. This section of the book is organized around fundamental resources, including fossil fuels, nuclear energy, geothermal and tidal energy, and direct and indirect solar energy. The survey of energy sources ends with a chapter that considers prospects for hydrogen-based energy, using hydrogen as either a chemical fuel or in nuclear fusion. Because fossil fuels dominate today's energy supply, there are two chapters dealing, first, with the fossil resource and fossil fuel technologies and, second, with the environmental impacts of fossil fuels. Whereas other textbooks have separate chapters on such energy-related issues as transportation, *Energy, Environment, and Climate* includes these topics in the appropriate energy chapters. For example, hybrid vehicles and combined-cycle power plants appear in the fossil-fuel chapters; fuel cells are discussed in the hydrogen chapter; and wind turbines are included in the chapter on indirect solar energy.

Four chapters on climate follow the section on energy sources. Chapter 12 describes the scientific principles that determine planetary climates, including the natural greenhouse effect in the context of planets Venus, Earth, and Mars. The chapter ends with a discussion of the nature of scientific theories and of certainty and uncertainty in science. Chapter 13 details the so-called "forcings"—both natural and anthropogenic—that can upset the energy balance that ultimately establishes Earth's climate. Chapter 14 documents the observations that suggest Earth is now undergoing unusually rapid climate change, and shows why scientists believe much of that change is attributable to human activities. Chapter 15 outlines projections of future climate, and includes a look at the workings of computer climate models and the role of climate feedbacks. The final chapter brings together the two main themes of the book—energy and climate—and explores how humankind might continue to enjoy the benefits of energy while minimizing climate-changing impacts.

Energy, Environment, and Climate is written primarily from a scientific perspective. However, questions of policy and economics are never far behind the science of energy and climate. The text therefore ventures occasionally into policy and economic considerations—although to a far lesser extent than a policy-oriented book would do. In particular, many chapters end with a section specifically dedicated to a policy-related issue that grows out of the science covered in the chapter.

Any serious study of energy and the environment has to be quantitative. We need to understand just how much energy we actually use and how much energy is available to us. It makes little sense to wax enthusiastic about your favorite renewable energy source if it can't make a quantitatively significant contribution to humankind's total energy supply. Assessment of environmental impacts, too, requires quantitative analysis: *How much* pollution does this energy source emit? *At what rate* are we humans increasing the atmospheric CO_2 concentration? *How long* will this nuclear waste remain dangerous? *How much* waste heat does this power plant dump into the river? *How much* CO_2 results from burning a gallon of gasoline? *What's exponential growth* and what are its consequences for future levels of energy consumption, environmental pollution, or carbon emissions? In dealing with such questions, this book doesn't shy away from numbers. At the same time it isn't a heavily mathematical text with equations on every page. Rather, the text attempts to build fluency with quantitative information—a fluency that means being able to make quick order-of-magnitude estimates, work out quantitative answers to simple "how much" questions, and to "read" the numerical information contained in graphs. The book doesn't require higher mathematics—there's no calculus here—but it does demand your willingness to confront quantitative data and to work comfortably with simple equations. Anyone with a solid background in high-school algebra can handle the material here. As for a science background, the text only assumes that the reader has some familiarity with high-school level chemistry and/or physics. Despite its scientific orientation, this book is written in a lively, conversational style that students have welcomed in my other textbooks.

Energy, Environment, and Climate helps reinforce qualitative and quantitative understandings with its end-of-chapter activities. **Chapter Reviews** summarize the big ideas presented in each chapter, invite the student to consider the meaning of new terms introduced in the chapter, and recap important quantitative information and equations. **Questions** probe the concepts behind energy sources, environmental impacts, and climate issues. **Exercises** involve calculations based on the material introduced in each chapter. Answers to the odd-numbered exercises are provided at the back of the book. **Research Problems** send the student to sources of contemporary data—usually Web-based—and allow for more detailed exploration of questions that may be related to energy and environmental issues in the student's home state or country. Given the discoveries quickly unfolding in this growing field, research problems may also ask the student to update data presented in the book or to look more deeply into quantitative data on global energy use and its impacts.

Energy, Environment, and Climate is illustrated with photos, line drawings, and graphs. Line drawings describe the workings of energy technologies, the flows of energy and material throughout the Earth system, climate models and feedback effects, pollution control and waste storage systems, and a host of other content that's best seen to be understood. Photos are largely of actual energy systems, presented to give a sense of the technologies and their scales. Graphs quantitatively describe everything from the breakdown of our energy use by source or by economic sector to projections of future global temperatures. Every graph is traceable to an authoritative source, and a list of credits and data sources documents these sources.

Also complementing the main text are tables displaying important energy- and climate-related quantities; some of the most useful also appear on the inside covers. An appendix tabulates relevant properties of materials, ranging from insulation R values of building materials to half-lives of radioactive isotopes to global warming potentials of greenhouse gases. A glossary defines all key terms that appear in the book, and includes acronyms and symbols for physical units and mathematical quantities. A list of suggested readings and authoritative websites is also provided. In addition, instructors teaching from *Energy, Environment, and Climate* will find supplementary resources at www.wwnorton.com/instructors, a password-protected website including color images of the figures appearing throughout the text.

Energy, Environment, and Climate is not a book of environmental advocacy or activism; it's much more objective than that. I have my own opinions, and I acknowledge that many—although not all—are in line with the views of the broader environmental movement. But I pride myself on independent thinking based on my own study of others' writings and research, and I'd like to encourage you to do the same. I'm also keenly aware that there is a stronger scientific consensus on some issues, particularly climate change, than either the popular media or the general public may realize. I'll be careful to base my scientific statements on the consensus of respected scientists and on peer-reviewed literature that's available to you and everyone else for direct examination. At the same time, I understand the uncertainties inherent in science, especially in an area as complex as the interconnected workings of the global environment. I openly state those uncertainties and quantify them whenever possible. That being said, I wouldn't be surprised or unhappy if the knowledge you gain from this book inspires you to work toward change in our collective patterns of energy consumption. I and the majority of my fellow scientists believe such actions are necessary in the coming decades if we're to avoid disruptively harmful environmental impacts.

No individual can be an expert on all the topics covered in a book like this one, and during the writing process I've been fortunate to be able to call on specialists in many fields. They've contributed to making this book more authoritative and timely than I, working alone, could have done. With appreciation, I acknowledge the individuals who have given their expert opinion, read drafts of individual chapters, or otherwise contributed advice and encouragement to this project:

Climate experts Gavin Schmidt (NASA Goddard Institute for Space Studies) and Michael Mastrandrea (Stanford University) reviewed the climate chapters and made many helpful suggestions. Dr. William Glassley (Lawrence Livermore National Laboratory and California Energy Commission) reviewed the sections on geothermal energy in Chapter 8; Dr. JoAnn Milliken (Acting Program Manager, U.S. Department of Energy Hydrogen Program) reviewed Chapter 11; Roger Wallace (Vermont Wood Energy) and Greg Pahl (Vermont Biofuels Association) reviewed sections of Chapter 10 on biomass. Others who offered advice include Dr. William Ruddiman (University of Virginia), Dr. Irina Marinov (Massachusetts Institute of Technology), Dr. Stephen Schneider (Stanford University), Dr. Michael Mann (Pennsylvania State University), Dr. Peter Vitousek (Stanford University), Dr. Robert Romer (Amherst College), Dr. Mark Heald (Swarthmore College), Dr. Gary Brouhard (Max Planck Institute), Elizabeth Rosenberg (Argus Media), and George Caplan (Wellesley College). My Middlebury colleagues Sallie Sheldon (biology), Jon Isham (economics), Jeffrey Munroe (geology), Chris Watters (biology), Bill McKibben (environmental studies), and Grace Spatafora (biology) were kind enough to share their expertise and encouragement. Finally, I thank my former student Peter Mullen for a thorough reading of the final manuscript, and I thank both Peter and Wendy Mullen for their support of this and other projects.

In addition to those acknowledged above, I am grateful to the following instructors of energy and/or climate courses who contributed reviews at the request of W. W. Norton. Their comments offered a blend of pedagogical and scientific expertise that have enhanced the readability, teachability, and authority of this textbook.

Dr. Robert L. Brenner (University of Iowa)
Dr. James Rabchuk (Western Illinois University)
Dr. Jonathan P. Mathews (Pennsylvania State University)
Dr. Sunil V. Somalwar (Rutgers University)

Dr. Cecilia Bitz (University of Washington)

Dr. F. Eugene Dunnam (University of Florida)

Dr. Ljubisa R. Radovic (Pennsylvania State University)

Dr. Dorothy Freidel (Sonoma State University)

Finally, special thanks to E. J. Zita (Evergreen State College) for her efforts in checking answers to the end-of-chapter exercises and preparing the online Solutions Manual for this text.

I'm honored to be publishing this book with W. W. Norton, and I am indebted to my editor, Leo Wiegman, for inviting me to write the first book in what we hope will become a substantial Norton list in environmental studies. Leo and his associates Sarah Mann, Lisa Rand, and Jennifer Cantelmi have helped guide this project from its inception through the production process. It has been a pleasure to work with all of them.

Finally, I thank my family for their support and patience through the long process of bringing this project to fruition.

Chapter 1

A CHANGING PLANET

Earth was born some 4.6 billion years ago, and our planet has been changing ever since. Agents of change include astrophysical, geological, chemical, and biological forces. Billions of years ago, primitive bacteria radically altered Earth's atmosphere and chemistry. Hundreds of millions of years ago, life emerged from the oceans and transformed the land surface. Just a few million years ago, our human species evolved and began its own process of environmental modification. We humans have now become sufficiently plentiful and technologically active that we, too, are having a global impact on Planet Earth.

1.1 Earth's Beginnings

The time is some 4.6 billion years ago; the place a vast cloud of interstellar gas and dust about two-thirds of the way out from the center of the Milky Way galaxy. Most of the material in the cloud is hydrogen and helium, the latter formed in the first thirty minutes after the universe began in a colossal explosion we call the Big Bang. But there are smaller amounts of oxygen, carbon, nitrogen, silicon, iron, uranium, and nearly all the other elements. These were formed by nuclear reactions inside massive stars that exploded and spewed their contents into the interstellar medium. Much of the matter in the cloud has been through multiple cycles of star formation, evolution, and explosion. With each cycle, the matter has become richer in the more complex elements beyond hydrogen and helium.

Gravitational attraction between the gas and dust that make up the cloud causes it to shrink, and—like ice skaters who come together, join hands, and spin—the shrinking cloud begins to rotate. As it rotates, it flattens into a disk, with all the matter in essentially the same plane. The collapse is remarkably rapid, taking only about 100,000 years.

A massive accumulation develops at the disk's center, and under the crushing pressure of gravitational attraction, its temperature rises. Eventually the central mass becomes so hot that hydrogen nuclei—protons—join, through a series of nuclear reactions, to produce helium. This process liberates vast amounts of energy. The Sun is born! The newborn Sun is about 30 percent fainter than it is today, but its energy output is still equivalent to some 300 trillion trillion hundred-watt lightbulbs (that's 3×10^{26} watts). Nuclear "burning" in the Sun's core can sustain the star for 10 billion years, during which time it will grow slowly brighter.

Farther out in the disk, dust particles occasionally collide and stick together. Mutual gravitation attracts more material, and small clumps form. These, too, collide and the clumps grow; in a mere million years the largest have reached kilometer sizes. The more massive clumps exert stronger gravitational forces, so they attract additional matter and grow still larger. After another hundred million years or so, the nascent Solar System has planet-size accumulations of matter, including the newborn Earth. But large numbers of smaller chunks continue in the mix, and they bombard the protoplanets mercilessly, cratering their surfaces and heating them. Occasional larger collisions occur, too; when Earth is a mere 50 million years old, a Mars-size object plows into the young planet, leaving it molten and so hot that it glows for a thousand years like a faint star. Material ejected in the collision condenses to form Earth's Moon.

In the young Earth, heavier elements sink toward the center, forming Earth's core, and lighter elements float to the surface, eventually forming a solid crust. Gases escape to form a primitive atmosphere of mostly carbon dioxide and nitrogen, although hydrogen may have been abundant as well. Chunks of interplanetary matter continue their relentless bombardment, heating and reheating the planet. But eventually Earth and its fellow planets have swept up much of the interplanetary stuff, and the bombardment tapers off, although occasional Earth-shaking impacts occur throughout the planet's history. Water vapor condenses to form primeval oceans. The time is now about 4 billion years ago.

The structure of Earth today reflects the basic processes from those early times. Its center is a solid inner core, mostly iron, at a temperature of many thousands of degrees Celsius. Surrounding this is an outer core of liquid iron, whose motions generate the magnetic field that helps protect us surface dwellers from high-energy cosmic radiation. Covering the core is the mantle—a hot, thick layer that's solid on short time scales but fluid over millions of years. On top of the mantle sits the thin solid crust on which we live. Thermally driven motions in the mantle result in continental drift, rearranging the gross features of Earth's surface over hundreds of millions of years, and giving rise to volcanic and seismic activity. Figure 1.1 takes a cross-sectional look at our planet.

1.2 Early Primitive Life

Sometime between 4.2 and 3.5 billion years ago, interactions among natural chemical substances, driven by readily available energy, led to the emergence of primitive life. The oldest unambiguous evidence for life consists of fossil algae from 3.5 billion years ago, with less direct evidence going back 3.8 billion years. Some would argue that life arose

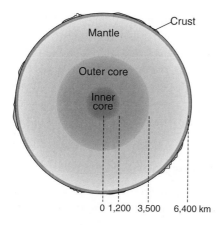

Figure 1.1
Structure of Earth's interior. The crust is not shown to scale; its thickness varies from about 5 to 70 kilometers. The distances indicated are radii measured from Earth's center.

within a few hundred million years of Earth's origin, perhaps as early as 4.2 billion years ago, but the violent bombardment and geological activity of Earth's first few hundred million years have obliterated any firm evidence for very early life. Nevertheless, even the 3.5-billion-year age of the earliest fossils shows that life has been a feature of Planet Earth for most of its history.

We don't know for sure how life developed. Today some biogeologists regard the formation of life as a natural continuation of the processes that differentiated Earth into its core, mantle, and crust. In this view, the first life probably arose deep underground and was fueled by a chemical disequilibrium resulting from Earth's internal heat. Primitive bacterial life of this sort could be common in the universe, with habitats ranging from Earth-like planets to the satellites of distant worlds. The more advanced forms of life we know on Earth, however, probably require specialized conditions, in particular a habitable planetary surface.

The earliest life-forms didn't change much over the billions of years. In fact, some early fossil bacteria and algae are strikingly similar to their modern counterparts. There's a good reason for this: These simple organisms are generalists, capable of surviving under a wide range of environmental conditions. In that sense they're highly successful, and there's little pressure on them to evolve. In contrast, most highly evolved organisms are specialists, surviving in narrow ecological niches and subject to continuing evolutionary pressures or, worse, extinction as the environment changes.

Photosynthesis

The earliest organisms extracted energy from their chemical surroundings, energy that came ultimately from Earth's interior heat. Some of those so-called chemotrophs are still at work in thermal vents on the ocean floor and at other subsurface locations. But at some point, organisms near the ocean's surface developed the ability to capture sunlight energy and use it to build organic molecules from carbon dioxide and water. This is the process of **photosynthesis**. These organic molecules represent stored solar energy, available to the photosynthesizing organisms and to others that prey on them. But there is—

or rather was, at first—a downside to photosynthesis: The process released a new chemical compound, the gas oxygen (O_2), into Earth's atmosphere. Oxygen is highly reactive, destructive of many chemical compounds, and therefore was toxic to the early life that had begun, inadvertently, to pollute its environment with this new substance.

Pinning down when photosynthesis first began is almost as hard as timing the origin of life itself. Solid fossil evidence shows photosynthetic bacteria dating to 2.7 billion years ago, but the process may have originated as much as a billion years earlier.

1.3 Evolution of Earth's Atmosphere

The histories of life and the atmosphere on Planet Earth are inextricably intertwined, so it's appropriate to pause here and focus on the atmosphere itself. Gases released from Earth's interior gave the young planet an atmosphere that was largely carbon dioxide (CO_2) and nitrogen (N_2), with trace amounts of methane (CH_4), ammonia (NH_3), sulfur dioxide (SO_2), and hydrochloric acid (HCl). Earth's gravity was not sufficient to hold onto hydrogen and helium, so these lighter gases escaped to space. Water vapor (H_2O) was probably a significant atmospheric component very early on, before most of it condensed to form the oceans.

Over roughly Earth's first 2 billion years, the levels of methane, ammonia, and carbon dioxide declined slowly. The details of this early atmospheric history are sketchy, but it's believed that geochemical and biological removal accounted for the decline in atmospheric CO_2. In the geochemical process, CO_2 dissolves in atmospheric water droplets to form carbonic acid (H_2CO_3). Rain carries the acid-laden droplets to the planet's surface, where the carbonic acid reacts with exposed rocks. The effect is to remove CO_2 from the atmosphere and sequester it in Earth's crust. In biological removal, early photosynthetic organisms at the ocean surface took CO_2 from the atmosphere and, when they died and sank to the deep ocean, sequestered the carbon in sediments that eventually became sedimentary rocks. The relative importance of geochemical versus biological CO_2 removal is not clear, and scientists are still arguing over these and other mechanisms. But it's clear that over billions of years CO_2 went from being a major atmospheric component to a gas present only in trace amounts.

Atmospheric nitrogen in the form N_2 is largely nonreactive, so it did not experience significant removal. As a result, Earth's atmosphere from about 3.5 to 2.5 billion years ago was probably almost entirely nitrogen.

Meanwhile photosynthetic organisms were at work taking in CO_2 and water, producing energy-storing organic molecules, and releasing the then-toxic gas oxygen. The oxygen content of the atmosphere rose. At first the rise was slow because the highly reactive O_2 combined with iron and other substances in the oceans and surface rocks. But eventually the planet was pretty much fully "rusted," and then—most likely around 2 billion years ago—atmospheric oxygen began to increase significantly. By 1.5 billion years ago, atmospheric oxygen may have reached its current concentration of around 20 percent. Most of the rest is still nitrogen.

A planetary atmosphere containing free oxygen is unusual. Of all the bodies in our Solar System, only Earth shows significant atmospheric oxygen. Because it's so reactive, oxygen in the form of O_2 soon disappears unless it's somehow replenished. On Earth, that replenishment occurs through photosynthesis. Both the origin and continued existence of our oxygen-rich atmosphere are the work of living organisms. Surely this global modification of Earth's atmosphere ranks as one of life's most profound impacts on our planet. Incidentally, many astrobiologists believe that finding the signature of oxygen in a distant planet's atmosphere might strongly suggest the presence of life.

Structure of the Atmosphere

Like the planet itself, Earth's atmosphere has several distinct layers. At the bottom, extending from the surface to an altitude that varies between about 8 and 18 kilometers, is the **troposphere**. Some 80 percent of the atmospheric mass lies within the troposphere, and it's in the troposphere that most phenomena of weather occur. The temperature generally declines

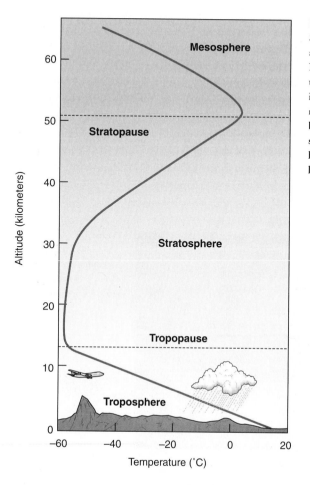

Figure 1.2
Structure of Earth's atmosphere, showing a typical temperature profile. Nearly all weather occurs in the troposphere, whereas the stratosphere is important in absorbing solar ultraviolet radiation. Tropopause altitude varies between about 8 and 18 kilometers. Not shown is the thermosphere, a region of high temperature but near vacuum that lies above the mesosphere.

with increasing altitude, although particular meteorological conditions may alter this trend in the lower troposphere. A fairly sharp transition, the **tropopause**, marks the upper limit of the troposphere. Above is the **stratosphere**, which extends upward to some 50 kilometers. The stratosphere is calmer and more stable than the troposphere; only the tops of the largest thunderstorms penetrate into its lowest reaches. The stratosphere contains the well-known **ozone layer** that protects us surface dwellers from harmful ultraviolet radiation. The formation of ozone (O_3) requires life-produced oxygen, so here's another way in which life helped modify Earth's environment, in this case making the land surface a safe place to live. The absorption of solar ultraviolet radiation causes the stratospheric temperature to increase with altitude. Only the troposphere and stratosphere suffer significant impacts from human activity, and these two layers also play the dominant role in Earth's climate. Above the stratosphere lie the **mesosphere** and **thermosphere**, where the atmosphere thins gradually into the near vacuum of space. There is no abrupt endpoint at which the atmosphere stops and space begins. Figure 1.2 shows the structure of Earth's atmosphere, including a typical temperature profile.

1.4 Aerobic Life

Back to the discussion of life—because, again, the evolution of life and atmosphere are inextricably linked. Although oxygen was toxic to the life-forms that originally produced it, evolution soon led to new forms that could use oxygen in their energy-releasing metabolic processes. In the new oxygen-based metabolism, the process of **aerobic respiration** combines organic molecules with oxygen, producing CO_2 and water and releasing energy. The result is a cycling of oxygen back and forth between life and atmosphere, alternating between the chemical forms CO_2 and O_2. I'll have much more to say about this when I describe the carbon cycle in Chapter 13.

Figure 1.3
The Cambrian period, from about 550 to 490 million years ago, produced an enormous diversity of marine life-forms, shown here in an artist's conception.

Because oxygen is so reactive, aerobic respiration releases energy at a greater rate than the **anaerobic respiration** that took place—and still occurs today—in the absence of oxygen. Aerobic respiration therefore helped facilitate the evolution of larger, more complex, and more mobile life-forms exhibiting new behaviors.

An important behavior that emerged about a billion years ago was sexual reproduction. The organized intermingling of genetic information from two distinct individuals led immediately to much greater diversity of life and an acceleration of the evolutionary process. Soon thereafter, about 850 million years ago, the first multicelled organisms appeared. The period from about 550 to 490 million years ago then produced a tremendous diversification of multicelled life-forms (Fig. 1.3). At this point, life was still a strictly marine phenomenon, but around 400 million years ago plants had begun to colonize the land, beginning another of life's major alterations to the planet. Animals soon followed and could take advantage of the food source represented by terrestrial plants. Amphibians, reptiles (including dinosaurs), birds, and mammals all appeared in the last 400 million years or so of Earth's 4.5-billion-year history. True humans occupy only about the last 2 million years (more about us below in Section 1.7). Figure 1.4 is a timeline of life's evolution on Earth.

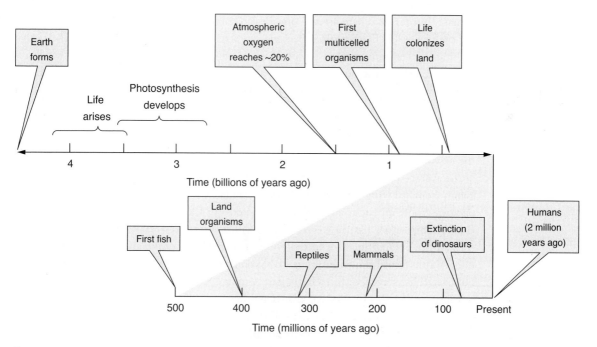

Figure 1.4
Some major events in the history of life on Earth. The origins of life and photosynthesis are uncertain, and the dates given for the last 500 million years represent the earliest definitive fossil evidence.

1.5   **Earth's Changing Climate**

Climate describes the average conditions that prevail in Earth's atmosphere—temperature, humidity, cloudiness, and so forth—and the resulting conditions at Earth's surface. Climate is distinct from weather, which describes immediate, local conditions. Weather varies substantially from day to day and even hour to hour, and changes regularly with the seasons. Climate, being an expression of average weather, changes on longer time scales. But change it does, and natural climate change has been a regular feature of Earth's history.

We've already seen that life and atmosphere are linked through Earth's life-produced atmospheric oxygen and the aerobic organisms that evolved to take advantage of it. Climate, too, is obviously linked with life because the climate of a region determines the kind of life that can survive there. Many different factors go into determining Earth's climate, but the two most important are light from the Sun and the composition of Earth's atmosphere. Sunlight provides the energy that warms our planet and drives the circulation of atmosphere and oceans. What's energy? I'll describe and quantify this all-important concept in the next three chapters, but for now suffice it to say that energy is the "stuff" that makes motion, action, change, and life possible.

I'll detail in Chapter 12 just how sunlight and atmosphere interact to establish Earth's climate, but for now here are the big ideas: (1) Sunlight brings energy to Earth, warming the planet. (2) Earth returns that energy to space, establishing an energy balance that maintains a constant average temperature. (3) Some gases in the atmosphere act like a blanket, blocking the outgoing energy and making Earth's surface temperature higher than it would be otherwise. These so-called **greenhouse gases** include especially water vapor and CO_2. Change either the rate at which Earth receives solar energy, or the concentration of atmospheric greenhouse gases, and you change Earth's climate.

We know from well-established astrophysical theories about how stars evolve that the newborn Sun was some 30 percent fainter than today. We have a very rough record of the average temperature at Earth's surface over the past 3 billion years—a record that comes from studying a variety of physical evidence containing information about temperature (more on such "ancient thermometers" in Chapter 14). The record shows that Earth's temperature has varied over that 3-billion-year interval, but remarkably, it hasn't tracked the gradual increase in the Sun's energy output. Take a look at this long-term temperature record shown in Figure 1.5; you can see that the temperature has fluctuated but overall has stayed within a fairly narrow range. Despite a fainter young Sun, throughout much of Earth's history our planet has been warmer than it is today.

Many factors influence climate, and they act on time scales ranging from years to billions of years (we'll consider some these influences in more detail in Chapters 12 to 15). But over the billions of years' time shown in Figure 1.5, scientists believe that geological processes regulate the concentration of atmospheric CO_2 and thus establish a relatively stable climate. The basic idea is simple: CO_2, as we saw earlier, is removed from the atmosphere by weathering of rocks; it's replenished by CO_2 escaping from Earth's

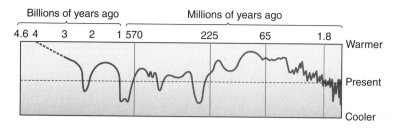

interior, especially through volcanoes. The chemical reactions that constitute weathering depend on temperature; the higher the temperature, the greater the weathering rate. And it's precipitation that brings CO_2 to Earth's surface in the form of the weathering agent carbonic acid. Precipitation, in turn, depends on how much water evaporates into the atmosphere—and that also increases with temperature. With increased temperature, then, both the rate of weathering reactions and the amount of precipitation increase. Those increases promote greater weathering, and thus remove more CO_2 from the atmosphere. Therefore the atmosphere acts less like an insulating blanket, and Earth's surface cools. This drop in temperature decreases the rate of weathering, and the continuing CO_2 emission from volcanoes gradually increases the atmosphere's CO_2 concentration. These two conditions enhance the insulating blanket, and Earth's surface warms.

What I've just described is a process of **negative feedback**. Earth warms, and the Earth-atmosphere system responds in a way that counters the warming. Earth cools, and the system responds to counter the cooling. This is *feedback* because a system, in this case the Earth and atmosphere together, responds to changes in itself. It's *negative* feedback because the response opposes the initial effect. (I'll describe and quantify many more climate feedbacks in Chapter 13.)

Scientists believe that the negative feedback process of CO_2 removal by rock weathering has acted over geologic time much like a household thermostat, regulating Earth's temperature within a fairly narrow range, even as the Sun's energy output gradually increased. Figure 1.6 shows this temperature-regulating effect.

Snowball Earth

Geological temperature regulation hasn't been perfect. Changes in volcanic activity, continental drift, variations in Earth's orbit, and many other factors have led to excursions toward warmer or cooler conditions, as suggested in Figure 1.5. Scientists have found evidence of dramatic climate swings that plunged Earth into a frozen "snowball" state that was followed by rapid warming. Perhaps as many as four such snowball episodes occurred in the period between about 750 and 580 million years ago. During that time Earth's continents were probably clustered near the equator, and the warm equatorial precipitation made for especially rapid removal of atmospheric CO_2 by rock weathering. Atmospheric CO_2 plunged, and ice advanced across the landless northern and southern

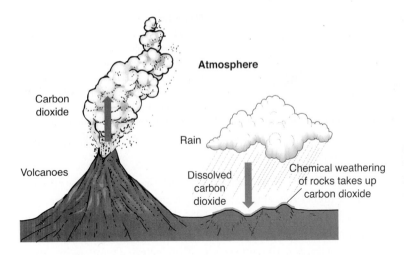

Figure 1.6
Over geologic time, removal of CO_2 by precipitation and chemical weathering of rocks balances volcanic CO_2 emissions. Carbon dioxide removal increases with temperature, providing a feedback that regulates Earth's temperature.

hemispheres. The larger expanses of ice reflected more of the incoming solar energy, cooling the planet further. The process overwhelmed the natural weathering thermostat, and soon the entire ocean was covered with ice. Starved of precipitation, land glaciers couldn't grow and thus some of the land remained ice-free (Fig. 1.7).

There's a problem here: Ice reflects most of the sunlight incident on it, so once Earth froze solid it would seem impossible for it ever to warm up and thaw again. But remember those volcanoes, which would continue to spew CO_2 from Earth's interior. Normally the atmospheric CO_2 concentration remains fairly constant, with CO_2 removal by weathering occurring at roughly the same rate as volcanic CO_2 emission. (Slight imbalances are what enable the rock-weathering thermostat, described above.) With the oceans frozen, however, there was no water to evaporate, precipitate, and cause rock weathering. But volcanism continued, driven by the planet's internal heat, so atmospheric CO_2 increased rapidly—and with it the insulating greenhouse effect and therefore Earth's surface temperature. Eventually equatorial ice melted, exposing dark ocean water to the strong

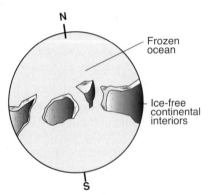

Figure 1.7
Earth at the height of a snowball episode. The continents are clustered near the equator, and the entire ocean is frozen to a depth of 1 kilometer. The lack of precipitation arrests glaciers and leaves some land ice-free.

tropical sunlight. The warming rate increased. Both theory and geological evidence suggest that the climate swung from extreme cold to stiflingly hot and wet in just a few centuries. Eventually the geological weathering feedback got things under control, and the climate returned to a more temperate state.

What happened to life during Earth's snowball episodes? It hunkered down beneath the kilometer-thick ice that covered the oceans, living off energy escaping from Earth's interior. Many single-celled organisms probably went extinct. Others, however, clustered around geothermal heat sources on the ocean floor, where they evolved in isolation and thus increased the overall diversity of living forms. In fact, some scientists credit snowball Earth episodes and their subsequent hot spells with engendering the huge proliferation of life during the Cambrian period around 500 million years ago (see Fig. 1.3).

1.6 Earth's Energy Endowment

The astrophysical, geological, and biological history I've just summarized has left Planet Earth with a number of natural sources of energy—sources that the planet taps to drive processes ranging from continental drift to photosynthesis. These energy sources are available to us humans, too, and ultimately they represent the only energy we have to power our industrial societies. Much of this book is about how we use that energy and the impacts our energy use has on the planet. Here I'll describe briefly the relatively few energy sources that constitute Earth's energy endowment.

Energy sources for Planet Earth take two forms, which I'll call *flows* and *fuels*. **Flows** are streams of energy that arrive at or near Earth's surface at a more or less steady rate, bringing energy whether it's needed or not. **Fuels**, in contrast, represent energy that's stored in one form or another—most commonly in the arrangements of matter that constitute molecules or atomic nuclei. Fuel energy usually remains stored energy until a deliberate act liberates it.

Energy Flows

By far, the dominant energy flow is sunlight, streaming across the 93 million miles of space between Sun and Earth in a mere eight minutes. Sunlight energy impinges on Earth's atmosphere at the rate of about 10^{17} watts. You'll get a solid feel—literally—for what a watt is in the next chapter; for now, suffice it to say that 10^{17} watts is the rate at which a thousand trillion ordinary 100-watt lightbulbs would use energy. This flow of solar energy represents 99.98 percent of all the energy arriving at Earth.

What happens to all that solar energy? Except for a nearly infinitesimal portion, it's soon returned to space. By "soon" I mean on time scales ranging from nearly instantaneous to the thousands of years it's stored in the longest-living plants; in any event, nearly all the solar energy enters into the Earth system and goes back out on very short time scales compared with Earth's geological history. This balance between incoming and outgoing energy determines Earth's climate (much more on this in Chapter 12).

Solar energy performs some useful roles between its arrival at Earth and its return to space. Figure 1.8 shows the big picture of Earth's energy flows, including not only solar energy but also additional flows that I'll discuss shortly. As the figure shows, some 31 percent of the incident solar energy is reflected immediately back to space, most of it from clouds and reflective particles in the atmosphere and some from ice, snow, light-colored deserts, and other surface features. Another 45 percent of the incoming solar

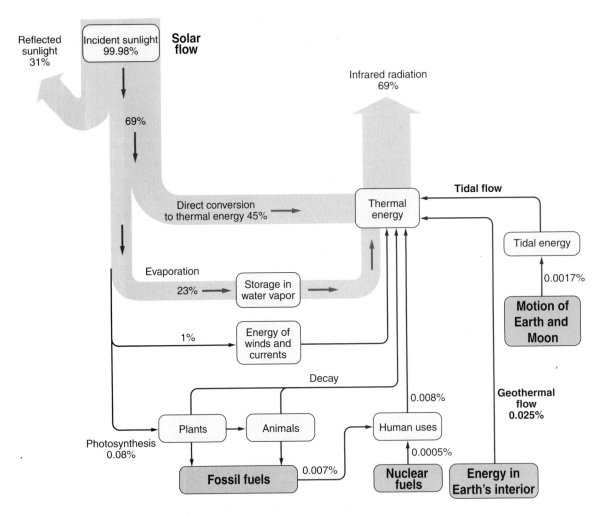

Figure 1.8
Earth's energy flows, nearly all of which are from the Sun. Other flows are geothermal and tidal energy. In addition, energy is stored in fossil and nuclear fuels. The 69 percent of incoming sunlight that's absorbed in the Earth-atmosphere system is almost perfectly balanced by the same amount of infrared radiation returned to space.

energy turns directly into thermal energy—what we call, loosely, "heat." That energy gets back to space almost immediately, through the process of radiation that I'll describe in Chapter 4. Another 23 percent of the incident solar energy goes into evaporating water, where it's stored in the atmosphere for periods of typically a few days as so-called latent heat—energy that's released when the vapor condenses back to liquid. A mere 1 percent ends up as the energy of moving air and water—the winds and ocean currents. Finally, a tiny 0.08 percent—less than 1 part in 1,000—of the incident solar energy is captured by photosynthetic plants and becomes the energy that sustains nearly all life. (I say "nearly all" because there are isolated ecosystems clustered around undersea thermal vents that thrive on energy from Earth's interior.) Organisms store some of this photosynthetically captured sunlight as chemical energy in their bodies, and it's this energy that may stick around a while until it all returns to space after death and decay.

Actually, not quite all. That infinitesimal portion I mentioned a couple of paragraphs ago gets buried without decay and becomes the energy stored in fossil fuels (more on these in the section "Fuels" on p. 14, and a lot more in Chapters 5 and 6).

Our planet's interior is hot—thousands of degrees Celsius at the core. The source of this **geothermal energy** is a mix of the primordial heat liberated in the gravitational accretion of matter that built our planet over 4 billion years ago, and ongoing energy release in the decay of long-lived natural radioactive elements, particularly uranium and thorium. Geologists aren't sure just how much of Earth's thermal energy is primordial and how much is due to radioactivity, but both sources are probably significant. The internal thermal energy causes the temperature to increase with depth into the planet; near the surface the rate of rise is typically about 25°C per kilometer, or roughly 70°F per mile.

This temperature difference between Earth's interior and the surface drives a steady flow of energy to the surface. On average, that flow amounts to some 0.087 watts arriving at every square meter of Earth's surface. The rate is much greater in regions of concentrated geothermal activity, where hot molten rock is closer to the surface. These include active and recently active volcanoes, as well as undersea rift zones like the Mid-Atlantic Ridge, where new ocean floor is forming as the continents move slowly apart. Geothermal heat, while geologically significant, contributes just 0.025 percent of the energy reaching Earth's surface, although the geothermal energy flow is second only to solar energy. The modest natural geothermal flow appears as a thin arrow in Figure 1.8. In Chapter 8 I'll describe how we humans can make use of geothermal energy.

The only other significant energy flow is **tidal energy**, which comes from the motions of Earth and Moon. Gravitational effects that I'll describe in Chapter 8 cause the oceans to slosh back and forth, giving most coastlines two high tides a day. The energy associated with the tides is dissipated, eventually ending up as heat, as the moving water interacts with the shores. That energy, which comes ultimately from the orbital motion of the Moon and the rotation of the Earth, accounts for less than 0.002 percent of the energy flowing to Earth. Although the tides play an important role in marine ecosystems and have been tapped in isolated instances to generate electrical energy (more on this in Chapter 8), the overall impact of tidal energy is very small. The natural tidal energy flow is shown in Figure 1.8.

Several other energy flows arrive at Earth. Although they have physical and cultural significance, they deliver negligible energy. For example, starlight (from stars other than the Sun) is an insignificant source of energy, but the presence of stars in the night sky has long stirred human imagination, inspired art and literature, and given astronomers throughout the ages a glimpse of the rich universe beyond our Solar System. The Sun itself puts out additional energy in the form of a wind of particles, flowing past Earth at some 400 kilometers per second. Although this energy is again negligible, especially strong bursts of solar particles can have significant impact on Earth's electrical and magnetic environment and may exert a subtle influence on climate by affecting the chemistry of the upper atmosphere. Cosmic rays—high-energy particles from beyond the Solar System—are yet another source. Again, the energy delivered is negligible, but cosmic rays play a role in the spontaneous mutations that drive biological evolution. Finally, our planet—and indeed the whole universe—is bathed in microwave radiation that originated some 300,000 years after the Big Bang, when the first atoms formed. Energetically, this radiation is totally insignificant, but it has given us profound insights into the origin of the universe itself.

So although Earth is connected to the greater universe through a variety of incoming energy flows, only sunlight, geothermal energy, and tidal energy are at all significant in terms of the energy that drives Earth's natural systems and human society. Of these three, sunlight is by far the dominant source of the energy arriving at Earth.

Fuels

Fuels are energy sources that Earth acquired long ago, in the form of substances whose molecular or nuclear configurations store energy. Most familiar are the **fossil fuels**, which supply the vast majority of the energy that drives human society. These substances—coal, oil, and natural gas—are relative newcomers to Earth's energy endowment. They formed over the past few hundred million years when once-living organic matter was buried before it had a chance to decay. Chemical and physical action then changed the material into fossil fuels, which contain the solar energy that ancient photosynthetic organisms had captured. Today, Earth's endowment of fossil fuels is considerably smaller than it was even a few decades ago. That's because we humans have already consumed somewhere around half of the readily available fossil fuel resources. Although fossil fuels continue to form today, they form at a rate millions of times slower than we consume them, so they're effectively a nonrenewable energy resource. Figure 1.8 has arrows showing both the natural formation of fossil fuels and the human consumption of fossil fuel energy.

In contrast to fossil fuel energy, nuclear fuels have been part of the Earth since the planet's birth. Natural uranium and thorium formed in stellar processes, and today their radioactive decay provides part of Earth's internal heat and thus part of the geothermal energy flow. Through nuclear reactions, uranium and thorium also constitute a more potent energy source. Today, we use uranium to generate some 17 percent of humankind's electrical energy. I'll explain how in Chapter 7, where I'll also describe the environmental benefits and the potential harm that accrue to our present-day use of nuclear energy. Earth's nuclear fuels formed before the planet itself, and there's no terrestrial process that can create them, so these are truly nonrenewable energy resources. Therefore Figure 1.8 shows a one-way arrow representing human consumption of nuclear fuels.

In principle, every atomic nucleus except that of iron represents a source of stored energy—a nuclear fuel. But so far we've only learned to tap that energy from a very few, very massive nuclei—most significantly, uranium—having the exceptional property that they readily split in two with a great release of energy. But we're exploring other approaches. At the other end of the nuclear mass scale, the hydrogen in seawater represents a vast nuclear energy source. Fused together to make helium, hydrogen releases so much energy that even a rare form of the element could supply humankind's present-day energy needs for 40 billion years—some ten times as long as the Sun will continue to shine. We know the process works because it's what powers the Sun itself. On a less positive note, we've also learned to release hydrogen's energy explosively in our thermonuclear weapons, or "hydrogen bombs." I'll describe the prospects for this vast energy resource in Chapter 11.

Hydrogen represents a potentially enormous resource in Earth's energy endowment, but one that we simply haven't yet learned to use. By the way, don't confuse hydrogen as a *nuclear* energy resource with the so-called hydrogen economy, which would use hydrogen gas (H_2) as a chemical fuel. There simply isn't any significant amount of H_2 in Earth's endowment, which is a big problem for the hydrogen economy. The hydrogen gas I'm talking about here is already "burned"—combined with oxygen to make water (H_2O). There's no more chemical energy to be extracted. But there's plenty of that harder-to-get-at nuclear energy. Figure 1.8 shows neither this hydrogen-stored nuclear energy nor any associated flow, because at present it isn't being released at all.

Since I'm being exhaustive about the energy sources available to us on Earth, I should mention that there might be sources outside of Earth's own endowment that humankind could someday use. For example, we might mine fuel substances from the Moon, where, among other things, we could find "unburned" hydrogen gas to use as a chemical fuel. Or we might harvest liquid methane (natural gas) on Saturn's moon Titan. But for now any thoughts of getting to such extraterrestrial energy reserves without expending far more energy than we would extract are pure fancy. That may change someday, but probably not any time soon.

A realistic picture of Earth's energy endowment, then, is this: We have available a substantial, continuous energy flow from the Sun, and much lesser flows from Earth's interior heat and from the tidal energy of the Earth–Moon system. Inside the Earth we have fossil fuels, which we're quickly depleting, and the nuclear fuels uranium and thorium that we know how to exploit. We also have a vast nuclear fuel resource in the hydrogen of seawater, but when and even if we'll learn how to use that one is a wide open question. That's it. When we talk about "energy alternatives," "renewable energy," and other popular solutions to energy shortages and energy-related environmental problems, there's no new, hidden, as-yet-undiscovered source. We either have to turn to one of the known sources that comprise Earth's energy endowment, or we have to use less energy.

1.7 The Human Era

Sometime around 5 million years ago, in eastern Africa, creatures evolved that had relatively large brains and walked erect on two feet. A number of human-like variants developed, some of which died out while others evolved further. About 2 million years ago—a drop in the bucket of geological time—humans of the genus *Homo* first appeared. We have firm evidence that our hominid ancestors were making primitive tools as long as 2.6 million years ago and that they had harnessed fire somewhere between 1.6 million and 780,000 years ago. Many factors distinguish human beings from other animals, but for the purposes of this book, the single most important distinction is that we make significant use of energy beyond what our own body's metabolism provides. In that context, the harnessing of fire is a seminal event in human history.

For much of our species' time on Earth, wood-fueled fire and the power derived from domesticated animals were our only energy sources beyond our own bodies. More recently we added oils extracted from animal bodies, tars and oils naturally seeping to Earth's surface, wind energy, water power, coal, and then oil, natural gas, and uranium. In rare instances we've also tapped the direct flows of solar, geothermal, and tidal energy.

Energy consumption is unquestionably a major factor in the advancement of human civilization. It has enabled us to manufacture goods, from the most primitive iron and bronze implements to the most sophisticated computer chips, spacecraft, and synthetic materials. It has enabled us to live comfortably far beyond the warm climates where our species evolved. Energy helps us transport goods for trade and ourselves for exploration, commerce, education, and cultural enrichment. Today, energy consumption enables the high-speed communication that binds us into one global community. As I'll show in the next chapter, measures of human well-being often correlate closely with energy consumption.

Perhaps most significantly, energy-intensive agriculture enables our planet to support a human population far beyond what would have been possible as recently as a few centuries ago. For most of our species' history, resource limitations and the hardships of the hunter–gatherer lifestyle kept the total world population under 10 million. By the beginning of the common era (c.e.; the year 1), the number of humans had reached about 200 million. By 1750, with the Industrial Revolution underway, that number had climbed to some 800 million; around 1800 it reached 1 billion. The entire nineteenth century added another half billion, and a full billion joined the world's population in the first half of the twentieth century, making the population just over 2.5 billion in 1950. By 2000 the total had passed 6 billion. Figure 1.9 summarizes human population growth over the past 12,000 years. Fortunately for our planet, the steep upswing shown in Figure 1.9 won't continue. As the figure also shows, the annual rate of growth as a percentage of population peaked in the 1960s at around 2 percent. It has been declin-

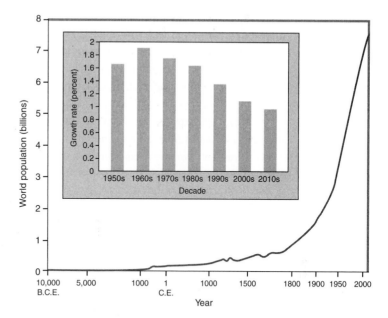

Figure 1.9

Human population growth over the past 12,000 years and projected to 2020. Although the world's population continues to grow, the inset shows that the growth rate as a percentage of population peaked in the 1960s. In absolute numbers, the maximum yearly growth occurred in the 1980s.

ing since, and by 2004 dropped to 1.2 percent. But the population itself continues to increase and will do so for some decades to come. Although some projections suggest population growth might continue beyond the year 2100, many show Earth's human population peaking at about 9 billion in the middle of the present century, then beginning a gradual decline to more sustainable levels. Many factors contribute to this welcome trend, but most are related to higher standards of living and particularly to education. Energy consumption helps make possible those higher standards.

A Global Impact

Ironically, the same energy consumption that brought some parts of human society an unprecedented level of comfort and material well-being also threatens natural Earth systems that maintain a supportive environment. Although our ancestors did plenty of local environmental damage—polluting waters, denuding hillsides, burning forests, perhaps driving species extinct—it's only in the past century that we humans have become so populous, so technologically vigorous, and so extravagant in our energy consumption that we've begun to alter Earth's environment on a planetary scale.

The title of this chapter suggests that change is a natural and ongoing feature of Earth's long history. And I've stressed that the agents of planetary change include life itself. Surely nothing we humans do is likely to be as dramatic as the work of those primitive photosynthetic organisms that gave Earth its oxygen atmosphere. Maybe not. But there are two issues here: One is the extent of the change; the other its speed. On the latter score, human-induced planetary change is almost unprecedented. The time scales for oxygen buildup in Earth's atmosphere and for the tectonic rearrangement of the continents is measured in hundreds of millions of years. The time scale for climate swings into and out of ice ages is thousands of years. But the time scale for the substantial human-caused increase in atmospheric CO_2 is about one hundred years. As I've outlined already in discussing the long-term history of Earth's climate, atmospheric CO_2 is a key factor in determining Earth's temperature. So here's a human-caused environmental change, resulting largely from combustion of fossil fuels, that has a direct bearing on global climate. And it's a change that's occurring on an unnaturally rapid time scale.

That last remark needs qualification: Although most natural changes in the global environment occur much more slowly than recent **anthropogenic**—that is, human-caused—changes, there are exceptions. Climatologists have detected extremely rapid global temperature fluctuations—on time scales as short as a few decades—as Earth came out of the last ice age. And as we saw earlier, the ending of a snowball Earth episode may have taken the planet from the deep freeze to a superhot, moist climate in a matter of a century or so. But none of those sudden changes occurred in the presence of a human civilization stretching the planet's resources in an effort to feed, transport, educate, and provide material goods for a burgeoning population. Our advanced human society can ill afford rapid environmental change, and yet such change is just what our very advancement has made increasingly likely.

Your Role

Given that you're reading this book, it's a good bet that you belong to one of the better-fed, better-educated, and more affluent subsets of human society. If that's true, then no matter how carefully you might try to maintain an environmentally sustainable lifestyle, you're in fact responsible for substantial rates of energy consumption, environmental contamination, CO_2 emission, resource depletion, and other affronts to the global environment. On the other hand, you probably also have the education, wealth, political savvy, and influence to help improve humankind's relationship with Planet Earth.

Chapter 1   **Chapter Review**

BIG IDEAS

1.1   Earth is 4.6 billion years old.

1.2   Life developed around 4 billion years ago. Photosynthetic life arose about 3.7 to 2.7 billion years ago.

1.3   Earth's early atmosphere consisted of carbon dioxide, nitrogen, methane, and other gases from the planet's interior. By about 2 billion years ago, photosynthetic plants had added significant amounts of oxygen. Today's atmosphere is largely nitrogen and oxygen. The lowest layer of the atmosphere is the **troposphere**.

1.4 Oxygen led to new, more complex, **aerobic** life-forms that could utilize food energy at a greater rate.

1.5 Earth's climate has evolved, regulated by factors including solar luminosity and atmospheric CO_2. Rock weathering and volcanism regulate CO_2 over the long term.

1.6 Earth's energy resources include **fuels** that store energy and **flows** that deliver continuous streams of energy. Sunlight is the dominant energy flow, and a small fraction of sunlight is trapped and stored in fossil fuels.

1.7 Humankinds' earliest ancestors evolved about 5 million years ago, and we are now sufficiently numerous and technological that our impact on Earth is comparable to that of natural processes. Much of that impact stems from our use of energy far in excess of what our own bodies can produce.

TERMS TO KNOW

aerobic respiration (p. 6)
anaerobic respiration (p. 7)
anthropogenic (p. 18)
climate (p. 8)
energy flow (p. 11)
fossil fuel (p. 14)
fuel (p. 11)
geothermal energy (p. 13)
greenhouse gas (p. 8)

mesosphere (p. 6)
negative feedback (p. 9)
ozone layer (p. 6)
photosynthesis (p. 3)
stratosphere (p. 6)
thermosphere (p. 6)
tidal energy (p. 13)
tropopause (p. 6)
troposphere (p. 5)

GETTING QUANTITATIVE

Earth forms: 4.6 billion years ago

Oxygen-containing atmosphere: ~2 billion years ago

Fraction of Earth's energy from the Sun: 99.98 percent

Human species evolves: ~5 million years ago

Humans harness fire: ~1 million years ago

Current human population: ~7 billion

QUESTIONS

1. What is the most significant change that life has made in Earth's environment?

2. Why can't we tell much about the possible origin of life on Earth before about 4 billion years ago?

3. Why did atmospheric oxygen concentration not rise immediately after life "invented" photosynthesis?

4. What's the difference between an energy flow and a fuel?

5. In discussing human population growth in Section 1.7, I noted that the annual percentage growth rate peaked in the 1960s, and yet population continues to increase. How is this possible?

6. The huge energy-consumption rate of modern industrial societies poses serious challenges to the environment. Yet energy consumption may help mitigate the growth of world population, which poses its own environmental challenges. Explain.

7. Although the percentage growth rate of world population peaked in the 1960s, it was 1988 when the world added the greatest number of people. Why the difference?

EXERCISES

1. Solar energy is incident at the top of Earth's atmosphere at the rate of about 1,368 watts on every square meter (W/m^2). This energy effectively falls on the cross-sectional area of the planet. From this fact, calculate the total rate at which solar energy arrives at Earth. You'll need to find Earth's radius; see the Physical Constants table on the inside back cover.

2. Use the answer to Exercise 1, along with the percentage of Earth's total energy supplied by the geothermal flow shown in Figure 1.8, to estimate the total geothermal energy flow.

3. In 1965 the world's population was about 3.4 billion and was growing at about 2 percent annually. In 1985 the population was 4.9 billion, growing at 1.7 percent, and in 2000 it was 6.1 billion, growing at 1.2 percent. In which of these three years did the actual number of people increase by the greatest amount? Show by calculating the number in each case.

4. If the 2004 world population growth rate of 1.2 percent were to continue, what would be the population in 2050? The 2004 population was about 6.4 billion.

5. Figure 1.8 shows that approximately 0.008 percent of the energy flow to Earth's surface is associated with humankind's consumption of fossil and nuclear fuels. Use the answer to Exercise 1 to determine the order of magnitude of that flow in watts.

RESEARCH PROBLEMS

(With research problems, always cite your sources!)

1. Prepare a plot of population versus time, like the one in Figure 1.9, but for your own country. Go back as far as you can find data. What factors might affect an individual country's population but not the world's?

2. Find the atmospheric composition of Earth and four other planets or planetary satellites in the Solar System. For each, list at least three of the most abundant gases and their percentage of concentration.

3. Washington State's Mount St. Helens volcano was active during the fall of 2004. Find the dominant gases emitted during this activity and give quantitative estimates of the gaseous emissions if you can find them.

Chapter 2

HIGH-ENERGY SOCIETY

You've just picked up this book, and now I'm going to ask you to put it down. That's because I want to give you a feel for what it means to be a member of a modern, industrialized, energy-intensive society. I mean "a feel" literally—a sense, in your own body, of the quantities of energy that you consume directly or that are consumed in your name. In the next two chapters we'll develop more rigorous understandings of energy and its measures, but an appreciation for energy and its role in human society is conveyed best by a literal feel for the energy associated with your own body.

So put down your book, and stand as you are able. Put your hands on your hips and start doing deep knee bends. Down and up, down and up—about once a second. Keep it going for a few minutes so you get a real sense of just how vigorously your body is working.

During this short exercise, your body is working at the rate of about 100 watts (W). This is the rate at which you are converting energy stored in chemical compounds derived from food into the mechanical energy associated with the motion and lifting of your body. If you're big or exercising more vigorously, the figure is a little higher. If you're small or exercising at a slower pace, the figure is a bit lower. But it won't be lower than several tens of watts, nor more than several hundred. One hundred watts is a nice round number, and we'll consider it typical of the energy output of the average human body. (In the next chapter you'll see how to calculate this figure for your knee-bend exercise.)

There are other paths to our 100-W figure. If you consider the daily caloric content of the typical human diet (calories being a measure of energy), you can show that it averages about 100 W (see Exercise 1 in Chapter 3). If you've ever turned a hand-cranked or foot-powered electric generator (Fig. 2.1), you probably found that you could comfortably keep a 100-W lightbulb lit for a moderate time. During strenuous activity, your body's energy output might be higher—some athletes can put out many hundreds or even thousands of watts for short periods—but 100 W remains a good figure for the average rate at which the human body expends energy.

Figure 2.1
Turning a hand-cranked electric generator, the average person can sustain a power output of about 100 W. Here the energy ends up lighting a 100-W lightbulb.

2.1   Energy and Power

I've quantified the energy output of the human body as being about 100 W. So what's a watt? One **watt** describes a *rate* of energy use. That's a rate as opposed to an *amount* of energy. What distinguishes our high-energy society is not so much the total amount of energy we use, but the amount we use each year, or each day, or each second, or whatever—in other words, the *rate* at which we use energy. The technical term for that rate is **power**. The watt is therefore a unit of power. So just what is a watt? The answer is in your muscles: From our knee-bend example, 1 W is about one-hundredth of the average power output of the human body. One kilowatt (kW), or 1,000 W, is then the equivalent of ten human bodies.

Power and energy are often confused, and it's not uncommon to read newspaper reports of a new power plant that produces "10 million watts every hour" or some such nonsense. Why is that nonsense? Because "every hour" or, more generally, "per time" is built into the meaning of power. Energy is the "stuff" we're talking about, and power—measured in watts—is the rate at which we use that "stuff."

Think of a car as something that produces "stuff," namely, miles of distance traveled. You're much more likely to ask how fast the car can go—that is, the *rate* at which it produces those miles—than how far it can go. If you know the rate, you can always figure out how far you can go in a given time.

Similarly, it makes sense to characterize energy-using devices and even societies by their power consumption—the rate at which they use energy. How much total energy a device or society actually uses depends on the time, so total energy is a less useful characterization.

If I say "This car can go 70 miles per hour," you're not inclined to ask "Is that 70 miles an hour each hour?" because "each hour" is built into the speed designation of 70 miles per hour. Similarly, if I say "This TV uses energy at the rate of 150 watts," it makes no sense to ask "Is that 150 watts every hour?" because the "per time" is implicit in the phrase 150 watts. How far the car goes, and how much energy the TV uses, are determined by how much time you choose to operate these devices. The only subtlety here is that we're explicit that speed is a rate; we say "miles per hour" or "meters per second" or whatever. But the "per time" is built right into the meaning of the term *watt*. Sometimes we're more explicit about energy rates; for example, the statement that the United States imports about 12 million barrels of oil per day is a statement about the rate of energy importation, and this 12 million barrels per day figure can be converted to watts if you know how much energy is in each barrel of oil (see Exercise 6 in Chapter 3).

If watts measure an energy rate, what measures the actual amount of energy? I introduce several energy units in the next chapter, but for now here's one that relates easily to the rate in watts: the kilowatt-hour (kWh). One kilowatt-hour is the amount of energy you use if you consume energy at the rate of 1 kW (1,000 W) for one hour. You would also use that much energy if you consumed one-tenth of a kilowatt—that is, 100 W—for 10 hours. So to produce 1 kWh of energy with your own body, you would have to do those knee bends for about 10 hours. You would use 1 kWh if you left a 100-W lightbulb on for 10 hours, or if you used a 1,000-W hair dryer for one hour. Electric bills generally show your monthly consumption of electricity in kilowatt-hours, so we tend to associate the kilowatt-hour with electrical energy. But it's a unit of energy, period, and therefore quantifies any kind of energy. If you understand that watts and kilowatts measure an energy rate, then the name *kilowatt-hour*—a rate multiplied by a time—makes it clear that the kilowatt-hour is a unit of energy, not rate.

Throughout this book I talk about both energy and power. Don't get them confused! The former is the rate, the latter is the "stuff."

By the way, I'll be employing phrases such as *using energy, producing energy, expending energy, importing energy, losing energy,* and the like. Quantified, most of these terms are usually expressed as a rate, measured in watts or other energy-per-time units. You may be aware that energy is a conserved quantity, so terms such as *producing energy* or *energy loss* aren't strictly accurate. What such terms actually describe is the conversion of energy from one form to another. When we "produce" electrical energy, for example, we're converting some other form of energy—usually the stored energy of fossil or nuclear fuels—into electricity. When we "lose" energy, it's usually being converted to the less useful form called, loosely, "heat." We'll deal extensively with energy conversion in subsequent chapters.

Example 2.1 Oil Heat

My home is heated with oil, and when it's on, my oil burner consumes 0.75 gallon of oil per hour. Very roughly, a gallon of heating oil, gasoline, diesel fuel, or other liquid petroleum product contains about 40 kWh of energy. What's the rate at which my oil burner consumes fuel energy?

Solution

That 0.75 gallon per hour amounts to (0.75 gal/h) × (40 kWh/gal) = 30 kWh/h. But what's 1 kilowatt-hour per hour? Simple: Since 1 kWh is the energy consumed in one hour when using energy at the rate of 1 kW, 1 kWh/h—a rate of energy use—is exactly the same thing as 1 kW. So my oil burner uses fuel energy at the rate of 30 kW. By the way, I bet you don't have electrical appliances that use energy at anywhere near this rate, which shows why those few homes with electric heat consume vastly more electrical energy than homes heated by other means.

2.2   Your Energy Servants

You now have a feel for the rate at which your own body can produce energy, about 100 W. So here's the big question for this chapter: At what rate do you, as a citizen of twenty-first-century industrialized society, use energy? More specifically, how many human bodies, producing energy at the rate of 100 W each, would it take to keep you supplied with energy?

Before you answer, think for a minute about your total energy consumption, including the energy that's used for you throughout our complex society. There's energy you use directly—the lightbulb you read by, the heat and air conditioning that keep you comfortable, the energy that runs your computer, the energy of the gasoline that powers your car, the electrical energy converted to sound in your stereo, the elevator that delivers you to your dorm room or apartment, the energy that boils water for a cup of tea or coffee, the refrigerator that stores your cold drinks, and so forth. Then there's energy used indirectly in your name. All those trucks on the highway are delivering goods, some of which are for you. Those energy-gobbling banks of open coolers at the supermarket hold frozen food, some of which is for you. And your food itself most likely comes from energy-intensive agricultural operations, with fuel-guzzling tractors, production and application of pesticides and fertilizers, and processing and packaging of the food. The airplanes overhead burn a lot of fuel; occasionally they transport you to distant places, and more regularly they supply fresh produce from across the planet or merchandise you ordered with overnight delivery. And speaking of merchandise, think of the factories churning out the products you buy, all of them using energy to process raw materials—to fabricate steel, cloth, silicon, and plastics; and to stamp, melt, mold, weld, rivet, stitch, and extrude the pieces assembled into finished products. Then there's the organizational infrastructure—the office buildings housing commercial, educational, health care, and government services, many with lights blazing around the clock; their air conditioners and heating systems running continuously; and the copiers, computers, and other office equipment on standby for instant use. There's a lot of energy being used in your name!

So how much? The answer, for the United States in the early twenty-first century, is that each of us is responsible for an average energy-consumption rate of roughly, in round numbers, 10,000 W or 10 kW. (The actual figure is between 11 and 12 kW, but we'll stick with the nice, round number—a slight underestimate.) Note, again, that I didn't really answer the question "How much energy do you use?" but rather the more meaningful question "At what rate do you use energy?" And again, don't ask if that's 10,000 W per day or per year or what, because 10,000 W describes the average *rate* at which energy is used in your name, around the clock, day in and day out.

I've now answered my first question, about the rate at which you use energy— 10,000 W. (This number is for citizens of the United States; later in this chapter we'll look at other countries.) The second question asks how many human bodies it would take to supply that energy. We've established that a single human body produces energy at an average rate of about 100 W. Since 100 × 100 = 10,000, this means it would take about 100 human bodies to supply your energy needs. So picture yourself with 100 "energy servants" working around the clock to supply you with energy (Fig. 2.2). If they work eight-hour shifts instead of all the time, you'd need to employ 300 servants. How much would you be paying those workers if the energy they produced were competitive with fossil fuel energy? The answer is shockingly low—see Exercises 1 and 2 at the end of this chapter.

So this is what it means to be a citizen of a modern, energy-intensive society: In the United States, you have the equivalent of about 100 energy servants working for you all the time. Imagine a room full of 100 people doing those deep knee bends around the clock, or cranking away on electric generators as in Figure 2.1, and you get the picture.

You

Some of your energy uses   Your energy servants

Figure 2.2
It would take 100 energy servants working around the clock to
supply the energy needs of the average American.

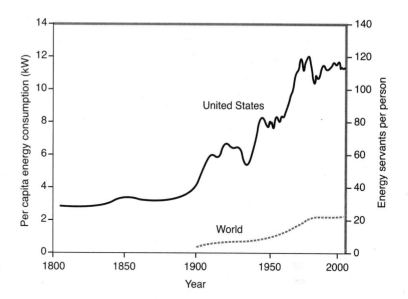

Figure 2.3
Per capita energy consumption rates since 1800. In the early twenty-first century, energy consumption in the United States is about 11 kW per capita, equivalent to some 110 energy servants working around the clock. This rate is about five times the world average.

Although "energy servant" isn't an official scientific unit, it's a handy way to get a quick feel for the magnitude of your energy consumption. Turn on your 1,000-W hair dryer, and picture ten energy servants leaping into action. Start up your 250-horsepower SUV and you've unleashed about 2,000 servants. (That's far more than your average of 100, but fortunately your SUV isn't running all the time. You'll have to borrow some servants from friends whose energy consumption is, temporarily, lower than average.) Just leaving your room light on while you're out means there's one poor servant doing deep knee bends the whole time you're gone. (Actually, it takes more like thirty servants for that hair dryer, and three for your room light; I'll show why in Chapter 4.)

We haven't always had 100 energy servants, of course. When humankind first evolved, each person had just one servant: themselves. As our species tamed fire and domesticated animals, that number grew. The discovery of coal and the use of waterpower further increased the number of servants. By the year 1800, the world average was about five servants, and by 2000 it had climbed to more than twenty (Fig. 2.3). But the distribution is uneven; today, citizens of many developing countries can still claim only a handful of energy servants, while North Americans have more than 100.

2.3 What Your Energy Servants Do

What, exactly, are all those 100 energy servants doing for you? Broadly, it's convenient to consider energy use in four sectors: residential, commercial, industrial, and transportation. How the energy servants divide their effort among those sectors varies from country to country and region to region. A geographically large country like the United States, with its transportation system emphasizing private automobiles, uses a greater

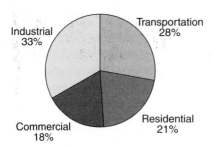

Figure 2.4

Energy consumption in the United States, by sector, in the first decade of the twenty-first century.

share of its energy on transportation than does a compact European country with a highly developed rail network. A cold country like Canada needs more energy for heat. And a developing country, with much lower energy consumption overall, devotes less of its limited energy to the industrial and transportation sectors and more to domestic tasks such as cooking.

Figure 2.4 shows the breakdown of energy use among the four sectors for the United States in the early twenty-first century. Since the average U.S. citizen has the equivalent of about 100 energy servants working around the clock, you can think of the numbers in Figure 2.4 either as percentages or as the total number of energy servants working in each sector. You can see from the figure that more than half of your servants work in the industrial and transportation sectors. Your thirty-three industrial servants produce the goods you consume, and the twenty-eight transportation servants move both you and your goods about. The residential and commercial servants heat and cool buildings and run lights, computers, stereos, copiers, refrigerators, and all manner of other modern conveniences. Most of your energy servants are working at tasks that may not be obvious to you, such as running machinery in factories that make your goods. If you decide to change your own patterns of energy consumption, you need to be aware not only of direct actions such as turning off lights or choosing a fuel-efficient car, but also of behavioral changes that might affect the number of energy servants working behind the scenes for you.

2.4   Who Are Your Energy Servants?

You don't really have 100 human servants turning cranks to generate electricity or pedaling like mad to propel your SUV. What you—and society—have are the natural energy resources I outlined in Chapter 1, and the devices that convert them into the forms of energy we find useful. Your car's engine is one such device; others include huge nuclear or coal-burning electric power plants, hydroelectric generators, wind turbines, boilers that burn fuel to heat buildings and to provide steam for industrial processes, solar panels that convert sunlight into electrical energy, and even systems that tap heat deep in the Earth. Behind all these devices are the energy resources themselves—fuels such as coal, oil, natural gas, wood, and uranium, and natural energy streams such as wind, flowing water, and sunlight. To ask "Who are they?" of your energy servants is to ask about these ultimate sources of the energy you use.

Again, the mix of servants varies from country to country, region to region. Places like Norway or the Pacific Northwest of the United States have abundant waterpower, and more of their energy servants are associated with running water. Policy decisions, such as France's choice of a nuclear path to energy independence, give prominence to particular energy sources. But averaged over a country as large as the United States, or for the world as a whole, a striking pattern emerges: The vast majority of the world's energy servants derive from the fossil fuels coal, oil, and natural gas.

Figure 2.5 shows the distribution of energy sources for the United States in the early twenty-first century. The figure makes it obvious that some 85 percent of U.S. energy comes from fossil fuels. Most of the rest is from nuclear and waterpower (hydro). A non-negligible 3 percent of U.S. energy comes from biomass—wood, corn-based alcohol fuels, and waste burned in "garbage to energy" plants that produce electric power. If you're an advocate of alternative energy sources, Figure 2.5 is sobering. Only one-half of 1 percent of U.S. energy comes from the "other" category that includes the environmentalists' favorites—solar and wind—as well as geothermal energy. In fact, some two-thirds of the "other" category represents geothermal energy, a resource that's not necessarily renewable, limited to a very few specific geographical locations, and presents a host of environmental problems of its own. The remaining one-third of the "other" category, a mere one-sixth of 1 percent of total U.S. energy consumption, comes from solar and wind, with wind supplying about two-thirds of that amount. (We'll explore geothermal energy in Chapter 8, solar energy in Chapter 9, and wind energy in Chapter 10.)

Once again, the fact that energy consumption in the United States is equivalent to about 100 energy servants means that, if you're in the United States, you can translate the percentages in Figure 2.5 directly into numbers of servants. So fully eighty-five of your servants are fossil fueled, eight are nuclear, and you share a single solar servant with eighteen fellow U.S. residents (that's from the solar third of the one-sixth of 1 percent that wind and solar together represent of the total).

Things aren't much different for the world as a whole, as Figure 2.6 suggests. The fossil fuel share is within a percentage point or two of the U.S. value, just over 85 percent. The world as a whole gets about twice as great a fraction of its energy from hydropower as does the United States, but less from biomass. And again, the "other"

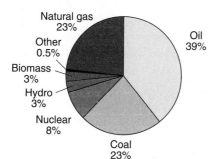

Figure 2.5
Sources of U.S. energy in the early twenty-first century. Fully 85 percent comes from fossil fuels. The "other" category includes geothermal, wind, and solar energy.

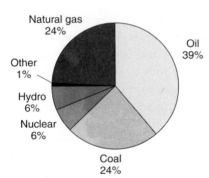

Natural gas
24%

Other
1%

Hydro
6%

Nuclear
6%

Coal
24%

Oil
39%

Figure 2.6
Sources of world energy in the early twenty-first century. More than 85 percent comes from fossil fuels. Here the "other" category includes biomass, geothermal, wind, and solar energy.

sources currently play a very minor role in world energy consumption, although both wind and solar energy are growing rapidly. Because average per capita energy consumption in the world as a whole is considerably less than in the United States, you can't think of the percentages in Figure 2.6 as representing equivalent numbers of energy servants.

You might wonder about the actual sizes of the pies in Figures 2.5 and 2.6. We'll quantify those values in the next chapter, but for now we'll just compare them. The U.S. share of the world's energy pie in the early twenty-first century is just about 25 percent, so the U.S. pie is one-fourth the size of the global pie.

The big picture that emerges from Figures 2.5 and 2.6 is this: The modern world runs on fossil fuels, with more than 85 percent of our energy coming from coal, oil, and natural gas. This fact establishes the substantial link between energy and climate that's implicit in the title of this book.

2.5 What Our Energy Servants Buy Us

What do we gain from the efforts of all those energy servants? Whether they're fossil, nuclear, or solar, all are working to provide us with the comforts, conveniences, material goods, and enriched lives that result from our exploitation of energy resources beyond our own bodies. Can we measure the benefits that result? Can we establish a relationship between energy consumption and quality of life?

Quality of life is a notoriously difficult thing to measure. Economists have traditionally avoided that difficulty by considering instead measures of economic activity. The presumption is that the higher material standard of living associated with greater economic activity translates for most people into richer, fuller, healthier, more comfortable lives. A commonly used economic indicator is a country's **gross domestic product** (GDP), a measure of total economic activity. The GDP includes such factors as personal consumption, government expenditures, private and public investment, product inventories, and net exports—the last being an increasingly important factor in today's global economy. Here we'll consider the link between GDP and energy. In the last section of this chapter, we'll look at alternatives to the GDP, which environmentally conscious economists are increasingly adopting as more appropriate measures for quality of life.

Energy Intensity

To explore a possible link between energy and GDP, I've located individual countries on a graph that plots the per capita energy-consumption rate (equivalent to the number of energy servants) on the horizontal axis, and per capita GDP on the vertical axis. Figure 2.7 shows the result for ten countries. Consider first the United States. Horizontally, it's located at a per capita energy consumption of just about 11 kW. That's the figure I introduced earlier and approximated as 10 kW, from which our 100 energy servants at 100 W each immediately followed. On the vertical axis, the United States is quite well off, with a GDP of over $30,000 per person. Australians use a little less energy each, and they make a little less money. (By the way, the GDP figures used here are what economists call *GDP ppp*, with "ppp" referring to "purchasing power parity." These figures reflect the actual value of goods and services that a given country's GDP can procure.) South Korea, Poland, Egypt, and the Congo are further down in both per capita energy consumption and GDP. However, all six countries lie pretty much on a straight line, as I've indicated in the figure. This means each uses roughly the same amount of energy to generate each dollar of GDP. This quantity—the ratio of energy use to GDP—is called **energy intensity**. In Figure 2.7, the six countries along the straight line have the same energy intensity. Many other countries would fall near this line, too, but some wouldn't. Countries such as Russia and Saudi Arabia, as shown in Figure 2.7, lie well

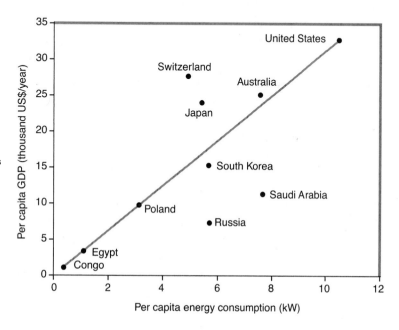

Figure 2.7
Per capita GDP (in thousands of U.S. dollars per year) versus per capita energy consumption for ten countries. The six countries that fall near the straight line have approximately the same energy intensity, or energy required per unit of GDP. Japan and Switzerland are more energy efficient, and Russia and Saudi Arabia less so. Multiplying the numbers on the horizontal axis by 10 gives the number of energy servants per capita.

below the line. They require more energy to produce a given amount of GDP, as you can see by comparing Saudi Arabia with Poland. Both have about the same per capita GDP, but Saudi Arabia uses a lot more energy to produce it. Therefore Saudi Arabia has the higher energy intensity. Russia's is even higher. On the other hand, Japan and Switzerland lie well above the line. Their citizens get more GDP for every unit of energy consumed. If you think of a country's economy as a machine that converts energy into GDP, then these countries are more efficient in their use of energy. Equivalently, they have lower energy intensity.

There are many reasons for the variations in energy intensity around the world. Small, compact countries need less energy for transportation, so they may have lower energy intensity. Tropical countries need less energy for heating, perhaps resulting in lower energy intensity. Military adventurism may lead to large energy consumption without GDP to show for it. Producing food for export with energy-intensive agriculture (as in the United States) raises the exporter's energy intensity and lowers that of the importer. Taxation, environmental regulation, commitment to public transit, and other policies aimed at improving energy efficiency, reducing air pollution, or mitigating climate change may all lead to lower energy intensity. In fact, energy intensity in many countries has fallen dramatically over the past century (Fig. 2.8). The fact that some countries continue to do better, using less energy per dollar of GDP, is a reminder that others may still have considerable room for improvement. Given that energy use has significant environmental as well as economic impacts, controlling both total energy consumption and energy intensity become important aspects of environmental policy.

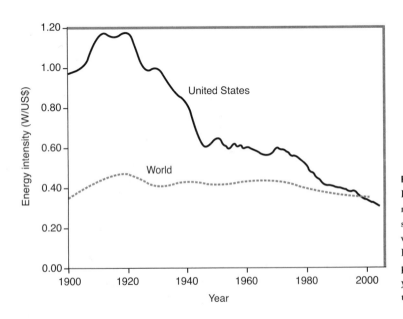

Figure 2.8
Energy intensity in the United States (and in many industrialized countries) fell substantially through the twentieth century, with less change in the world as a whole. Intensity is measured in watts of average power needed to produce $1 of GDP each year (based on the value of one U.S. dollar in the year 2000).

2.6   Policy Issue: Measuring Quality of Life

Is the GDP an appropriate measure for quality of life? Many think not. Because the GDP includes virtually all economic activity, money spent on such things as cleaning up oil spills, repairing hurricane damage, treating air pollution–induced emphysema, and building prisons all contribute to increasing the GDP. But oil-fouled waters, intense hurricanes, illness, and criminal activity clearly don't enhance the quality of life.

Some economists have developed indices that claim to provide more appropriate indications of quality of life, especially in the context of human and environmental health. One such measure is the United Nations' Human Development Index (HDI), which combines the traditional GDP with life expectancy, adult literacy, and school enrollment.

How does energy consumption relate to these broader quality-of-life indices? You can explore this question quantitatively for the HDI in Research Problem 6. For a variety of individual indices, an interesting pattern emerges: At first, increasing per capita energy consumption goes hand in hand with improvements in quality of life; but the improvements eventually saturate, meaning that additional energy consumption doesn't "buy" a higher quality of life. Figure 2.9 shows this effect for one commonly used quality-of-life indicator, namely life expectancy.

So do all those energy servants buy us a better life? Yes, up to a point. But beyond that point we don't gain much. And a quality-of-life indicator that accounts for negative factors such as increased pollution or the detrimental effects of climate change might well show a decline in quality of life with excessive energy consumption.

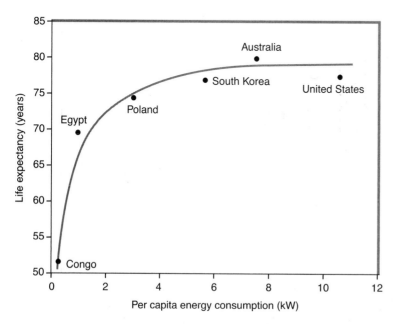

Figure 2.9
Life expectancy compared to energy consumption for the six countries that lie near the straight line in Figure 2.7. Only at very low energy-consumption rates is there a correlation; at higher energy-consumption rates the life expectancy curve saturates. Many other quality-of-life indicators show similar behavior in relation to energy consumption.

Chapter 2   Chapter Review

BIG IDEAS

2.1   **Power** measures the *rate* at which we use or produce energy. The standard unit for power is the **watt**, about one-hundredth of the power output of a typical human body.

2.2   Residents of modern industrial societies use far more energy than their own bodies can produce—about one hundred times as much in North America and twenty times as much averaged over the globe.

2.3   Energy consumption is distributed among four major sectors: residential, commercial, industrial, and transportation.

2.4   Fossil fuels are by far the world's dominant energy source.

2.5   Energy consumption is closely correlated with measures of economic well-being, particularly a country's **gross domestic product**. But there are exceptions, as different countries use energy more or less efficiently.

2.6   Measures of human well-being that take into account factors other than economics also show increases with energy consumption, but only up to a point, beyond which additional energy consumption does not appear to increase the quality of life.

TERMS TO KNOW

energy intensity (p. 31)   power (p. 23)
gross domestic product (p. 30)   watt (p. 23)

GETTING QUANTITATIVE

Typical power output of the human body: 100 W

Typical per capita energy-consumption rate, global average: just over 2 kW (20 energy servants)

Typical per capita energy-consumption rate, North America: just over 10 kW (100 energy servants)

Portion of world energy supplied by fossil fuels: just over 85 percent

QUESTIONS

1.   A newspaper article claims that a new power plant "produces 50 megawatts of energy each hour, enough to power 30,000 homes." Criticize this statement.

2.   Figure 2.3 suggests that the global average per capita energy consumption has risen about fourfold since the year 1800. You might find that change surprisingly low, given the huge number of new energy-dependent technologies developed since 1800. What factors might account for this apparent discrepancy?

3.   Find the energy-consumption rate for five electrical devices you use every day. This value—in watts—is generally listed somewhere on the device or its power adapter, if it has one.

4.   For what energy statistic do Egypt and the United States share a common value?

5.   Are there enough people in the world to supply all U.S. energy needs, if that energy were to come entirely from human muscle power?

EXERCISES

1.   Suppose you got all your energy from human servants turning hand-cranked electric generators and each producing energy at the rate of 100 W. If you paid those servants $8 per hour, a bit more than the U.S. minimum wage, what would be your cost for 1 kWh of energy? (The typical cost of 1 kWh is actually around 10¢.)

2.   What hourly wage would you have to pay your servants from Exercise 1 if the energy they produced were competitive with electrical energy produced by conventional means such as fossil fuels? Use the typical cost of 10¢ per kilowatt-hour.

3.   Figure 2.5 shows that 8 percent of U.S. energy comes from nuclear power plants. The output of those plants is electricity, and that nuclear-generated electricity represents 20 percent of U.S. electrical energy. Use these facts to determine the fraction of the total U.S. energy supply that is in the form of electricity.

4.   For the devices identified in Question 3, estimate the total time you use each device in a day, and use this quantity along with the power consumption to estimate the total energy each device uses in a day. (You can think of this quantity as an average power consumption for each device.)

5.   Your author's home uses about 450 kWh of electrical energy each month. Convert this to an average rate of electrical energy consumption in watts.

6.   Find the cost per kilowatt-hour of electricity in your area, and the cost of a gallon of gasoline. Considering that gasoline contains about 40 kWh per gallon, compare the price per kilowatt-hour for gasoline versus electrical energy.

7.   What is the approximate slope of the straight line shown in Figure 2.7? Be sure to include units. What is the meaning of this quantity?

8.   A country uses energy at the rate of 3.5 kW per capita, and its citizens enjoy an annual per capita GDP of $7,500. Calculate the energy intensity of this country, and determine

whether it uses energy more or less efficiently than the countries on the diagonal line in Figure 2.7.

RESEARCH PROBLEMS

1. Find five countries whose energy intensity is significantly greater than that of the United States. Is there an obvious explanation for their greater energy intensity? Do any of these countries have anything in common? Note: You may find territories of the United States listed as separate countries, but don't include them in your list.

2. China has undergone rapid industrialization in the past few decades, and its energy consumption has increased accordingly. What has happened to its energy intensity? Support your answer with a table or graph.

3. Find your state or country's per capita energy consumption and compare it with the U.S. average of roughly 11 kW per capita. Are there any subtleties in making this comparison?

4. Which of the U.S. or world energy-source categories in Figures 2.5 and 2.6 has changed most significantly over the past few decades? Support your answer with data.

5. France has pursued energy independence through reliance on nuclear power. Prepare a table or graph showing France's energy use by category, like the ones in Figures 2.5 and 2.6, and compare it with those figures.

6. Find values for the United Nations' Human Development Index (HDI) for the countries listed in Figure 2.7 and prepare a similar graph showing HDI versus energy consumption. Is there any correlation between the two? Is a saturation effect obvious? If you can't discern a trend, add more countries to your plot. You can find the data on the UN Development Programme's web site, where the annual *Human Development Report* is available.

ENERGY: A Closer Look

What, exactly, is this energy that we use at such a prodigious rate in modern society? You show your intuitive sense of the term *energy* when you speak of a person or a performance as having a "high energy level," when you find an event "energizing," or when, at the end of a long day, you're "low on energy." In all those cases, the word *energy* seems to be associated with motion, activity, or change. Although science can refine the meaning of energy, your intuitive sense is pretty good. Energy is the "stuff" that makes everything happen. Without energy, there would be no motion, and nothing would ever change.

Ultimately, energy is one manifestation of the basic "stuff" that makes up the universe. The other manifestation, perhaps more obvious, is matter. But there's only one kind of basic "stuff." As Einstein showed with his famous equation $E = mc^2$, energy and matter are in fact interchangeable. You can turn one into the other, with energy disappearing and matter materializing in its place, or vice versa, but the total amount of that basic substance—call it *mass-energy* for want of a better term—doesn't change. It's this property, called *conservation of mass-energy*, that makes it worthwhile to talk of mass-energy as a substance.

In our everyday world, however, the interchange of matter and energy is a very subtle effect, essentially immeasurable. So for us it's convenient to talk separately of matter and energy, and to consider that each is separately conserved. That's the approach I'll take almost everywhere in this book.

Energy can change from one form to another. For example, a car's engine converts some of the energy stored in gasoline into the energy of the car's motion, and then the car's brakes turn that energy into heat. But energy can't disappear into nothingness, nor can it be created. In addition to changing form, energy can also move from one place to another. For example, a house cools down as thermal energy inside the house flows out through the walls. But again the energy isn't gone; it's just relocated.

Although this treatment of energy and matter as distinct and separately conserved substances is justified in our everyday world, you should be aware that there are times and places where the interchange of matter and energy is so blatantly obvious that it can't be ignored. During the first second of the universe's existence, following the Big Bang, the average energy level was so high that matter particles could form out of pure energy, and matter could annihilate with antimatter to form pure energy. Neither energy nor matter was separately conserved. Physicists recreate the conditions of the early universe in the huge particle accelerators they use to probe the structure of matter, and here again neither matter nor energy is separately conserved. When two subatomic particles collide, for example, the result can be a whole spray of new particles created from the energy the particles brought to the collision. I'll have more to say about the interchangeability of matter and energy when we explore nuclear energy in Chapter 7.

3.1 Forms of Energy

Stand by the roadside as a truck roars by, and you have a gut sense of the vast energy associated with its motion. This energy of motion, **kinetic energy**, is perhaps the most obvious form of energy. But here's another, more subtle form: Imagine climbing a rock cliff, a process you can feel takes a lot of energy. You also know the danger you face, a danger that exists because the energy you put into the climb isn't gone, but could reappear as kinetic energy were you to fall. Or imagine that I lift a bowling ball and hold it over your head; again you're aware of a danger from an energy that isn't visibly obvious but that you know could reappear as kinetic energy of the ball's downward motion. This energy, associated with an object that's been lifted against Earth's gravity, is **potential energy**; specifically, **gravitational potential energy**. It's potential because it has the potential to turn into the more obvious kinetic energy, or for that matter into some other form. Stretch a rubber band or bungee cord, draw a bow or a slingshot, or compress a spring, and again you sense that there's stored energy. This is called **elastic potential energy** because it involves changing the configuration of an elastic substance. Figure 3.1 shows simple examples of potential and kinetic energy.

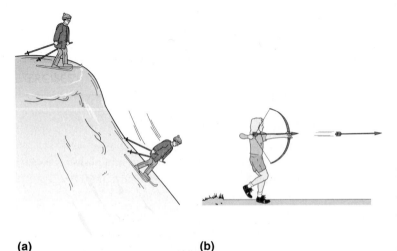

(a)　　　　　**(b)**

Figure 3.1
Potential and kinetic energy.
(a) Gravitational potential energy of the skier becomes kinetic energy as she heads down the slope. (b) Elastic potential energy stored in the bow becomes kinetic energy of the arrow.

Force and Energy

Are there other forms of energy? In the context of this book and its emphasis on the energy that powers human society, you might think of answers such as "coal," "wind energy," "solar energy," "waterpower," and the like. But here we'll take a more fundamental look at the different types of energy available to us. There's the kinetic energy associated with moving objects—essentially the same type of energy whether those objects are subatomic particles, trucks on a highway, Earth orbiting the Sun, or our whole Solar System in its stately 250-million-year circle around the center of the Milky Way galaxy. Then there's potential energy, which is intimately related to another fundamental concept—that of **force**. At the everyday level, you can think of a force as a push or a pull. Some forces are obvious, such as the pull of your arm as you drag your luggage through the airport, or the force your foot exerts when you kick a soccer ball. Others are equally evident but less visible, such as the gravitational force that pulls an apple from a tree or holds the Moon in its orbit, the force of attraction between a magnet and a nail, or the frictional force that makes it hard to push a heavy piece of furniture across the floor.

Today, physicists recognize just three fundamental forces that appear to govern all interactions in the universe. For our purposes I'm going to discuss the **gravitational force**, the **electromagnetic force**, and the **nuclear force**, although a physicist would be quick to point out that the electromagnetic and nuclear forces are simply aspects of more fundamental forces. A "holy grail" of science is to understand all three forces as aspects of a single interaction that governs all matter and energy, but we're probably some decades away from achieving that understanding. Figure 3.2 suggests realms and applications in which each of the fundamental forces is important.

Gravity seems familiar, since we're acutely aware of it in our everyday lives here on Earth. It's the force that keeps us on the planet, that pulls a falling apple to the ground, and that holds the Moon in orbit. Earth has no monopoly on gravity; it's the Sun's gravity

Figure 3.2

The fundamental forces and some applications relevant to this book. (a) Gravity governs the large-scale structure of the universe. It holds you to Earth and keeps a satellite in orbit. Gravitational potential energy is the energy source for hydroelectric power plants. (b) The electromagnetic force is responsible for the structure of matter at the molecular level; the associated potential energy is released in chemical reactions such as those that occur in burning fuels. Electromagnetism is also involved in the production and transmission of electrical energy. (c) The nuclear force binds protons and neutrons to make atomic nuclei. The associated potential energy is the energy source for nuclear power plants.

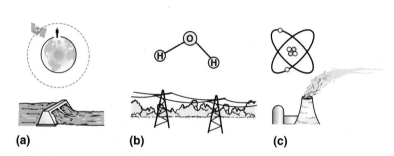

(a)   (b)   (c)

that tugs on Earth to keep our planet on its yearlong orbital journey. Actually, gravity is universal; it's a force of attraction that acts between every two pieces of matter in the universe. But it's the weakest of the fundamental forces and is significant only with large-scale accumulations of matter such as planets and stars.

The electromagnetic force comprises two related forces involving electricity and magnetism. The electric force acts between matter particles carrying the fundamental property we call electric charge. The magnetic force also acts between electric charges, but only when they're in relative motion. In this sense magnetism is intimately related to electricity. The two are complementary aspects of the same underlying phenomenon, which we call electromagnetism. That complementarity has much to do with the ways we generate and transport electrical energy. Given the importance of electromagnetism in our energy technologies, I'll say a lot more about it in Section 3.2.

The nuclear force binds together the protons and neutrons that form atomic nuclei. It's the strongest of the three forces—a fact that accounts for the huge difference between nuclear and chemical energy sources, as I'll describe in Chapter 7. We can thank nuclear forces acting deep inside the Sun for the stream of sunlight that supplies nearly all the energy arriving at Earth.

Forces can act on matter to give it kinetic energy, as when an apple falls from a tree and the gravitational force increases its speed and hence its kinetic energy, or when, under the influence of the electric force, an electron "falls" toward a proton to form a hydrogen atom and the energy ultimately emerges as a burst of light. Alternatively, when matter moves against the push of a given force, energy is stored as potential energy. That's what happened when I lifted that bowling ball over your head a few paragraphs ago, or when you pull two magnets apart, or when you (or some microscopic process) yank that electron off the atom, separating positive and negative charge by pulling against the attractive electric force. So each of the fundamental forces has associated with it a kind of potential energy.

What happened to all those other kinds of forces, like the push of your hand or the kick of your foot, or the force in a stretched rubber band, or friction? They're all manifestations of one of the three fundamental forces. And in our everyday lives, the only forces we usually deal with are gravity and the electromagnetic force. Gravity is pretty obvious: We store gravitational energy any time we lift something or climb a flight of stairs. We gain kinetic energy from gravity when we drop an object, take a fall, or coast down a hill on a bicycle or skis. We exploit the gravitational force and gravitational energy when we generate electricity from falling water. All the other forces we deal with in everyday life are ultimately electromagnetic, including the forces in springs, bungee cords, and rubber bands, and the associated potential energy. More significantly for our study of human energy use, it also includes the energy stored in the food we eat and in the fuels we burn. The energy stored in a molecule of gasoline, for example, is associated with arrangements of electric charge that result in electromagnetic potential energy. (In these cases the energy is essentially all electrical energy; magnetism plays no significant role in the interactions among atoms that are at the basis of chemistry and chemical fuels.) In this book, we're also concerned with one form of energy that isn't either gravitational or electromagnetic—namely, the nuclear energy that we use to generate electricity and that our star uses to make sunlight.

3.2 Electrical Energy: A Closer Look

Obviously, electricity involves energy. You probably know that electricity, as we commonly think of it, is a flow of electrons through a wire. That flow is called **electric current**, and what's important here is that the electrons carry **electric charge,** a fundamental electrical property of matter. Electrons carry negative charge, and protons carry equal but opposite positive charge. Because electrons are much lighter, they're usually the particles that move to carry electric current. You might think that the energy associated with electricity is the kinetic energy of those moving electrons, but this isn't the case. The energy associated with electricity, and with its close cousin, magnetism, is in the form of invisible **electric fields** and **magnetic fields** created by the electric charges of the electrons and protons that make up matter. In the case of electric current, electric and magnetic fields surround a current-carrying wire and act together to move electrical and magnetic energy along the wire. Although a little of that moving energy is inside the wire, most is actually in the space immediately around it! Strictly speaking, the energy moving along a current-carrying wire isn't electrical energy but electromagnetic energy, a term that reflects the intimate relationship between electricity and magnetism.

Making Electricity

Electrical energy, as associated with electric current flowing in wires, is an important and growing form of energy in modern society (Fig. 3.3). In the United States, for example, about 40 percent of our overall energy consumption goes toward making electricity, although only about a third of that actually ends up as electricity, for important reasons that I'll discuss in the next chapter. Electrical energy plays an increasingly important role for two reasons. First, it's versatile: Electrical energy can be converted with nearly 100 percent efficiency to any other kind of energy—mechanical energy, heat, light, or whatever you want. Second, electrical energy is especially easy to transport. Thin wires made from electrically conducting material are all it takes to guide the flows of electrical energy over hundreds and even thousands of miles.

How do we produce electrical energy? In principle, any process that forces electric charges apart will do the trick. In **batteries**, chemical reactions separate positive and negative charge, transforming the energy contained within individual molecules into the energy associated with distinct regions of positive and negative charge—the two terminals of the battery (Fig. 3.4). Hook a complete circuit between the terminals, such as a lightbulb, motor, or other electric device, and current flows through it, converting electrical energy into light, mechanical energy, or whatever.

In some batteries, the chemical reactions go only one way, and when the original chemicals have given up all their energy, the battery is "dead" and must be discarded

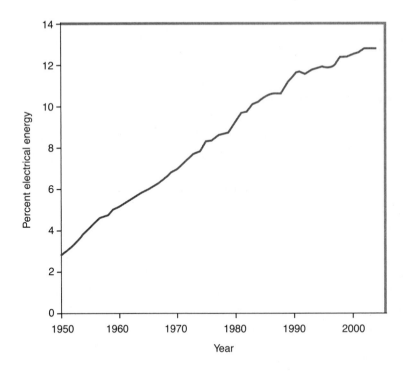

Figure 3.3
End-use electrical energy consumption in the United States has increased rapidly and is now around 13 percent of total U.S. energy consumption. But inefficiencies in power plants mean that it takes nearly 40 percent of the total U.S. energy supply to generate the 13 percent that's consumed as electricity.

(or, better, recycled). In other batteries, the chemical reactions can be reversed by forcing electric current through the battery in the direction opposite its normal flow. In this mode, electrical energy is converted into chemical energy, which is the opposite of what happens in the battery's normal operation. Such batteries are, obviously, **rechargeable**. Rechargeable batteries power your cell phone and laptop computer, and it's a large rechargeable battery that starts your car or, if you have a gas-electric hybrid, provides some of the car's motive power. Batteries are great for portable devices, but they're hardly

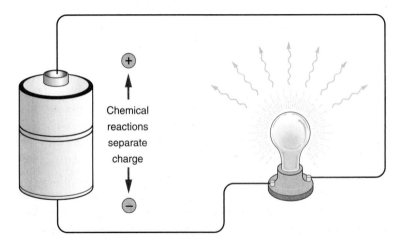

Figure 3.4
Chemical reactions in a battery separate positive and negative charge. Connecting an external circuit, such as the lightbulb shown here, allows the battery to deliver electrical energy as charge flows through the circuit.

up to the task of supplying the vast amounts of electrical energy that we use in modern society. Nor are the batteries available today a serious alternative to the internal-combustion engine as a source of energy for transportation, as shown by the demise of all-electric cars, first in the early twentieth century and again following a brief revival at the end of that century.

A cousin of the battery is the **fuel cell**, which, conceptually, is like a battery whose energy-containing chemicals are supplied continually from an external source. Today fuel cells find use in a number of specialized applications, including increasingly as backup sources of electrical energy in the event of power failures. And they, unlike conventional batteries, have substantial potential for powering transportation vehicles. I'll describe their operation and potential role in transportation in Chapter 11.

So where does most of our electrical energy come from, if not from batteries? It originates in a phenomenon that reflects the intimate connection between electricity and magnetism. Known since the early nineteenth century and termed **electromagnetic induction**, this fundamental phenomenon entails the creation of electrical effects from *changing* magnetism. Wave a magnet around in the vicinity of a loop of conducting wire, and an electric current flows in the wire. Move a magnetized strip past a wire coil and you create a current in the coil, which is what happens when you swipe your credit card to pay for your groceries. And on a much larger scale, rotate a coil of wire in the vicinity of a magnet and you have an electric generator that can produce hundreds of millions of watts of electric power. It's from such generators that the world gets the vast majority of its electric power (Fig. 3.5).

Remember, however, that energy is conserved, so an electric generator can't actually *make* energy; it simply converts mechanical energy into electrical energy. Recall Figure 2.1, which shows a person turning a hand-cranked electric generator to power a lightbulb. The person turning the generator is working hard because the electrical energy to light the bulb comes from her muscles. The whole sequence goes something like this:

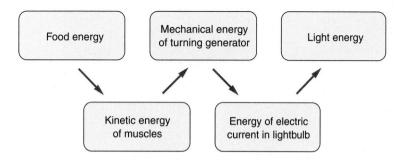

Had the lightbulb been off, and not connected in a complete circuit to the generator, the generator would have been very easy to turn. Why? Because turning it wouldn't produce electrical energy, so the turner wouldn't need to supply any energy. So how does the generator "know" to become hard to turn if the lightbulb is connected? Ultimately,

(a)

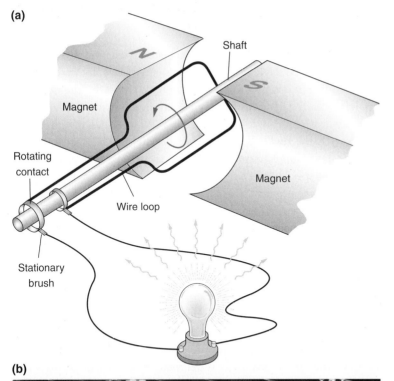

Shaft

N

Magnet

S

Magnet

Rotating
contact

Wire loop

Stationary
brush

(b)

Figure 3.5

(a) A simple electric generator consists of a
single wire loop rotating between the poles
of a magnet. The stationary brushes allow
current to flow from the rotating contacts to
the external circuit. A practical generator has
many loops, each with many turns of wire.
(b) This large generator produces 650 MW
of electric power.

the answer lies in another fundamental relationship between electricity and magnetism: Electric current—*moving* electric charge—is what gives rise to magnetism; that's how an electromagnet works. So when current is flowing, there are *two* magnets in the generator—the original, "real" magnet that was built into the generator, and the electromagnet arising from current flowing through the rotating coil. These two magnets repel each other, and that's what makes the generator hard to turn.

I'm going into this level of physics detail because I want you to be acutely aware of what happens every time you turn on a lightbulb, a stereo, a TV, a hair dryer, or an electric appliance: More current flows, and somewhere an electric generator gets harder to turn. Despite my fanciful energy servants of Chapter 2, the generators that produce our electric power aren't cranked by human hands. Most are turned by high-pressure steam created using heat from the combustion of fossil fuels or the fissioning of uranium, some are turned by flowing water, and a tiny fraction are driven by geothermal steam or wind (Fig. 3.6). So when you turn on that light or whatever, a little more fossil fuel has to be burned, or uranium fissioned, or water let through a dam, to satisfy your demand for energy. The electric switches you choose to flip on are one of your direct connections to the energy-consumption patterns I described in Chapter 2, and the physical realization of that connection is in the electromagnetic interactions that make electric generators hard to turn.

There are a handful of other approaches to separating electric charge and thus generating electrical energy, but none plays a significant role in global energy production today.

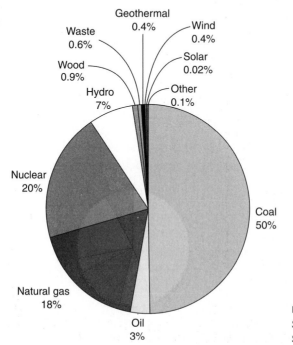

Figure 3.6
Sources of electrical energy in the United States.

Heat applied to a junction of two different electrically conducting materials results in a current flow, but this process is too inefficient for large-scale energy production. It's used on spacecraft exploring the outer Solar System, where sunlight is too weak to be a viable energy source, and in other specialized and limited applications. Closer to the Sun, the energy in sunlight can be made to separate charge and thus drive substantial electric currents in specialized semiconductor devices called **photovoltaic cells**. Today photovoltaics power most near-Earth spacecraft, from TV satellites to the Hubble Space Telescope, and are increasingly used in remote terrestrial applications ranging from weather-sensing buoys deployed at sea to water pumping in remote villages of the developing world. At present, however, photovoltaics play an insignificant role in global electrical energy production, but this could change dramatically, as I'll discuss in Chapter 9.

Stored Electrical Energy

Electrical energy is associated not only with electric currents and the batteries, generators, and photovoltaic devices that drive them; it also exists in every configuration of electric charges. Since matter is made up of electrons and protons, all matter contains electrical energy. Just how much depends on how the electric charges are arranged. We can create stored electrical energy, for example, by putting positive charge on one metal surface and negative charge on a separate, nearby metal surface. Configurations such as this one store the energy that powers a camera flash or represents information saved in your computer's memory. A larger but similar example is a lightning storm, in which electrical energy is stored in the electric fields associated with layers of charge that build up in violent storm clouds. That energy is released—converted to heat, light, and sound—by lightning discharges. In these and the cases I'll describe next, by the way, the term *electrical energy* is more appropriate than *electromagnetic energy*; magnetism plays a very small role in the energy storage associated with configurations of electric charge, especially when the movement of charge is not significant.

Far more important than the energy of a lightning storm or energy-storage technology is the electrical energy stored at the microscopic level in the configurations of electric charge that we call molecules. That's right: Molecules, made up of anywhere from a few to a few thousand atoms, are arrangements of the electric charges that make up their constituent atoms. How much energy is stored in a given molecule depends on its physical structure, which changes in the chemical reactions that rearrange atoms. Most fuels, including in particular the fossil fuels that I introduced in Chapter 1, are substances that store energy in the electric fields associated with their molecular arrangements of electric charges. Gasoline, for example, has molecules consisting of carbon and hydrogen. When gasoline burns, its molecules interact with atmospheric oxygen to form carbon dioxide (CO_2) and water (H_2O). The total electrical energy stored in the CO_2 and H_2O molecules is less than what was in the original gasoline and oxygen, so the process of burning gasoline releases energy. We'll see in Chapter 5 how we harness that energy.

So fuels store energy, ultimately, as electrical energy of molecular configurations. We generally call that stored energy **chemical energy**, because chemistry is all about the interaction of atoms to make molecules. But the origin of chemical energy is in the electrical nature of matter. By the way, not all fuels are chemical. In Chapter 1, I mentioned the nuclear fuels that power our reactors and bombs; they derive their energy from fields associated with electrical and nuclear forces that act among particles within the atomic nucleus. I'll continue to use the term *fuel* broadly to mean a substance that stores energy in its microscopic structure, at either the atomic/molecular level (chemical fuel) or at the nuclear level (nuclear fuel).

Magnetism is intimately related to electricity and it, too, can store energy. Magnetic energy plays a lesser role than electricity in our everyday lives, although it is associated with the conversion of kinetic energy to electrical energy in electric generators, and vice versa in electric motors. Magnetic energy storage is important in some technological devices, and it plays a major role in the giant eruptive outbursts that occur on our Sun and sometimes hurl high-energy particles toward Earth. The brilliant auroral displays visible at high latitudes are the result of these particles interacting with Earth's own magnetic field and atmosphere. Although stored magnetic energy is of less practical importance than stored electrical energy, be aware that any flow or movement of what you might want to call electrical energy must necessarily involve magnetism, too. This is the case for the current-carrying wire that I discussed at the beginning of this section (Section 3.2), and it's especially true of another means of energy transport that I'll describe next.

Electromagnetic Radiation

What's the ultimate source of the energy that powers life on Earth? The Sun, of course. As shown in Figure 1.8, our star accounts for 99.98 percent of the energy reaching Earth's surface. We'll see in Chapters 7 and 11 how the Sun's energy arises from nuclear reactions deep in the solar core, but here's the important point for now: Sunlight carries energy across the 93 million miles of empty space between the Sun and Earth, so light itself must be a form of energy. Light—along with radio, TV, and microwaves; infrared, ultraviolet, and X rays; and the penetrating nuclear radiation known as gamma rays— are a kind of energy-in-transit called **electromagnetic radiation**. Light and other electromagnetic radiation are made up of **electromagnetic waves**, structures of electric and magnetic fields in which change in one field regenerates the other to keep the wave moving and carrying energy through empty space (Figures 3.7 and 3.8). I'll occasionally talk of "light energy" or "the energy of electromagnetic radiation" as though these were yet another form of energy, but ultimately, light and other electromagnetic radiation are a manifestation of the energy contained in electric and magnetic fields.

Electromagnetic waves originate in the accelerated motion of electric charges— electric currents in antennas producing radio waves; the jostling of charges in a hot object, causing it to glow; atomic electrons jumping between energy levels to produce light; atomic nuclei rearranging themselves and emitting gamma rays. And when electromagnetic waves interact with matter, it's ultimately through their fields producing forces on electric charges—a TV signal moving electrons in an antenna; light exciting nerve impulses in the receptor cells of your eye; sunlight jostling electrons in a solar collector to produce heat, or in the chlorophyll of a green plant to produce energy-storing sugar.

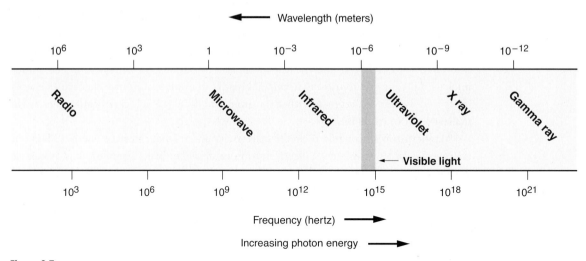

Figure 3.7

The electromagnetic spectrum. Electromagnetic waves are characterized by their frequency in hertz (wave cycles per second) or their wavelength in meters (the distance between wave crests). The two are inversely related. Note the highly nonlinear scale, with each tick mark representing a factor of 1,000 increase or decrease in frequency or wavelength. The Sun's energy output is mostly in the visible and adjacent infrared portion of the spectrum, with a little ultraviolet.

Although it usually suffices to think of electromagnetic radiation in terms of waves, at the subatomic level the interaction of radiation with matter involves quantum physics. Specifically, the energy in electromagnetic waves comes in "bundles" called **photons**, and it's the interaction of individual photons with electrons that transfers electromagnetic wave energy to matter. For a given frequency f of electromagnetic wave, there's a minimum possible amount of energy, corresponding to one photon. In terms of frequency or, equivalently, wavelength, that energy is given by

$$E = hf = \frac{hc}{\lambda} \text{ (photon energy)} \tag{3.1}$$

Here E is the photon energy in joules (J); f is the frequency in hertz, or cycles per second; and λ is the wavelength in meters. The constants c and h are, respectively, the speed of light ($c = 3.00 \times 10^8$ m/s) and **Planck's constant** ($h = 6.63 \times 10^{-34}$ J·s), which

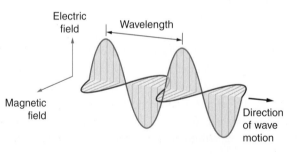

Figure 3.8

Structure of an electromagnetic wave. The wave carries energy in its electric and magnetic fields, which are at right angles each other and to the direction in which the wave is traveling.

describes the fundamental "graininess" of nature at the quantum level. We'll find the concept of quantized energy and photons especially useful in Chapter 9's description of photovoltaic solar energy conversion.

3.3 Quantifying Energy

A common excuse against increased use of solar energy is that there isn't enough of it. That excuse is wrong, as I'll make clear in Chapter 9, but you can't very well make an argument against it unless you have a quantitative answer to the question "How much is there?" You may be enthusiastic about wind energy, but unless you can convince me that the wind resource is sufficient in quantity for wind to make a significant contribution to our energy needs, I'm unlikely to share your enthusiasm. A noted environmentalist claims that buying an SUV instead of a regular car wastes as much energy as leaving your refrigerator door open for seven years. Is this right? Or is it nonsense? You can't make a judgment unless you can get quantitative about energy.

As these examples suggest, we often need to use numerical quantities to describe amounts of energy and rates of energy use. In Chapter 2, I introduced a basic unit of power, or rate of energy use—the watt (W). I explained what a watt is by quantifying the rate at which your own body can produce energy, about 100 W. I also introduced the prefix *kilo,* meaning 1,000, hence the kilowatt (kW, or 1,000 W). I emphasized the distinction between power—the rate of energy use—and actual amounts of energy. And I introduced one unit that's often used to describe amounts of energy, namely the kilowatt-hour (kWh). Again, here's the relationship between power and energy: Power is a rate and can be expressed in energy units per time units. Equivalently, energy amounts can be expressed in power units multiplied by time units; hence the kilowatt-hour as an energy unit.

Suppose you use energy at the rate of 1 kW for two hours. Then you've used (1 kW) × (2 h) = 2 kWh of energy. Or suppose you consume 60 kWh of energy in two hours. Then your energy-consumption rate is (60 kWh)/(2 h) = 30 kW. Notice how the units work out in this second calculation: The hours (h) cancel, giving, correctly, a unit of power, in this case the kilowatt. Even simpler: If you use 1 kWh of energy in one hour, your rate of energy consumption is 1 kWh/h, or just 1 kW. That is, a kilowatt-hour per hour is exactly the same as a kilowatt.

Energy Units

Still, the kilowatt-hour, although a perfectly good energy unit, seems a bit cumbersome. Is there a simple unit for energy itself that isn't expressed as a product of power with time? There are, in fact, many. In the International System of Units (SI, as it's abbreviated from the French Système International d'Unités), the official energy unit is the **joule** (J), named for the British physicist and brewer James Joule (1818–1889), who explored and quantified the relationship between mechanical energy and heat. SI is the standard unit system for the scientific community (and for most of the everyday world beyond the United States). If this were a pure science text, I would probably insist on expressing all energies in joules from now on. But energy is a subject of interest to scientists, economists, policymakers, investors, engineers, nutritionists, corporate executives, athletes, and many others, and as a result there are a great many different energy units in common use. I'll describe a number of them here, but I'll generally limit most of the discussion in this book to the joule, the kilowatt-hour, and only occasionally other units.

So what's a joule? Since you already know in your muscles what a watt is, I can easily describe what a joule is: 1 J is 1 watt-second (W·s). That is, if you use energy at the rate of 1 W for one second, then you've used 1 J of energy. Use energy at the rate of 1 kW for one second, and you've used (1,000 W) × (1 s) = 1,000 W·s = 1,000 J, or 1 kilojoule (kJ). Use energy at the rate of 1 kW for a full hour—3,600 seconds—and you've used (1,000 W) × (3,600 s) = 3.6 million W·s, or 3.6 megajoules (MJ; here the prefix *mega* stands for million). But 1 kW for one hour amounts to 1 kWh, so 1 kWh is 3.6 MJ.

If you'd like a more direct feel for a joule of energy, consider that this book weighs around 2 pounds. Lift it about 4 inches and you've expended about 1 J of energy, which is now stored as gravitational potential energy. In Section 3.4 I'll explain how I did this calculation, and I'll use a similar calculation to verify the 100-W figure from the knee-bend exercise in Chapter 2.

The joule is the official energy unit of the scientific community, but there are plenty of others. Table 3.1 lists a number of them, along with their equivalents in joules. A unit often used in biology, chemistry, and nutrition is the **calorie** (cal). One calorie is defined as the amount of energy it takes to raise the temperature of 1 gram of water by 1 degree Celsius (°C). The calorie was named before the connection between energy and heat was understood (more on this in Chapter 4). It was Joule himself who established this connection; today, we know that 1 cal is 4.184 J. (There are actually several definitions for the calorie, which vary slightly. I'm using what's called the *thermochemical calorie*.) You're probably most familiar with the calorie as something to be conscious of if you don't want to gain weight. The caloric value of food is indeed a measure of the food's energy content, but since your body stores unused food energy as chemical energy in molecules of fat, it's also an indication of potential weight gain. The "calorie" you see listed on a food's nutritional label is actually 1,000 cal, correctly called a *kilocalorie* (kcal) or sometimes a *large calorie* and written with a capital C: Calorie. Obviously, 1 kcal is then 4,184 J or 4.184 kJ. By the way, "low-calorie soda" might be a meaningful description for an American, but in other countries it's "low-joule soda," making obvious the point that calories and joules measure the same thing (Fig. 3.9). That thing is energy.

Since the calorie is a unit of energy, calories per time is a unit of power, convertible to watts. I'll let you show (in Exercise 1 at the end of this chapter) that the average human diet of 2,000 kcal per day is roughly equivalent to 100 W, thus providing another confirmation of Chapter 2's value of 100 W for the typical power output of the human body.

In the English system of units, no longer used in England but only in the United States and a very few other countries, the analog of the calorie is the **British thermal unit** (Btu), defined as the amount of energy needed to raise the temperature of 1 pound of water by 1 degree Fahrenheit (°F). Table 3.1 shows that 1 Btu is 1,054 J, or just over 1 kJ. You could also calculate this conversion knowing the relationship between grams and pounds, and the Celsius and Fahrenheit scales (see Exercise 2). A power unit often

Table 3.1 Energy and Power Units

Energy units	Joule equivalent*	Description
joule (J)	1 J	Official energy unit of the SI unit system; equivalent to 1 W · s or the energy involved in applying a force of 1 newton over a distance of 1 meter.
kilowatt-hour (kWh)	3.6 MJ	Energy associated with 1 kW used for one hour.
gigawatt-year	3.16 PJ	Energy produced by a typical large (1 gigawatt) power plant operating full-time for one year.
calorie (cal)	4.184 J	Energy needed to raise the temperature of 1 gram of water by 1°C.
British thermal unit (Btu)	1,054 J	Energy needed to raise the temperature of 1 pound of water by 1°F, very roughly equal to 1 kJ.
quad (Q)	1.054 EJ	Quad stands for quadrillion Btu, or 10^{15} Btu, and is roughly equal to 1 exajoule (10^{18} J).
erg	10^{-7} J	Energy unit in the centimeter-gram-second system of units.
electron volt (eV)	1.6×10^{-19} J	Energy gained by an electron dropping through an electric potential difference of 1 volt; used in atomic and nuclear physics.
foot-pound	1.356 J	Energy unit in the English system, equal to the energy involved in applying a force of 1 pound over a distance of 1 foot.
tonne oil equivalent (toe)	41.9 GJ	Energy content of 1 metric tonne (1,000 kg, roughly 1 English ton) of oil.
barrel of oil equivalent (boe)	6.12 GJ	Energy content of one 42-gallon barrel of oil.

Power units	Watt equivalent	Description
watt (W)	1 W	Equivalent to 1 J/s.
horsepower (hp)	746 W	Unit derived originally from power supplied by horses; now used primarily to describe engines and motors.
Btu per hour (Btu/h, or Btuh)	0.293 W	Used primarily in the United States, usually to describe heating and cooling systems.

*See Table 3.2 for SI prefixes.

used in the United States to describe the capacity of heating and air conditioning systems is the Btu per hour (Btu/h, but often written, misleadingly, as simply Btuh). As Table 3.1 shows, 1 Btu/h is just under one-third of a watt. My household furnace is rated at 112,000 Btu/h; Example 3.1 shows that this number is consistent with its fuel consumption rate of about 1 gallon of oil per hour. The British thermal unit is at the basis of a unit widely used in describing energy consumption of entire countries, namely the **quad** (Q). One quad is 1 quadrillion Btu, or 10^{15} Btu. The United States' rate of energy consumption in the early twenty-first century, for example, is just about 100 Q per year, a figure that accounts for approximately one-fourth of humankind's total yearly energy consumption.

Figure 3.9
"Low joule" describes this diet soft drink from Australia.
Joules and calories measure the same thing, namely energy.

Example 3.1 Home Heating

Assuming home heating oil contains about 40 kWh of energy per gallon, determine the approximate rate of oil consumption in my home furnace when it's producing heat at the rate of 112,000 Btu/h.

Solution

We have the oil's energy content in kilowatt-hours, so we need to convert that 112,000 Btu/h into compatible units, in this case kilowatts. Table 3.1 shows that 1 Btu is 1,054 J, or 1.054 kJ, and one hour is 3,600 seconds, so the heat output of the furnace becomes

$$(112,000 \text{ Btu/h})(1.054 \text{ kJ/Btu})(1/3600 \text{ h/s}) = 33 \text{ kJ/s} = 33 \text{ kW}$$

Notice how I was careful to write out all the units and to check that they multiplied together to give the correct final unit:

$$(\text{Btu/h})(\text{kJ/Btu})(\text{h/s}) \rightarrow \text{kJ/s} = \text{kW}$$

In any numerical problem like this one, it's essential that the units work out correctly. If they don't, the answer is wrong, even if you did the arithmetic correctly. Note in this case that the unit for hour (h) is in the denominator, so I had to multiply by hours per second (h/s) to convert the time unit to seconds. Since there are 3,600 seconds in one hour, there's $\frac{1}{3,600}$ of an hour in each second, hence the factor 1/3,600 h/s.

So my furnace produces heat at the rate of 33 kW, or 33 kWh/h. Since there are 40 kWh of energy in a gallon of oil, the furnace would need to burn a little over three-quarters of a gallon per hour (33/40) if it were perfectly efficient. But it isn't, so that 112,000 Btu/h heat output requires close to 1 gallon of oil per hour.

Table 3.1 lists several other energy units that we'll have little use for in this book, but that often appear in the scientific literature. The erg is the official unit of the centimeter-gram-second system of units (as opposed to the SI meter-kilogram-second units). The electron volt is a tiny unit (1.6×10^{-19} J) used widely in nuclear, atomic, and molecular physics. Finally, the foot-pound is an English unit, being the energy expended when you push on an object with a force of 1 pound as the object moves 1 foot.

The last two energy units in Table 3.1 deserve special mention because they're based not directly on energy but on oil. The **tonne oil equivalent** is the energy content of a metric ton (tonne) of typical oil, while the **barrel of oil equivalent** is the energy content of a standard 42-gallon barrel. The values given for the tonne oil equivalent and barrel of oil equivalent in Table 3.1 are formal definitions; the actual energy content of oil varies somewhat, depending on its source and composition.

Units for energy and other quantities are often modified with prefixes that indicate multiplication by various powers of ten; every three powers gets a new name. We've already met kilo (k, meaning 1,000 or 10^3) and mega (M, meaning 1 million or 10^6). Others you've surely heard of are giga (G, meaning 1 billion or 10^9), milli (m, meaning 1/1,000 or 10^{-3}), micro (μ, the Greek letter "mu," meaning one-millionth or 10^{-6}), and nano (n, meaning one-billionth or 10^{-9}). Table 3.2 lists others, all formally part of the SI unit system. Many are probably less familiar to you, but they often prove useful in quantifying the huge rates of energy consumption in the industrialized world. The United States' 100 Q per year, for example, can be expressed in SI as being about 100 exajoules per year or one-tenth of a zettajoule per year; converting the U.S. energy-consumption rate of 100 Q per year to watts (Exercise 3) shows that the total U.S. energy-consumption rate is about 3 terawatts (3 TW, or 3×10^{12} W). I'll use SI prefixes

Box 3.1   Tons and Tonnes

American readers are familiar with the **ton**, an English unit equal to 2,000 pounds and useful for describing large weights. (The pound—and weight itself—is actually a measure of force, used loosely for mass as well.) But most of the world talks in terms of **metric tons**, also called **tonnes**. One tonne is 1,000 kilograms. Since 1 kilogram is 2.2 pounds, 1 tonne is therefore 2,200 pounds—close enough, for our purposes, to an English ton. So when I refer to *tonnes*, you're welcome to think in terms of *tons*. The two are interchangeable to within 10 percent.

Table 3.2 SI Prefixes

Multiplier	Prefix	Symbol
10^{-24}	yocto	y
10^{-21}	zepto	z
10^{-18}	atto	a
10^{-15}	femto	f
10^{-12}	pico	p
10^{-9}	nano	n
10^{-6}	micro	μ
10^{-3}	milli	m
$10^{0} (= 1)$	—	—
10^{3}	kilo	k
10^{6}	mega	M
10^{9}	giga	G
10^{12}	tera	T
10^{15}	peta	P
10^{18}	exa	E
10^{21}	zeta	Z
10^{24}	yotta	Y

routinely throughout this book, and you can find them here and inside the front cover. Note that the symbols for SI prefixes that multiply by less than 1 are in lowercase, while those that multiply by more than 1 (except for the kilo) are capitalized.

Back to Table 3.1, which also lists several units for power. Among those in common use is **horsepower** (hp), a holdover from the day when horses supplied much of the energy coming from beyond our own bodies. One horsepower is 746 W, or about three-quarters of a kilowatt. So a 400-hp car engine can, in principle, supply energy at the rate of about 300 kW (most of the time the actual rate may be much less, and very little of that energy ends up propelling the car; more on this in Chapter 5).

Fuels—those substances that store potential energy in the configurations of molecules or atomic nuclei—are characterized by their energy content, expressed as energy

Table 3.3 Energy Content of Fuels

Fuel	Typical energy content (varies with fuel source)	
	SI units	Other units
Coal	29 MJ/kg	7,300 kWh/ton 25 MBtu/ton
Oil	43 MJ/kg	40 kWh/gallon 138 kBtu/gallon
Gasoline	44 MJ/kg	36 kWh/gallon
Natural gas	55 MJ/kg	30 kWh/100 cubic feet 1,000 Btu/cubic foot
Biomass, dry	15–20 MJ/kg	13–17 MBtu/ton
Hydrogen gas (H_2) burned to produce H_2O	142 MJ/kg	320 Btu/cubic foot
Uranium, nuclear fission Natural uranium Pure U-235	 580 GJ/kg 82 TJ/kg	 161 GWh/tonne 22.8 TWh/tonne
Hydrogen, deuterium-deuterium nuclear fusion Pure deuterium Normal water	 330 TJ/kg 12 GJ/kg	 13 MWh/gallon, 350 gallons gasoline equivalent per gallon water

contained in a given mass or volume. Table 3.3 lists the energy contents of some common fuels. Some of these quantities find their way into alternative units for energy and for energy-consumption rate, as in the tonne oil equivalent and barrel of oil equivalent listed in Table 3.1. Related power units of millions of barrels of oil equivalent per day often describe the production of fossil fuels, and national energy-consumption rates are sometimes given in millions of barrels of oil equivalent per year. Another handy figure to use in considering fossil fuel energy is to approximate the energy content of a gallon of petroleum product—oil, kerosene, gasoline—as being about 40 kWh (the exact amount varies with the fuel, and the amount of useful energy obtained depends on the efficiency of the energy-conversion process; more on this in the next chapter). Finally, Table 3.3 hints at the huge quantitative difference between chemical and nuclear fuels; just compare the energy per kilogram of petroleum with that of uranium!

You'll find Tables 3.1 through 3.3 sufficiently useful that they're printed inside the front cover for easy reference, along with other useful energy-related information.

3.4 Energy and Work

Push a stuck car out of the mud, a lawnmower through tall grass, or a heavy trunk across the floor, and in all cases you're doing a lot of work. **Work** has a precise scientific meaning: It's a quantity equal to the force you apply to an object multiplied by the distance over which you move that object. Work is essentially a measure of the energy you expend as you apply a force to an object. This energy may end up as increased kinetic energy of the object, as when you kick a soccer ball and set it in motion, or it may end up as potential energy, as when I lifted that bowling ball over your head. One caveat: You do work only when the force you apply to an object is in the direction of the object's motion. If the force is at right angles to the motion, no work is done, and there's no change in the object's energy. If the force is opposite the motion, then the work is negative and you take energy away from the object. Figure 3.10 illustrates these possibilities.

So work is a measure of energy supplied to an object by mechanical means—that is, by applying a force, such as pushing or pulling. Again, it's the product of force times the distance the object moves:

$$W = Fd. \tag{3.2}$$

Here W is the work, F the force, and d the distance. Equation 3.2 applies in the case where the force and the object's motion are in the same direction. In the English system commonly used in the United States, force is measured in pounds and distance in feet; hence the English unit of work is the foot-pound (see Table 3.1). In the SI system, the unit of force is the **newton** (N), named in honor of the great physicist Isaac Newton, who formulated the laws governing motion. One newton is roughly one-fifth of a pound (1 pound = 4.48 N). The SI unit of distance is the meter, so work is measured in newton-meters (N·m). And what's a newton-meter? Since work is a measure of energy transferred to an object, 1 N·m is the SI unit of energy, namely the joule. So another

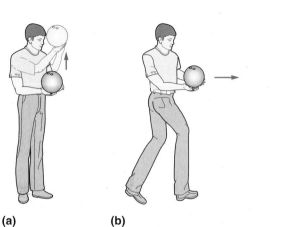

Figure 3.10
Force and work. (a) When you lift a bowling ball, you apply a force in the direction of its motion. You do work on the ball, in this case increasing its gravitational potential energy. (b) When you carry the ball horizontally, you apply a force to counter its weight, but you don't do work on the ball because the force is at right angles to its motion. (c) When you apply an upward force to stop a falling ball, the force is opposite the ball's motion, so here you do negative work that reduces the ball's kinetic energy.

(a)   **(b)**   **(c)**

way of understanding the joule is to consider it the energy supplied by exerting a force of 1 N on an object as the object moves a distance of 1 m.

An object's **weight** is the force that gravity exerts on it. Near Earth's surface, the strength of gravity is such that an object experiences a force of 9.8 N for every kilogram of mass it possesses. We designate this quantity g, the strength of gravity near Earth's surface. The value of g is then 9.8 newtons per kilogram (N/kg), which is close enough to 10 N/kg that I'll frequently round it up. (If you've had a physics course, you probably know g as the acceleration of gravity—the rate at which an object falling near Earth's surface gains speed. My definition here is equivalent, but more useful for thinking about energy.) We can sum all this up mathematically: If an object has mass m, then its weight is given by

$$F_g = mg \tag{3.3}$$

Here I've designated weight as F_g, for "force of gravity," because I've already used the symbol W for work. If m is in kilograms and g in newtons per kilogram, then the weight is in newtons.

To lift an object at a steady rate, you have to apply a force that counters the force of gravity. That is, the force you apply is equal to the object's weight, which Equation 3.3 shows is simply the product mg. Suppose you lift the object a height h. Using h for the distance in Equation 3.2, we can then combine Equations 3.2 and 3.3 to get an expression for the work you do in lifting the object:

$$W = mgh \tag{3.4}$$

This work ends up being stored as gravitational potential energy. Since the gravitational force "gives back" stored potential energy, Equation 3.4 also describes the energy gained, perhaps as kinetic energy, when an object with mass m falls a distance h.

Earlier I suggested that you could get a feel for the size of a joule by lifting your 2-pound book about 4 inches. Two pounds is about 1 kg, so your book's weight is about 10 N (here I multiplied 1 kg by the approximate value for g, namely 10 N/kg). Four inches is about 10 centimeters (cm), or one-tenth of a meter, so Equation 3.2 gives $W = (10\ \text{N}) \times (0.10\ \text{m}) = 1.0\ \text{J}$. Or just derive it from Equation 3.4:

$$W = mgh = (1\ \text{kg})(10\ \text{N/kg})(0.10\ \text{m}) = 1\ \text{J}$$

So there it is: Lift your 2-pound book 4 inches and you've done one 1 J of work, giving the book 1 J of gravitational potential energy.

What's your weight? My mass is about 70 kg, so according to Equation 3.3 my weight is (70 kg) $\times$ (10 N/kg) = 700 N. With 4.48 N per pound, this is equivalent to (700 N)/(4.48 N/lb) = 156 lb. That's just what my U.S. scale reads. Suppose I start doing those knee bends from Chapter 2. I've marked the position of the top of my head on a chalkboard when standing and again when at the lowest point of my knee bend, and the marks are 17 cm apart (0.17 m) (Fig. 3.11). So as I come up out of each knee bend, I raise most of my weight by 17 cm. Most? Yes: My feet stay on the ground and

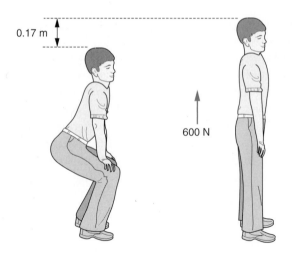

0.17 m

600 N

Figure 3.11
Rising out of a knee bend requires that I apply a 600-N force over a distance of 0.17 m, resulting in 100 J of work done. Repeating once per second gives a power output of 100 W.

my legs participate in only some of the motion. So the weight that's actually raised is somewhat less than my actual weight; suppose it's about 600 N. Then each time I rise to my full height, the work I do is $W = (600\ \text{N}) \times (0.17\ \text{m}) \simeq 100\ \text{N·m} \simeq 100\ \text{J}$. If I do those knee bends once every second, that's 100 J/s, or 100 W. Again, this is a rough figure; you could quibble with my choice for just what fraction of my weight I actually raise, and surely the once-per-second is only approximate. But there it is: While doing those knee bends, my body expends energy at the rate of roughly 100 W.

You might wonder what happens as I lower my body when bending my knees. In that downward motion I do negative work, or the force of gravity does work on me. If my muscles were like springs, this energy would be stored as elastic potential energy and be available for the next upward motion, with the result that my overall energy output would be considerably less. But my muscles aren't springs, and most of that energy is lost in heating the muscles and other body tissues. There is a little bit of springlike energy storage in muscles, but it isn't very significant with the relatively slow muscle contractions and extensions of this knee-bend exercise.

Example 3.2   Mountain Run!

(a) How much work do I do in running up a mountain with a vertical climb of 2,500 feet (760 m)? Express the answer in both joules and calories. (b) If the run takes 50 minutes, what's the average power involved?

Solution

My mass is 70 kg, g is 9.8 N/kg, and I'm climbing 760 m; Equation 3.4 then gives

$$W = mgh = (70\ \text{kg})(9.8\ \text{N/kg})(760\ \text{m}) = 521\ \text{kJ}$$

With 1 kcal = 4.184 kJ, this amounts to (521 kJ)/(4.184 kJ/kcal) = 125 kcal. This is the bare minimum amount of work it would take to scale the mountain because my body is far from 100 percent efficient, and I do work against frictional forces in my own muscles even when walking horizontally. So I probably burn up a lot more than the 125 "calories"—actually kilocalories—implied by this answer.

Expending those 521 kJ over 50 minutes gives an average power of

$$P = \frac{521 \text{ kJ}}{(50 \text{ min})(60 \text{ s/min})} = 0.174 \text{ kJ/s} = 174 \text{ W}$$

Again, the actual rate of energy expenditure would be a lot higher, although it might take rather more than 50 minutes to make a 2,500-foot climb.

3.5   Work and Kinetic Energy

In the examples of the preceding section, we did work that resulted in stored gravitational potential energy. But that happens only when the force we apply acts against another force, as in lifting an object upward against the force of gravity. In cases where there is no opposing force, the work then goes into increasing an object's kinetic energy. Examples include pushing a fellow skater or a hockey puck on ice, kicking a ball on a smooth horizontal surface, accelerating a car from a stoplight, or any other case in which opposing forces, including friction, are absent or negligible or themselves do no work.

An important result [that follows from Newton's laws of motion] is the work-energy theorem, which states that the net work done on an object ends up increasing the object's energy of motion—its kinetic energy. The theorem specifically identifies kinetic energy K with the quantity $\frac{1}{2} mv^2$, where m is the object's mass and v (for velocity) is its speed:

$$K = \frac{1}{2} mv^2 \quad \text{(kinetic energy)} \tag{3.5}$$

The increase in kinetic energy is numerically equal to the net work done on the object. By "net work" I mean the work associated with all the forces acting on the object. When you lift an object with constant speed, for example, you exert an upward force that just balances gravity. The net force is zero, the net work done is zero, and the kinetic energy doesn't change. But when you push an object along a horizontal surface where friction is negligible, you do work that ends up accelerating the object; both the speed v and kinetic energy $\frac{1}{2} mv^2$ increase.

Note that the kinetic energy depends on an object's speed *squared*, which means kinetic energy increases rapidly with speed. For example, doubling the speed (a factor of 2) requires quadrupling the energy (a factor of 2^2, or 4). That's one reason why driving at high speeds is particularly dangerous, more so than if energy increased in direct proportion to speed (Fig. 3.12). And it's also why accelerating rapidly to high speeds is a rather inefficient use of fuel.

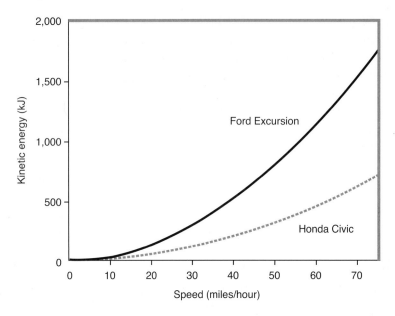

Figure 3.12
Kinetic energy increases with the square of the speed. Shown here is kinetic energy versus speed for a Honda Civic compact car (empty weight 1,268 kg) and a Ford Excursion SUV (empty weight 3,129 kg), each with a 68-kg driver. The weight difference accounts for much of the difference in fuel efficiency between these vehicles.

Example 3.3 Takeoff Power!

A fully loaded Boeing 767-300 jetliner has a mass of 180,000 kg. Starting from rest, it accelerates down the runway and reaches a takeoff speed of 270 kilometers per hour (km/h) in 35 seconds. What engine power is required for this takeoff roll?

Solution

The plane gains kinetic energy during the takeoff roll, and we know how long the roll takes, so we can calculate the energy per time, or power. Equation 3.5 gives energy in joules if the mass is in kilograms and speed is in meters per second. So first we convert that 270 km/h into meters per second:

$$(270 \text{ km/h})(1,000 \text{ m/km})(1/3,600 \text{ h/s}) = 75 \text{ m/s}$$

Then the plane's kinetic energy at takeoff is

$$K = \frac{1}{2}\,mv^2 = \left(\frac{1}{2}\right)(180,000 \text{ kg})(75 \text{ m/s})^2 = 5.1 \times 10^8 \text{ J} = 510 \text{ MJ}$$

The plane is on the runway for 35 seconds, so the engines must be supplying power at the rate of 510 MJ/35 s = 14.6 MJ/s or 14.6 MW. With 1 hp being 746 W, that's about 20,000 hp.

3.6   The Role of Friction

We've now seen that doing work on an object generally results in either stored potential energy or increased kinetic energy (or both if, for example, you simultaneously lift and accelerate an object). But there's one case where doing work increases neither potential nor kinetic energy. That's when you do work solely to overcome friction. When I lifted that bowling ball over your head, you were worried because you knew that the gravitational force would "give back" the stored potential energy if I let go of the ball. But if you push a heavy trunk across a level floor, you don't have to worry about the trunk sliding back and doing damage once you let go. Why not? Because the energy you expended pushing the trunk didn't end up as stored potential energy. But energy is conserved, so where did it go? Most of it went into the form of energy we call, loosely, "heat." Unlike the case of lifting an object or compressing a spring, that energy became unavailable for conversion back to kinetic energy. To convince yourself that friction turns mechanical energy into heat, just try rubbing your hands rapidly together!

The **frictional force** is fundamentally different from forces like gravity or the force in a spring, in that work done against friction doesn't end up as stored potential energy. Rather, it ends up in a form—"heat"—that isn't particularly useful. This inability to recover the energy "lost" to friction is an important limitation on our efforts to use energy wisely and efficiently. For example, despite our best engineering efforts, some two-thirds of the mechanical energy extracted from gasoline in a typical automobile is lost to friction in the engine, transmission, and tires. Similar "losses" occur even in cases where friction isn't present, including additional losses in automobile engines, electric power plants, and other energy-extracting devices. We'll take a closer look at these situations in the next chapter.

3.7   The Art of Estimation

In this chapter I've thrown around a lot of numbers and a few equations, because it's important to be able to calculate, and calculate accurately, when dealing quantitatively with energy. But it's equally important to be able to make a quick, rough, "back of the envelope" estimate. This kind of estimate won't be exactly right, but it should be close enough to be useful. Many times that's all you need to appreciate an important energy concept or quantity. Below are a couple of examples of estimation. Note how readily I've approximated quantities to nice, round numbers or guessed quantities I wasn't sure of.

Example 3.4   Lots of Gasoline

What's the United States' annual gasoline consumption?

Solution

How many cars are there in the United States? I don't know, but I do know that there are around 300 million people. Suppose there are half as many cars—150 million. How far does a typical car go in one year? Certainly more than a few thousand miles, but (for most of us), not 50,000 or 100,000 miles. So suppose it's about 10,000 miles. And what's the average fuel efficiency of a car in the United States? Unfortunately, that's been falling in recent years, and it's in the low twenties of miles per gallon. Let's call it just 20 miles per gallon. Then the total amount of gasoline used is

$$(150 \times 10^6 \text{ cars})(10^4 \text{ miles/car/year})\left(\frac{1}{20 \text{ miles/gallon}}\right) = 8 \times 10^{10} \text{ gallons/year}$$

Note my use of scientific notation to make dealing with big numbers easier. I rounded the final result to just one digit because this is an estimate; I might as well have rounded further to 10^{11} gallons per year because the numbers going into my estimate were so imprecise. How did I know to put the 20 miles per gallon in the denominator? Two ways: First, it makes the units come out right, as they must. I was asking for gallons per year, so I better do a calculation that gives gallons per year. Second, and a bit more physically meaningful, I know that higher fuel efficiency should result in lower fuel consumption, so the fuel efficiency had better be in the denominator. More cars and more miles driven increase consumption, so they go in the numerator.

How good is this estimate? According to the U.S. Energy Information Administration, daily gasoline consumption in the United States during 2005 was 385 million gallons per day. Multiplying by 365 days per year gives 1.4×10^{11} gallons. My estimate, just under 10^{11} gallons per year, is a bit low but certainly in the ballpark, and I promise you that I didn't look up the actual figure until after making the estimate. By the way, the number of cars registered in the United States in the early 2000s was about 140 million, so my estimate there was quite good.

Example 3.5   A Solar-Powered Country?

The average rate at which solar energy reaches Earth's surface is about 240 W/m²—that is, 240 watts on each square meter of surface area (this figure accounts for night and day, reflection off clouds, and other factors). If we had devices capable of capturing this energy with 100 percent efficiency, how much area would be needed to supply all U.S. energy needs?

Solution

At what rate does the United States use energy? The U.S. population is about 300 million, and as we saw in the preceding chapter, U.S. residents consume energy at the rate of about 10 kW per capita. So the area needed would be:

$$(300 \times 10^6 \text{ people})(10^4 \text{ W/person})\left(\frac{1}{240 \text{ W/m}^2}\right) = 10^{10} \text{ m}^2$$

where I rounded 300/240 to 1, and put the watts per square meter in the denominator for the same reason I did the miles per gallon in the previous example. There are 1,000 meters or 10^3 m in 1 km, so there are 10^6 m^2 in 1 km^2 (picture a square that is 1,000 m on a side; it contains a million little squares that are each 1 m $\times$ 1 m). So that 10^{10} m^2 is 10^4 km^2. For comparison, that's roughly one-sixth the area of California's Mojave Desert. Of course we don't have solar energy conversion devices that are 100 percent efficient, so considerably more area would actually be needed—perhaps ten times as much with efficiency and infrastructure taken into account. Still, the required land area is remarkably small. See Research Problem 4 for more on this point.

Wrapping Up

This chapter examines basic energy concepts at the most fundamental level; in that sense it's both essential for and yet most removed from the main themes of this book, namely human energy use and its environmental impacts. You now understand that energy manifests itself either as the kinetic energy associated with motion, or as potential energy associated with the various forces that occur in nature. You understand conceptually how energy is stored as gravitational potential energy or as chemical or nuclear energy in fuels. And you know how to talk fluently about energy and power in a variety of unit systems. Finally, you can calculate potential and kinetic energies in simple systems, determine power consumption, and make quick estimates of energy-related quantities.

Still, something may seem missing. Have we really covered all forms of energy? In one sense, yes. But in another, no: We haven't said much about that form of energy called, loosely, "heat." That's a topic with big implications for our human energy consumption, and it's the subject of the next chapter.

Chapter 3   Chapter Review

BIG IDEAS

3.1   Many forms of energy are instances of **kinetic energy** or **potential energy**. Kinetic energy is the energy of moving objects, while potential energy is stored energy.

3.2   Electricity and magnetism play a crucial role in energy storage and energy technologies. The energy of chemical fuels is ultimately stored in the electric fields associated with molecular configurations. The electric power we use is produced through the process of **electromagnetic induction**. Electromagnetic energy is also carried by **electromagnetic radiation**, of which light is an important example.

3.3 The watt (W) is the standard unit of power, and the joule (J) is the corresponding energy unit. One watt is 1 joule per second (J/s). Other common energy units include the kilo-watt-hour (kWh), calorie (cal), and British thermal unit (Btu). Multiples of units are expressed with standard prefixes, and conversion factors relate energy measurements in different units.

3.4 You do **work** when you apply a force to an object as it moves, provided the force acts in the direction of the object's motion. Work is the product of the force and the distance the object moves. Doing work on an object increases its energy. For example, applying a force to lift an object results in an increase in its gravitational potential energy.

3.5 The net work done on an object—the work done by all forces acting on the object—increases that object's kinetic energy, which is given by $K = \frac{1}{2} mv^2$, where m is the object's mass and v its speed.

3.6 Friction is a force that dissipates energy, turning the energy of motion into less useful forms. Friction can limit our ability to use energy efficiently.

3.7 Energy is a quantitative subject. However, you can learn a lot about energy with quick, simple estimates of numerical quantities.

TERMS TO KNOW

barrel of oil equivalent (p. 53)
battery (p. 41)
British thermal unit (p. 50)
calorie (p. 50)
chemical energy (p. 46)
electric charge (p. 41)
electric current (p. 41)
electric field (p. 41)
elastic potential energy (p. 38)
electromagnetic force (p. 39)
electromagnetic induction (p. 43)
electromagnetic radiation (p. 47)
electromagnetic wave (p. 47)
force (p. 39)
frictional force (p. 61)
fuel cell (p. 43)
gravitational force (p. 39)

gravitational potential energy (p. 38)
horsepower (p. 54)
joule (p. 49)
kinetic energy (p. 38)
magnetic field (p. 41)
newton (p. 56)
nuclear force (p. 39)
photon (p. 48)
photovoltaic cell (p. 46)
Planck's constant (p. 48)
potential energy (p. 38)
quad (p. 51)
rechargeable battery (p. 42)
ton, tonne, metric ton (p. 53)
tonne oil equivalent (p. 53)
weight (p. 57)
work (p. 56)

GETTING QUANTITATIVE

Energy of a photon: $E = hf = \dfrac{hc}{\lambda}$ (Equation 3.1; p. 48)

Planck's constant: $h = 6.63 \times 10^{-34}$ J·s

Speed of light: $c = 3.00 \times 10^8$ m/s

Energy and power units: see Table 3.1

Energy content of fuels: see Table 3.3

Work: $W = Fd$ (Equation 3.2; p. 56)

Force of gravity: $F_g = mg$ (Equation 3.3; p. 57)

Work done lifting object of mass m a distance h: $W = mgh$ (Equation 3.4; p. 57)

Kinetic energy: $K = \dfrac{1}{2}mv^2$ (Equation 3.5; p. 59)

QUESTIONS

1. Why is it harder to walk up a hill than on level ground?

2. Describe qualitatively the relationship between force and potential energy.

3. Table 3.3 shows that hydrogen has a higher energy content per kilogram than natural gas, but a lower energy content per cubic foot. How can this be consistent?

4. You jog up a mountain and I walk. Assuming we weigh the same, compare (a) our gravitational potential energies when we're at the summit, and (b) the average power each of us expends in climbing the mountain.

5. How many (a) megajoules are in 1 exajoule; (b) petagrams in 1 gigatonne (1 tonne = 1,000 kg); (c) kilowatt-hours in 1 gigawatt-hour?

EXERCISES

1. The average daily human diet has an energy content of about 2,000 kcal. Convert this 2,000 kcal per day into watts.

2. Using appropriate conversions between pounds and grams, and degrees Celsius and Fahrenheit, show that 1 Btu is equivalent to 1,054 J.

3. Express in watts the U.S. energy-consumption rate of approximately 100 Q per year.

4. There are two ways to calculate the power output of a car when you know that (1) the car has a 250-horsepower engine and (2) the car gets 20 miles per gallon when traveling at 60 miles per hour: (a) Convert the horsepower rating into watts. (b) Calculate the gasoline consumption rate and, using Table 3.3's energy equivalent for gasoline, convert the result to a power in watts. Comparison of your results shows that a car doesn't always achieve its engine's rated horsepower.

5. An oil furnace consumes 0.80 gallons of oil per hour while it's operating. (a) Using the approximate value of 40 kWh per gallon of petroleum product, find the equivalent power consumption in watts. (b) If the furnace runs only 15 percent of the time on a cool autumn day, what is the furnace's average power consumption?

6. The United States imports about 12 million barrels of oil per day. (a) Consult the tables in this chapter to convert this quantity to an equivalent power, measured in watts. (b) Suppose we wanted to replace all that imported oil with energy produced by fission from domestic uranium. How many 1,000-MW nuclear power plants would we have to build?

7. Assuming that 1 gallon of crude oil yields roughly 1 gallon of gasoline, estimate the decrease in daily oil imports (see preceding question) that we could achieve if the average fuel efficiency of U.S. cars and light trucks, now around 21 miles per gallon, were raised to the 50 miles per gallon typical of a modern hybrid car. Assume the average vehicle is driven about 10,000 miles per year and that there are about 200 million cars and light trucks operating in the United States.

8. In the text I cited an environmentalist's claim that buying an SUV instead of a regular car wastes as much energy as leaving your refrigerator door open for seven years. Let's see if that's right. A typical refrigerator consumes energy at the rate of about 400 W when it's running, but it usually runs only about a quarter of the time. If you leave the door open, however, the refrigerator will run all the time. Assume that an average SUV's fuel efficiency is 15 miles per gallon, an average car gets 25 miles per gallon, gasoline contains 40 kWh of energy per gallon, and you drive the vehicle 15,000 miles per year. Calculate how long you would have to leave your refrigerator door open to use as much extra energy as the difference between the car and SUV in the first year you own the vehicle.

9. I want to expend 100 "calories" (that is, 100 kcal) of energy working out on an exercise machine. The readout on the machine says I'm expending energy at the rate of 270 W. How long do I need to exercise to expend those 100 kcal?

10. You buy a portable electric heater that claims to put out 10,000 Btuh (meaning 10,000 Btu/h). If it's 100 percent efficient at converting electrical energy to heat, what is its electrical energy-consumption rate in watts?

11. A car with a mass of 1,700 kg can go from rest to 100 km/h in 8.0 seconds. If its energy increases at a constant rate, how much power must be applied to achieve this magnitude of acceleration? Give your answer in both kilowatts and horsepower.

12. An energy-efficient refrigerator consumes energy at the rate of 280 W when it's actually running, but it's so well insulated that it runs only about one-sixth of the time. You pay for that efficiency up front: It costs $950 to buy the refrigerator. You can buy a conventional refrigerator for $700. However, it consumes 400 W when running, and it runs one-fourth of the time. Calculate the total energy used by each refrigerator over a 10-year lifetime and then compare the total costs—purchase price plus energy cost—assuming electricity costs 10¢ per kilowatt-hour.

RESEARCH PROBLEMS

1. Find a value for the current population of the United States, and use your result along with the approximate U.S. annual energy-consumption rate of 100 Q per year to get an approximate value for the per capita U.S. energy-consumption rate in watts.

2. Choose a developing country and find the values for its current population and its total annual energy consumption in quads. Using these numbers, calculate the per capita energy-consumption rate in watts.

3. Find the official EPA (Environmental Protection Agency) fuel efficiency for the car you or your family drive. Estimate your yearly mileage, and determine the amount of fuel you would save each year if you switched to a 50-mile-per-gallon hybrid.

4. Make the assumption that the solar-collector area calculated in Example 3.5 should be increased by a factor of 10 to account for the inefficiency of the solar cells and to allow room for infrastructure. What fraction of the total area of (a) the state of New Mexico and (b) the continental United States would then be needed? (c) Compare your answer to part (b) with the fraction of land that's now under pavement.

Chapter 4

ENERGY AND HEAT

In Chapter 3 I introduced kinetic energy, gravitational and elastic potential energy, electrical and magnetic energy, and nuclear energy. Are there other forms of energy? Think about this question and you'll surely answer "heat." In one sense you're right that heat is a new and different form of energy, but not in another sense. The issues surrounding the energy we commonly call *heat* are both sufficiently subtle and of such importance to human society's energy appetite that we need to explore them in more detail.

4.1 Heat and Internal Energy

You probably think of heat as the energy contained in a hot object; a bathtub full of hot water, for example, seems to have a lot of heat. But strictly speaking, that's not what heat means. The energy in the bathtub or other hot object is **internal energy**, also called **thermal energy**. It consists of kinetic energy associated with the random motions of the atoms and molecules that make up the object. So it's a form of energy that we've already discussed. The only difference between thermal energy and the kinetic energy of, say, a moving car is that thermal energy is random, with the individual molecules moving in different directions at various speeds. In contrast, all parts of the moving car participate in the same motion. Both are examples of kinetic energy, but the difference between random and directed motion is a big one, with serious consequences for human energy use.

So what's heat? Strictly speaking, **heat** is energy that is flowing as a result of a temperature difference. The term *heat* always refers to a *flow* of energy from one place to another, or one object to another. And not just any flow, but a flow that originates because of a temperature difference. If I take a can of gasoline out of the garage and carry it to my lawnmower, that's a flow of energy. But it's not heat, because the flow of energy from garage to lawnmower is caused not by a temperature difference but by my carrying the can. Similarly, and perhaps surprisingly, the flow of energy from your microwave oven to the cup of water you're boiling is not heat. There's no temperature difference involved, but rather a carefully engineered flow of electro-magnetic energy. On the other hand, the loss of energy from your house on a cold day, associated with the flow of energy through the walls and windows, is heat. That's because this flow is driven by the temperature difference between inside and outside. If the inside and outside of the house were at the same temperature, there would be no energy loss. If it were warmer outside, the flow would be into the house. In or out, that flow is properly described as *heat* or *heat flow*. Other examples include the heat radiating from a hot woodstove or lightbulb filament, the transfer of energy from a hot stove burner into a pan of water, and the movement of hot air from Earth's surface to higher levels of the atmosphere. All are heat flows because they occur as a result of a temperature difference.

You might think I'm splitting hairs with this distinction. After all, a microwave oven heats water just as well as a stove burner, so is there really a significant differ-ence between the two processes? There is, and that's precisely the point. You can raise temperature with heat flow, but there are other ways to do it that don't involve a temperature-driven energy transfer. The microwave is just one example; I could also heat the water by agitating it vigorously with a spoon. The spoon is no hotter than the water, so this energy flow isn't heat. Instead, it's a transfer of kinetic energy as the moving spoon does work on the molecules of water. Another example: Push on the plunger of a bicycle pump, and you raise the temperature of both the pump and the air inside it. This doesn't happen because of heat transfer from a hotter object, but because you supplied mechanical energy that ended up as internal energy of the pump and the air. By the way, an experiment involving the agitation of water is what allowed James Joule to determine the quantitative relationship between heat and mechanical energy. That relationship is further developed in the so-called first and second laws of thermodynamics.

The **first law of thermodynamics** embodies this distinction between heat and mechanical energy flows. The first law says that the change in an object's internal energy is the sum of the mechanical work done on the object and the heat that flows into it. The first law is essentially a generalization of the law of conservation of energy to include heat. It doesn't matter where an object's internal energy comes from, whether by heat transfer or mechanical energy or a combination of the two. A cup of hot water is just that; whether its temperature increased because of heat flow or because someone shook it violently makes no difference. By the way, heat or mechanical energy can flow either into or away from an object, and the first law handles both cases with the appropriate use of positive and negative signs.

4.2 Temperature

If heat is energy flowing because of a temperature difference, then what's temperature? The answer is simple: **Temperature** measures the average thermal energy of the mole-cules in a substance. That is, temperature measures the average of the quantity $\frac{1}{2}mv^2$

(kinetic energy) associated with individual molecules. The word *average* is important, because thermal motions are random in both direction and speed. Figure 4.1 suggests this kinetic-energy meaning of temperature.

You're probably familiar with several temperature scales. The Fahrenheit scale, used mainly in the United States, has water freezing at 32°F and boiling at 212°F. The Celsius scale, used in science and in everyday life throughout most of the rest of the world, has water freezing at 0°C and boiling at 100°C. But neither scale is truly fundamental, because neither measures zero energy at the zero-degree mark. The **Kelvin scale**, the official temperature scale of the SI unit system, has the same size degree gradations as the Celsius scale, but its zero is at the true zero of energy, about −273°C. This quantity is called **absolute zero**, because molecular kinetic energy can't be less than zero. The temperature unit on the Kelvin scale is the kelvin (not degrees kelvin or °K, but just K). A temperature change of 1 K is the same as a change of 1°C, but the actual values on the Kelvin and Celsius scales differ by 273. Another absolute scale, used primarily by engineers in the United States, is the Rankine scale. Its degrees are the same size as Fahrenheit degrees, but its zero, like that of the Kelvin scale, is at absolute zero. We'll find the absolute property of the Kelvin scale useful in some areas of energy and climate studies. On the other hand, the Celsius scale is more familiar to nonscientists and often more appropriate for expressing everyday temperatures, so I'll be going back and forth between the two. Again, where temperature *differences* are concerned, there's no distinction between the Kelvin and Celsius scales. Figure 4.2 compares the four temperature scales.

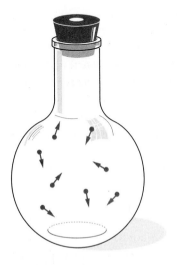

Low temperature

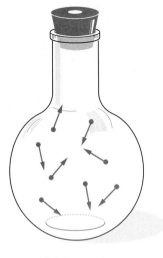

High temperature

Figure 4.1
Temperature measures the average molecular kinetic energy. At lower temperatures (left), the molecules of a gas have lower kinetic energy and move more slowly; at higher temperatures (right), the molecules have higher kinetic energy and move more quickly.

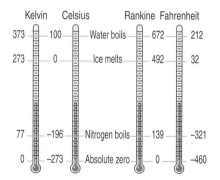

Figure 4.2
Four temperature scales compared: Kelvin, Celsius, Rankine, and Fahrenheit.

4.3 Heat Transfer

Several different physical mechanisms produce the temperature-difference-driven energy flow that we call heat. We refer to these as *heat-transfer mechanisms*, although strictly speaking, "heat transfer" is redundant because heat *means* energy that's flowing from one place to another.

Conduction

Think of a cool pan sitting on a hot stove burner. Because of the higher temperature, the molecules in the stove burner have more kinetic energy. Where these molecules collide with those in the pan, they transfer kinetic energy. As a result, the pan warms. That's a microscopic description of heat transfer by **conduction**, the direct sharing of kinetic energy by molecular collisions between materials in contact. Another example is the flow of heat through the walls of a house. Air molecules in the house collide with the wall, transferring kinetic energy. Molecules at the indoor edge of the wall transfer their energy to molecules deeper within the wall, and eventually the energy gets to the outdoor edge of the wall where it's transferred to the exterior air. The temperature difference between the warmer inside and cooler outside ensures that collisions, on average, transfer energy in the direction from warmer to cooler.

Conduction is an important process in the study of energy and climate. Conduction is largely responsible for energy losses from buildings, and the vast energy resources we devote to heating attest to these losses. Conduction also transfers energy from the sunlight-warmed surface of the Earth into the atmosphere, after which another heat-transfer mechanism takes over.

In the case of buildings, we would like to avoid energy loss by conduction through walls. We do so with insulating materials, which, by virtue of their structure, are poor conductors of heat. A given material is characterized by its **thermal conductivity**, designated k. Table 4.1 shows thermal conductivities of some common materials. The heat flow through a given piece of material is determined by its thermal conductivity, its area,

Table 4.1 Thermal Conductivities of Selected Materials

Material	Thermal conductivity, (W/m·K)	Thermal conductivity, (Btu·inch/hour·ft²·°F)
Air	0.026	0.18
Aluminum	237	1,644
Concrete (typical)	1	7
Fiberglass	0.042	0.29
Glass (typical)	0.8	5.5
Rock (granite)	3.37	23.4
Steel	46	319
Styrofoam (extruded polystyrene foam)	0.029	0.2
Water	0.61	4.2
Wood (pine)	0.11	0.78
Urethane foam	0.019	0.13

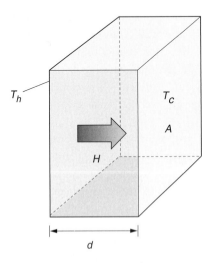

Figure 4.3
Conductive heat flow H through a slab of material with conductivity k is given by Equation 4.1. T_h and T_c are the temperatures on the warmer side and the cooler side of the slab, respectively. A is the slab area, and d is its thickness.

its thickness, and the temperature difference from one side of the piece to the other. Figure 4.3 and Equation 4.1 show how to determine this heat flow:

$$H = kA \, \frac{T_h - T_c}{d}$$

(4.1)

Here H is the heat flow, measured in watts; k is the thermal conductivity; A and d are the area and thickness, as shown in Figure 4.3; and T_h and T_c are the temperatures on the hot and cold sides of the material, respectively.

Example 4.1   A Better Refrigerator

Assume it's 3°C inside your refrigerator, while the surrounding kitchen is at a typical room temperature of 20°C (68°F). The refrigerator is essentially a rectangular box 150 cm high by 76 cm wide by 64 cm deep. Determine the rate at which heat flows into the refrigerator if it's insulated on all sides using 4-cm-thick urethane foam.

Solution

Heat flows into the refrigerator as a result of thermal conduction. Equation 4.1 describes conduction quantitatively; to determine the heat flow, we need to know the area, the temperature difference, and the thickness and thermal conductivity of the insulating material. We're given the two temperatures and the insulation material (urethane foam), so we can get the thermal conductivity from Table 4.1; it's 0.029 W/m·K. We know the insulation thickness (4 cm), but we need to calculate its area, which is the total of the refrigerator's top and bottom, front and back, and two side areas. You can see from Figure 4.4 that the total area is then

$$2[(76 \text{ cm})(150 \text{ cm}) + (76 \text{ cm})(64 \text{ cm}) + (150 \text{ cm})(64 \text{ cm})] = 51{,}728 \text{ cm}^2 \simeq 5.2 \text{ m}^2$$

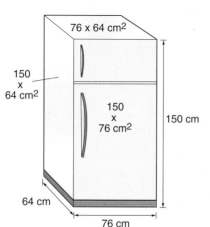

Figure 4.4
The refrigerator from Example 4.1 has a total area of 5.2 m², and heat flows through its insulated walls at the rate of 64 W.

where the last step follows because there are 100 cm in 1 m, and thus $100 \times 100 = 10^4$ cm^2 in 1 m^2. Using this area in Equation 4.1, along with the temperatures, insulation thickness, and thermal conductivity, we get

$$H = kA \, \frac{T_h - T_c}{d} = (0.029 \text{ W/m·K})(5.2 \text{ m}^2) \, \frac{20°C - 3°C}{0.04 \text{ m}} = 64 \text{ W}$$

To keep its interior cold, the refrigerator's mechanism needs to remove energy at this rate. That's why it's plugged into the wall and consumes electricity. However, its average electrical power consumption may be less than 64 W, for reasons I'll describe later in Section 4.9.

Building Insulation and *R* Value

The thermal conductivities of the materials comprising a building's walls and roof determine the rate of heat flow out of or into the building, and therefore its energy requirements for heating and cooling. Building (or retrofitting) with insulating materials that have low conductivity is a critical step in reducing overall energy consumption, especially the nearly 40 percent that occurs in the commercial and residential sectors.

If you go to your local building-supply outlet and buy a roll of fiberglass insulation or a slab of rigid foam, you'll find it marked with its *R* **value**. This quantity measures the material's resistance to the flow of heat and is closely related to thermal conductivity (k). However, where k is an intrinsic property of a given material, R characterizes a specific thickness of material. The relationship between the two is $R = d/k$ where d is the thickness. The R value is almost always measured in English units; although rarely stated explicitly, those units are ft^2·°F·h/Btu. How on Earth are we to make sense of that? Simple: R is a measure of resistance to heat flow, and its inverse is a measure of how easy it is for heat to flow. And the inverse of R has the following units, as you can see from inverting that mouthful of units for R: British thermal units per hour per square foot per degree Fahrenheit. Take, for example, a 1-inch-thick slab of "blueboard," a foam insulation often used on basement walls (Fig. 4.5). Its R value is 5, which means each square foot of blueboard loses one-fifth of a Btu of energy every hour for every degree Fahrenheit temperature difference across the insulation. So if it's 70°F inside your basement and 45°F in the ground outside, and if your basement wall area totals 1,000 square feet, then the rate of energy loss through the wall is

$$(1/5 \text{ Btu/h/ft}^2/°F)(1,000 \text{ ft}^2)(70°F - 45°F) = 5,000 \text{ Btu/h}$$

Incidentally, that's about 1,500 W (see Exercise 1), so picture fifteen energy servants hunkered down in your basement working away just to produce the energy that's lost through your basement wall. Of course, things would be a lot worse without the insulation, and in any event on a cold day there's a lot more energy being lost through the walls and windows upstairs.

Figure 4.5
Worker installing "blueboard" rigid foam insulation on exterior foundation walls. This material provides an R value of 5-per-inch thickness, some forty times that of the concrete foundation itself.

One virtue of the R value is that the R values of different materials making up a composite structure such as a house wall can simply be added together to give an overall R value. Table 4.2 lists R values for some common materials used in construction, although you could calculate some of these from Table 4.1 using $R = d/k$. Note that an 8-inch-thick concrete wall rates at roughly $R = 1$, so it's a lousy insulator compared with our 1-inch foam. But we would have had a more accurate account of the heat loss through the basement wall had we added the concrete's $R = 1$ to the foam's $R = 5$, for a total R value of 6. Because the R value scales with thickness, you can easily determine the R value for any material appearing in Table 4.2, even if your particular thickness isn't listed in the table.

Calculating the total R value for a composite wall is a bit subtle (Fig. 4.6); in addition to the building materials, there's usually a layer of "dead" air against the wall, which provides additional insulation. Table 4.2 lists the R value of this dead air for both interior and exterior wall surfaces, but be aware that these figures depend on the air being still. A strong wind blowing outside replaces slightly warmed air with cold air, removing that extra insulating value. Dead air actually contributes more than glass to the insulating properties of windows and is especially effective when trapped between the panes of a double- or triple-glazed window. Replacing air with

Table 4.2 *R* Values ($ft^2 \cdot °F \cdot h/Btu$) of Some Common Building Materials

Air layer		
Adjacent to inside wall	0.68	
Adjacent to outside wall, no wind	0.17	
Concrete, 8-inch	1.1	
Fiberglass		
3.5-inch	12	
5.5-inch	19	
Glass, 1/8-inch single pane	0.023	
Gypsum board (Sheetrock), 1/2-inch	0.45	
Polystyrene foam, 1-inch	5	
Urethane foam, 1-inch	7.5	
Window (*R* values include adjacent air layer)		
Single-glazed wood	0.9	
Standard double-glazed wood	2.0	
Argon-filled double-glazed with low-E coating	2.9	
Argon-filled triple-glazed with low-E coating	5.5	
Wood		
1/2-inch cedar	0.68	
3/4-inch oak	0.74	
1/2-inch plywood	0.63	
3/4-inch white pine	0.96	

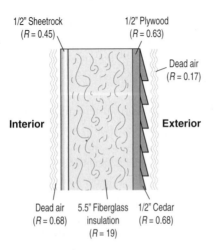

1/2" Sheetrock (*R* = 0.45)

1/2" Plywood (*R* = 0.63)

Dead air (*R* = 0.17)

Interior

Exterior

Dead air (*R* = 0.68)

5.5" Fiberglass insulation (*R* = 19)

1/2" Cedar (*R* = 0.68)

Figure 4.6

The *R* value of a composite wall is the sum of the *R* values of its individual components. Here the *R* value is 21.6, nearly all of it from the fiberglass insulation.

argon gas further enhances the insulating qualities of a window, as do "low-E" coatings that reduce energy loss by infrared radiation (more on radiation heat loss later in this section). Premium windows cost a lot, but over the lifetime of a building they can pay for themselves many times over in fuel costs, as well as reduce the building's environmental impact. Table 4.2 includes some windows, from the most basic to the most advanced; the difference in R value is dramatic.

Example 4.2   Saving on Heat

Imagine you've bought an old house that measures 36 feet wide by 24 feet deep and is 17 feet from the ground to the second-floor ceiling. The walls have no insulation, and the R value of the wall materials and dead air is only 4. The second-floor ceiling has 3.5 inches of fiberglass insulation over 1/2-inch gypsum board. The attic under the sloping roof is uninsulated, a common practice to avoid problems with condensation and rot. This lack of insulation also makes this example easier, since the house can be considered a rectangular box.

(a) On a winter day when it's 10°F outside and 68°F inside, what is the rate of heat loss through the walls and ceiling? Neglect heat flow through the floor (a reasonable approximation because the ground is much warmer than the outside air) and through the windows (less reasonable, because windows generally account for a lot of heat loss; see Exercise 9). (b) How much oil would you burn in a winter month with this average temperature? Assume an oil furnace that is 80 percent efficient at converting the chemical energy of oil into heat in your house. (c) At $2.20 per gallon, how much would it cost to heat this house for a month? (d) How much would you save on your monthly heating bill if you put 3.5-inch fiberglass insulation in the walls and raised the ceiling insulation to 11 inches of fiberglass?

Solution

Since they have different R values, we need to calculate separately the areas of ceiling and walls to determine the individual heat losses. The ceiling measures (24 ft) $\times$ (36 ft) = 864 ft^2. Table 4.2 shows the ceiling's R value is 13.3, consisting of the 3.5-inch fiberglass insulation ($R = 12$) plus the gypsum board ($R = 0.45$), the inside air layer ($R = 0.68$), and the outside layer ($R = 0.17$). This means every square foot of ceiling loses 1/13.3 Btu per hour for every degree Fahrenheit of temperature difference. Here we have 864 square feet of ceiling area and a temperature difference of 68°F − 10°F = 58°F (I'm assuming the unheated, ventilated attic is at the outdoor temperature, which may be a slight underestimate.) Then the heat loss through the entire ceiling is

$$(1/13.3 \text{ Btu/h/ft}^2/°F)(864 \text{ ft}^2)(58°F) = 3{,}768 \text{ Btu/h}$$

Now for the walls: The total area of the equal-sized front and back and the two identical sides is $(2)[(36 \text{ ft} \times 17 \text{ ft}) + (24 \text{ ft} \times 17 \text{ ft})] = 2{,}040 \text{ ft}^2$. We're given the total R value of 4, so the heat-loss rate through the walls is

$$(1/4 \text{ Btu/h/ft}^2/°F)(2{,}040 \text{ ft}^2)(58°F) = 29{,}580 \text{ Btu/h}$$

So the total loss rate—walls and ceiling—is 33,348 Btu/h, which I'll round to 33 kBtu/h. (Again, this is a very rough calculation, neglecting windows, losses through the floor, losses due to air infiltration, etc.) There are $24 \times 30 = 720$ hours in a 30-day month, so this amounts to a monthly heat loss of (33 kBtu/h)(720 h/month) = 24 MBtu/month, where again I rounded and changed to the more convenient prefix *mega* (M) instead of *kilo* (k).

On to the oil consumption and cost. With a furnace that is 80 percent efficient, you need oil with an energy content of 24 MBtu/0.80 = 30 MBtu. Table 3.3 lists oil's energy content at 138 kBtu/gallon, or 0.138 MBtu/gallon, so we'll need (30 MBtu)/(0.138 MBtu/gal) = 217 gallons of oil to heat the house for a month. At $2.20 per gallon, that will cost $477.

Putting 11-inch fiberglass in the ceiling gives twice the insulating value of the 5.5-inch fiberglass listed in Table 4.2, or an *R* value of 38 for the insulation alone; adding the gypsum board and dead air on both sides gives a total *R* value of 39.3. Adding 3.5-inch fiberglass to the walls increases the walls' *R* value by 12, to 16. Areas and temperature differences remain the same, so we can simply repeat the above calculations with the new *R* values. The result is a heat-loss rate of 8,670 Btu/h, about one-fourth of the earlier value. Everything else goes down proportionately, so you need only 56 gallons of oil for the month, at a total cost of $123. You save $477 − $123 = $354 on your monthly oil bill, and you produce a lot less carbon dioxide and pollutants.

So the insulation is a good investment. Exercise 9 explores this situation in the more realistic case of a house with windows (when you decide to upgrade those, too).

Example 4.3 Heat Loss and Solar Gain

On a sunny winter day, a window is situated so that during the midday hours it admits solar energy at the rate of approximately 300 watts per square meter of window surface—equivalently, 95 Btu per hour per square foot. For the best and worst windows listed in Table 4.2, find the minimum outdoor temperature for which that window provides the house with a net energy gain. Assume the indoor temperature is fixed at 70°F.

Solution

We know the rate at which each square foot of window gains energy: 95 Btu/h. We want to know when the rate of heat loss is equal to this gain. Since the inverse of the *R* value gives the heat-loss rate in Btu per hour per degree Fahrenheit per square foot, multiplying the temperature difference (call it ΔT) by $1/R$ gives the heat loss per square foot. Now, we want to know when the rate of energy gain exceeds the heat-loss rate. The break point occurs when they're just equal, so we set the heat-loss rate $\Delta T/R$ equal to the rate at which the window gains energy:

$$\frac{\Delta T}{R} = 95 \text{ Btu/h/ft}^2$$

Solving for ΔT gives $\Delta T = (95 \text{ Btu/h/ft}^2)(R)$. The single-glazed window listed in Table 4.2 has an R value of 0.9, giving $\Delta T = (95 \text{ Btu/h/ft}^2)(0.9 \text{ ft}^2 \cdot {}^\circ\text{F} \cdot \text{h/Btu}) = 86^\circ\text{F}$ (note how everything works out when the units of the R value are stated explicitly). With an interior temperature of 70°F, this means the outdoor temperature can be as low as $70^\circ\text{F} - 86^\circ\text{F}$, or a chilly -16°F, and even this relatively inefficient window still provides a net energy gain. (Of course, that's only true during the midday hours on sunny days.) I won't bother to repeat the calculation for the better window, since it provides an energy gain down to temperatures far below anything you would encounter. By the way, the better-insulated window probably admits a little less solar energy than the single-pane model, but it more than compensates for that with far lower heat loss at night and on cloudy days. In fact, many high-quality windows are designed to reduce solar gain as well as heat loss in order to cut energy use for air conditioning during the summer.

Convection

Conduction carries energy from a hot stove burner into a pan of boiling water, but from there the vigorous boiling motion is what carries energy to the water's surface. That bulk motion of the water (as opposed to the random motions of individual molecules) is called **convection**. In general, convection is the bulk motion of a fluid, driven by temperature differences across the fluid. Convection generally occurs in the presence of gravity, because as fluid becomes less dense when heated, it rises, gives up its energy, then sinks to repeat the convective circulation (Fig. 4.7).

Home heating systems often exploit convection. Baseboard hot water pipes or old-fashioned radiators heat the air around them, which then circulates about the room in a large-scale version of the convection patterns shown in Figure 4.7. Convection carries heated air near Earth's surface higher into the atmosphere; on a planetary scale, such convection produces large-scale atmospheric circulation patterns that transfer energy from the tropics toward the poles (Fig. 4.8). On a smaller atmospheric scale, the violence of a thunderstorm is closely related to vigorous convective motions within the storm; the sudden rush of wind you often feel just before a thunderstorm hits is due to the downward-flowing part of a convection flow hitting the ground and spreading in all directions. And on a much larger scale, convection is responsible for the transport of energy from the Sun's interior to just below its surface—energy that becomes the sunlight that ultimately powers life on Earth.

Cool

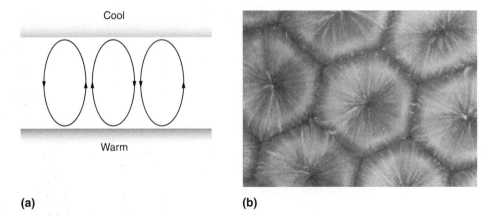

Warm

(a)   **(b)**

Figure 4.7
Convection. (a) Warm fluid rises, gives up its thermal energy, and
then sinks to form individual cells of convective motion. (b) A top
view of convection cells in a laboratory experiment shows striking
regularity, with fluid rising in the centers of the hexagonal cells and
sinking at the edges.

Convection is also responsible for energy loss. Although air is a good thermal insulator, its convective motion between the two panes of a double-glazed window can still
result in heat loss. For that reason, a narrower air space that inhibits convection is actually better than a wider one. Insulating materials such as fiberglass and foams work
because the interstices between glass fibers or tiny bubbles in the foam trap air or gases
and prevent convective motions from developing.

The strength of convective flows, and therefore the rate of energy transfer, depends
on the temperature difference across a convecting fluid. But fluid motion is complicated,
and there's no simple formula that applies in all cases of convection.

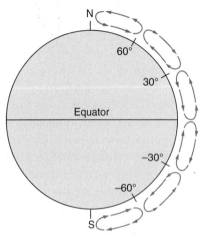

Figure 4.8
Large-scale convective motions transport energy from Earth's
warm equatorial regions to the cold polar regions. Coupled
with Earth's rotation, these convective motions also give rise
to the prevailing winds that blow from east to west at low
latitudes and from west to east in the temperate zones.

What I've been describing here is **natural convection**, so-called because it occurs naturally when there's a vertical temperature difference. There's also **forced convection**, which occurs when we use fans or pumps to move fluids. Hot-air and hot-water heating systems generally use forced convection to move fluid from the furnace or boiler into a living space where natural convection in the air then takes over. Forced convection also moves heat-transfer fluids through solar energy systems and through the boilers of fossil fuel and nuclear power plants. So forced convection is an important process in many energy technologies.

Radiation

If you stand near a hot woodstove, you can feel heat coming directly from the stove to your skin. Around a campfire, even on a frigidly cold evening, you're warmed by heat coming directly from the fire. And when you lie on the beach, your body is heated by energy direct from the Sun. All these cases involve heat transfer by **radiation**—in particular, electromagnetic radiation.

As long as a material is above absolute zero, its atoms and molecules are in random thermal motion, moving about in gases and liquids or vibrating in place in solids. And because atoms and molecules consist of electrically charged particles, those random motions result in electromagnetic radiation. So all objects, as long as they're above absolute zero temperature, emit energy in the form of electromagnetic radiation. They may also gain electromagnetic energy from their surroundings, so there isn't necessarily a net energy loss. If an object is hotter than its surroundings, it emits more energy than it absorbs; if it's cooler, it gains energy. So the net flow of electromagnetic energy to or from an object is driven by the temperature difference between the object and its surroundings. This temperature-driven flow of electromagnetic radiation therefore qualifies as a third mechanism for heat transfer.

The total energy radiated increases rapidly with an object's temperature—specifically, as the fourth power of the absolute temperature measured on the Kelvin scale. It also depends on the object's surface area and on a property of the object called its **emissivity**. Emissivity is a number between 0 and 1 that describes a material's efficiency as an emitter of electromagnetic radiation. A well-known law of physics states that a given material's efficiency as an emitter is equal to its efficiency as an absorber; that is, materials that are good at absorbing radiation are equally good at emitting it. A black object (which looks black because it absorbs all colors of light incident on it) is, in principle, a perfect absorber. Therefore it's also a perfect emitter, and it has an emissivity of 1. A shiny material such as metal reflects nearly all the light incident on it, and therefore it's both a poor absorber and a poor emitter, with an emissivity close to zero. You may have noticed that the inside of a thermos vacuum bottle is shiny; this gives it a low emissivity to cut down on energy loss by radiation. The vacuum between the walls of the bottle ensures that there's no conduction or convection either. That's why your coffee stays hot for hours. Those low-E windows I mentioned earlier are actually coated with a thin film that reduces emissivity and thus lowers energy loss by radiation.

Although the efficiency of absorption and emission are equal, both may vary with the wavelength of the radiation. However, they vary by the same amount so they stay equal at any given wavelength. An object that looks black to the eye has high absorption and therefore high emissivity for visible light, but it might behave very differently for infrared, ultraviolet, or other forms of electromagnetic radiation. The difference between emissivity for visible and infrared wavelengths has great practical significance; for example, low-E windows manage to minimize heat loss via infrared radiation without cutting down too much on the visible light they admit. And that same difference between visible light and infrared emissivities plays a major role in climate, especially in the greenhouse effect and the role our fossil-fueled society plays in altering climate (much more on this in Chapter 12).

Putting together fourth-power dependence on temperature with emissivity gives the **Stefan-Boltzmann law** for radiation:

$$P = e\sigma A T^4 \tag{4.2}$$

Here P is the power, in watts, emitted by an object at temperature T; A is the object's surface area; and e is its emissivity. The symbol σ (the Greek letter "sigma") is the **Stefan-Boltzmann constant**, a universal quantity whose value, in SI units, is 5.67×10^{-8} W/m^2/K^4. Here I'm using the symbol P rather than the H of Equation 4.1 because Equation 4.2 gives, strictly speaking, only the power radiated *from* an object. To find the net heat transfer by radiation would involve subtracting any radiation coming from the object's surroundings and being absorbed by the object. In many cases where radiation is important, however, the object in question is significantly hotter than its surroundings. Then that strong T^4 temperature dependence means the energy emitted far exceeds the energy absorbed, and in such cases Equation 4.2 provides an excellent approximation to the net heat transfer by radiation.

Example 4.4 A Red-Hot Stove

A hot stove burner with nothing on it loses energy primarily by electromagnetic radiation, which is why it glows red hot. A particular burner radiates energy at the rate of 1,500 W and has a surface area of 200 cm^2. Assuming emissivity of 1, what is the temperature of the burner?

Solution

Equation 4.2 describes the rate of energy loss by radiation, here 1,500 W. In this example, the only thing in Equation 4.2 that we don't know is the temperature, which is what we're trying to find. It's straightforward algebra to solve Equation 4.2 for the quantity T^4; raising the result to the 1/4 power (or taking the fourth root) then gives the answer. But one caution: Equation 4.2 works in SI units, which means using square meters for area. We're given the area in square centimeters; since 1 cm is 10^{-2} m, 1 cm^2 is 10^{-4} m^2. Thus our 200-cm^2 area amounts to 200×10^{-4} m^2. So here's our calculation for the burner temperature:

$$T^4 = \frac{P}{e\sigma A}, \text{ or}$$

$$T = \left(\frac{P}{e\sigma A}\right)^{1/4} = \left(\frac{1,500 \text{ W}}{(1)(5.67 \times 10^{-8} \text{ W/m}^2\text{/K}^4)(200 \times 10^{-4} \text{ m}^2)}\right)^{1/4} = 1,070 \text{ K}$$

or about 800°C.

Equation 4.2 shows that the total radiation emitted by a hot object increases dramatically as its temperature rises. But something else changes, too—namely, the wavelength range in which most radiation is emitted (Fig. 4.9). At around 1,000 K, the stove burner of Example 4.3 glows red, although, in fact, most of its radiation is invisible infrared mixed with some visible light. A lightbulb filament heats up to about 3,000 K, and it glows with a bright yellow-white light, although as Figure 4.9 shows, it's still emitting mostly infrared and is therefore a rather inefficient light source. Put the bulb on a dimmer and turn down the brightness, though, and you'll see its color changing from yellow-white to orange and then to a dull red as the filament temperature drops. The Sun's surface is at about 6,000 K, so its light is whiter than that of the lightbulb (which is why photographs taken in artificial light rather than sunlight sometimes show strange colors). On the other hand, a much cooler object such as Earth itself, with a tem-

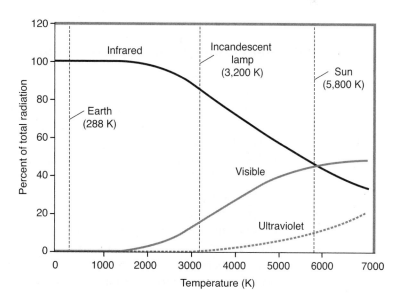

Figure 4.9
Percentage of radiation emitted in different wavelength ranges, as a function of temperature. The infrared curve represents longer wavelengths; the ultraviolet curve represents shorter wavelengths. The visible spectrum is defined as wavelengths in the range of 380 to 760 nanometers.

perature around 300 K, emits essentially all of its radiation at infrared and longer wavelengths. This distinction between the 6,000-K Sun radiating primarily visible light and the 300-K Earth radiating primarily infrared plays a major role in establishing Earth's climate. Finally, the entire universe is at an average temperature of only 2.7 K (about −270°C), and it radiates primarily in the microwave region of the electromagnetic spectrum. Studying that microwave emission gives astrophysicists detailed information about the structure and evolution of the universe.

Example 4.5 Earth's Radiation

Earth's radius (R_E) is 6,370 km, or 6.37 Mm. Assuming a temperature of 288 K and an emissivity of 1, what is the total rate at which Earth loses energy by radiation?

Solution

This is a straightforward application of Equation 4.2, with the surface area of the spherical Earth given by $4\pi R_E^2$. So we have:

$$P = e\sigma A T^4 = (1)(5.67 \times 10^{-8} \text{ W/m}^2/\text{K}^4)(4\pi)(6.37 \times 10^6 \text{ m})^2(288 \text{ K})^4 = 2.0 \times 10^{17} \text{ W}$$

Because of Earth's relatively cool temperature, this power is essentially all in the form of infrared radiation. It's no coincidence that the number here is in the same order-of-magnitude ballpark as the 10^{17} W I gave in Chapter 1 as the total solar energy flow to Earth. In fact, there's an almost perfect balance between incoming sunlight and outgoing infrared, and that's what determines Earth's temperature. I'll have much more to say on this topic in Chapters 12 and 13.

4.4 Heat Capacity and Specific Heat

When you put a small amount of water in a pan and heat it to boiling so you can steam vegetables, it doesn't take long to reach the boiling point. But put a big pot of water on the same burner and bring it to a boil so you can cook spaghetti, and it takes a lot longer. The reason is obvious: After all, energy is going into the water at the same rate in both cases (you're using the same burner), but there's a lot more water to be heated for the spaghetti. We describe this difference by saying that the larger amount of water has a larger **heat capacity**, which is the energy needed to cause a given temperature rise.

If, instead of water, you were heating oil for deep-fat frying, you would find the oil warms faster than an equal amount of water. In this case we're describing a more fundamental quantity, a property not of a particular amount of material but of the material itself. This property is the material's **specific heat**, which measures the heat capacity on a per-unit-mass basis. In the SI system, specific heat has the units of joules per kilogram per kelvin ($J/kg/K$). Because 1 K and 1°C are the same as far as temperature *differences*

Table 4.3 Specific Heats (J/kg/K) of Some Common Materials	
Aluminum	900
Concrete	880
Glass	753
Steel	502
Stone (granite)	840
Water	
Liquid	4,184
Ice	2,050
Wood	1,400

are concerned, this can equally well be written J/kg/°C. Water's specific heat of 4,184 J/kg/K is especially large, which is why it takes a long time to heat water, and why large lakes heat or cool slowly and thus exert a moderating effect on local climate. By the way, if the figure 4,184 sounds familiar, it's because that's also the number of joules in a kilocalorie. The calorie was initially defined as the heat needed to raise the temperature of 1 gram of water by 1°C, so the specific heat of water is simply 1 cal/g/°C. Converting calories to joules and grams to kilograms gives the figure 4,184 J/kg/°C. Table 4.3 lists the specific heats of some common materials.

The definition of specific heat as the energy per kilogram per kelvin (or degree Celsius) shows immediately how to calculate the energy needed to raise the temperature of a given mass m of material by an amount ΔT (here Δ is the capital Greek letter "delta," commonly used to signify "a change in . . . "). We simply multiply the specific heat (commonly given the symbol c) by the mass involved and by the temperature change:

$$Q = mc\Delta T \tag{4.3}$$

where Q is the energy required, in joules.

Example 4.6 Meltdown!

A cooling-water leak causes an emergency shutdown of a nuclear power plant's 2-gigawatt (GW) reactor. But the decay of radioactive material in the reactor continues to release energy at one-tenth the plant's normal power output, or 200 megawatts (MW). The plant's emergency cooling system dumps 4 million kg of water (that's 4,000 metric tons), initially at 50°C, into the reactor. How hot does that water get in one hour?

Solution

This example is a good test of your fluency in working quantitatively with energy and power. We're given power, or the *rate* at which the radioactive material is putting out energy. But to find the temperature rise ΔT in Equation 4.3, we need the total energy. We're asked for that rise at the end of one hour. Given that the power output is 200 MW, in a full hour the radioactive material will have produced 200 megawatt-hours (MWh) of energy. Because the specific heat in Table 4.3 involves joules, not megawatt-hours, we have to convert. Remember that 1 J is 1 W·s, so if we convert hours to seconds, we're there. There are 3,600 seconds in one hour, so our 200 MWh amounts to:

$$Q = (200 \text{ MWh})(3,600 \text{ s/h}) = 720,000 \text{ MW·s} = 720,000 \text{ MJ}$$

This can be written more neatly as 720 GJ or 7.2×10^{11} J. Now we can solve Equation 4.3 for the temperature rise:

$$\Delta T = \frac{Q}{mc} = \frac{7.2 \times 10^{11} \text{ J}}{(4 \times 10^6 \text{ kg})(4,184 \text{ J/kg/°C})} = 43°C$$

Since the water started at 50°C, it's now at 93°C—perilously close to boiling. Fortunately, as we'll see in Example 4.7, it's going to take a long time to boil all that water away.

4.5   Phase Changes and Latent Heat

Supplying heat to an object usually raises its temperature. Usually, but not always. Sometimes the energy goes into a change of state instead, as when ice melts or water boils. What's happening is that the added energy is breaking the bonds that hold molecules into the rigid structure of a solid or in close contact in a liquid. Different materials change state at different temperatures and require different amounts of energy to make those changes.

The energy required to change the state of a unit mass of material (1 kg in SI units) is the **heat of transformation**—specifically, the **heat of fusion** for the liquid-to-solid transformation and the **heat of vaporization** for the change from liquid to gas. In this book we're interested mostly in water, for which the heats of fusion and vaporization at the melting and boiling points are, respectively, 334 kJ/kg and 2.257 MJ/kg. At more typical atmospheric temperatures where evaporation takes place, water's heat of vaporization is somewhat larger—approximately 2.5 MJ/kg. Those are big numbers, and they tell us that it takes lots of energy to melt ice and even more (nearly seven times as much) to vaporize an equal amount of water.

The adjective *latent* is sometimes added in front of *heat* in the heats of transformation. That's because a substance in, say, the gaseous state contains more energy than the same substance in the liquid state, even if they're at the same temperature. The extra energy is latent in the higher-energy state and can be released in a change of state.

Specifically, water vapor contains more energy than liquid water. When water vapor in the atmosphere condenses, energy is released. Evaporation of ocean waters and the subsequent rise of humid air into the atmosphere is an important energy-transfer process in weather and climate. Hurricanes are powered by the energy released through the condensation of water vapor that rises into the atmosphere above warm tropical oceans. On a larger scale, transfer of energy by water vapor rising in the atmosphere plays a significant role in Earth's global energy balance. In the context of weather and climate, we refer to the energy associated with water vapor as **latent heat**, again because the energy is latent and can be released by condensation to liquid water. Latent heat contrasts with **sensible heat**, which, in the language of weather and climate, means the internal kinetic energy I introduced at the beginning of this chapter. This is heat you can feel, hence *sensible* heat.

Example 4.7   Back to the Meltdown

Once the emergency cooling water of Example 4.6 reaches the boiling point (only a little over an hour into the crisis), how long will it take to boil away completely?

Solution

The heat of vaporization of water is 2,257 kJ/kg, and in Example 4.6 we have 4 million kg of water. So the total energy required is $(2{,}257 \text{ kJ/kg}) \times (4 \times 10^6 \text{ kg}) = 9.0 \times 10^9$ kJ, or 9,000 GJ. The reactor is putting out energy at the rate of 200 MW, or 0.20 GW—that is, 0.20 GJ per second. So it's going to take

$$\frac{9{,}000 \text{ GJ}}{0.20 \text{ GJ/s}} = 45{,}000 \text{ s}$$

With 3,600 seconds in an hour, that's about 12 hours. All the while, the water stays at a safe 100°C. So there's a reasonable amount of time to deal with the crisis.

4.6 Energy Quality

A 2-ton car moving at 30 miles per hour has, in round numbers, about 200 kJ of kinetic energy. So does 1 teaspoon of gasoline. And the extra internal energy stored in a gallon of water when heated by 10°C is also around 200 kJ. Are all of these energies equivalent? Are they equally valuable? Equally useful?

The answer is no. To understand why, consider what it would take to convert the energy of either the water or the gasoline to the kinetic energy of a car. The energy in the water is the random kinetic energy of individual H_2O molecules. Somehow we would have to get them all going in the same direction and then impart their energy to the car. To use the gasoline, we'd have to burn it, again producing the random kinetic energy of molecules, and then somehow get all those molecules moving in the same direction so they could impart all their energy to the car.

How likely is it that the randomly moving molecules in water or in combustion products of the gasoline will organize themselves so they're all going in the same direction? If there were only two or three molecules, we might expect, on rare occasion, that random collisions among the molecules or their container would result in all of them going in the same direction. But the number of molecules in a macroscopic sample of material is huge—on the order of 10^{23} or more. It's extraordinarily unlikely that these molecules will ever find themselves all moving in the same direction. So unlikely, in fact, that it probably wouldn't happen spontaneously even if you waited many times the age of the universe.

The extreme improbability of those random molecular motions ever being in the same direction means that it is, in practice, impossible to convert all the internal energy of a heated material into kinetic energy. However, it's easy to go the other way. When our car brakes, its 200 kJ of directed kinetic energy ends up as random internal energy associated with the frictional heating of the brakes (that's true unless the car is a gas-electric hybrid; more on this in Chapter 5). We have here an **irreversible process**: Once directed kinetic energy of a macroscopic object is converted to heat—that is, to random internal energy associated with an increased temperature—then it's impossible to turn it all back into kinetic ordered energy (Fig. 4.10).

Irreversible processes are common. For example, beat an egg, intermixing the white and yolk; even if you reverse the beater, you'll never see the white and yolk separate from the scrambled mix. Take two cups of water, one hot and one cold, and put them together inside an insulated box. A while later you have two cups of lukewarm water. Wait as long as you want and you'll never see them spontaneously revert to one hot and one cold cup. That's true even though the amount of energy in the two lukewarm cups is exactly the same as in the original hot/cold situation. (I put them in an insulated box so they could exchange energy with each other but not with the outside world, thus maintaining the same total energy.)

There's something ordered about the kinetic energy of a moving car, with all its molecules sharing the common motion of the car. There's something ordered about an unbeaten egg, with all its yolk molecules in one place and its white molecules in another. And there's something ordered about cups of hot and cold water, with more of the faster-moving molecules in one cup and more of the slower-moving ones in the other. But these ordered states gradually deteriorate into states with less order. So do many other ordered

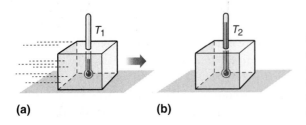

(a)   **(b)**

Figure 4.10
An irreversible process. (a) A block slides along a surface, gradually slowing. Friction converts the kinetic energy of its directed motion to random thermal energy of its molecules, indicated by the higher temperature when it comes to rest (b). The reverse process, from (b) to (a), is so highly improbable that it's essentially impossible.

states you can think of, such as the state of your room after you clean it. It gradually gets messy, and the messy room never spontaneously gets neater.

There's a simple reason for this tendency to disorder—namely, of all the possible ways to arrange the molecules of an egg, or the stuff in your room, or the molecular energies of two cups of water, there are far more ways that correspond to disordered states. You could arrange the books on your shelf in different ways, but if you toss them randomly on the floor, there are hugely more possible arrangements. Because of the vastly greater number of disordered states, systems naturally tend to end up disordered. Ordered states are few, and therefore special, and it takes an effort to put a system into an ordered state.

That systems tend naturally toward disorder is a general statement of the **second law of thermodynamics**. The second law places serious restrictions on our ability to harness energy, and this is what I'll emphasize from now on. But be aware that the second law is broader, describing the general tendency to disorder—a tendency that occurs simply because there are more disordered states.

Back to our 200 kJ of energy in the moving car, the gasoline, and the hot water. The second law of thermodynamics says that we can't turn all the internal energy of the water or the gasoline into the directed kinetic energy of a car. We can, on the other hand, turn all the kinetic energy of the car into internal energy. So there's a very real sense in which the kinetic energy of the car is more valuable, more versatile, more useful. Anything we can do with 200 kJ of energy, we can do with the kinetic energy of the car. We can't do just anything with the gasoline or the hot water. In particular, we can't turn it all into kinetic energy, so the energy of the gasoline and hot water is less valuable.

I'm not talking here about different *amounts* of energy (we have 200 kilojoules in all three cases), but about different *qualities* of energy. The kinetic energy is of higher quality, because it's the most versatile. And the gasoline's energy is of higher quality than that of the heated water. Why? Because when I burn the gasoline I can convert its energy to internal energy of far fewer molecules than are in a gallon of water. Those molecules each have, on average, more energy than the molecules in the water, meaning a higher temperature. Because there are fewer of these high-temperature molecules, the number of disordered states, although still fantastically huge, is nevertheless smaller than in the case of the water. This translates, for reasons I'll discuss in the next section, into our being able to convert more of the gasoline's energy to directed kinetic energy.

4.7 Entropy, Heat Engines, and the Second Law of Thermodynamics

Entropy is a measure of disorder. A moving car represents a state of low entropy. Convert the car's kinetic energy to random thermal energy, and entropy has increased, even though the total energy remains the same. The combustion products of our teaspoon of gasoline have higher entropy than the moving car, but less entropy than the hot water. There's actually a hierarchy of energy qualities, from the highest quality and lowest entropy to lower qualities and higher entropy. At the top are mechanical energy—the potential and kinetic energies of macroscopic systems. Of equal quality is the directed energy associated with electric current, meaning we can convert electrical energy to any other form with, in principle, 100 percent efficiency. Lower in quality and higher in entropy is the energy of objects at high temperature. Still lower in quality and even higher in entropy is energy associated with lower temperatures. Figure 4.11 shows this hierarchy of energy qualities.

The second law of thermodynamics can be restated in terms of entropy: The entropy of a closed system can never decrease. The "closed system" designation is important, for reasons we'll see in Section 4.9. And in almost all cases, entropy in fact increases. So in general, energy quality deteriorates. That's why the fact that energy is conserved doesn't really do us much good; as we use energy, it degrades in quality and becomes less useful.

We're now ready to understand in more detail why it is that we can't extract as mechanical energy all the energy contained in fuels. We *can* burn fuel directly and convert all its energy to heat—that is, to internal energy. But this is low-quality energy. What we can't do is turn it all into high-quality mechanical energy or electricity.

Heat Engines

A **heat engine** is a device that converts random thermal energy into mechanical energy. Examples include the gasoline engine in your car, the boilers and steam turbines in electric power plants, and jet aircraft engines. Although these engines differ in their design and operation, they're conceptually similar: All extract random thermal energy from a hot substance and deliver mechanical energy. In an **internal-combustion engine,** the fuel burns inside the engine and the combustion products become the hot substance. This is what happens in the cylinders of your car engine. In an **external-combustion engine**, burning fossil fuel or fissioning uranium heats a separate substance, usually water. Either way, we can consider that we start with a hot substance whose high temperature is sustained by burning a fuel. More on the details of real engines in Chapter 5; for now, we're just interested in the essential concept of an engine as a device that converts random thermal energy into mechanical energy.

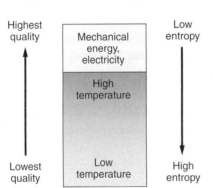

Figure 4.11
Quality associated with a given quantity of energy is highest for mechanical energy and electricity, and lowest for low-temperature thermal energy. Low entropy corresponds to high quality, and vice versa.

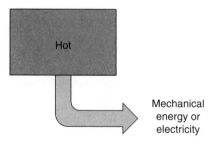

Figure 4.12
Conceptual diagram of an ideal heat engine, which extracts thermal energy from a hot substance and delivers an equal amount of high-quality mechanical energy or electricity. Such a device is impossible.

Figure 4.12 shows a conceptual diagram of what an ideal heat engine would be like. It would extract energy from a hot material and deliver it all as mechanical energy. But the second law of thermodynamics tells us that can't happen. In fact, another statement of the second law is this: *It is impossible to build a perfect heat engine.* What a real engine does is to extract energy from a hot substance and deliver only *some* of it as mechanical energy and the rest as heat transferred to its surroundings. Figure 4.13 shows a conceptual picture of a real heat engine. The real engine is less than 100 percent efficient, because the mechanical energy it delivers is less than the energy it extracts from the hot substance (which in turn is kept hot by the burning fuel). The efficiency, *e*, is just the ratio of mechanical energy delivered to total energy extracted:

$$e = \frac{\text{mechanical energy delivered}}{\text{energy extracted from fuel}} \tag{4.4}$$

So how efficient can we make a heat engine? Maybe we can't convert all the fuel energy to mechanical energy, but why can't clever engineers get the efficiency of Equation 4.4 close to 1 (i.e., to 100 percent)? Early in the nineteenth century a young French engineer named Sadi Carnot pondered this question. He designed a simple type of engine involving a fluid inside a cylinder closed with a moveable piston. Placed in contact with

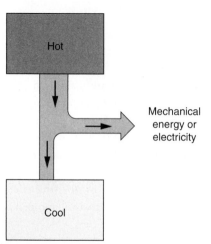

Figure 4.13
Conceptual diagram of a real heat engine, which extracts thermal energy from a hot substance and converts only some of it into high-quality energy. The rest is rejected to a cool substance, usually the surrounding environment. Can you estimate the efficiency of this engine?

a hot substance, the fluid would push on the piston and deliver mechanical energy. But then the fluid would have to be cooled to its original state, so the cylinder would be placed in contact with the ambient environment. In the process, heat would flow to the environment—this heat being energy extracted from the fuel but *not* delivered as mechanical energy. Although the engineering details of the **Carnot engine** are not exactly those of a modern gasoline engine or a steam power plant, the general principles are similar. More important, Carnot proved that no engine, no matter how cleverly designed, could be more efficient than his engine.

So what is the efficiency of Carnot's engine? I won't go through the derivation, which is done in most introductory physics texts. The result depends on just two numbers: the temperature of the hot substance (T_h) and the temperature of the cool substance (T_c) to which the waste heat flows. This efficiency, often called the **thermodynamic efficiency**, is given by

$$e = 1 - \frac{T_c}{T_h} \tag{4.5}$$

Here the temperatures must be expressed on an absolute scale (Kelvin or Rankine).

We're generally stuck with the low temperature; it's that of Earth's ambient environment, around 300 K. And T_h is the hottest temperature we can achieve in our heat engine, which is set by the materials from which the engine is made. As a practical matter, for example, the maximum steam temperature in a fossil-fueled power plant is around 650 K; for a nuclear plant, where water is heated in a single large pressure vessel, it's around 570 K. These lead to thermodynamic efficiencies of

$$e_{\text{fossil}} = 1 - \frac{T_c}{T_h} = 1 - \frac{300 \text{ K}}{650 \text{ K}} = 0.52$$

or 52 percent, and

$$e_{\text{nuclear}} = 1 - \frac{T_c}{T_h} = 1 - \frac{300 \text{ K}}{570 \text{ K}} = 0.46$$

or 46 percent. T_h is somewhat higher for an automobile engine, but the low temperature of the warm engine block is higher, and the thermodynamic efficiency comes out about the same, a little under 50 percent.

Be aware that Equation 4.5 gives the absolute maximum efficiency that's theoretically possible. Given the temperatures T_h and T_c, no amount of cleverness on an engineer's part can increase the efficiency over the Carnot limit expressed in Equation 4.5. And the overall efficiency can be a lot lower if friction and other losses sap energy, or if energy is diverted to pumps, pollution control systems, or other devices essential to the operation of an engine or power plant. By the time those losses are considered, most older electric power plants are somewhere around 30 to 40 percent efficient, meaning that roughly two-thirds of the energy extracted from fuels is dumped to the environment as waste heat.

Example 4.8   Ocean Thermal Energy Conversion

Ocean thermal energy conversion (OTEC) is a scheme for extracting energy from the temperature difference between the warm surface waters of tropical oceans and the cool water deeper down. Solar heating of the surface water would take the place of a burning fuel. Although there are many engineering details to work out, OTEC has potential as an energy source, particularly for tropical island nations (more on OTEC in Chapter 10). However, as this example shows, OTEC is not particularly efficient.

Typical tropical ocean surface temperatures are around 25°C, while several hundred meters down the temperature remains at about 5°C. What is the thermodynamic efficiency of a heat engine operated between these two temperatures?

Solution

Equation 4.5 gives the thermodynamic efficiency, but it requires absolute temperatures. Since 0°C is 273 K, our hot and cold temperatures of 25°C and 5°C become, respectively, 298 K and 278 K. Then Equation 4.5 gives

$$e = 1 - \frac{T_c}{T_h} = 1 - \frac{278 \text{ K}}{298 \text{ K}} = 0.067$$

or only 6.7 percent. However, there's no fuel to pay for, so this low efficiency is not necessarily unacceptable.

Thermal Pollution

We need to dispose somehow of the waste heat produced in our heat engines. In a car, some of the heat goes out the tailpipe as hot gases; much of the rest goes out through the radiator, a device designed explicitly to dump waste heat to the surrounding air. Power plants are almost always built on the shores of large bodies of water, giving them a place to dump waste heat. But the associated temperature rise of the water can affect the aquatic ecology. To avoid such **thermal pollution**, water from the power plant is often cooled through contact with air before being returned to a river, lake, or bay. Those huge concrete cooling towers you probably associate with nuclear power (but which are used in fossil-fueled power plants as well) serve that purpose; the cloudlike plumes emerging from the towers consist of water droplets condensed from water that was cooled by evaporation inside the towers (Fig. 4.14). So vast is the amount of waste heat produced in power generation that most of the rainwater falling on the continental United States eventually makes its way through the cooling systems of electric power plants.

Our understanding of heat engines illustrates a point I made earlier when I described the cups of hot and cold water that gradually became two cups of lukewarm water, despite no overall loss of energy. Given the original cups of hot and cold water, we could have

Figure 4.14
The cooling tower at this nuclear power
plant transfers waste heat to the atmosphere.

run a heat engine and extracted some useful, high-quality mechanical energy. But if we
let the cups become lukewarm we could no longer do that, despite all the original energy
still being in the water. It's in that sense that the quality of energy in the lukewarm
water is lower; none of it is available for conversion to mechanical energy.

4.8 Energy Quality, End Use, and Cogeneration

To propel a car, truck, or airplane, we need kinetic energy—that is, mechanical energy
of the highest quality. To operate a computer, TV, or washing machine, we also need
the highest-quality energy, this time in the form of electricity. But heating water for a
shower or producing steam for processing food only requires lower-quality thermal energy.
These different **end uses** have different energy-quality requirements. We can produce
low-quality energy simply by burning a fuel and transferring the energy as heat to water
or whatever. The second law of thermodynamics places no restrictions on this process.
So it makes a lot more sense, for example, to heat your hot water with a gas flame than
with electricity. That's because each joule of electrical energy you use represents two
additional joules that were rejected to the environment as waste heat back at the elec-

tric power plant. Energy from the gas, on the other hand, transfers to the water with close to 100 percent efficiency (practical heaters can exceed 85 percent efficiency).

So why can't we use all that waste heat for low-quality energy purposes? Increasingly, we do. Whole cities in Europe are sometimes heated with energy that's the byproduct of electric power generation. Increasingly, institutions in the United States are choosing a similar energy path called **cogeneration**. In this process, fossil fuels are burned to produce steam, which turns a turbine and generates electricity just as in any electric power plant. But the waste heat is then used for heating buildings or for industrial processes. In the United States today, only about 5 percent of electric utilities' production involves cogeneration, so we're still a long way from using most of the waste heat that's forced on us by the second law of thermodynamics. This means there's considerable additional potential for energy savings through cogeneration.

Primary Energy

The second law of thermodynamics can lead to confusion when we try to quantify energy consumption. If I ask about the energy used to run your car, do I mean the energy content of the gasoline you burned, or only the actual kinetic energy that was put into the car's motion? In this case the answer is perhaps obvious: You burn gasoline right there

---

Box 4.1   Cogeneration at Middlebury College

Like many institutions, Middlebury College heats its buildings using a central steam-production plant. Steam is delivered through heavily insulated underground pipes to the various campus buildings, where it can be used directly for heating or to operate chillers for summer air conditioning. Over the years, Middlebury has used oil, coal, and wood as the fuel for the steam-production boilers.

The high steam temperatures and pressures generated in operating the boilers are greater than required for heating buildings; thus they constitute a higher-quality form of energy than what's needed. To make use of this high-quality energy, Middlebury has installed turbine-generators that extract high-quality energy to generate electricity and pass the "waste heat" on to the steam distribution system. This is a classic example of cogeneration, in this case installed retroactively on what was a purely thermal energy system. Today the college operates several such generators, with a combined electrical generation capability of 1.7 MW (Fig. 4.15). Although the turbine-generators don't run all the time at capacity, they do produce some 15 percent of the college's total electrical energy consumption.

The newest turbine-generator operates at an inlet temperature of 406°F and an outlet temperature of 259°F, which indicates the decrease in steam energy that goes toward making electricity. A pressure drop from 250 pounds per square inch at the inlet to 20 pounds per square inch at the outlet also signals energy extraction and the decrease in energy quality across the turbine-generator.

Figure 4.15
This cogeneration unit at Middlebury College produces nearly 1 MW of electric power.

in your car, and it's clearly the total energy of the gasoline that you're responsible for consuming, despite the fact that most of that energy was simply dumped to the environment as waste heat.

Electrical energy presents a murkier situation. Your house is miles from the power plant that produces your power, and you tend to think only about what's going on at your end of the wire. When you turn on your air conditioner and consume, say, 5 kWh of electrical energy, should you count that as your total energy consumption? Or should you recognize that the electricity came from a power plant with an overall efficiency of only about 30 percent, in which case your total energy consumption is more like 15 kWh?

That 15 kWh—the total energy extracted from the fuel used to make your electricity—is called **primary energy**. Primary energy includes the substantial amount of fuel energy that the second law of thermodynamics keeps us from converting to electricity or other high-quality energy. But primary energy is a more honest accounting of the actual energy consumed to provide us with what's generally a lesser amount of useful energy. In discussions of world energy use, you'll often see the term **total primary energy supply** (TPES) used to describe a country's energy use. Again, *primary* means that we're accounting for energy right from the source, not just the end use.

In the case of electric power plants, we often distinguish the thermal power output—the primary energy, or total energy extracted from fuel, including what's dumped as waste heat—from the electrical energy that is the plant's useful product. You'll sometimes see the designation MWth, for *megawatts thermal,* and MWe, for *megawatts electric,* to distinguish between the two. Most power plants are specified by their electrical output; after all, electrical energy is what they're trying to produce and sell. A typical large power plant might have an output of 1,000 MWe (equivalently, 1 GWe). If its efficiency is 33 percent, then it extracts energy from fuel at three times this rate, so its thermal output is 3,000 MWth. Incidentally, you might want to look back at Example 4.6 and decide whether that 2-GW nuclear plant is a 2-GWe or a 2-GWth plant. In the former case, it would be a very large plant indeed; in the latter, it would be of only modest size.

The electricity picture is further complicated by the mix of different sources of electrical energy, and even of the different efficiencies of thermal power plants. As a result, there's no single conversion between 1 kWh of electrical energy consumption and the corresponding primary energy. In the Pacific Northwest of the United States, for example, much of the electricity comes from nonthermal hydropower (nonthermal, that is, unless you consider Earth's water cycle as a natural heat engine). In the Pacific Northwest, 1 kWh of electrical energy therefore corresponds, on average, to only a little more than 1 kWh of primary energy. In New England, which gets much of its electricity from nuclear power plants that are roughly 33 percent efficient, the ratio is more like a factor of 3 or so. In fact, for the United States as a whole, the total primary energy consumed to generate electricity in the early twenty-first century is some 2.9 times the total amount of electrical energy produced. Accounting for electricity consumed within power plants and lost in electrical energy transmission raises the average primary energy consumed to nearly 3.2 kWh for every 1 kWh of electrical energy delivered to the end user. Again, think of your energy servants: For every one you've got working for you running anything electrical, there are on average two more back at the power plant laboring away just to produce waste heat.

4.9 Refrigerators and Heat Pumps

Equation 4.5 shows that we can increase the thermodynamic efficiency of a heat engine either by raising the maximum temperature or by lowering the minimum temperature. The materials the engine is made of set the maximum temperature. But why can't we just refrigerate the cooling water used in a power plant or car engine to lower the minimum temperature? This won't work, and the reason is again the second law of thermodynamics. How do we cool things? With a **refrigerator**, which, conceptually, is like a heat engine run in reverse. With a heat engine, we extract heat from a hot substance, turn some of it into mechanical energy or electricity, and reject the rest as waste heat to a cool substance. With a refrigerator, we extract energy from a cool substance and transfer it to a hot substance (Fig. 4.16). An ideal refrigerator would do exactly that, with no other energy needed. But this is impossible, as my example of the two cups of lukewarm water suggests. They never become, spontaneously, one hot and one cold cup.

In fact, another statement of the second law is this: *It is impossible to build a perfect refrigerator.* Equivalently, *heat doesn't flow spontaneously from a cooler to a hotter object.* The key word here is *spontaneously.* You can make heat flow from cool to hot, but only if you also supply high-quality energy. That's why you have to plug in your refrigerator. And that's why you can't make a more efficient heat engine by teaming it up with a refrig-

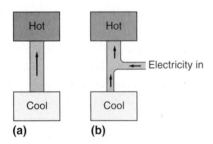

Figure 4.16

(a) A perfect refrigerator would transfer energy spontaneously from cool to hot, but this is impossible according to the second law of thermodynamics. (b) A real refrigerator requires a high-quality electrical or mechanical energy input to effect the energy transfer from cool to hot.

erator; at best you'll use up the extra energy running the refrigerator, and in reality you'll lose even more energy to waste heat.

Put a room-temperature soda in the refrigerator and it will cool down. The energy extracted from the soda comes out the back of the fridge as heat—and it looks like you have a more organized state, with one cooler thing (the soda) and one warmer thing (the room the fridge is in). So has entropy decreased, in violation of the second law? No: It takes electricity—energy of the highest quality—to operate the refrigerator and transfer energy from the cooler to the hotter side. This high-quality energy is degraded to heat, which is dumped out of the back of the refrigerator, along with the energy extracted in cooling the soda. In a perfectly efficient refrigerator, the entropy increase associated with this energy degradation would just compensate for the entropy decrease in creating the more organized state of warm and cold. Any real refrigerator has inefficiencies associated with irreversible thermodynamic processes, and as a result entropy inevitably increases. You can make a similar argument for Planet Earth; surely the emergence of life and of civilization represents increasing organization and therefore lower entropy. But Earth isn't a closed system. It gets its energy from the Sun, and degradation of that high-quality solar energy more than compensates for the entropy decrease associated with organized life.

There is a way in which refrigerators can actually help us save energy. A **heat pump** is basically a refrigerator or air conditioner run in reverse. In the winter it extracts thermal energy from the cool outside air (or from the ground in cold climates) and delivers it to the warmer interior of the house. Of course, heat won't flow spontaneously in that direction, so the heat pump needs a source of high-quality energy, namely electricity. But a well-designed heat pump operating from a cool source that isn't too cold can transfer a lot more thermal energy than it uses as electricity, and therefore heat pumps have considerable potential for energy savings. I'll discuss heat pumps further in Chapter 8. Incidentally, the fact that a heat pump can transfer more thermal energy than it uses in high-quality energy is why the refrigerator of Example 4.1, which removes thermal energy from its interior at the rate of 64 W, doesn't consume electrical energy at that rate.

Energy Overview

Energy is a big subject, and Chapters 3 and 4 have just skimmed the surface. But you now have enough background to understand quantitatively how we power our high-energy society, what constraints the fundamental laws of nature place on our use of energy, and what steps we might take to reduce our energy use or mitigate its environmental impact. In particular, you should be fluent in the various units used to measure energy and power, you should recognize the different fundamental kinds of energy, and you should be keenly aware of the restrictions the second law of thermodynamics places on our ability to make use of thermal energy. Along the way, you've learned to calculate the effects of energy flows as they cause changes in temperature or in the physical states of substances, and in that context you've learned to evaluate quantitatively the important role of insulation in making energy-efficient buildings. You're now ready to apply your understanding of energy to the various sources of society's energy and to the complex flows of energy that govern Earth's climate.

Box 4.2 Energy Flows in the United States

Chapter 2 introduced humankind's energy sources, and Chapter 3 quantified our energy consumption. In this chapter we've seen how the principles of physics limit our ability to use all that energy. A dramatic example is the flow of energy from sources to end uses in the United States. Figure 4.17 depicts this complex set of energy flows, and shows that more than half the primary energy is wasted—much of it through second-law and other inefficiencies in electric power generation.

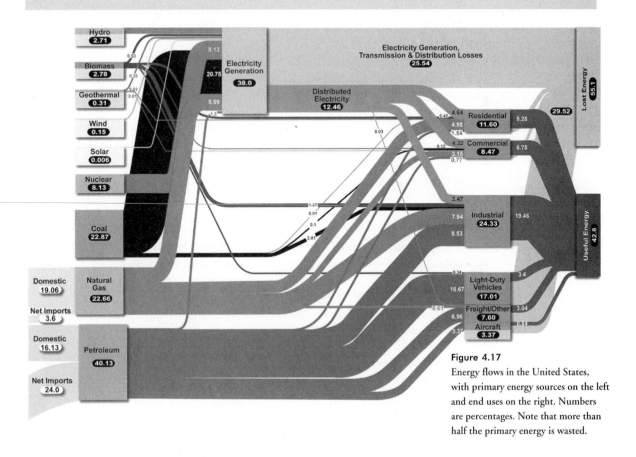

Figure 4.17
Energy flows in the United States, with primary energy sources on the left and end uses on the right. Numbers are percentages. Note that more than half the primary energy is wasted.

Chapter 4   Chapter Review

BIG IDEAS

4.1   **Heat** is energy that flows because of a temperature difference. Heat usually raises a material's **internal energy**, the kinetic energy of random molecular motion. The **first law of thermodynamics** says that the change in internal energy is the sum of the mechanical work done on an object and the heat that flows into it.

4.2   **Temperature** measures the average thermal energy of molecules. Common temperature scales include Celsius (°C) and Fahrenheit (°F), as well as the Kelvin scale (K), which is based on absolute zero.

4.3   Heat-transfer mechanisms include **conduction**, important in building insulation; **convection**, important in circulating heat within buildings as well as in energy loss associated with windows; and **radiation**, especially important for objects such as planets and stars that are surrounded by vacuum.

4.4   **Heat capacity** and **specific heat** describe the amount of heat needed to effect a given temperature change.

4.5   **Heats of transformation** describe the energy per unit mass required to change a material's state. On melting or vaporizing, the heat of transformation is stored as **latent heat**.

4.6   Energy has *quality* as well as quantity. The highest quality energies include mechanical and electrical energy. Internal energy is of lower quality. The **second law of thermodynamics** prohibits the transformation of low-quality energy into high-quality energy with 100 percent efficiency.

4.7   The second law of thermodynamics is ultimately about the tendency toward more disorder. **Entropy** is a measure of that disorder. The second law limits our ability to produce high-quality energy efficiently through methods involving heat. As a result, **heat engines** inevitably reject as waste heat some of the energy they extract from their fuels.

4.8   Energy efficiency dictates that we not use high-quality energy like electricity for low-quality uses such as heating. In the efficient process of **cogeneration**, waste heat from electricity generation is put to use for heating.

4.9   **Refrigerators** and **heat pumps** transfer heat from cooler to hotter systems. Heat pumps can provide efficient heating by using small amounts of high-quality energy to transfer larger amounts of heat.

TERMS TO KNOW

absolute zero (p. 70)

Carnot engine (p. 92)

cogeneration (p. 95)

conduction (p. 71)

convection (p. 79)

emissivity (p. 81)

end use (p. 94)

entropy (p. 89)

external-combustion engine (p. 90)

first law of thermodynamics (p. 69)

forced convection (p. 81)

heat (p. 68)

heat capacity (p. 84)

heat engine (p. 90)

heat of fusion (p. 86)

heat of transformation (p. 86)

heat of vaporization (p. 86)

heat pump (p. 98)

internal-combustion engine (p. 90)

internal energy (p. 68)

irreversible process (p. 88)

Kelvin scale (p. 70)

latent heat (p. 87)

natural convection (p. 81)

primary energy (p. 96)

radiation (p. 81)

refrigerator (p. 97)

R value (p. 74)

second law of thermodynamics (p. 89)

sensible heat (p. 87)

specific heat (p. 84)

Stefan-Boltzmann constant (p. 82)

Stefan-Boltzmann law (p. 82)

temperature (p. 69)

thermal conductivity (p. 71)

thermal energy (p. 68)

thermal pollution (p. 93)

thermodynamic efficiency (p. 92)

total primary energy supply (p. 96)

GETTING QUANTITATIVE

Conductive heat flow: $H = kA \dfrac{T_h - T_c}{d}$ (Equation 4.1; p. 73)

Stefan-Boltzmann law for radiation: $P = e\sigma AT^4$ (Equation 4.2; p. 82)

Energy required for temperature change: $Q = mc\Delta T$ (Equation 4.3; p. 85)

Efficiency of an engine: $e = \dfrac{\text{mechanical energy delivered}}{\text{energy extracted from fuel}}$ (Equation 4.4; p. 91)

Thermodynamic efficiency: $e = 1 - \dfrac{T_c}{T_h}$ (Equation 4.5; p. 92)

QUESTIONS

1. Describe the difference between heat and thermal energy.

2. My neighbor tells me that his new wood stove "puts out a lot of heat." Is this a correct use of the term *heat*? Discuss.

3. Solid glass is not a particularly good insulator, but fiberglass—a loose agglomeration of thin glass fibers—is an excellent insulator widely used in buildings. What gives fiberglass its insulating quality?

4. You can bring water to a boil fairly quickly, but it takes a long time to boil it away. What two properties of water determine this fact?

5. Some modern medical thermometers work by measuring the infrared emission from the eardrum, and they give nearly instantaneous readings. What's the principle behind such thermometers?

6. Large lakes tend to exert a moderating effect on the surrounding climate. Why?

7. What is the essential idea behind the second law of thermodynamics?

8. Explain, in the context of the second law of thermodynamics, why heating water with electricity is not an energy-efficient thing to do.

9. In the United States, a significant portion of the energy we produce is in the form of high-quality, low-entropy electricity. But many end uses of energy tend more toward low- and medium-quality thermal energy, so we end up using some of our high-quality energy for low-quality uses. Why is this an inefficient use of energy resources?

10. My local nuclear power plant is rated at 650 MWe. What does the lowercase *e* mean here?

11. Power plants are slightly more efficient in the winter than in the summer, to the extent that power-plant ratings are sometimes given separately for the two seasons. Why the difference?

EXERCISES

1. Verify that the 5,000 Btu/h heat flow through the basement wall described in Section 4.3 is equivalent to about 1,500 W and therefore fifteen energy servants.

2. My solar hot-water system has a 120-gallon storage tank with a surface area of 4.3 m². It's insulated with 5-cm-thick urethane foam, and the water in the tank is at 76°C. It's in the basement, where the ambient temperature is 15°C. What is the rate of heat loss from the tank?

3. You're considering installing a 4-foot-high by 8-foot-wide picture window in a north-facing wall (no solar gain) that now has a total *R* value of 23. As a compromise between initial

cost and energy savings, you've chosen a standard double-glazed window like the one listed in Table 4.2. As a result of installing this window, how much more oil will you burn in a 6-month winter heating season with an average outdoor temperature of 25°F and an indoor temperature of 68°F? Assume your oil furnace is 80 percent efficient, as in Example 4.2.

4. The electric heating element in a 40-gallon water heater is rated at 5 kW. After your sister takes a shower, the water temperature drops to 90°F. How long will it take the heater to bring the temperature back up to 120°F? The density of water is 3.79 kg per gallon. Neglect heat loss, which in a well-insulated heater should be a lot less than that 5-kW power input.

5. (a) Calculate the energy needed to bring a cup of water (about 250 grams) from 10°C to the boiling point. Then find the time it takes to heat this water (b) in a 900-W microwave oven that transfers essentially all of its microwave energy into the water and (c) in a 1-kg aluminum pan sitting on a 1,500-W electric stove burner that transfers 75 percent of its energy output to the water and the pan. Assume the pan, too, starts at 10°C and has to be heated to water's boiling point.

6. Assuming your electricity comes from a power plant that is 33 percent efficient, find the actual primary energy consumed in heating the water as described in both parts (b) and (c) of Exercise 5. Assume that the microwave oven uses electrical energy at the rate of 2.4 kW to produce its 900 W of microwave energy. Remember also that you have to heat the aluminum pan on the stove. Compare your answers with the energy needed solely to heat the water.

7. Lake Champlain, bordered by New York, Vermont, and Quebec, has a surface area of 1,126 km². (a) If it freezes solid in the winter with an average ice thickness of 0.5 m, how much total energy does it take to melt the ice once spring comes? The density of ice is about 920 kg/m³ and the heat of fusion is 334 kJ/kg. (b) If this energy is provided by sunlight shining on the lake at an average rate of 200 W/m² but with only half that amount absorbed, how long will it take the lake to absorb the energy needed to melt it?

8. A house measures 42 feet wide by 28 feet deep. It sits on a 4-inch-thick concrete slab. The ground below the slab is at 48°F, and the interior of the house is at 68°F. (a) Find the rate of heat loss through the slab. (b) By what factor does the heat-loss rate drop if the slab is poured on top of a 2-inch layer of polystyrene insulation?

9. Rework Example 4.2, now assuming that the house has fourteen windows, each measuring 45 inches high by 32 inches wide. When you buy the house, they're single-pane windows, but when you upgrade the insulation, you also upgrade the windows to the best you can buy (see Table 4.2). Remember that the windows replace some of the wall area. Neglect any solar gain through the windows, which would increase your savings even more.

10. The surface of a woodstove is at 250°C (that's degrees Celsius, not kelvins). If the stove has a surface area of 2.3 m² and an emissivity of 1, at what rate does it radiate energy to its surroundings?

11. The filament of a typical lightbulb is at a temperature of about 3,000 K. For a 100-W bulb, what is the surface area of the filament? You can assume that the entire 100 W goes into heating the filament, and that the filament has an emissivity of 1.

12. Assuming a low temperature of 300 K, at what high temperature would you have to operate a heat engine for its thermodynamic efficiency to be 80 percent?

13. A 2.5-GWth nuclear power plant operates at a high temperature of 570 K. In the winter the average low temperature at which it dumps waste heat is 268 K; in the summer it's 295 K. Determine the thermodynamic efficiency of the plant in each season, and then derive from it the maximum possible electric power output.

14. A jet aircraft engine operates at a high temperature of 750°C, and its exhaust temperature—the effective low temperature for the engine—is a hot 350°C. What is the thermodynamic efficiency of this engine?

15. Show that an R value of 1 (that is, 1 ft$^2\cdot$°F$\cdot$h/Btu) is equivalent in SI units to 0.176 m$^2\cdot$K/W (equivalently, m$^2\cdot$°C/W).

RESEARCH PROBLEMS

1. Find the average temperature for a particular winter month in the region where your home is located. From your family's heating bills, estimate as best you can the amount of oil or gas consumed to heat your house during this month. Estimate the surface area of your house and, on the assumption that all the energy content of your heating fuel escaped through the house walls and roof, estimate an average R value for the house.

2. Locate the nearest thermal power plant to your home or school and describe what type it is, what kind of fuel it uses, and any other details. From this information, determine (a) its electric power output and (b) its thermal power output, and then (c) calculate its actual efficiency. If you're in the United States, you may find useful the U.S. Department of Energy's listings entitled "Existing Electric Generating Units in the United States."

3. Locate a microwave oven and find its rated microwave power and its actual electric power consumption. From these determine its efficiency at converting electrical energy into microwaves. Then time how long it takes to heat a known amount of water from a known initial temperature until it reaches the boiling point. Assuming that all the microwave energy ends up in the water, use the result of your experiment to calculate a value for the microwave power. How does it compare with the oven's rated power?

4. Find specifications on your car or your family's: its mass, its engine's power rating in horsepower, and its acceleration (0 to some speed in how many seconds?). Compare the horsepower rating with the actual average power needed to give the car its kinetic energy at that final speed.

FOSSIL FUEL ENERGY

The time is 100 million years ago, the place is a steamy swamp. Earth is a much warmer planet than today, thanks to a level of atmospheric CO_2 more than three times what it is now, and more than four times what it was before we humans began dumping CO_2 into the atmosphere. Our swamp could be almost anywhere, because although the tropics in this ancient world are just a few degrees Celsius warmer than today, the poles are some 20°C to 40°C warmer and therefore ice-free. Dinosaurs thrive even north of the Arctic Circle. However, there are plenty of places in today's world where our swamp couldn't be. That's because sea level is some 200 meters higher than at present, so some of today's land is under water.

The tropical heat and abundant CO_2 promote rapid plant growth. Plants take in CO_2 from the atmosphere and water from the ground. Using energy from sunlight, they rearrange the simple CO_2 and H_2O into sugars and other complex molecules that store this solar energy in the electric fields associated with the new molecular configuration. This, of course, is the process of photosynthesis. The energy-storing molecules produced in photosynthesis become the energy source for the plants that made them, for the herbivorous animals that eat those plants, for the carnivorous animals that eat the herbivores, and for the insects and bacteria that decompose all of these organisms when they die.

5.1 The Origin of Fossil Fuels

Coal

Back to our particular swamp. Giant ferns surround stagnant, acidic waters. An herbivorous dinosaur munches contentedly on the vegetation, ambling slowly through muddy ooze composed largely of dead plant matter. The acidic conditions limit the decay of this organic material, and as more and more generations of dead plants (and eventually our dinosaur's body) accumulate, they push the older material ever deeper. Eventually the material forms a soft, crumbly, brown substance called **peat**. Peat contains some of the energy that was originally fixed by photosynthesis when the plants were alive. Indeed, peat is still forming today, in places ranging from the Florida Everglades to the bogs of Scotland, Canada, Scandinavia, and Russia. When dried, peat provides fuel for warming houses and for cooking.

But our peat doesn't meet that fate. Rather, geologic processes bury it ever deeper. Temperature and pressure increase, and the material undergoes physical and chemical changes. Much of the oxygen and hydrogen escape, leaving a substance rich in carbon. The final result is the black solid we call **coal**. Coals vary substantially in type, depending on where they form and how far evolved they are. The most evolved—bituminous and anthracite coal—are largely carbon. However, coals contain significant quantities of other elements, including sulfur, mercury, and even uranium.

Coal also contains energy. The carbon atoms in coal aren't combined with oxygen as they are in CO_2. As such they, together with oxygen in the air, represent a state of higher energy than does CO_2. Combining the carbon with oxygen in the process of combustion releases that excess energy. Coal is quite a concentrated energy source; the highest-quality anthracite contains some 36 MJ/kg, or about 10,000 kWh per ton. (The 29-MJ/kg figure in Table 3.3 is an average over different types of coal.)

So coal's energy is ultimately sunlight energy trapped by long-dead plants. This direct link with once-living matter makes coal a fossil fuel. Some lower-grade coals actually contain recognizable plant fossils (Fig. 5.1).

Coal is still forming today, although under current conditions that happens much more slowly than it did when the dinosaurs lived. But the rate of coal formation has always been agonizingly slow. Nearly all living matter decays promptly after death, releasing the CO_2 that plants removed from the atmosphere not long before. Nearly all the stored energy is released, too, and is quickly radiated to space. But as I discussed in Chapter 1 when I described Earth's energy endowment, a tiny fraction of living material escapes immediate decay and is buried to form coal and other fossil fuels. The fossil fuels we burn today are the result of hundreds of millions of years of growth, death,

Figure 5.1
A fossil leaf in low-grade coal.

escape from decay, and physical and chemical processes that transform once-living matter into these fuels.

Oil and Natural Gas

Here's another ancient scene: This time we're in a warm, shallow ocean. A steady "rain" of dead organic matter falls to the ocean floor. Most of this consists of tiny, surface-dwelling organisms called, collectively, **plankton**. Plankton includes animal species (zooplankton) and plant species (phytoplankton), the latter using photosynthesis to capture the solar energy that is the basis of marine food chains. The organic material reaching the ocean floor joins the inorganic matter carried into the sea from rivers. The result is a loose mix of ocean-floor sediments.

In time, the pressure of the overlying water compacts the sediments. Natural cements and the crystallization of inorganic minerals eventually convert the sediments into solid material. Over geologic time, ocean-floor sediments are buried and subject to high pressures and temperatures. The result is the layered solid we call sedimentary rock. During the formation of sedimentary rock, organic material may be "cooked," first to a waxy substance called **kerogen** and eventually to a combination of organic liquids and gases called, collectively, **petroleum** (this term is often used for liquids alone, but formally *petroleum* refers to both liquids and gases). Trapped in petroleum is some of the energy that was captured long before by photosynthetic plankton on the ocean surface.

Petroleum occurs throughout many sedimentary rock formations, but usually not in sufficient quantities to make its extraction economically feasible (at least not at present; more on this later). But being fluid, some petroleum migrates to collect in porous rocks such as sandstones or limestones. Overlying layers of impermeable rock may trap the petroleum, resulting in a concentrated source of fossil energy (Fig. 5.2). Fast-forward to the present, and you're likely to find an energy company drilling to extract the trapped fuel.

Like coal, petroleum represents fossil energy. And as with coal but even more so, we're removing petroleum from the ground at a rate millions of times greater than the rate at which it forms.

5.2 The Fossil Fuels

Coal and petroleum are the fossil fuels that power modern society. As I showed in Chapter 2, these fuels provide nearly 90 percent of the energy we humans consume. A few other fossil products, especially peat and the tarry solid called *bitumen,* are also used for energy or other applications such as paving roads. But coal and petroleum are by far the dominant fossil fuels.

Composition

We've seen that coal consists largely of carbon, although lower grades contain significant quantities of hydrocarbons (molecules made of hydrogen and carbon) and contaminants such as sulfur and mercury. Petroleum is even more of a mix. As with coal, the exact composition of petroleum varies with where and under what conditions it was

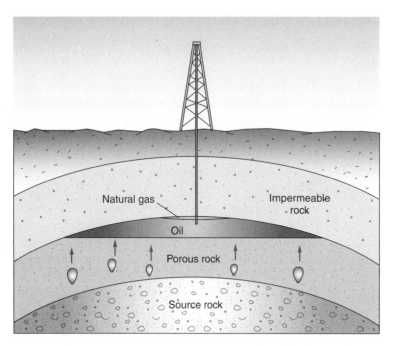

Natural gas

Impermeable rock

Oil

Porous rock

Source rock

Figure 5.2

Formation of an oil deposit in a geological structure called an *anticline,* in which rock layers have bent upward. The oil originally forms in the source rock, then migrates upward through the porous layer to collect at the top, where it is trapped by the overlying impermeable rock. Natural gas may collect above the oil.

formed. Most significant is the ratio of gas to liquid. The gaseous component of petroleum is **natural gas,** which consists largely of methane (CH_4, the simplest organic hydrocarbon) and other light, gaseous hydrocarbon molecules. The liquid component of petroleum is **crude oil**. Crude oils vary widely in composition, but all contain a mix of hydrocarbons. Typically, crude oil is about 85 percent carbon by weight, and most of the rest is hydrogen. Sulfur, oxygen, and nitrogen are also present in significant quantities.

Refining

We **refine** crude oil to produce a variety of fuel products, ranging from the heaviest oils for industrial boilers, to fuel oil used in home heating, to diesel oil that powers most trucks and some cars, to jet aircraft fuel, to gasoline for our cars. Each of these products contains a mix of different molecules, so no single chemical formula describes a given fuel. Molecules typically found in gasoline, for example, include heptane (C_7H_{16}), octane (C_8H_{18}), nonane (C_9H_{20}), and decane ($C_{10}H_{22}$). These molecules consist of long chains of carbon with attached hydrogens; for every n carbons, there are $2n + 2$ hydrogens (Fig. 5.3). Higher grades of gasoline have a greater ratio of octane to heptane, which makes them burn better in high-performance engines. (This is the origin of the term *high octane* you see at the gas pump.) The energy content of refined fuels varies somewhat, but a rough figure is about 45 MJ/kg for any product derived from crude oil. By the way, refining is itself an energy-intensive process; some 8 percent of total U.S. energy

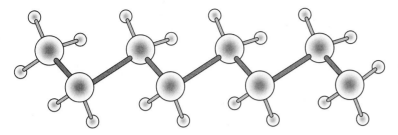

Figure 5.3
Structure of the octane molecule (C_8H_{18}), a typical component of gasoline. The larger dark circles are the eight carbon atoms, each bonded to two hydrogen atoms, with an extra hydrogen at each end of the molecule.

consumption is used to run oil refineries. Figure 5.4 shows the typical range of refinery products. The actual quantitative breakdown varies with demand; for example, the percentage of crude oil refined into fuel oil for heating increases in the winter.

Other Products from Fossil Fuels

We don't burn all the products we make from petroleum. About half of the "other" category in Figure 5.4, which is around 7 percent of all crude oil, is turned into road-making products, lubricants, and petrochemical feedstocks. The latter are oil derivatives that go into making a vast array of the products we use today—everything from plastics to medicines, ink to contact lenses, insecticides to toothpaste, perfumes to fertilizers, lipstick to paint, false teeth to food preservatives. As oil grows scarce, it will become increasingly valuable as a raw material for manufacturing products, giving us all the more incentive to look elsewhere for our energy. Our descendants a century hence may well look back and ask, incredulously, "Did they actually burn such valuable stuff?"

History of Fossil Fuel Use

Humankind has used coal as an energy resource, albeit in limited quantities, for as long as two thousand years. Some evidence suggests that around 1000 B.C.E. the Chinese were burning coal to smelt copper; archeologists have found coal cinders from 400 C.E. in what was Roman-occupied Britain; and written accounts of coal use begin

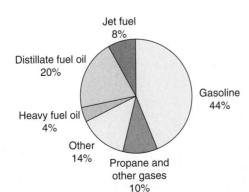

Jet fuel
8%

Distillate fuel oil
20%

Heavy fuel oil
4%

Other
14%

Propane and
other gases
10%

Gasoline
44%

Figure 5.4
Typical distribution of refinery products in the United States. Heavy fuel oil is used primarily in industrial boilers. Distillate fuel oil includes diesel fuel and heating oil. The "other" category includes lubricants, asphalt and road oil, kerosene, petroleum coke (a coal-like solid used as an industrial fuel), waxes, and feedstocks for petrochemicals. Percentages shown are on a per volume basis.

appearing in the thirteenth century. Coal was a widely used heating fuel in seventeenth-century England, so much so that the famous St. Paul's Cathedral was blackened from coal smoke even before it was completed. Coal consumption in Europe expanded greatly with the beginning of the industrial revolution and James Watt's eighteenth-century improvements to the steam engine. By the 1830s, coal mining thrived in the eastern United States, supplying coal for industry and for steam locomotives on the newly developed railroads.

Petroleum, too, has a long history. Liquid petroleum seeps naturally to Earth's surface in some locations, and as early as 5000 B.C.E. humans exploited this resource for a variety of purposes, including medicinal use in ancient Egypt. Oil became a military resource when Persians and Arabs developed oil-soaked arrows and other incendiary weapons. By the twelfth century, use of oil for illumination had spread from the Middle East to Europe. Indians of North America discovered oil seeps in what are now New York and Pennsylvania, and they used the oil for medicines. Whale oil was a favored fuel for lighting purposes until the nineteenth century, but the decline of whale populations drove the need for new liquid fuels, including petroleum. In 1859, Edwin Drake drilled the first well specifically designed to extract oil from the ground. Drake's well, at Titusville in northwestern Pennsylvania, struck oil at a mere 69 feet and marked the beginning of an ever-expanding quest for oil worldwide. Today, oil is surely the single most important natural resource for modern society.

Although isolated uses of natural gas date back thousands of years, much of the natural gas emerging from oil wells in past times was simply vented to the atmosphere or burned, uselessly, at the well site in a process known as *flaring* (Fig. 5.5). Flaring is increasingly illegal, and today natural gas is sometimes reinjected into the ground for future extraction when it cannot be economically distributed from a particular site.

From the modest beginnings outlined here, the fossil fuels have become the predominant energy source for humankind. Figure 5.6 shows quantitatively the historical growth in fossil fuel consumption.

Figure 5.5
Gas flaring at oil fields in Prudhoe Bay, Alaska.

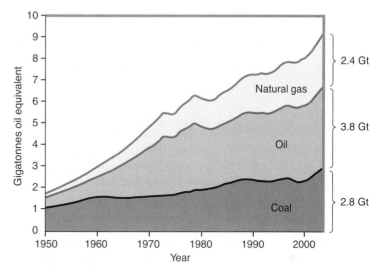

Figure 5.6

World fossil fuel consumption since 1950. The height of each shaded area represents the amount for one of the three fuels, so the top curve is the total fossil fuel consumption. Thus the graph shows that, in the final year plotted, fossil fuels equivalent to 9 gigatonnes of oil (Gtoe) were consumed globally: 2.8 Gtoe of coal, 3.8 Gtoe of oil, and 2.4 Gtoe of natural gas.

5.3 Energy from Fossil Fuels

When we burn a fossil fuel, we bring fuel into contact with atmospheric oxygen and raise the temperature so that chemical reactions occur between the fuel and oxygen. In sustained combustion, as in a coal-fired power plant or a gas-fired home water heater, heat released in the burning process sustains the combustion. In a gasoline engine, an electric spark provides the heat to ignite the gasoline-air mixture in each cylinder, whereas in a diesel engine, ignition results from compression and consequent heating of the fuel-air mixture as the piston rises in the cylinder. In any event, chemical reactions rearrange the atoms of fuel and oxygen, resulting in new chemical products and simultaneously releasing energy as heat.

Products of Fossil Fuel Combustion

Because fossil fuels consist largely of hydrocarbons, complete combustion produces mostly carbon dioxide gas (CO_2) and water vapor (H_2O). Both are nontoxic, commonplace substances already present in the atmosphere, but both are also climate-changing greenhouse gases. Although water vapor is actually the dominant, naturally occurring greenhouse gas in Earth's atmosphere, recent changes in atmospheric composition are largely the result of CO_2 produced from fossil fuels.

Because the various fossil fuels have different chemical compositions, their combustion results in differing amounts of CO_2 emission. Coal is mostly carbon, so CO_2 is the main product of coal burning. Oil consists of long hydrocarbon molecules with roughly twice as many hydrogen as carbon atoms (recall Fig. 5.3). So when oil burns, a good deal of the energy released comes from the formation of H_2O and so, for a given amount of energy, less CO_2 is emitted. Finally, natural gas is largely methane (CH_4), so it contains still more hydrogen and produces even less CO_2. That's why a shift from coal and oil to

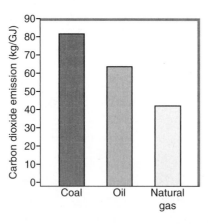

Figure 5.7
Carbon dioxide emission per gigajoule of energy released in the combustion of the three fossil fuels. Natural gas produces just over half the CO_2 of coal, making it a more climate-friendly fuel.

natural gas would slow but not halt the process of anthropogenic climate change. Figure 5.7 shows quantitatively the carbon emissions per unit of energy released in burning each of the three main types of fossil fuel. Unfortunately, we're going to run out of natural gas (and oil) long before we run out of coal, so switching to gas isn't a long-term solution. However, it could help buy some time until we start making large-scale use of carbon-free energy sources.

Carbon emission from fossil fuel burning is increasingly recognized as a major global environmental problem because of its potential for changing Earth's climate (this is the subject of Chapters 12 to 15). But CO_2 is not some incidental, unfortunate by-product of fossil fuel combustion. Rather, in a very real sense, it's exactly what we want to produce (along with water) when we burn fossil fuels. Carbon dioxide represents the lowest possible energy state of carbon in Earth's oxygen-containing atmosphere, so by making CO_2, we've extracted the most energy possible from our fossil fuels.

Example 5.1 Coal versus Gas

An older coal-burning power plant produces electrical energy at the rate of 1 GWe, typical of a large power plant. The plant is 35 percent efficient, meaning that only 35 percent of the coal's energy ends up as electricity. Estimate the rate at which the plant emits CO_2, and determine the reduction in CO_2 emissions if the coal plant were replaced with a modern gas-fired facility with 44 percent efficiency.

Solution

Figure 5.7 shows that coal burning yields CO_2 emissions of 90 kg per gigajoule (GJ) of energy produced. This figure refers to the thermal energy content of the fuel; at 35 percent efficiency, a 1 GWe power plant would need to produce thermal energy at the rate of 1 GW/0.35 = 2.86 GWth (I've added the *th* to remind you of the difference between electric and thermal power). Since 1 GW is 1 GJ per second, this 2.86-GWth rating, coupled with CO_2 emissions of 90 kg/GJ, gives a CO_2 emissions rate of

$$(2.86 \text{ GJ/s}) \times (90 \text{ kg/GJ}) = 257 \text{ kg/s}$$

for the coal-burning plant. That's more than 500 pounds of CO_2 every second!

A gas-fired plant that is 44 percent efficient, in contrast, would require a thermal power of 1 GW/0.44 = 2.27 GWth to produce that same 1 GW of electric power. For natural gas, Figure 5.7 shows 50 kg of CO_2 per gigajoule of energy, so the emissions rate for the gas-fired plant becomes

$$(2.27 \text{ GJ/s}) \times (50 \text{ kg/GJ}) = 114 \text{ kg/s}$$

The increased efficiency and, more significant, the lower CO_2 emissions per unit of energy produced from natural gas combine to give the gas-fired plant a total CO_2 emissions rate of less than half that of the coal-fired plant.

However, fossil fuels aren't pure hydrocarbon, and even if they were they wouldn't necessarily burn completely to CO_2 and H_2O. So a host of other substances result from fossil fuel combustion. Many of these are truly undesirable, and they are often unneces-

---

Box 5.1   Carbon versus Carbon Dioxide

Example 5.1 asked for the rate of CO_2 emission from a single coal-fired power plant, and the answer was 257 kg of CO_2 per second. You could use that figure to project an estimate of, say, the total rate of CO_2 emissions from all our coal combustion, or even from all fossil fuels. In describing anthropogenic emissions from fossil fuel consumption, different authors in different contexts may express the emission rate either as an amount of CO_2 per unit time, as I did in Example 5.1, or in carbon (C) per unit time. The latter is in some sense more general, because we do emit carbon in other forms (CH_4 [methane], for example). When you're reading about flows of carbon or CO_2, whether natural or anthropogenic, be sure you know which is being specified.

What's the relationship between CO_2 emissions when specified as CO_2 and as C? Simple: Carbon has an atomic weight of 12, and oxygen's atomic weight is 16, so the weight of the CO_2 molecule is $12 + (2 \times 16) = 44$—that is, 44/12, or 3.67 times that of a carbon atom. This means that a CO_2 emission rate specified as, say, 7 gigatonnes (Gt) of carbon per year is equivalent to $(7 \text{ Gt}) \times (44/12) = 26$ Gt of CO_2. (This 7 Gt per year isn't just an arbitrary figure; it's roughly the annual total of global anthropogenic carbon emissions from fossil fuel combustion.) In short, to convert from carbon to CO_2, just multiply by 44/12. No need to learn this figure; you should be able to arrive at it yourself as needed.

sary in that their production isn't required in the energy-release process. These substances comprise pollution. Pollution from fossil fuels is such a major environmental problem that I devote all of Chapter 6 to it.

5.4 Fossil Energy Technologies

Nearly all technologies for extracting energy from fossil fuels involve burning the fuel (we'll see one exception, the fuel cell, in Chapter 11). Burning releases energy as heat, which we can use directly to heat water, air, or other substances and thus use nearly all the energy contained in the fuel. But heated substances represent low-quality energy, as I described in Chapter 4. If we want high-quality energy, such as electricity or energy of motion, then we need to run a heat engine that converts *some* of the heat released from burning fuel into high-quality energy. The second law of thermodynamics prohibits us from converting *all* of that heat to high-quality energy.

The ingenuity of engineers, tinkerers, and inventors from James Watt and his predecessors on to the present day has resulted in a bewildering array of heat-engine designs. But, as Sadi Carnot showed, none can escape the fundamental efficiency limitations of the second law of thermodynamics, and none can be more efficient than Carnot's early-nineteenth-century design for a theoretical engine with the maximum possible efficiency. This doesn't mean we can't continue to improve real heat engines, but it does mean that such improvements can at best approach Carnot's ideal limit, not reach it. We won't explore heat engine designs in detail here, but it is helpful to distinguish two general types, external-combustion engines and internal-combustion engines.

External-Combustion Engines

The first steam locomotives had a long, cylindrical boiler heated by a wood or coal fire. The water boiled to steam, and steam pressure drove pistons that turned the locomotive's wheels (Fig. 5.8). Because the fuel burned in an external firebox, which subsequently heated the water, these were called *external-combustion engines*. With their efficiencies measured in single digits, steam locomotives were never very efficient, and they came nowhere near Carnot's theoretical limit. Steam automobiles were produced in the early twentieth century, but they, too, were not at all efficient. External-combustion engines for transportation vehicles are now a thing of the past.

However, the external-combustion engine lives on as the prime mover in most electric power plants. In a typical fossil-fueled plant, fuel burns in a boiler consisting of a chamber containing banks of water-filled pipes. This arrangement allows hot combustion gases to surround the pipes, resulting in efficient heat transfer. The water boils to high-pressure steam, which drives a fanlike **turbine** connected to an electric generator (recall the discussion of electromagnetic induction in Chapter 3). The spent steam, now at low pressure, still contains energy in the form of its latent heat. It

Figure 5.8
This steam locomotive, built around 1930, is an example of an external-combustion engine. The efficiencies of steam locomotives were low, typically in the single digits.

moves through a **condenser**, where the steam-containing pipes contact a flowing source of cooling water, usually from a nearby river, lake, or ocean. The steam condenses back to water and returns to the boiler. In the process, it gives up its latent heat to the cooling water. This energy is usually dumped to the environment as waste heat, either directly to the cooling water source or to the atmosphere via the large **cooling towers** typically seen at power plants. Figure 5.9 is a diagram of such a **thermal power plant**, with photos of several main components. Incidentally, this description applies to nuclear plants as well, with the exception that (in the design used in the United States) the water occupies a single large pressure vessel in which fissioning uranium fuel is immersed.

Thermal power plants are large, stationary structures. Their performance is optimized for a particular turbine speed and power output. For that reason they can be more efficient than vehicle engines, which have to function over a range of speed and power. The fuel requirements for a large fossil fuel power plant are staggering: A 1-GW plant may consume ten or more 110-car trainloads of coal each week! (See Fig. 5.10 and Example 5.2).

| Example 5.2 | Coal Train!

Coal is probably the most important commodity carried almost exclusively by rail. A typical modern coal train, called a *unit train,* consists of approximately one hundred carloads of coal, each with a capacity of about 100 tonnes. Consider a coal-fired power plant that produces electrical energy at the rate of 1 GWe and is 33 percent efficient. Estimate the number of unit coal trains needed each week to keep this plant supplied with fuel.

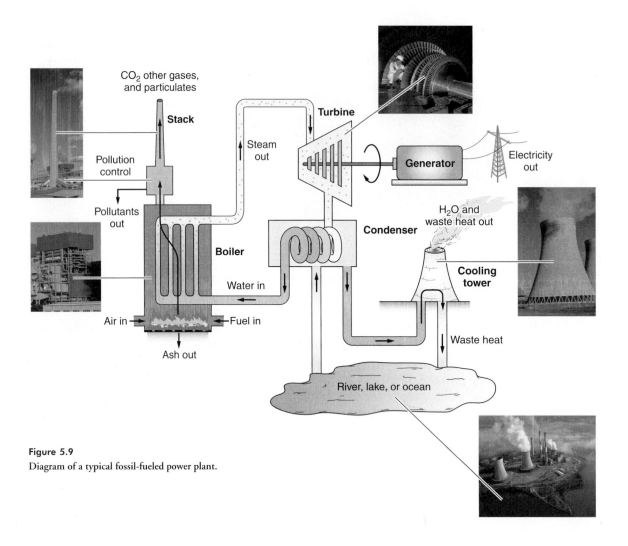

Figure 5.9
Diagram of a typical fossil-fueled power plant.

Solution

Table 3.3 shows that coal's energy content averages 29 MJ/kg; for this rough estimate I'll round that to 30 MJ/kg. Recall that 1 tonne (metric ton) is 1,000 kg. So our unit train, with one hundred 100-tonne coal cars, carries 10^4 tonnes, or 10^7 kg of coal. Therefore, the energy carried in a unit train is

$$(10^7 \text{ kg})(30 \text{ MJ/kg}) = 30 \times 10^7 \text{ MJ} = 3 \times 10^5 \text{ GJ}$$

Figure 5.10
A 110-car coal train arrives at a Kansas power plant. This particular plant consumes fourteen such trainloads of coal each week.

where the last conversion follows because 1 GJ (10^9 J) is 1,000 MJ. Our power plant is putting out electrical energy at the rate of 1 GWe, or 1 GJ per second, but it's only 33 percent efficient, so it uses fuel energy at three times this rate. Therefore the plant consumes fuel energy at the rate of 3 GW, turning 1 GW into electricity and dumping the rest as waste heat. How long does it take to go through a unit train? The unit train carries 3×10^5 GJ, and the plant consumes fuel energy at the rate of 3 GW or 3 GJ/s. So a unit train is going to last for a time given by

$$\frac{3 \times 10^5 \text{ GJ}}{3 \text{ GJ/s}} = 10^5 \text{ s}$$

Now, for purposes of estimation, 10^5 seconds is just about one day (24 hours times 60 minutes per hour times 60 seconds per minute gives 86,400 seconds per day). So our power plant needs about seven trainloads a week! For many power plants, these unit trains act like conveyer belts, shuttling back and forth between the power plant and a coal mine that may be hundreds of miles distant. You can explore the fuel needs of a similar oil-fired power plant in Exercise 2.

In Chapter 4, I showed how the second law of thermodynamics prohibits any heat engine from extracting as useful work all the energy contained in a fuel. In particular, the efficiency of a heat engine depends on the maximum and minimum temperatures within the engine. For a typical fossil-fueled power plant, with maximum steam temperature of 650 K, we calculated this maximum thermodynamic efficiency at just over 50 percent. But real plants fall well short of this theoretical maximum. Friction robs us of some of the energy, and so do deviations from Carnot's ideal engine cycle. In addition, a significant part of the power plant's output goes into running pumps, pollution control systems, cooling towers, and other essential equipment. All that extra energy eventually ends up as heat, adding still more to the waste required by the second law of thermodynamics. The net result is that a typical coal-fired steam-turbine power plant has an overall efficiency of around 35 percent. That's where I get the rough rule of thumb that two-thirds of the energy extracted from fuel is discarded as waste heat. We'll see later in this chapter how more advanced systems can achieve higher efficiencies.

Internal-Combustion Engines

In an internal-combustion engine, the products of fuel combustion themselves provide the pressures that ultimately produce mechanical motion. There is no transfer of heat to a secondary fluid, as in the steam boiler of an external-combustion power plant. In **continuous-combustion engines**, fuel burns continually and generally produces rotary motion directly. Jet aircraft engines are a good example; so are devices used in some advanced power plants. In **intermittent-combustion engines**, fuel burns periodically, usually in a cylindrical chamber containing a movable piston.

Intermittent-combustion engines are the workhorses of our surface transportation fleet. More than a century of engineering has brought this technology to a high level of sophistication, performance, and efficiency. Nearly all engines used in cars and trucks, as well as in many ships, railroad locomotives, and smaller aircraft, are intermittent-combustion engines in which rapid burning of the fuel in a cylinder produces hot, high-pressure gas that pushes on a piston. The piston is connected to a **crankshaft**, with the result that the back-and-forth motion of the piston becomes rotary motion of the crankshaft. Most modern engines, except those used in smaller devices such as lawn-mowers, have more than one cylinder. The physical layout of the cylinders, the strategies for initiating and timing the combustion, and the type of fuel all vary among different engine designs. Here I'll mention just a few variations on the basic intermittent-combustion engine.

Spark-ignition engines include the gasoline engines used in most cars in the United States. Fuel is injected into the cylinders at an appropriate point in the cycle and then, at the optimum instant, an electric spark ignites the fuel. In modern automobile engines a computer controls the spark timing to maximize performance, or to maximize fuel efficiency, or to minimize pollution. Figure 5.11 shows a cutaway diagram of a typical modern gasoline engine. Used in today's automobiles, such engines are typically about 20 percent efficient at converting the energy content of gasoline into mechanical energy, but some of this energy is lost to friction in other mechanical components, leaving only

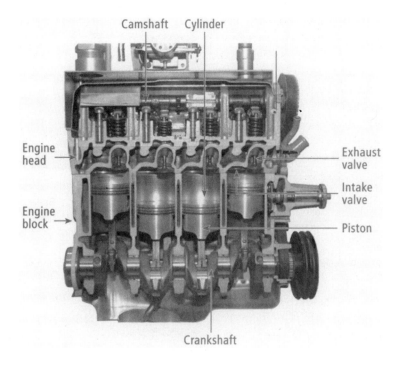

Camshaft Cylinder

Engine head

Engine block →

Exhaust valve

Intake valve

Piston

Crankshaft

Figure 5.11

Cutaway diagram of a four-cylinder gasoline engine. The up-and-down motion of the pistons is converted to rotary motion of the crankshaft. Valves admit the fuel–air mixture and release exhaust gases.

about 15 percent of the original fuel energy available at the wheels. Figure 5.12 shows energy flows in a typical automobile.

Compression-ignition engines include the diesel engine, used universally in large trucks, buses, and locomotives, as well as in about half the cars sold in Europe since the turn of the twenty-first century. If you've pumped a bicycle tire, you know that compressing air in the pump makes it hot; what you're doing is turning mechanical work into thermal energy. In a diesel engine, the rising piston compresses air in the cylinder to such an extent that it's hot enough to ignite fuel, which is injected into the cylinder at just the right instant. Diesel engines burn a refined oil that's heavier than gasoline and results in significantly more particulate pollution (more on this in Chapter 6). However, higher compression—implying higher temperature and thus a higher thermodynamic efficiency—along with other design features make the diesel engine considerably more fuel efficient than its gasoline-powered cousin. This is the reason for Europe's ongoing boom in diesel car sales. The few diesel-fueled cars sold in the United States in the early twenty-first century have fuel efficiencies comparable to those of the newer gas-electric hybrids. Greater fuel efficiency means that diesel cars emit less CO_2 per mile of driving and thus have a smaller impact on climate. However, their high compression means that diesel engines must be more ruggedly built, which makes them heavier and more expensive than their gasoline counterparts.

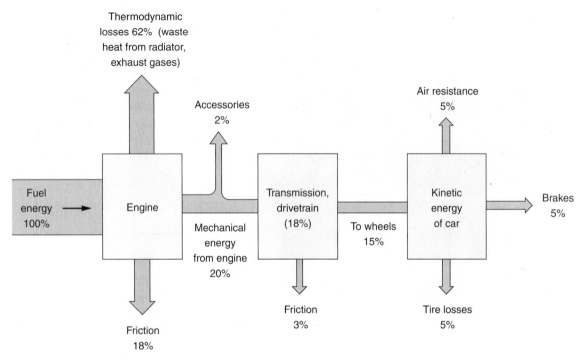

Figure 5.12

Energy flows in a typical gasoline-powered car. Thermodynamic losses and friction leave only about 15 percent of the fuel energy available at the wheels, all of which is dissipated by air resistance, tire friction, and braking. The power needed for accessories runs the air conditioning, lights, audio system, and vehicle electronics.

Hybrid Vehicles

Recently, gasoline engines have been combined with electric motors to make **gas-electric hybrid** cars and light trucks. In these vehicles, motive power to the wheels may come directly from the gasoline engine, from the electric motor, or from a combination of the two (Fig. 5.13). The gasoline engine also runs a generator that charges a battery, and the electric motor uses the energy stored in this battery. This means that all of the car's energy comes ultimately from burning gasoline; hybrid cars aren't plugged into the electric power grid. When a hybrid car brakes, the wheels turn the electric motor, which functions as a generator to charge the battery and thus store what was the car's kinetic energy as chemical energy in the battery. In a gasoline-only car, this energy would be dissipated as useless heat through friction in the brakes. Such **regenerative braking**, coupled with selective use of the electric motor and a smaller than usual gasoline engine, all serve to give hybrids substantially better fuel efficiency than standard gasoline-engine vehicles (Fig. 5.14).

Hybrids come in several flavors. **Series hybrids**, not now widely available, use a gasoline engine solely to generate electricity; motive power comes entirely from an electric

(b)

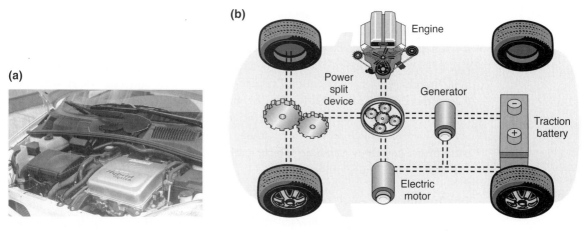

(a)

Figure 5.13

(a) Under the hood of a Toyota Prius. At left is the gasoline engine, at right the high-voltage electric power control unit. (b) Diagram of Toyota's Hybrid Synergy Drive. Energy flows among the different components depending on the driving situation. In braking, for example, the car's motion drives the electric motor as a generator and stores energy in the battery. The Prius engine is rated at 57 kW (76 horsepower) and the electric motor at 50 kW (67 horsepower).

motor. In **parallel hybrids**, a gasoline engine is the prime source of motive power, assisted on occasion by an electric motor; these are often called *mild hybrids*. Parallel systems range from those in vehicles with merely souped-up starter motors providing only a modest hybrid effect to systems such as Honda's Insight hybrid that make extensive use of electric power. More sophisticated are series/parallel or "full" hybrids that can use either the

Figure 5.14

Fuel efficiencies of conventional, hybrid, and diesel vehicles for the 2006 model year. The Honda Insight and and the Toyota Prius are hybrids only. The Honda Civic, Honda Accord, and Ford Escape have hybrid and conventional models. The Volkswagen Jetta is available in conventional gasoline and diesel models in the United States; there is no VW hybrid. Can you tell which hybrid optimizes performance instead of fuel efficiency?

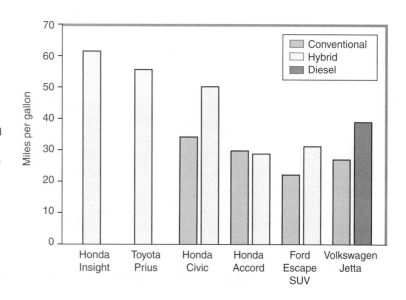

gasoline engine or the electric motor, or both simultaneously. Toyota's Hybrid Synergy Drive (also licensed by Ford) is such a system.

Whatever the technological details, you should recognize that today's hybrid vehicles are powered entirely by fossil fuel. Their use of an electric motor may make them more fuel efficient than conventional vehicles, but the source of the electric power is ultimately the gasoline engine (although that may change with so-called plug-in hybrids). And I say "*may* make them more fuel efficient" because hybrid technology can also be used to increase vehicle performance without much improvement in fuel mileage. In the performance-conscious American vehicle market, this is already beginning to happen, as one example in Figure 5.14 suggests.

Gas Turbines

The second major class of internal-combustion engine is the continuous-combustion engine, of which the **gas turbine** is the most widely used example. In a gas turbine, fuel burns to produce hot, pressurized gas that rushes out of the engine exhaust and, on the way, drives a rotating turbine. The turbine may, in turn, drive an electric generator, airplane propeller, or other device. Alternatively, in a jet aircraft engine, the turbine simply helps compress the outflowing gases to provide the thrust that gives the jet its motive power (Fig. 5.15). Incidentally, the smaller propeller-driven aircraft used in commercial aviation are usually powered by gas turbines and are called *turboprops* to distinguish them from even smaller planes with propellers driven by internal-combustion engines. To complicate things further, some internal-combustion engines employ devices called *turbochargers,* which are essentially gas turbines placed in the exhaust flow. The turbine extracts some of the energy from waste heat that would normally have gone out the tailpipe and uses it to pump air into the engine at a greater rate than would otherwise be possible. However, a turbocharged engine has to work harder to push exhaust through

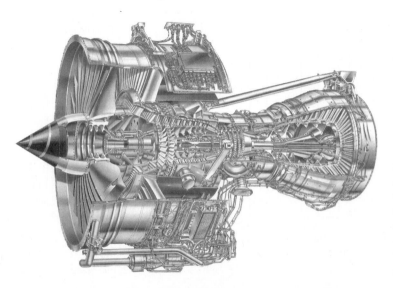

Figure 5.15
A jet aircraft engine is an example of a continuous-combustion gas turbine. At left are a fan and compressor that pressurize the air before it enters the combustion chamber; combustion products then expand to turn the turbine and compressors, providing the jet thrust. Exhaust is at the right.

the turbine. Turbochargers generally enhance engine performance without contributing excess weight. Although they can, in principle, increase fuel efficiency, turbochargers are normally used to achieve enhanced performance, greater acceleration, and higher top speed—all of which lead to higher fuel consumption.

Gas turbines produce lots of power for a given weight, so they're ideal for aircraft engines. But because of their high exhaust temperature (T_c in Equation 4.5), they're not very efficient and, in particular, gas turbines don't make a lot of sense in stationary applications such as power plants. Nevertheless, electric utilities often maintain small gas-turbine-driven generators for so-called peaking—that is, supplying extra power during short periods of high demand—because gas turbines, unlike external-combustion steam power plants, can start and stop very quickly as needed.

Combined-Cycle Systems

Despite their low second-law efficiency, gas turbines increasingly play a role in newer and very efficient **combined-cycle power plants**. In these plants, a gas-turbine engine turns an electric generator. But instead of being dumped to the environment, the hot exhaust from the turbine goes through a so-called heat recovery boiler where some of its energy is used to boil water that then runs a steam turbine to power another generator. Modern combined-cycle power plants can achieve overall efficiencies approaching 60 percent. Ultimately, the second law of thermodynamics limits the efficiency of a combined-cycle plant just as it does any other heat engine. But the second law doesn't care about the details of the mechanical system used to extract energy, so all that really matters, at least in principle, are the highest and lowest temperatures in the entire combined system. The high temperature of the gas turbine (typically 1,000 to 2,000 K; T_h in Equation 4.5) coupled with the lower final temperature from the steam portion of the combined-cycle plant (T_c in Equation 4.5) makes for a higher thermodynamic efficiency than the gas turbine alone. Most combined-cycle power plants use natural gas as their fuel, and as Figure 5.7 suggested, this further reduces their carbon emissions. Figure 5.16 is a diagram of a typical combined-cycle power plant.

Example 5.3   A Combined-Cycle Power Plant

A combined-cycle power plant has a gas turbine operating between maximum and minimum temperatures of 1,450°C and 500°C, respectively. Its 500°C waste heat is used as the high temperature for a conventional steam turbine cooled by river water at 7°C. Find the thermodynamic efficiency of the combined cycle, and compare it with the efficiencies of the individual components if they were operated independently.

Solution

This is a straightforward calculation of thermodynamic efficiencies from Equation 4.5. First, though, we need to convert all temperatures to absolute units—that is, kelvins. Adding the 273 K difference between degrees Celsius and kelvins, we have a highest

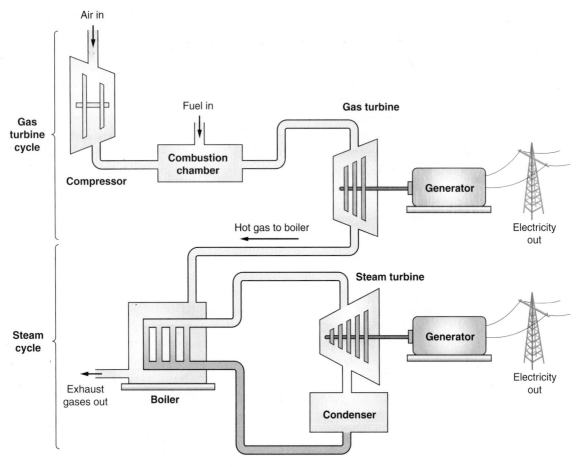

Figure 5.16
Diagram of a combined-cycle power plant. The steam section is similar
to the one illustrated in Figure 5.9, although details of the cooling and
exhaust systems aren't shown. Hot gas from the gas turbine replaces
burning fuel in the steam boiler.

temperature of $1{,}450 + 273 = 1{,}723$ K, an intermediate temperature of $500 + 273 = 773$ K, and a lowest temperature of $7 + 273 = 280$ K.

The thermodynamic efficiency is determined by the second law alone, through Equation 4.5; the details of the heat engine don't matter. So the second-law efficiency for the combined-cycle plant is

$$e_{\text{combined}} = 1 - \frac{T_c}{T_h} = 1 - \frac{280 \text{ K}}{1{,}723 \text{ K}} = 0.84 = 84 \text{ percent}$$

Friction and other losses would reduce this figure substantially, but a combined-cycle plant operating at these temperatures might have a practical efficiency near 60 percent.

Meanwhile, the gas turbine and steam-cycle sections alone have efficiencies given by

$$e_{gas} = 1 - \frac{T_c}{T_h} = 1 = \frac{773\ K}{1{,}723\ K} = 0.55 = 55\ \text{percent}$$

and

$$e_{steam} = 1 - \frac{T_c}{T_h} = 1 = \frac{280\ K}{773\ K} = 0.64 = 64\ \text{percent}$$

Here's a case where the whole is better than either of its parts! In general, the combined-cycle efficiency can be written $e_{combined} = e_{gas} + (1 - e_{gas})e_{steam}$, as you can show for the numbers here and prove in general in Exercise 5.

5.5   Fossil Fuel Resources

Where are the fossil fuels, and how large is the fossil energy resource? We've seen that fossil fuels began forming hundreds of millions of years ago under fairly common circumstances, primarily acidic swamps for coal and ancient seas for petroleum. But geological conditions resulting in significant, economically recoverable concentrations of fossil fuel are less common. Consequently, the distribution of fossil fuel resources around the globe is quite uneven—a fact that has blatantly obvious economic and geopolitical consequences.

Before we proceed, I need to clarify some vocabulary: The term **resource** refers to the total amount of a fuel in the ground—discovered, undiscovered, economically recoverable, or not economically recoverable. Obviously, estimates of fossil fuel resources are necessarily uncertain, because they include fuel we haven't yet discovered. **Reserves**, in contrast, describe fossil fuels that we're reasonably certain exist, based on geological and engineering studies, and that we can recover economically with existing technology. As prices rise and technology improves, reserves may grow even without new discoveries, because it becomes feasible to exploit what were previously uneconomical sources of fossil fuels.

Coal Resources

Coal is the most widely distributed and most abundant of the fossil fuels. Nevertheless, some countries have much more coal than others, as Figure 5.17 shows. The United States is the leader, with more than one-fourth of the world's known coal reserves. Other major coal reserves are in Russia, China, Australia, and India. Total world reserves amount to some 10^{12} tonnes, about half of which is the higher-quality bituminous and anthracite coal. Energy content varies substantially with coal quality, but a rough conversion suggests that the world's 10^{12}-tonne reserve is equivalent to about 25,000 exajoules (EJ) of energy or, again roughly, 24,000 quads (Q) (see Exercise 6).

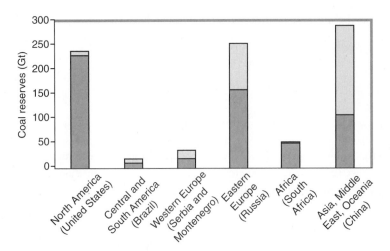

Figure 5.17
World coal reserves in gigatonnes (billions of metric tons). The full height of each bar gives the total coal reserves for the indicated continent, and the lower part of each bar indicates the total reserves for the listed country, which has the most coal in that continent.

Much of the world's coal reserves lie deep underground, and this coal is extracted using traditional mining methods. In the western United States, where coal is closer to the surface, strip mining techniques remove the surface layers to expose coal seams. The coal is extracted and the land then restored to some semblance of its original state. Both deep mining and strip mining have significant environmental impacts, which I discuss in Chapter 6.

Oil Resources

Citizens of today's world, and especially of the United States, cannot help but be acutely aware of the uneven global distribution of petroleum reserves. Figure 5.18 shows that the politically volatile Middle East holds some 65 percent the world's crude oil; Saudi Arabia

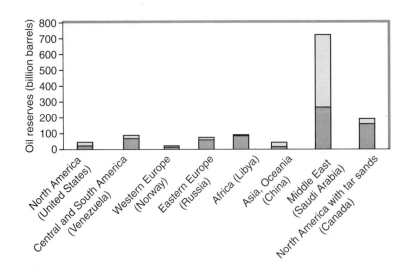

Figure 5.18
World oil reserves, presented as for coal in Figure 5.17, with the Middle East broken out separately. North America is shown twice: The United States has the largest conventional reserves in North America (left), but if the Canadian tar sands are included, Canada has almost as much oil as Saudi Arabia (right).

alone accounts for more than a quarter of the world's known conventional oil reserves. The United States, by far the largest oil consumer at about one-fourth of total world oil consumption, has only about 2 percent of the global reserves. That discrepancy, along with ever-increasing demand, explains why U.S. dependence on foreign oil sources has grown dramatically in recent decades, from less than 10 percent of total petroleum consumption in 1950 to nearly 70 percent in the early twenty-first century (Fig. 5.19).

Estimates of total world oil reserves are approximately 1 trillion barrels (authoritative sources give estimates ranging from about 1 trillion to more than 1.2 trillion barrels). You might contemplate that figure in light of a world oil consumption rate of 25 billion barrels per year in the early 2000s, rising at more than 1 percent per year. As you can show in Exercise 7, this trillion-barrel reserve translates into a little over 6,000 EJ, or a little under 6,000 Q. It's instructive to consider this figure in light of the world's total energy consumption rate of, very roughly, 10^{13} W (see Exercise 8).

Oil extraction technology has evolved substantially since Drake struck Pennsylvania oil at a mere 69 feet in 1859. By the 1930s, the deepest wells went straight down some 3 km (about 2 miles), and today some wells' depths exceed 5 km (about 3 miles). An equally significant advance is the development of so-called directional drilling, which allows drills to penetrate rock vertically, horizontally, and diagonally to tap multiple oil deposits from a single wellhead (Fig. 5.20). By the middle of the twentieth century, oil drilling had begun to move offshore. Today more than 30 percent of the world's oil comes from offshore drilling stations, including platforms mounted on the ocean floor as well as on floating structures. Today's offshore rigs can work in waters approaching 2 miles deep, below which their drill strings may penetrate several additional miles (Fig. 5.21).

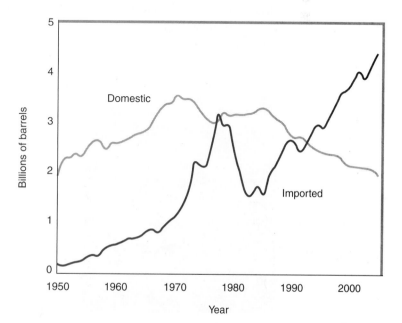

Figure 5.19

U.S. oil imports have risen substantially to compensate for declining domestic production.

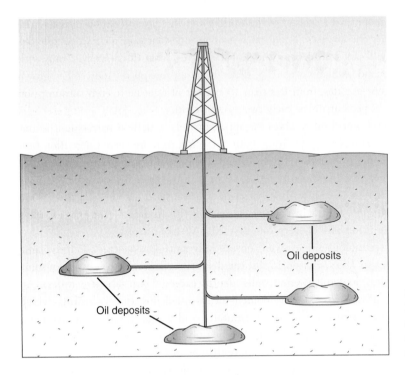

Figure 5.20
Directional drilling taps multiple oil deposits from a single wellhead. This technology not only produces more oil, but also reduces the environmental impact at the surface.

Natural Gas Resources

Most natural gas occurs in conjunction with crude oil, so its distribution is similar to that of oil. Being gaseous, however, it's harder to handle and transport than liquid oil. But for environmental reasons and because of its availability, natural gas is playing an ever-larger role in the world's energy supply. In the United States, natural gas is com-

Figure 5.21
Offshore drilling rigs produce some 30 percent of the world's oil.

monly measured in cubic feet; in this unit, the world's reserves are estimated to be around 6 quadrillion cubic feet, which amounts to another 6,000 EJ, or 6,000 Q—roughly the same energy content as the world's oil reserves (see Exercise 9).

Unconventional Fossil Resources

There's a lot more fossil fuel than what's included in the proven reserves of coal, oil, and natural gas; most of it is oil and gas locked in geological structures where extraction has not been technologically or economically feasible. But as conventional supplies run short and thus become more expensive, and as technology advances, the world will inevitably turn to these unconventional fossil resources.

I began this chapter with a description of fossil fuel formation through a variety of physical and chemical processes, and I pointed out that today's economically recoverable oil occurs when liquid petroleum migrates and concentrates in geological "traps." But much of the fuel remains more dispersed, sometimes at an early stage in the formation process, in the very rocks where it originated. **Oil shale** is rock that contains a significant amount of kerogen, the waxy substance that represents a step on the way to liquid crude oil. Shales that are particularly rich in kerogen can burn directly and have seen limited use as fuels for centuries. To produce true oil from oil shale involves pulverizing the rock and applying heat and pressure to complete what nature has left undone. The process is expensive and, because huge amounts of rock are involved, it has major environmental impacts. However, the oil shale resource is extensive and not heavily concentrated in one part of the world. The United States alone may have over 1 trillion barrels of oil locked in oil shale, which is equivalent to the proven world reserves of liquid crude oil. However, the energy content isn't equivalent, because large amounts of energy go into processing the oil shale. Nevertheless, it's clear that oil shale might become a significant contributor to our fossil fuel supply, if we can surmount numerous technological, economic, and environmental issues.

Tar sands are another unconventional fossil fuel resource. These are sand deposits into which liquid oil has leaked. Over the eons, the lighter components have evaporated, leaving a heavy tar. The Canadian province of Alberta is home to a large portion of the world's known tar sands, equivalent to some 1.6 trillion barrels of oil, which is more than the world's reserves of liquid crude. With some 10 percent or more of this resource considered economically extractable, mining operations in Alberta are already underway (Fig. 5.22). With Alberta's extractable tar sands considered, the conventional oil-reserve picture in Figure 5.18 changes dramatically as Canada becomes second only to Saudi Arabia. As with oil shale, however, extracting liquid oil from tar sands is an expensive, energy-intensive, and environmentally damaging process.

Natural gas, too, has unconventional sources. Like oil, it's present in some rock formations. It also occurs in coal seams, where it's responsible for explosive mining accidents. And under the extreme pressures at the bottom of deep oceans, methane molecules can become trapped in icelike crystals to form **methane clathrates**, also known as **methane hydrates**. Although the amount of this trapped methane is unknown, it could

(a)

(b)

Figure 5.22
(a) Tar sands from Alberta, Canada. (b) It takes 2 tons of sand to make a barrel of oil. This shovel suggests the enormous scale of tar-sand extraction.

be as much as one hundred times the conventional natural gas reserves. At present, though, we lack the technology to extract or even accurately locate this fuel resource. Methane hydrates may also play a disturbing role in sudden climate change by releasing large quantities of the greenhouse gas methane into the atmosphere.

5.6   When Will We Run Out?

Fossil fuels formed over hundreds of millions of years, and we're using them at a vastly greater rate than they're now forming. So it's clear we'll run out sometime, but when?

The simplest way to estimate that time is to divide a fossil reserve by the rate at which we're using it, yielding a number known as the **reserve/production ratio** (R/P ratio). But this approach is inaccurate for several reasons. First, known reserves don't account for as-yet undiscovered resources, and even if there were no undiscovered resources, economic and technological factors would continue to expand the reserve as, for example, marginal oil fields or unconventional sources such as oil shale and tar sands become economically viable. On the other hand, our fossil fuel consumption rate is growing, so using today's rate overestimates the remaining supply. Finally, the real question isn't when we'll run out completely (when we pump the last drop of oil from the ground) but when and whether production of fossil fuels begins to drop even as demand continues to rise.

Example 5.4   Running Out: A Naïve Approach

Use Figure 5.6 to estimate the current rate of world oil consumption, expressed in power units. Use that result in conjunction with known oil reserves to estimate the R/P ratio, which is a naïve estimate of the time remaining until those reserves run out.

Solution

Figure 5.6 shows an early-twenty-first-century global oil consumption rate close to 4 Gt oil equivalent per year; with Table 3.1's conversion factor of 41.9 GJ per tonne oil equivalent, that works out to be about 170 EJ per year. With world oil reserves at some 6,000 EJ, this means we have about

$$\frac{6,000 \text{ EJ}}{170 \text{ EJ/year}} = 35 \text{ years}$$

until we run out.

The naïve approach of Example 5.4 neglects future growth in oil consumption. Of course, no one can predict the future, but we can extrapolate from past experience, which shows world oil consumption increasing since the mid-1990s at somewhat over 1 percent per year. This increase means that our 35-year result in Example 5.4 may be an overestimate of how long it will take to consume all known oil reserves. When we account for the mathematics of exponential growth (see Box 5.2 and Exercise 11), we find that we only have about 27 years of oil left, if the current percentage growth rate holds.

Box 5.2   Exponential Growth

One day, a gardener noticed a lily pad growing in her garden pond. The next day there were two lily pads, the following day four, and so forth, with the number doubling each day. At first the lily pads covered only a small portion of the pond, but as their incessant doubling continued, the gardener grew alarmed. When the pond was half covered, she decided to take action. At that point, how much time did she have before the pond would be covered?

The answer, of course, is one day.

This story illustrates the power of exponential growth—growth not by a fixed amount but by a fixed percentage of what's already there (Fig. 5.23). In the case of the lily pond, the rate is 100 percent per day, meaning that each day the total population of lily pads increases by 100 percent of what there was before.

Many quantities increase exponentially. Money in the bank earns interest at a fixed percentage of what's already there. As your money grows, so does the actual amount of interest you receive. Put a bacterium in a petri dish with all the food it needs, and soon you'll have 2, then 4, 8, 16, 32, 64, and so on. For centuries the growth of the human population could be described as exponential—although only approximately, because the percentage rate actually increased until the 1960s, after which it has fallen. And the growth in world energy consumption, or in the consumption of a given fossil fuel, has often been approximately exponential.

Unchecked, exponential growth always leads to huge values, but it can't go on forever. Shortages of resources—money, food, oil, whatever—eventually halts the exponential growth and results in a transition to slower growth, decline, or even collapse of the numbers that were once growing so rapidly. Because it starts slowly and then increases dramatically, exponential growth can "sneak up" on us and leave little time for an orderly transition to the postexponential phase. That's our lily-pond gardener's plight, and it's one of the big concerns whenever resource use grows exponentially.

Mathematically, exponential growth is characterized by a function with time in the exponent. If at time $t = 0$, we start with a quantity N_0, whether it be dollars or people or barrels of oil per year, and if that quantity grows at some rate r, specified as the fraction by which the quantity increases in whatever time unit we're using, then at a later time t we'll have a quantity N given by

$$N = N_0 e^{rt} \tag{5.1}$$

Here e is the base of natural logarithms, with an approximate value of 2.71828 . . . , but this value isn't important here because your scientific calculator has the exponential function built in.

When exponential growth describes resource consumption, and therefore N represents the consumption rate, then the calculus-savvy can show that the total amount of the resource used up in some time t—the cumulative total of the increasing yearly consumption—is given by

$$\text{Cumulative consumption} = \frac{N_0}{r} (e^{rt} - 1) \tag{5.2}$$

where N_0 is the initial consumption rate. I used Equation 5.2 in making that 27-year estimate for the time remaining until oil might run out. I set the annual growth rate to 2 percent—that is, $r = 0.02$, typical of the recent past (see Research Problem 7). I used $N_0 = 170$ EJ per year from Example 5.4 as the initial oil-consumption rate, and I solved for the time t when the cumulative oil consumption reaches the 6,000-EJ contained in the total reserves. Exercises 11 through 16 explore exponential growth in the context of fossil fuel reserves, including the derivations of Equation 5.2 and of an explicit expression for the time.

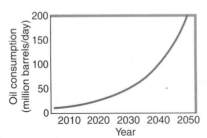

Figure 5.23

China's 7.5 percent annual increase in oil consumption is an example of exponential growth. If this rate were to hold steady through 2050, China's current consumption of roughly 7 million barrels per day would grow nearly thirty times, to more than 200 million barrels per day—more than double the current *global* daily consumption rate.

But again, that's not the whole story. As the supplies dwindle, oil will get very expensive and we'll surely reduce our consumption. Reserves may last a bit longer, or the actual reserve may be larger, making that 27-year figure an underestimate. But near the end, whenever it comes, we'll be in a real energy crisis if we haven't found an alternative to oil. "By then" is not an abstract concept; it means something on the order of a few decades from now, and if you're a typical college student, it means the prime of midlife for you. This is your problem, not something for later generations to worry about.

Hubbert's Peak

So, again, the issue isn't when the last drop of oil disappears, but when we begin to feel the pinch of dwindling reserves. In 1956 Shell Oil Company geophysicist M. King Hubbert made a remarkable prediction: He suggested that oil production in the United States, then the world's dominant oil producer, would peak in the late 1960s or early 1970s. Oil companies, government agencies, and economists alike dismissed Hubbert's prediction, but he was right. Oil production in the United States peaked in 1970 and has been declining ever since.

Hubbert based his estimate on a simple idea—that the extraction of a finite resource such as oil starts out slowly, rises with increasing demand, eventually levels off as the finite resource becomes harder to extract, and then declines. In short, production should follow, roughly, a bell-shaped curve (Fig. 5.24). This curve shows production in, say, billions of barrels of oil per year or some similar unit—that is, it shows a rate of oil extraction. The total amount of oil extracted up to a given time is the area under the curve from the starting point to the time in question (if you've had calculus, you'll recognize this as a statement about the definite integral). Given a perfectly symmetric bell-shaped curve, we reach the peak—**Hubbert's peak**—when we've used up half of the oil reserve.

Now here comes the crunch: Oil consumption has been rising more or less steadily for decades, with no obvious end in sight. Production has been rising, too, to keep up with demand. Although some oil is stored for future use (as in the U.S. Strategic Petroleum Reserve; see Research Problem 1), the amount going into and out of storage is small, so in practice we use oil at essentially the same rate at which it's pumped from the ground. But after Hubbert's peak, production necessarily declines. In the case of U.S. production, which peaked in 1970, we've made up the increasing difference between production and consumption with ever-rising oil imports (recall

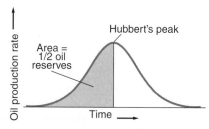

Figure 5.24

An idealized bell-shaped curve for oil production known as Hubbert's peak. Production peaks when half the resource has been exhausted; thereafter, the production rate declines as the remaining oil reserves become more difficult and expensive to extract.

Fig. 5.19). But globally, the world's oil reserves are all we have. Unless we can begin to slow our global oil consumption well before the world reaches its Hubbert's peak, we may be in big trouble.

So when do we reach the global Hubbert's peak for oil? Serious estimates range from a few years hence to about 2040 (Fig. 5.25). A few pessimists believe we're already at or even past the peak now, and some optimists think that technological advances and new oil discoveries will push Hubbert's peak beyond the turn of the next century. Hubbert's bell-shaped curve is only an approximation, and peak estimates are sensitive to the exact form one assumes. There's also the question of the ultimate oil reserve; the U.S. Geological Survey, for example, estimates that actual reserves may be two to four times that trillion-barrel or 6,000 EJ figure I've been using. But others point to an ongoing decline in new oil discoveries as evidence that the currently known reserves may be all we have. Those variations make less difference than you might think, as you can show in Exercise 12. In any event, it's fair to say that a broad consensus has the crunch coming in a matter of a few decades.

The world's dependence on oil has created a huge economic and technological infrastructure dedicated to oil production, supply, and consumption. Think of all the gasoline stations, all the oil-heated homes, all the jet airplanes, all the lubricating oils used in the machinery of the world's factories, all the tankers plying the seas to deliver imported oil to all the refineries, and you get the picture of a vast oil enterprise. There's an enormous inertia associated with that enterprise, and it's going to be very difficult to turn it around on a time scale of mere decades.

The Cornucopian View

Not everyone subscribes to the impending crunch implicit in Hubbert's idea. So-called **cornucopians** represent the extreme opposite in a wide range of opinions about the future of oil and other finite resources. Cornucopians believe that human ingenuity will remain

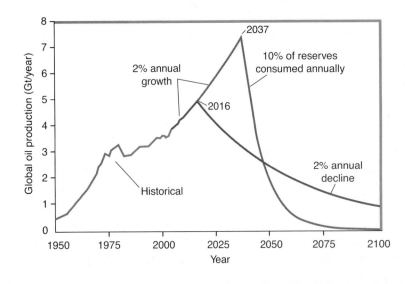

Figure 5.25
Two scenarios for Hubbert's peak, both of which assume recoverable reserves of 3 trillion barrels and 2 percent growth in oil production until the peak. One scenario assumes a symmetric curve, with a 2 percent annual decline in oil production following its peak in 2016. The other scenario assumes that we'll use 10 percent of the remaining reserves each year after the peak occurs, yielding a steeper decline. The areas under the two curves must each equal the total oil reserves, so the peak must occur earlier in the more slowly declining scenario. The sharp peaks probably aren't realistic.

always a step ahead of Hubbert's peak, continuing to find new reserves, advancing technologies to make currently uneconomical resources viable, or sliding smoothly to an economy based on alternative fuels or altogether new energy sources. In this view, the finite nature of a given resource is simply irrelevant.

Cornucopians point to past energy predictions that have been wildly off the mark. The 35-year R/P ratio we found in Example 5.4 sounds scary, but a cornucopian would note, correctly, that if we had calculated this ratio in 1980, the result would have been just over 25 years, meaning that we would now be out of oil. In 1945 it looked like we had just 20 years of oil left!

What happened? How did we go through decades of oil consumption and end up with what looks like a longer oil future? We discovered more oil, expanding the known reserves. We improved extraction technologies, making new sources accessible and enhancing the yield from older wells. And despite our SUV craze and other energy follies, we became more efficient in our energy use. Energy intensity in the United States, for example, dropped by half between 1960 (0.6 W per dollar of GDP) and 2005 (about 0.3 W per dollar), as I showed in Figure 2.8. Although our total annual energy consumption increased substantially in that time, the increase was a lot less than it would have been without efficiency improvements.

So is Hubbert's peak an alarmist notion? Not in the context of Hubbert's correct prediction that U.S. oil production would peak in the early 1970s. And there's good evidence that the rate of new oil discovery is dropping significantly. The issue may not be so much whether the idea of Hubbert's peak is valid or not, but the time until we reach that peak if we stay our present course. If that time is long, then there's a reasonable chance of making a smooth transition to a different energy regime and thus avoiding peak altogether. But if the time is short—and a few decades is probably short on the time scale needed to change our vast energy infrastructure—then Hubbert's peak really does represent a looming energy crunch.

Coal and Natural Gas Supplies

How about coal and natural gas? As with any finite resource, their production, too, might experience a Hubbert's peak. For gas, this will probably happen a decade or two after oil's peak. Exactly when depends in part on how vigorously the world tries to curtail its carbon emissions by substituting natural gas for coal and oil.

Coal is a different story. A naïve calculation like the one we did for oil in Example 5.4 yields several hundred years of remaining supply (see Exercise 10). But growth in coal consumption, which we might expect as oil and gas supplies falter, could reduce that number. Also recall that Hubbert's peak comes when we've extracted only half the world's coal, not when we're down to the last chunk. But still, with coal we don't face the urgency we do with oil; the time scale to peak is a century or more, compared with a few decades. Furthermore, resource limitations may turn out to have little relevance in the case of coal, because coal's deleterious environmental impacts—especially on climate—may prove to be the dominant factor limiting our use of this abundant fossil fuel. In that case, Hubbert's peak would be irrelevant for coal, much of which may remain forever in the ground.

5.7   Policy Issue: Carbon Tax or Cap-and-Trade?

Clearly, it's in our best interest to wean ourselves gradually from fossil fuels. This conclusion follows obviously from the preceding section, and it will be reinforced when we look at the environmental impacts of fossil fuels in Chapter 6 and at climate change in Chapters 12 to 16. But how are we to reduce fossil fuel consumption in a world where energy use is closely tied to our standard of living (recall Fig. 2.7), and where the developing world seeks to emulate the developed world's material well-being?

One proposed approach is a carbon tax that would be levied on fuels and based on the amount of carbon a given fuel puts into the atmosphere. The carbon tax is often proposed as a way of reducing the climate impact of fossil fuel combustion, because it would encourage shifts from coal and oil to natural gas, as well as away from fossil fuels altogether. But unless it's designed carefully, a carbon tax could be regressive, having a greater impact on low-income consumers than on the well-to-do. Proponents of a carbon tax have suggested redistributing some of the tax revenue based on income, but that goal may conflict with another: using carbon-tax revenues to fund research into nonfossil energy sources.

Although carbon-based taxation is a new and relatively untried idea, taxes on fuel itself have long been with us. European countries, in particular, use the gasoline tax as an instrument of environmental policy, explicitly encouraging lower fuel consumption, minimizing environmental impacts, and guiding consumers to fuel-efficient vehicles. The gasoline tax is one big reason why, as I noted earlier, half the new cars sold in Europe are fuel-efficient diesels. European gas taxes are considerable, amounting to some 60 to 80 percent of the total fuel cost. For the United States, that number is only about 30 percent. Figure 5.26 compares gasoline prices in the United States and other industrialized countries. An increase in the low U.S. gasoline taxes, as sensible as it would seem,

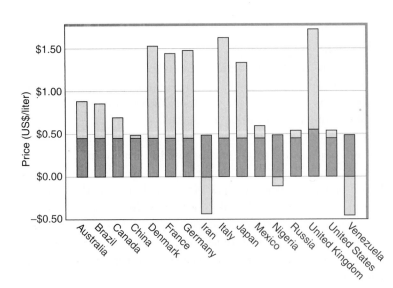

Figure 5.26
Gasoline prices vary dramatically as a result of different tax policies. European countries have the highest prices because of taxes intended to encourage energy efficiency. The dark portion of each bar shows the basic gasoline cost; the lighter area indicates taxes. Some major oil producers subsidize gasoline rather than taxing it, hence the negative taxes given for Iran, Nigeria, and Venezuela-dark part = .46

has been politically unachievable for many years. The last increase, in 1993, amounted to a mere 4.3 cents—less than the typical variations in gasoline prices from week to week.

An alternative to the carbon tax is a so-called cap-and-trade scheme, in which a governing body sets limits on carbon emissions, then issues allowances to individual emitters that permit a certain amount of carbon emissions. These allowances can either be granted free at the outset or auctioned to raise money to be used, for example, for research into energy alternatives. If a company produces less carbon than it's allowed, it can sell its allowances on an open market. Thus the cap-and-trade scheme provides a market-based incentive to reduce emissions. The European Union has adopted cap-and-trade to help meet its goals under the Kyoto Protocol, and today you can track the European Carbon Index—the price of carbon allowances as traded on the open market—just like any other commodity or stock. The United States, which did not join the Kyoto Protocol, has only a small, voluntary carbon emissions market, the Chicago Climate Exchange. Research Problem 8 explores the cost of carbon allowances on the European market and the Chicago Climate Exchange.

Should we increase the gasoline tax? Should we adopt a carbon tax? Should we require mandatory participation in a cap-and-trade market? Those are questions of policy, not science. But an intelligent answer requires understanding the science behind fossil fuels—their origin, their uses, their availability, and their environmental impact. This chapter discussed the first three of these; Chapter 6 focuses on the last.

Chapter 5   Chapter Review

BIG IDEAS

5.1   Coal, oil, and natural gas are fossil fuels that contain stored solar energy captured by plants tens to hundreds of millions of years ago.

5.2   The fossil fuels are hydrocarbons; coal is mostly carbon, whereas oil and natural gas have larger amounts of hydrogen. Today we **refine** the fossil fuels into products ranging from heavy oils to jet fuel to gasoline to the feedstocks for making plastics.

5.3   Burning fossil fuels produces largely CO_2 and water vapor. For a given amount of energy, coal produces the most CO_2 and natural gas the least.

5.4   We extract energy from fossil fuels with **external-combustion engines** such as electric power plants and with **internal-combustion engines** for transportation. Their efficiencies are limited by the second law of thermodynamics and by design factors. Modern technologies such as gasoline-electric hybrid vehicles and combined-cycle power plants provide substantial efficiency gains.

5.5 Fossil fuel **resources** are limited. **Reserves** describe the amount of fuel we're reasonably confident is in the ground. Unconventional sources such as tar sands might increase fossil reserves, but extracting fuel from these sources is expensive, energy intensive, and environmentally disruptive.

5.6 Petroleum reserves will probably last a matter of decades, coal a few centuries. Resource limitations become severe when production peaks and then declines even as demand continues to increase. Whether we reach the impending peak crisis or slide into a new energy regime depends on our energy policy and the prospect of increasing reserves.

5.7 Both the impending shortages and the environmental impacts of fossil fuels argue for an energy future less dependent on these fuels. A carbon tax is one approach to reducing fossil fuel consumption.

TERMS TO KNOW

coal (p. 106)
combined-cycle power plant (p. 123)
compression-ignition engine (p. 119)
condenser (p. 115)
continuous-combustion engine (p. 118)
cooling tower (p. 115)
cornucopian (p. 134)
crankshaft (p. 118)
crude oil (p. 108)
gas-electric hybrid (p. 120)
gas turbine (p. 122)
Hubbert's peak (p. 133)
intermittent-combustion engine (p. 118)
kerogen (p. 107)
methane clathrate (p. 129)
methane hydrate (p. 129)

natural gas (p. 108)
oil shale (p. 129)
parallel hybrid (p. 121)
peat (p. 105)
petroleum (p. 107)
plankton (p. 107)
refine (p. 108)
regenerative braking (p. 120)
reserve (p. 125)
reserve/production ratio (p. 130)
resource (p. 125)
series hybrid (p. 120)
spark-ignition engine (p. 118)
tar sands (p. 129)
thermal power plant (p. 115)
turbine (p. 114)

GETTING QUANTITATIVE

Coal energy content: averages 29 MJ/kg, ~10,000 kWh per ton

Energy content of petroleum products: ~45 MJ/kg, ~40 kWh/gallon

CO_2 emissions from 1-GW coal-fired power plant: ~260 kg/s, ~500 pounds per second

Mass ratio, CO_2 to carbon: 44/12, or 3.67

Efficiency of typical gasoline engine: 20 percent, fuel energy to mechanical energy

Efficiency of typical coal-fired power plant: 35 percent, fuel energy to electricity

Efficiency of typical combined-cycle power plant: 60 percent, fuel energy to electricity

Coal reserves: 10^{12} tonnes, ~25,000 EJ or ~25,000 Q

Oil reserves: ~10^{12} barrels, ~6,000 EJ or 6,000 Q

Natural gas reserves: ~6,000 EJ or 6,000 Q

World energy consumption rate: ~10^{13} W

$$\text{R/P ratio} = \frac{\text{reserves}}{\text{consumption rate}} \text{ (units of time)}$$

Exponential growth: $N = N_0 e^{rt}$ (Equation 5.1; p. 132)

Cumulative consumption: $\dfrac{N_0}{r} (e^{rt} - 1)$ (Equation 5.2; p. 132)

QUESTIONS

1. Explain, in terms of basic chemistry, why natural gas produces less CO_2 per unit of energy generated than do the other fossil fuels.

2. Fossil fuels are continuously forming, even today. So should we consider coal, oil, and natural gas to be renewable energy sources? Explain.

3. Explain how combined-cycle power plants manage to achieve a high thermodynamic efficiency.

4. Why can't a clever engineer design a heat engine with 100 percent efficiency?

5. In the context of fossil fuels, what's the difference between reserves and resources?

6. How is it possible for oil reserves to increase without the discovery of any new oil fields?

7. In the cornucopian view, the finite nature of a natural resource need not be an issue. How is this possible? Back up your answer with the example of coal.

8. How does the occurrence of Hubbert's peak for U.S. oil production in 1970 relate to the increasing U.S. dependence on imported oil?

EXERCISES

1. Complete combustion of a single octane molecule results in how many molecules of CO_2 and how many of H_2O? How many oxygen (O_2) molecules are needed for this combustion?

2. Repeat Example 5.2 for the case of a 1-GWe oil-fired power plant that is 41 percent efficient. Assume the oil is delivered by rail, in 100-car trains with each tank car holding 600 barrels of oil, and determine the number of such trains needed each week. (In practice, an oil-fired power plant would be more likely to be supplied from an oil pipeline.) Hint: Table 3.1's last energy entry could save you some unit conversions.

3. A combined-cycle power plant operates with a maximum temperature of 1,080°C in its gas turbine, then dumps waste heat at 10°C from its steam turbine. What is the overall thermodynamic efficiency of this plant?

4. A combined-cycle power plant has a thermodynamic efficiency of 81 percent. It dumps waste heat from its steam cycle at 5°C. (a) What is the maximum temperature in its gas turbine? If the gas turbine stage alone has 63 percent thermodynamic efficiency, what are (b) the intermediate temperature and (c) the thermodynamic efficiency of the steam cycle alone?

5. Show that the efficiency of a combined-cycle power plant is given by $e_{combined} = e_{gas} + (1 - e_{gas})e_{steam}$, where e_{gas} and e_{steam} are the efficiencies of the gas turbine and steam cycle alone.

6. World coal reserves amount to about 520 Gt of anthracite and bituminous coal with an average energy content of about 32 MJ/kg, and about 470 Gt of lower-grade coal with an average energy content of about 18 MJ/kg. Verify the text's figure of about 25,000 EJ for the energy equivalent of the world's coal reserves.

7. Use Table 3.1 to show that a world oil reserve of 1 trillion barrels contains roughly 6,000 EJ or 6,000 Q of energy.

8. World oil reserves amount to about 6,000 EJ, and humankind uses energy at the rate of roughly 10^{13} W. If all our energy came from oil and assuming there was no growth in energy consumption, estimate the time left until we would exhaust the 6,000-EJ reserve.

9. Verify the statement made in the text that the world's natural gas reserve of some 6 quadrillion cubic feet has an energy content of about 6,000 EJ.

10. Use this chapter's estimate of coal reserves, along with current world coal consumption of 5 Gt per year, to make a naïve estimate of the time remaining until coal runs out.

11. (a) Use Equation 5.2 to verify my 27-year estimate for the time that proven oil reserves will last with a 2 percent annual growth in world oil consumption. How would the time estimate change if the growth rate (b) increased to 3 percent, (c) dropped to 1 percent, or (d) became negative at −0.5 percent, indicating a decline in consumption? (Equation 5.2 still applies in this case.)

12. A rough figure for proven world oil reserves is the 1 trillion barrels (10^{12} barrels or 6,000 EJ) that I gave in the text, which led to Example 5.4's naïve estimate of 35 years' remaining oil supply if there's no change in the world oil consumption rate. However, the U.S.

Geological Survey estimates that recoverable oil reserves may actually be somewhat more than double that figure. Approximating the figure as 2 trillion barrels or 12,000 EJ would obviously double the naïve estimate of 35 years. But what would it do to my 27-year estimate, which assumes a 2 percent annual growth in oil consumption? Study Box 5.2 for help with this exercise.

13. In the early twenty-first century, China consumes oil at the rate of about 7 million barrels per day, increasing by 7.5 percent annually. The United States consumes about 21 million barrels per day, increasing by about 1.4 percent annually. If these rates hold, how long before China overtakes the United States in oil consumption?

14. Show that Equation 5.2 implies that the time to reach a cumulative consumption C is

$$t = \frac{\ln(rC/N_0 + 1)}{r}$$

15. If you've had calculus, derive Equation 5.2. Remember that N here is the annual rate of consumption (which is a function of time) and that you can get the total consumption from time 0 to time t by integrating the consumption rate over this time interval.

16. Show that Figure 5.23's assumptions about exponential growth in China's oil consumption would result in China's having consumed roughly the world's entire 1-trillion-barrel oil reserve by 2050.

RESEARCH PROBLEMS

1. How much oil is in the U.S. Strategic Petroleum Reserve? For how much time could the reserve supply the oil we're now importing from other countries?

2. Research advanced combined-cycle power plant designs and find an example of a contemporary high-efficiency design. Give the manufacturer, the total power output, and the overall efficiency. From the highest and lowest temperatures, calculate the thermodynamic efficiency—the theoretical maximum for any heat engine operating between these temperatures. Compare this result with the actual overall efficiency.

3. Find the composition of a typical gasoline, listing the major chemical components and the percentage of each.

4. Find the latest edition of the U.S. Department of Energy's *Annual Energy Review* (available online). Under the "International Energy" section, find data you can use to produce an updated version of the oil production curve in Figure 5.24. Is Hubbert's peak evident yet?

5. Find as many different estimates as you can of current world oil reserves. List each, along with its source.

6. Use a spreadsheet or other software to fit a curve to the data found in Research Problem 4. If your software permits, use a Gaussian curve (the standard bell curve); if not, use a quad-

ratic (a second-order polynomial). From either the graph or its equation (which your software should be able to provide), determine when the curve you've produced reaches its peak. This is a simple way to estimate Hubbert's peak for oil.

7.   Consult the latest edition of the U.S. Department of Energy's *Annual Energy Review* (available online), and under the "International Energy" section, find data on world oil consumption. Plot the data from the early 1990s to the present, and fit an exponential function (which will involve the term e^{rt}, where t is the time in years). Compare the growth rate r of your exponential fit with the 1.4 percent figure used in Box 5.2.

8.   Find the current prices of allowances to emit a ton (or tonne; see Box 3.1) of carbon on the European Carbon Index and the Chicago Climate Exchange.

ENVIRONMENTAL IMPACTS OF FOSSIL FUELS

Fossil fuels dominate today's global energy supply, so it's important to understand the environmental impact of these fuels. Furthermore, fossil fuels are among the most environmentally deleterious ways we have to produce energy, so their impact on the environment is doubly significant.

In this chapter I interpret the term *environmental impact* broadly, to include not only such obvious factors as air pollution and landscape desecration, but also direct and indirect effects on human health and safety and on quality of life. However, I leave the details of the most significant global impact of fossil fuel consumption—climate change—for later chapters focusing on that big subject.

Environmental impacts result, obviously, from the combustion of fossil fuels and the associated release of combustion products to the atmosphere. Extraction, transportation, and refining of fossil fuels carry additional environmental and health implications. Even the thermodynamically mandated waste heat can affect local ecosystems.

6.1 Is Carbon Dioxide a Pollutant?

Before we plunge into these myriad environmental impacts, let me remind you that the combustion of fossil fuels, whatever other implications it may have, necessarily entails the production of CO_2 and water. That's because these fuels are, chemically, hydrocarbons, and extracting the stored energy means combining the fuel's carbon and hydrogen with atmospheric oxygen to make CO_2 and H_2O. If we're going to burn fossil fuels, then we're necessarily going to make CO_2. Unfortunately, this necessary by-product has an undesirable consequence, namely climate change. But given that nearly 90 percent of our energy comes from fossil fuels, we're stuck with CO_2 as an inevitable by-product of our current energy use.

However, fossil fuels aren't pure hydrocarbon, and even if they were they wouldn't necessarily burn completely to CO_2 and water. A host of other substances result from fossil fuel combustion, many of which are undesirable and even unnecessary in that their production isn't required in the energy-release process. These substances constitute **pollution**.

There's considerable confusion, in the minds of both the public and of policymakers, about the distinction between CO_2 and other emissions from burning fossil fuel. Some would lump them all together as "pollution"—an approach that might be strategically valuable because it would allow government environmental agencies to regulate CO_2. But such an approach confuses the essential role of CO_2 in fossil fuel energy production with the inessential role of other emissions. Traditionally, pollutants have been considered substances that are toxic, carcinogenic, or otherwise directly harmful to living things and the immediate environment. Nontoxic CO_2, essential for plant life and an ingredient in soda and beer, has not usually been considered a pollutant in that sense. Furthermore, CO_2 is a natural atmospheric constituent, albeit in a relatively low concentration. But as we learn more about the long-term effect of CO_2 on global climate, the view of CO_2 as an innocuous gas is changing. Indeed, a landmark 2007 decision by the U.S. Supreme Court paved the way for regulation of CO_2 as a harmful pollutant.

In this book I will continue to distinguish CO_2 from traditional pollutants. That's not because CO_2 is less harmful than the others, but because it's an *essential* product of fossil fuel combustion. When I talk about a "low-emission vehicle" for example, I mean one that produces low levels of traditional pollutants; it may or may not have low carbon emissions. When I talk about a "dirty" fuel, I mean one that produces lots of traditional pollutants; it may or may not also produce lots of CO_2. You'll often find terms like *low emission* and *dirty* used sloppily, sometimes referring to traditional pollutants and sometimes to CO_2 emissions. It's better to maintain a clear distinction. For that reason I treat CO_2 emissions and their climatic effects separately from other environmental implications of fossil fuel use. I deal with conventional pollutants in this chapter, then discuss CO_2 in the chapters on climate. Box 6.1 makes an obvious and practical statement about this distinction.

6.2 Air Pollution

The harmful and largely inessential by-products of fossil fuel combustion that I'm calling *pollution* result from several distinct sources. Some, such as emissions of sulfur and mercury, arise from substances other than carbon and hydrogen that occur in fossil fuels. Others, such as carbon monoxide (CO) and a host of complex and often carcinogenic compounds, result from incomplete combustion. A third category includes the nitrogen oxides, which form when usually inert atmospheric nitrogen combines with oxygen at the high temperatures present in the burning of fossil fuels. Even before combustion, leakage of fuel vapors can also contribute to air pollution. Finally, derivative pollutants, including the complex mix called *smog,* form as a result of atmospheric chemical reactions among primary pollutants.

Box 6.1   Gas or Diesel?

Although diesel cars are common in Europe, they're rare in the United States. Among the few diesels available in the U.S. market are the Volkswagen TDI (turbo direct injection) models. The TDI engine is designed to minimize common diesel problems, including odor, smoke, and hard starting. Furthermore, the diesel offers substantially greater fuel efficiency than comparable gasoline models; for example, the manual-transmission gasoline and diesel models of the New Beetle shown in Figure 6.1 average, respectively, 28 miles per gallon and 42 miles per gallon. (Vehicles with automatic transmissions get even lower mileage than those with manual transmissions; see Exercise 2.)

Which New Beetle should you buy? Fewer gallons burned in the diesel translate into substantially lower CO_2 emissions, so from a climate perspective the diesel is preferable. Indeed, the U.S. Environmental Protection Agency gives the New Beetle's diesel model a greenhouse gas score of 9 out of a possible 10; however, it has an air pollution score of only 2. That difference—very good in one index, very poor in the other—illustrates directly the distinction I'm making between CO_2 and traditional pollutants. (The gasoline model rates 6 on both scales.)

In California, home to the strictest air pollution standards in the United States, you don't have the choice suggested in Figure 6.1. You can buy a gasoline-powered Beetle engineered specifically for the California market that does better on pollution (9 out of 10) without sacrificing fuel efficiency. But you can't buy the diesel model in California, because its emissions of traditional pollutants don't meet state standards.

Particulate Matter

Solid material that doesn't burn comes through the combustion process in the form of tiny particles that are carried into the air along with the stream of gaseous combustion products. Other particles result from incomplete combustion; soot from a candle flame is a familiar example. Still others form in the atmosphere from reactions among gaseous pollutants. Such **particulate pollution** is especially significant with the combustion of coal and heavier liquid fuels, including diesel oil. You've surely seen the cloud of black smoke emitted when a large truck starts up; this is particulate pollution. Less obvious are tiny sulfur-based particles emitted from coal-burning power plants; these produce atmospheric haze that, among other things, is responsible for significant loss of visibil-

Figure 6.1
Which New Beetle should you buy?

ity in scenic areas of the American West. Heavier particulate matter soon falls to the ground or is washed out of the air by rain, but lighter particulates can remain airborne for long periods.

Particles less than 10 microns (10 μm, or one-millionth of a meter) in size reach deep into the lungs and are therefore especially harmful to human health. Breathing particulate-laden air may cause or exacerbate respiratory diseases such as asthma, emphysema, and chronic bronchitis. Indeed, a careful study done for the Clean Air Task Force in 2004 using the methodology of the U.S. Environmental Protection Agency (EPA) suggested that some 24,000 premature deaths per year in the United States may be the result of particulate air pollution from fossil-fueled power plants. Of these, the study concluded that nearly 17,000 deaths could be prevented each year by the adoption of policy measures that were proposed, but not passed, in the U.S. Congress. This number of preventable air pollution–related deaths is higher than the U.S. annual homicide rate in the early twenty-first century. In addition, the Clean Air Task Force study cites far higher numbers of respiratory diseases and lost work days associated with power-plant particulate emissions—the latter exceeding 3 million per year! Figure 6.2 summarizes some of the study's findings.

Reducing Particulate Pollution The technology to reduce particulate pollution is readily available and is in widespread use. Long gone, at least in wealthier countries, is the pall of black smoke that once hung over industrialized areas (Fig. 6.3). The very tall smokestacks of fossil-fueled power plants are themselves crude pollution control devices; by injecting emissions into the atmosphere at great heights where winds are strong, they ensure that pollutants, including particulates, are dispersed throughout a larger volume of air and are carried away from the immediate area.

Several techniques remove particulate pollution from the gases produced in fossil fuel consumption. These, in turn, are based on distinct physical principles. **Filters** are per-

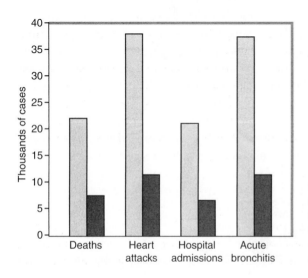

Figure 6.2
Some estimated health effects of fossil-fueled power plant pollution in the United States, in thousands of cases annually, from the Clean Air Task Force study. The tall bars reflect the number of total cases, and the short bars give projected numbers had the U.S. Congress passed Senate Bill 366, "The Clean Power Act."

Figure 6.3
Air pollution hangs over New York City in this photo from the 1950s.

haps the most obvious particulate-reduction devices. They employ fine fabric or other meshes to trap particulate matter while permitting gases to pass through. In large-scale industrial and power-plant applications, fabric filters often take the form of tube-shaped bags, with hundreds or thousands of bags in an enclosure called a **baghouse**. Filtration can be highly effective, removing some 99.5 percent of all particulate pollution. However, frequent mechanical shaking of the bags is necessary to remove the particulate material that would otherwise block the filter, and bag filters are expensive, require frequent maintenance, and don't work at high flue-gas temperatures.

Cyclones use the principle of inertia—that an object tends to remain in a state of uniform motion unless a force acts on it. In a cyclone, flue gas is forced to travel in a spiral vortex something like the inside of a tornado. The lighter gas molecules have low inertia, so they readily "go with the flow." But the more massive particulates have high inertia, so they have trouble following the tight, high-speed spiral motion of the gas. These particulates strike the inside walls of the cyclone, then drop into a collection hopper (Fig. 6.4a). You may have noticed the characteristic upside-down cone shape of cyclones on industrial buildings (Fig. 6.4b). Cyclones typically remove between 50 and 99 percent of particulate matter, depending on particle size. Given their inertial operation, you can understand why they're more effective with larger particles. Among the least expensive particulate-control devices, cyclones are used on their own in dust-producing industrial applications, and often as pretreatment for flue gas before it enters more efficient pollution-control devices.

Electrostatic precipitators offer a more sophisticated approach. In these devices, a high voltage is applied between a thin, negatively charged wire and positively charged metal plates. The strong electric field in the vicinity of the wire ionizes gas molecules, resulting in free electrons that attach to particles in the gas stream. The negatively charged

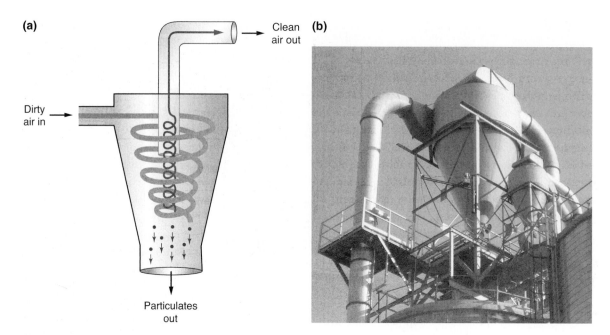

Figure 6.4

Cyclone separator for reduction of particulate air pollution.

(a) A cutaway diagram shows the spiral air path within the device.

(b) Cyclones atop an industrial facility.

particles are driven onto the positively charged metal plates (Fig. 6.5). Periodically, the plates are mechanically shaken, and the particulate matter falls into a hopper where it collects and is eventually trucked away. This so-called **fly ash** collected from coal-burning power plants actually contains metals that may be economically worth extracting; alternatively, fly ash is used in the manufacture of cement, ceramics, and even some plastics.

Electrostatic precipitators can remove more than 99 percent of particulate matter. Precipitators installed in U.S. factories and power plants since 1940 have reduced sub-10-μm particulate emissions by a factor of 5, even as energy and industrial output grew substantially. But this clean air comes at a cost: Electrostatic precipitators and other particulate removal systems typically consume some 2 to 4 percent of a power plant's electrical energy output, which is one of the many reasons why actual power-plant efficiencies are well below the theoretical limits imposed by the second law of thermodynamics.

Sulfur Emissions

Sulfur is present in coal and, to a lesser extent, in oil. Unless it's removed before combustion—a technologically feasible but expensive process—the sulfur burns to produce sulfur dioxide (SO_2). However, the native sulfur content of fuels varies substantially, so sulfur reduction also can be accomplished by changing the fuel source. For example, coal

(b)

(a)

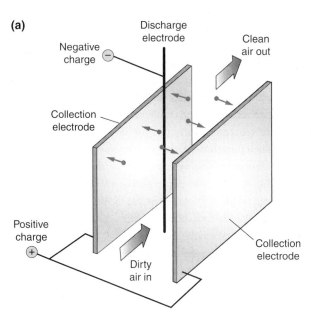

Figure 6.5
Electrostatic precipitators are among the most effective devices for
particulate pollution control. (a) The discharge electrode is charged
to a high negative voltage, which negatively charges particles in the
gas stream. The particles are then attracted to the positively charged
collection electrodes, from which they are removed by mechanical
shaking. (b) Electrostatic precipitators are large devices that consume
a lot of energy.

from the western United States typically contains less than 1 percent sulfur, whereas coals
from the Midwest may contain 4 to 5 percent sulfur.

Sulfur dioxide itself is detrimental to human health. As usual, the health burden falls
most heavily on the very young, the very old, and those already suffering from respiratory diseases. Acute episodes of high SO_2 concentrations have resulted in outright mortality; in a 1952 air pollution incident in London, some four thousand people perished
when atmospheric SO_2 rose sevenfold.

Once in the atmosphere, SO_2 undergoes chemical reactions that produce additional
harmful substances. Oxidation converts SO_2 to sulfur trioxide (SO_3), which then reacts
with water vapor to make sulfuric acid (H_2SO_4):

$$2SO_2 + O_2 \rightarrow 2SO_3$$
$$SO_3 + H_2O \rightarrow H_2SO_4$$

The second reaction can take place either with water vapor or in liquid water droplets. In the liquid water, sulfuric acid dissolves into positive hydrogen ions and negative sulfate ions (SO_4^-). Sulfate ions can join with other chemical species to make substances that, if the water evaporates, remain airborne as tiny particles known as **sulfate aerosols**. These particles, typically well under 1 μm in size, are very efficient scatterers of light. As a result, they reflect sunlight and thus lower the overall energy reaching Earth. Consequently, as you'll see in Chapter 13, sulfate aerosols have a cooling effect on climate.

Acid Rain   When water droplets laden with sulfuric acid fall to Earth, the result is **acid rain**, which has significantly greater acidity than normal rain. Acidity is measured on the **pH scale**; a pH of 7 is neutral and anything less than 7 is acidic. Technically, pH is the negative of the logarithm of the concentration of hydrogen ions in a solution; in pure water that concentration is 1 part in 10 million, or 10^{-7}, for a pH of 7. Natural atmospheric CO_2 dissolves to some extent in rainwater, forming carbonic acid (H_2CO_3) and making normal rain somewhat acidic, with a pH of 5.6. Rain with a pH lower than this value is considered acid rain. By comparison, vinegar has a pH of about 3, and most sodas have a pH around 4 because of their dissolved CO_2. In addition to forming dissolved sulfuric acid in rainwater, sulfate aerosols also contribute directly to environmental acidity when they land on the ground or fall into surface waters.

Acidity with a pH level below about 5 has a detrimental effect on aquatic life. Reproduction in fish falters, with death or deformity widespread among young fish. Amphibians and invertebrates suffer similarly, with the result that highly acidic lakes have little animal life. The northeastern United States has been particularly affected by acid rain, much of it from coal-burning power plants in the Midwest, and so has much of northern Europe. Many high-altitude lakes, even those in protected wilderness areas, have pH levels well below 5 and are considered "dead." The effect of acid rain depends in part on geology; lakes or soils on limestone rock suffer less harm because the limestone can neutralize the excess acidity, but granite and quartz-based rocks lack this buffering effect. Figure 6.6 maps the acidity of precipitation in the United States.

Acid precipitation affects not only water quality; it also damages terrestrial vegetation, especially high-altitude trees, and causes increased corrosion and weathering of buildings. Limestone and marble are particularly susceptible to damage from acid precipitation, as are some paints.

Reducing Sulfur Emissions   Because sulfur pollution forms initially as the gas SO_2, it's not removed by particulate pollution controls. The most effective sulfur-removal technique now in widespread use is called **flue gas desulfurization**, also known as **scrubbing**. In a so-called wet scrubber, flue gases pass through a spray of water containing chemicals, usually calcium carbonate ($CaCO_3$) or magnesium carbonate. In a dry scrub-

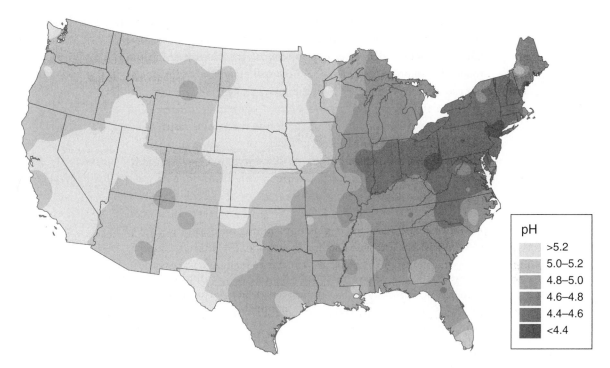

Figure 6.6
Values of pH in precipitation falling on the United States in the
early twenty-first century. Lower pH corresponds to greater acidity.
Excess acidity in the Northeast is largely the result of sulfur
emissions from coal-burning power plants in the Midwest.

ber, the flue gas contacts pulverized limestone, which is largely calcium carbonate. Either
way, chemical reactions yield solid calcium sulfate ($CaSO_4$), which drops out of the gas
stream or is removed along with other particulates:

$$2SO_2 + 2CaCO_3 \rightarrow 2CaSO_4 + 2CO_2$$

Scrubbers remove some 98 percent of sulfur from the flue gases, but again there's a price.
The monetary cost of scrubbers can be 10 to 15 percent of the total cost of a coal-burning
power plant, and the energy cost may be as much as 8 percent of the plant's total electric power output. This is in addition to the 2 to 4 percent for particulate removal. Much
of that energy goes into reheating the flue gas after its contact with water in a wet scrubber, which is necessary to restore buoyancy so the flue gas can rise up the smokestack
and disperse in the atmosphere.

Sulfur-removing scrubbers are widely used in Europe and Japan. In the United States,
however, electric utilities can meet sulfur emissions regulations with a variety of less-

effective strategies, including switching to lower sulfur fuels or trading so-called emissions credits. Recent interpretation of the New Source Review provisions of the U.S. **Clean Air Act** has allowed enhancements costing as much as 20 percent of a power plant's total value to proceed without the need to upgrade sulfur emission controls. The resulting excessive sulfate aerosol pollution plays a major role in the health impact study I discussed in the section "Particulate Matter" above, and the health effects documented in that study would be reduced dramatically if all U.S. power plants met the standards required of new plants.

Finally, sulfur removal results in its own environmental problems, namely the need to dispose of large amounts of sulfurous waste. Example 6.1 explores just how much of this waste is produced.

Example 6.1   Sulfurous Waste

Consider a 1-GWe power plant burning coal with a sulfur content of 3 percent by weight. Estimate (a) the amount of coal burned per day, assuming 33 percent efficiency, and (b) the amount of $CaSO_4$ produced daily if all the sulfur in the coal is converted to $CaSO_4$ in the plant's scrubbers.

Solution

Table 3.3 shows that the energy content of coal is about 30 MJ/kg. Our 1-GWe power plant is 33 percent efficient, so it goes through coal energy at the rate of 3 GW to produce its 1 GW of electric power (recall the distinction between GWe—the electric power output of a plant—and the total thermal power extracted from fuel and measured in GWth). That 3 GW is 3,000 MW, or 3,000 MJ/s. At 30 MJ/kg, our plant therefore uses coal at the rate of

$$\frac{3,000 \text{ MJ/s}}{30 \text{ MJ/kg}} = 100 \text{ kg/s}$$

That's a tonne of coal every 10 seconds! Since the coal contains 3 percent sulfur by weight, this amounts to 3 kg/s of sulfur. But the scrubbers combine each sulfur atom with one calcium atom and four oxygen atoms to make $CaSO_4$. Refer to the periodic table in the Appendix (p. 487), and you'll find that sulfur's atomic weight is 32, whereas calcium's is 40 and oxygen's is 16. This means that, for every 32 units of sulfur by weight, we get an additional 40 units from the calcium and $4 \times 16 = 64$ units from the oxygen; equivalently, for every 1 unit of sulfur we get 40/32 units of calcium and 64/32 units of oxygen. So our 3 kg/s of sulfur translates into

$$(3 \text{ kg/s})\left(1 \text{ (S)} + \frac{40}{32} \text{ (Ca)} + \frac{64}{32} \text{ (O)}\right) = 13 \text{ kg/s}$$

of $CaSO_4$. There are 86,400 seconds per day (figure that out!), so the power plant's scrubbers produce

$$(13 \text{ kg/s})(86,400 \text{ s/day}) = 1,100,000 \text{ kg/day}$$

That's about 1,000 tonnes per day of $CaSO_4$! Actually, the $CaSO_4$ generally ends up in a water-based slurry totaling perhaps four times this mass.

Carbon Monoxide

Carbon monoxide results from the incomplete combustion of fossil fuels or other carbon-containing materials, such as tobacco, charcoal, or wood. Carbon monoxide formation is enhanced when combustion occurs with insufficient oxygen present, as in indoor burning or the smoldering of a cigarette. But all fossil fuel combustion produces some CO. Although CO occurs naturally in the atmosphere at low levels, its concentration in industrialized areas and traffic-congested cities can be ten to one thousand times greater.

Carbon monoxide's dominant health effect results from the gas's affinity for hemoglobin, the molecule in red blood cells that carries oxygen to the body's tissues. Carbon monoxide binds 240 times more readily to hemoglobin than does oxygen, and it binds more tightly than oxygen, making it hard to displace. A person breathing CO-contaminated air therefore carries a substantial amount of so-called carboxyhemoglobin (COHb), which reduces the blood's oxygen-carrying ability. At very high concentrations, CO causes death from lack of oxygen in body tissues. At lower concentrations, oxygen deprivation damages tissues and affects a variety of physiological functions. Carboxyhemoglobin levels of 6 percent can cause irregular heart rhythm in cardiovascular patients, increasing the risk of mortality. By comparison, nonsmokers normally carry COHb levels of 1 percent or lower; for smokers, the normal level averages 4 percent and can be twice that. Concentrations in the 5 percent range can also affect the brain, increasing reaction time and decreasing ability at tasks requiring high coordination. Some studies have suggested that automobile accidents occur more frequently among drivers with high COHb levels resulting from prolonged exposure to traffic pollution.

Control of CO emissions is easier with stationary sources such as power plants, where combustion takes place under steady conditions—in particular, precisely controlled and optimized fuel/air ratios. Careful engineering with attention to adequate mixing of fuel and air greatly reduces the incomplete combustion that causes CO formation. Mobile sources are another story: With the intermittent combustion that occurs in gasoline and diesel engines, and with engines operating under a wide range of speeds and loads, CO production is inevitable. In the United States, cars and trucks produce about half of all CO emissions, while other mobile sources, such as planes, boats, and farm equipment, are responsible for roughly another 20 percent (Fig. 6.7).

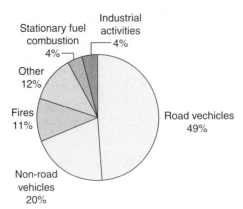

Figure 6.7
More than two-thirds of all CO emissions in the United States come from vehicles of all types. Non-road vehicles include recreational vehicles, boats, trains, aircraft, farm equipment, and lawnmowers. Nearly half the emissions in this category are from lawn and garden equipment.

CO emissions would be higher if it weren't for the widespread use of pollution control measures. In cars, most CO reduction occurs in **catalytic converters** placed in the exhaust system. These devices essentially "burn" CO to CO_2 with the aid of a chemical catalyst that allows the reaction to proceed at lower temperatures than in combustion. Although it facilitates the reaction, the catalyst itself is not consumed. Another approach is to reformulate gasoline by adding oxygen-containing chemicals that help to ensure complete conversion of all carbon to CO_2. Such additives include alcohols, both methanol and corn-derived ethanol, and methyl tertiary-butyl ether (MBTE), a substance that reduces CO production but has been linked to water pollution when fuel leaks into the ground.

Because CO represents a higher energy state than CO_2 (the latter being the product of complete combustion), CO in the atmosphere is naturally oxidized to CO_2 on a fairly short time scale of about a month, which means CO never has time to become widely distributed and remains essentially a local pollutant. However, the oxidation of atmo-spheric CO to CO_2 involves a complex sequence of reactions that affect other atmospheric chemicals and may have implications for climate. Also, microorganisms in the soil rapidly absorb CO, helping to remove it from the air in all but the most heavily paved urban areas. In any event, CO emissions don't stay around for long, and the atmospheric concentration of CO fluctuates substantially according to the time of day, traffic conditions, and specific location.

Nitrogen Oxides

Earth's atmosphere is predominantly nitrogen (N_2). In that form, nitrogen is relatively inert, meaning it doesn't participate readily in chemical reactions. But at the high temperatures typical of fossil fuel combustion (especially over 1,100°C), nitrogen combines with oxygen to form a variety of compounds, collectively designated as the nitrogen oxides (NO_x, often pronounced "nox"). Among these are nitric oxide (NO), also known as nitrogen monoxide; nitrogen dioxide (NO_2); dinitrogen pentoxide (N_2O_5); and nitrous oxide (N_2O), also known as "laughing gas" and commonly used as a dental anesthetic. Figure 6.8 shows that fossil-fueled electric power plants, industry, and transportation are all significant sources of NO_x emissions.

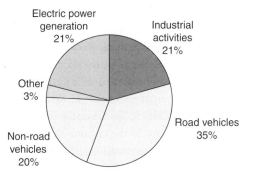

Figure 6.8
Vehicles, industry, and power generation all contribute significantly to NO_x emissions.

Most of the nitrogen that forms NO_x comes not from fuel, but from the air itself. So there's no way to prevent NO_x emissions by pretreating fuel. Nor is it practical to remove nitrogen from the air supplied for combustion (after all, nitrogen constitutes some 80 percent of the atmosphere). In motor vehicles, the same catalytic converters that oxidize CO to CO_2 also convert much of the NO_x into the normal, harmless atmospheric components N_2 and O_2. A variety of additional techniques can help reduce nitrogen oxide production in stationary sources such as power plants and industrial boilers, including modifications to the combustion chamber and flame geometry, adjustments in the fuel and air flows, and recycling of flue gases through the combustion chamber. The latter reduces combustion temperature and with it the amount of NO_x produced. Ironically, reducing the temperature reduces the thermodynamic efficiency, lowering the useful energy available from a power plant.

Nitric oxide, the dominant component of NO_x, is not particularly harmful in itself, but NO_2 is toxic, and long-term exposure may result in lung damage. The reddish-brown color of NO_2 is responsible for the dirty brown appearance of polluted urban air. It doesn't stop there, however, because nitrogen oxides undergo further chemical reactions in the atmosphere. Oxidation, for example, converts NO to the more harmful NO_2. More important are **photochemical reactions**, driven by the energy of sunlight, the result of which is **photochemical smog**.

Photochemical Smog

The atmosphere above urban areas contains a complex soup of chemicals, many resulting directly or indirectly from fossil fuel combustion in both stationary and moving sources. Among the most important for smog formation are the nitrogen oxides that we've just discussed; hydrocarbons from incomplete combustion as well as from the production, handling, and storage of gasoline; and evaporation of organic solvents in paints, inks, dry-cleaning fluids, and similar chemicals.

Trapping of these and other pollutants occurs under weather conditions known as **inversion**. To understand this phenomenon, keep in mind that air temperature usually decreases with altitude, a reflection of the fact that the atmosphere is largely transparent and therefore doesn't absorb much energy from sunlight. Instead, Earth's surface heats up and transfers energy to the adjacent atmosphere. The rate at which temperature declines with altitude is called the **lapse rate**, and it's normally around 6.5°C per kilometer. Now consider a blob of air at Earth's surface that's somehow heated. It might, for example, be the air over a black, sunlight-absorbing surface such as a parking lot, or it might be air mixed with hot gases from fossil fuel combustion. The warmed air is less dense than its surroundings, so it rises just like a hot-air balloon. As it rises, the air expands and cools, expending its internal energy as it pushes against the pressure of the surrounding air. Under normal conditions, this cooling isn't as great as the lapse rate that describes the cooling with altitude of the surrounding air (Fig. 6.9a), so the blob remains warmer and less dense than its surroundings and continues to rise, carrying away any embedded pollutants. In an inversion, though, the atmospheric temperature either increases with height or decreases more slowly than a rising blob of air would cool (Fig. 6.9b,c). As a result, such a blob soon becomes cooler and thus denser than its surroundings, and it sinks back down. Under such conditions the air is stable, with the layers nearest the ground trapped in place. Trapped with the air are any pollutants it contains. Nearby mountains and other topographical conditions can exacerbate this trapping. A classic example of a smog-prone area is the Los Angeles basin, where mountains help trap polluted air and the cool Pacific waters encourage inversion conditions (Fig. 6.10).

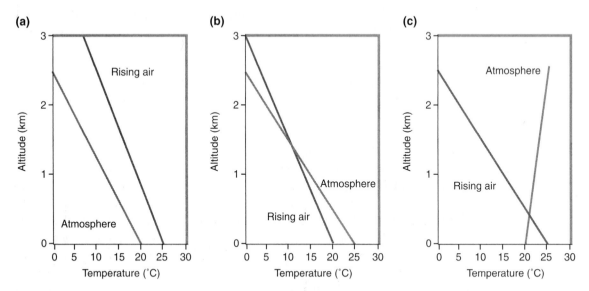

Figure 6.9
(a) Under normal conditions, atmospheric temperature decreases more rapidly than that of a rising blob of air. Because the rising air stays warmer than the surrounding air, it continues to rise, carrying pollutants high into the atmosphere. (b) Here the atmospheric temperature decreases more slowly, and a rising blob of air quickly comes to equilibrium at about 1.5 km. (c) In a true inversion, temperature increases with altitude, and rising air reaches equilibrium at only half a kilometer. In both (b) and (c) the air is stable, and pollutant-laden air remains trapped near the ground.

Figure 6.10
Los Angeles under normal and inversion conditions.

Sunlight shining on polluted air trapped by an inversion carries enough energy to induce chemical reactions involving pollutant molecules. One important reaction is the decomposition of nitrogen dioxide into nitric oxide and an isolated oxygen atom:

$$NO_2 + \text{solar energy} \rightarrow NO + O$$

Now, atomic oxygen is very reactive, and it can join molecular oxygen (O_2) to make another highly reactive chemical, **ozone** (O_3):

$$O + O_2 \rightarrow O_3$$

You probably know ozone as the gas in the stratosphere that protects Earth's surface from the harmful effects of ultraviolet radiation. But ozone in the lower atmosphere is harmful to health. (Ozone is also a climate-changing greenhouse gas; more on that in Chapter 13.) Furthermore, the high reactivities of ozone and monatomic O can lead to a great deal of chemical mischief, resulting in still more harmful substances. This is because O_3 and O facilitate reactions among hydrocarbons and nitrogen oxides that result in a host of so-called peroxyacyl nitrates, or PANs. Again, these are highly reactive substances. In humans they cause eye irritation as well as respiratory problems. PANs and other reactive molecules can also injure plants and damage materials such as paints, fabrics, and rubber.

The photochemical nature of smog is evident in a plot showing the levels of smog precursors and ozone in an urban area over a 24-hour period. At night, levels of all these substances are low. Morning rush hour quickly increases nitrogen oxides and hydrocarbons. As sunlight strengthens through the morning, NO and hydrocarbons decline and ozone increases. Finally, by midday all these precursor molecules have declined through the formation of more complex molecules, including PANs. Figure 6.11 shows pollut-ant concentrations measured throughout a typical 24-hour period in Los Angeles.

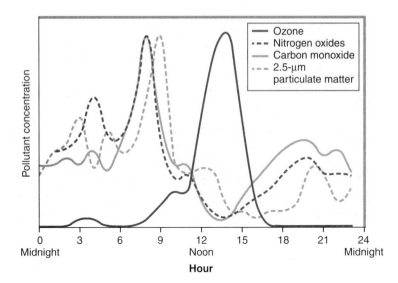

Figure 6.11
Hourly concentrations of several pollutants for Los Angeles on a midweek day in January. Notice how ozone peaks after the others since it's a secondary pollutant formed by photochemical processes. Can you identify the effects of the morning and evening rush hours? The concentrations are scaled so they all peak at the same height on the graph.

The best way to minimize photochemical smog is to reduce emissions of smog precursors, especially nitrogen oxides and hydrocarbons. I discussed catalytic converters and other approaches to NO_x reduction in the preceding section. Because hydrocarbon emissions often result from evaporation of fuels and other products, careful handling to avoid spills, leakage, and vapor escape can reduce these emissions. Cars, for example, have vapor recovery systems as part of their pollution control systems; these capture gasoline vapors in a charcoal canister and recycle them back into the fuel system rather than letting them escape to the atmosphere. And increasingly, vapor release in the fueling process itself is reduced through the use of special gasoline nozzles (Fig. 6.12). Paints, paint thinners, and other solvents can be reformulated to reduce their volatility (tendency to vaporize). Widespread use of water-based paints further reduces hydrocarbon emissions.

Heavy Metals

Fossil fuels contain a variety of heavy metals that are released as pollutants when fuels burn. Other metals enter the environment through wear in engines, turbines, and other machinery. In their elemental form (i.e., not combined with other elements to make chemical compounds), the heavier metals inhibit the actions of biological enzymes and therefore have a

Figure 6.12
The rubber bellows on this gasoline nozzle seals the gasoline tank opening to prevent leakage of hydrocarbon vapors that are precursors of photochemical smog.

wide range of toxic effects in humans. Especially significant are the deleterious impacts on brain development in young children. Lead, once a common component of gasoline and paints, is a particularly serious contaminant whose intentional use has been substantially curtailed; nevertheless, levels of lead remain high in urban air and as deposited particulate matter on urban surfaces. Another significant heavy metal is mercury, which again is particularly harmful to developing organisms. Mercury occurs naturally in coal, and coal-burning electric power plants are the dominant source of this widespread pollutant. When mercury pollution lands in surface waters, it's concentrated in the food chain and ends up at high levels in predatory fish. As a result, fish in pristine lakes of the central and eastern United States are so highly contaminated that pregnant women are advised not to eat freshwater fish of certain species, and for the general populace, warnings to eat no more than one meal of freshwater fish per month are common. A particularly disturbing study shows that one-fifth of all Americans may have mercury levels exceeding EPA recommendations of no more than 1 part per million. (This standard applies to hair, which incorporates mercury from the blood; the study sampled hair from fifteen hundred individuals). No other pollutant even comes close to mercury for violating federal standards so widely. The dominant source of this mercury contamination is coal-burning power plants. Figure 6.13 maps the extent of mercury pollution in the United States.

Radiation

It might come as a surprise to learn that fossil fuel combustion is a source of radioactive materials. Coal, especially, contains small quantities of uranium and thorium, as well as their radioactive decay products, such as radium. Although most coal contains uranium at about the 1-part-per-million (ppm) level, this figure varies widely; one sample of North Dakota lignite coal measured 1,800 ppm of uranium. Radioactive constituents of coal end up in the air or in fly ash recovered from particulate-removal devices. Although radiation from fossil fuels is not a major concern, it's interesting to note that the level of radiation emitted in coal burning can be higher than that emitted in the normal operation of a nuclear power plant. As a result, people living near some coal-burning power plants may be subjected to as much as five times more radiation than those living near nuclear plants (although this comparison is a bit misleading, as I'll show in Chapter 7).

6.3 Other Environmental Impacts of Fossil Fuels

The environmental impacts of fossil fuels begin long before combustion. Since fossil fuels form underground, we generally drill, dig, mine, or otherwise penetrate the Earth to recover them. After we extract them from the ground, we have to transport and process these fuels before they're ready for combustion.

Coal Extraction

When you think of coal extraction, you probably picture miners toiling away deep underground. This is the traditional method, but today some 60 percent of all coal

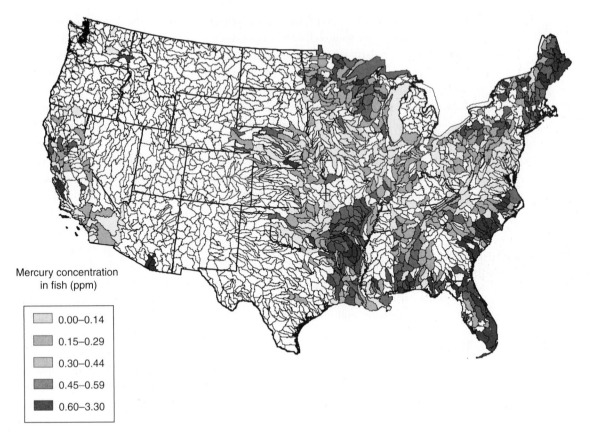

Mercury concentration
in fish (ppm)

	0.00–0.14
	0.15–0.29
	0.30–0.44
	0.45–0.59
	0.60–3.30

Figure 6.13
Average mercury concentration in U.S. fish, in parts per million
(ppm), plotted by watershed region. The EPA maximum permissible
concentration is 0.3 ppm, and advisories start at 0.15 ppm.

comes from **strip mining**, in which surface layers are removed to reach the coal. In
the United States, the proportion of strip-mined coal is likely to increase because this
coal, which is largely from the American West, has a significantly lower sulfur content than coal from the East and Midwest. In strip mining, a relatively thin layer
(typically tens of feet) of rock and soil is removed to expose coal seams that are typically 50 to 100 feet thick. A related method, used in the Appalachian mountains of
the eastern United States, is so-called **mountaintop removal**. In this relatively new
technique, the top thousand feet or so of a coal-containing mountain is removed by
blasting, the coal is extracted, and the remaining rubble is deposited in nearby valleys. Figure 6.14 shows the aftermath of mountaintop removal; in recent decades,
hundreds of peaks have met similar fates.

Figure 6.14
A mountaintop-removal coal operation
encroaches on the West Virginia community
below.

Obviously, surface mining by stripping or mountaintop removal results in major environmental alteration. In the United States, the Surface Mining Control and Reclamation Act
of 1977 requires that strip-mined land be restored to its original biotic productivity. Whether
that is realistically possible in the arid West is unclear. Serious erosion damage can occur
before plants have a chance to reestablish themselves. Complete restoration of mountainous
topography after mountaintop removal is obviously impractical. The removal process often
buries streams, altering natural water flows and resulting in water pollution and additional
erosion. With all types of coal mining, acidic runoff from mines adds further insult to surface waters; such **acid mine drainage** has rendered many Appalachian streams lifeless.

Large-scale use of mountaintop removal began only in the 1980s, and two decades
later some 700 miles of Appalachian streams had been buried in the resulting rubble.
Public outcries in the 1990s sent the practice into a decline, until just after the turn of
the century when an industry-friendly administration introduced a subtle change in federal environmental regulations. This change—a substitution of the word *fill* for *waste* to
describe the rubble from mountaintop removal—was all it took to reaccelerate the pace
of mountaintop removal.

Even old-fashioned underground mining has serious implications for the environment
and for human health. Despite tightened health and safety regulations, mining remains
among the most dangerous occupations in the United States. China, where rapid industrialization drives a growing appetite for coal, averages more than a dozen deaths each
day from coal-mining accidents. And miners around the world continue to suffer from
black lung disease and other infirmities brought on by exposure to dust-laden air in
underground mines. Environmental dangers persist long after mining ends; for example,

fires in abandoned coal mines can burn for years or even decades, threatening natural and human communities that have the misfortune to be located above the mines (Fig. 6.15). Coal fires are also a significant source of air pollution, and they produce enough CO_2 to worry scientists concerned about climate change.

Oil and Natural Gas Extraction

Conventional extraction of oil and natural gas is, at least in principle, less messy than coal mining. These fluid substances come from deep in the ground and are "mined" through narrow, precision-drilled wells. High pressures in the geologically trapped petroleum require careful handling, however, and an occasional "gusher" spews oil and natural gas into the environment—although modern drilling techniques have greatly reduced such wasteful and polluting events. Leakage of natural gas from drilling sites remains a concern, and "flaring"—the intentional burning of unusable gas—continues in some parts of the world, although it's largely banned in Europe and North America. Oil leaks and spills from offshore drilling rigs have the potential to disrupt marine ecosystems. The existence of oil reserves in environmentally sensitive areas pits environmentalists against the oil industry and its political allies; witness the debates surrounding construction of the Trans-Alaska oil pipeline in the 1970s and drilling in the Arctic National Wildlife Refuge in the 2000s. And in many parts of the world, oil exploration and extraction are done with little regard for their impacts on the environment or local human populations. Finally, deliberate ignition of oil wells during the 1991 Persian Gulf War briefly produced air pollution at ten times the rate of all U.S. industrial and power-generating activities, and pollution from these oil fires was detected halfway around the world (Fig. 6.16).

Figure 6.15
An underground coal seam fire has been burning for decades inside this Colorado mountain. Occasionally it bursts to the surface, igniting brush fires like the one shown here.

Figure 6.16
Oil wells burning in Kuwait after the 1991
Persian Gulf War.

The potential environmental impacts of unconventional petroleum extraction may be orders of magnitude greater than those of conventional drilling. Extracting oil from oil shale requires grinding up masses of rock. Exploiting the coal sands of Alberta, Canada, has turned vast areas into moonscapes (Fig. 6.17). If the world's demand for oil continues to grow, the unconventional sources of petroleum will surely become major environmental battlegrounds.

Figure 6.17
Extraction of oil from tar sands entails
massive environmental degradation.

Transportation

All fossil fuels contain on the order of 30 to 60 MJ of energy per kilogram. The world's population consumes energy at the rate of around 470 EJ per year, most of it from fossil fuels. This means we have to burn a great many kilograms of fossil fuels. We have to get all those kilograms from mines and oilfields to the power plants, cars, homes, and factories where we're going to burn them, so there must be a huge mass of fossil fuels in transit at any given time. You can estimate how much in Exercise 3; the answer is something approaching 100 million tons! We can't move that much material around without significant environmental impact.

One immediate consequence is that we have to use even more fuel to run all those tanker ships, trains, trucks, and pipelines. In fact, a small but significant fraction of our fossil fuel consumption goes into transporting the fuels themselves. For Alaskan oil that travels by pipeline and ship to ports on the U.S. West Coast, for example, the energy required to move the oil amounts to about 1 percent of the oil's energy content. A supertanker carrying oil from the Middle East to the United States, however, can use the equivalent of 15 percent of its cargo's energy (see Exercise 16). This additional energy use compounds the environmental impacts of fossil fuel combustion, but fuel transport also has its own unique impacts on the environment and on human well-being.

Oil spills from supertanker accidents are perhaps the most dramatic examples of environmental damage from fossil fuel transport (Fig. 6.18). As tankers have grown in size, so have the spills; the largest have involved more than a quarter million tons of oil. Some spills are caused when supertankers run aground or collide with other vessels, while others result from fires, hull failures, or accidents during loading and unloading. With 37,000 tonnes spilled, the Exxon *Valdez* accident in 1989 was relatively small, but it occurred in an ecologically sensitive Alaskan bay and caused widespread environmental damage. Oil spills block sunlight from reaching photosynthetic organisms, reduce dissolved oxygen needed to sustain life, and foul the feathers of birds and the gills of fish (Fig. 6.18b). Economic impacts result when oil piles up on beaches. Because supertanker accidents involve large oil releases that occur over short periods and affect relatively small areas, their specific impact can be very large. In the past few decades, tightened regulations and technological fixes such as the use of double-hulled oil tankers have greatly reduced both the number and size of spills (Fig. 6.19). Although plenty of smaller spills occur, it's the big headline-grabbing accidents that are responsible for most of the spilled oil; in actuality, the total amount of oil spilled worldwide varies substantially from year to year.

Although oil spills provide the most dramatic examples of environmental degradation resulting from marine oil transport, a lesser-known but significant effect is the spread of exotic species to coastal ecosystems around the globe. This occurs when ballast water, used to provide stability, is pumped from ships when they take on more oil. Ironically, legislation mandating separate ballast and oil chambers to prevent oil discharge into the oceans has actually enhanced shipboard conditions for the survival of invasive species.

(a)

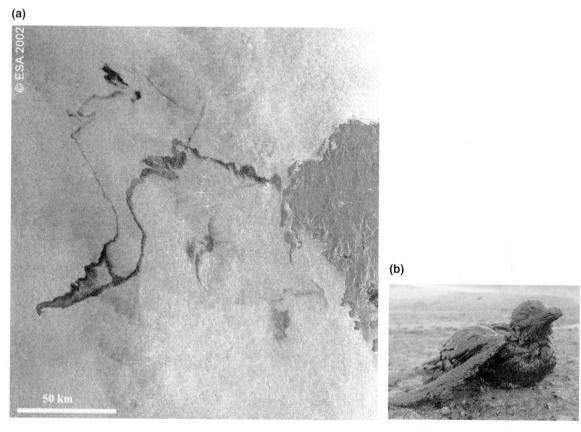

(b)

Figure 6.18
(a) Satellite image of oil spreading into the Atlantic Ocean from a
stricken oil tanker off the Spanish coast in 2002. (b) Oil-soaked bird
on a Spanish beach.

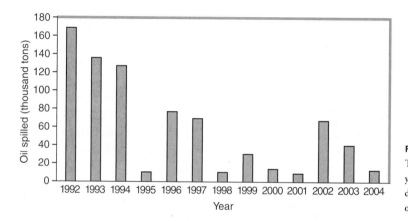

Figure 6.19
The total quantity of spilled oil varies from
year to year, but there has been a general
downward trend despite greater quantities of
oil being transported at sea.

Many coastal ecosystems around the world have been transformed dramatically in recent decades as invasive exotic species out-compete native organisms.

On land, crude oil is transported by pipeline to refineries, where it's processed into a wide variety of fuels, lubricants, and other products. Refineries themselves are major pollution sources, as I'll describe in the next section. Liquid and gaseous fuels leave refineries by pipeline, truck, train, or ship. We hear regularly of small-scale accidents involving tanker trucks or rail cars; both the flammability and environmental toxicity of petroleum fuels make these serious incidents that too often result in loss of life. Pipelines can rupture or leak, causing spills or, worse, explosions. Typical of a large-scale pipeline accident is the rupture of a 34-inch pipeline near Cohasset, Minnesota, in 2002. This event spilled a quarter million gallons of oil into a marsh and resulted in economic losses of nearly $6 million. Investigators determined the cause of this accident to be a crack that developed when the pipeline was improperly loaded for transport to its installation site. Smaller-scale incidents occur frequently when excavating equipment accidentally hits buried pipelines. An event in Wilmington, Delaware, in 2003 is typical: An excavator struck a small natural gas line with a backhoe. The line didn't break at the site of the incident, but in the basement of a nearby building. Gas accumulated, triggering an explosion that destroyed two residences and resulted in fourteen injuries. In some impoverished countries, pipeline leaks are intentional and can have disastrous consequences. In 1998 in Nigeria, more than one thousand people died while attempting to scavenge gasoline from a ruptured pipeline, which exploded in an inferno of liquid fire.

Transport of natural gas is especially tricky. Unlike oil, it can't be poured into ships, tanker cars, or trucks. Gas is so diffuse that it must be compressed substantially for economical transport. This process takes energy, and it results in dangers associated with high pressure. Pressure exacerbates any tendency to leakage, which is an explosive danger in confined spaces and a climate-change worry when natural gas escapes to the open atmosphere. Once a purely domestic product, natural gas is now transported across the open ocean in special tanker ships that can maintain the gas in liquid form at very low temperatures (about $-160°C$, or $-260°F$). This liquefied natural gas (LNG) has a volume only 1/600 of its gaseous state. LNG tankers pack a large amount of energy (see Exercise 4), and although LNG transport has an enviable safety record, there's plenty of worry among communities near LNG docking centers about the impact of an explosion or a terrorist attack on LNG tankers or storage facilities.

Coal, being a solid, presents less serious transportation challenges. The dominant mode of coal transport is by so-called unit trains of about one hundred cars each, which I introduced in Example 5.2. A typical unit train carries about 10,000 tonnes (10 kilotons, or 10^7 kg) of coal. Coal-train derailments are not infrequent accidents, although spilled coal, unlike oil, rarely causes significant environmental damage. However, every year several hundred people die in collisions between vehicles and trains at grade crossings; given that coal accounts for nearly one-third of all train car loadings, we can infer that roughly one hundred people die each year in the United States from grade-crossing accidents involving coal trains.

I mention some of these seemingly mundane consequences of fossil fuel transport because the outcomes—in terms of environmental damage or loss of human life—are often greater than other things we tend to worry much more about. You may be anxious about terrorism or nuclear power accidents or airplane crashes, but do you worry about being hit by a coal train? Yet the number of people killed in coal-train accidents—just one tiny aspect among many death-dealing consequences of fossil fuel use—likely exceeds all deaths from some more widely publicized dangers, such as nuclear power.

Fossil Fuel Processing

Fossil fuels don't come from the mine or well ready to burn. Even coal needs washing, crushing, sizing, and other processing, while crude oil undergoes extensive refining. Among industrial polluters in the United States, refineries emit the greatest amount of volatile organic compounds—chemicals such as the carcinogen benzene and the neurotoxin xylene—which are harmful in their own right and also contribute to smog production. Refineries are the second-largest industrial source of sulfur dioxide emissions and are third in industrial production of nitrogen oxides. A typical refinery produces some 10,000 gallons of waste each day in normal operation; overall, refineries emit some 10 percent of all pollutants required to be reported to the federal government, despite their constituting less than 1 percent of the industries required to report emissions. Anyone who has driven through the oil-refining regions of New Jersey or Texas has experienced first-hand the degradation in air quality that accompanies large-scale refining operations.

Thermal Pollution

I've emphasized that the second law of thermodynamics requires that much of the energy released in burning fossil fuels remains as low-grade heat. Somehow this energy must be disposed of, either by using it for heating or simply dumping it to the environment. The waste heat from large-scale electric power plants presents special problems, since it's enough to alter substantially the temperature of the surrounding environment. All power plants require enormous volumes of water to cool and condense the steam from their turbines; this is where most of the second-law waste heat is extracted. Power plants are almost always built on rivers, lakes, or seashores for a ready supply of cooling water.

Simply dumping the now-heated water back into its natural source usually won't do; this constitutes thermal pollution, whose associated temperature rise can substantially alter the local ecology (although sometimes this effect can be put to advantage, encouraging the growth of commercially desirable aquatic species). Most of the time the water is first run through an air-based cooling system that drops its temperature significantly, effectively transferring the power plant's waste heat to the atmosphere. The huge concrete cooling towers at large power plants accomplish this transfer (recall Fig. 4.14). In so-called **wet cooling towers**, cooling water comes into direct contact with air, resulting in evaporative cooling (ultimately associated with the latent heat discussed in Chapter 4). The evaporated water often recondenses to form clouds of steam rising from the cooling towers. **Dry cooling towers** are used where water is scarce. In these, water remains in sealed pipes while air blows past to extract the heat without evaporating any water. In arid climates, power plants may use closed cooling systems in which water is continuously cycled without the need for a large body of water. Even where water is available, some plants use open-cycle cooling in the winter, when ample cold water is available, and closed-cycle cooling in the drier, hotter summer months. Exercise 15 illustrates the magnitude of the cooling-water problem faced by large power plants.

6.4 Policy Issue: The Clean Air Act

As this chapter clearly suggests, air pollution is (aside from climate change) probably the greatest single environmental impact of the world's voracious appetite for fossil fuels. In the United States, though, despite several decades' growth in fossil fuel consumption, air quality is generally better than it was in the 1970s. Largely responsible for that welcome environmental news is the set of **Clean Air Act Amendments** of 1970, 1977, and 1990. These amendments substantially strengthened the 1967 Air Quality Act, which itself was a successor to the 1963 Clean Air Act and the 1965 Motor Vehicle Pollution Act. Still earlier, the 1955 Air Pollution Control Act spelled out the first nationwide air quality policy. Individual cities got into the act much earlier, with Chicago and Cincinnati

Table 6.1 National Ambient Air-Quality Standards

Pollutant	Primary standards	Averaging times	Secondary standards
Carbon monoxide	9 ppm (10 mg/m^3) 35 ppm (40 mg/m^3)	8 hours 1 hour	None None
Lead	1.5 μg/m^3	Quarterly average	Same as primary
Nitrogen dioxide	0.053 ppm (100 μg/m^3)	Annual	Same as primary

Particulate matter
  10 μm
  2.5 μm
	150 μg/m^3
	15.0 μg/m^3
	35 μg/m^3
		24 hours
		Annual
		24 hours
			Same as primary

Ozone	0.08 ppm 0.12 ppm	8 hours 1 hour	Same as primary Same as primary
Sulfur oxides	0.03 ppm 0.14 ppm —	Annual 24 hours 3 hours	— — 0.5 ppm (1,300 μg/m^3)

enacting pollution regulations in the late nineteenth century. And as early as 1306 in London, England's King Edward I banned the burning of so-called sea coal, a particularly dirty-burning coal found in coastal outcroppings. Many environmentalists would argue that this history of advancing progress in air quality took a backward turn in the early 2000s with the policies of a U.S. administration friendly to the energy industry.

One caveat here: As I emphasized at the beginning of the chapter, by "clean air" I mean air free of the traditional pollutants, not including CO_2. The Clean Air Act and its amendments say nothing about CO_2, and the level of that climate-change agent has risen steadily and substantially during the time of modern air-quality policy.

The Clean Air Act identifies six substances that it calls **criteria pollutants**, and it establishes national standards for their maximum concentrations in ambient air. These **National Ambient Air-Quality Standards** (NAAQS) include **primary standards**, designed to protect human health, and **secondary standards**, associated with more general welfare that includes environmental quality and protection of property. Note the emphasis on standards for air quality: The NAAQS do not directly address emissions of pollutants, but rather their concentrations in the ambient air. This leaves states and municipalities somewhat free to choose policies that ensure their air meets the national standards. In addition, however, the Clean Air Act Amendments require the EPA to establish emissions standards for a number of pollutants and their sources.

Table 6.2 Emissions Standards

Source	Pollutant	Emissions standard	Comments
Coal-fired power plant	SO_2	516 grams/gigajoule (g/GJ) fuel energy	
Oil-fired power plant	SO_2	86 g/GJ	
Gas-fired power plant	NO_x	86 g/GJ	
All power plants	PM	13 g/GJ	PM = particulate matter
Cars	CO	3.4 grams/mile (g/mi)	
	NO_x	0.4 g/mi	Tier 1 1994; must meet standard for first 50,000 miles
		0.07 g/mi	Tier 2 fleet average 2004–2009
	PM	0.08 g/mi	Tier 1
		0.01 g/mi	Tier 2
SUVs, light trucks	CO	4.4 g/mi Tier 1	Vehicle category LDT2 (light duty truck 2): medium-weight SUVs and light trucks
		3.4 g/mi Tier 2	
	NO_x	0.7 g/mi	Tier 1
		0.07 g/mi	Tier 2 fleet average
	PM	0.08 g/mi	Tier 1
		0.01–0.02 g/mi	Tier 2

Table 6.1 shows the six criteria pollutants and their ambient air-quality standards, both primary and secondary (seven pollutants are listed because the NAAQS distinguishes two size ranges for particulate matter). All are substances I've addressed to some extent in this chapter. Note that each standard has an associated averaging time, or in some cases two times, which means a pollutant level can exceed its standard for short periods as long as the average over the specified time meets the standard.

Table 6.1 specifies the acceptable quality of the ambient air; it says nothing about emissions restrictions to achieve that quality. Table 6.2, in contrast, shows some individual emissions standards for a variety of pollutants and sources. These are drawn from separate regulations, including New Source Performance Standards for power plants and Vehicle Emissions Standards for motor vehicles. Emissions standards are not set in stone, but evolve over time. Figure 6.20, for example, shows the evolution of vehicular emissions standards for nitrogen oxides from the 1970s. These standards progress through several levels, called *tiers*. The new Tier 2 vehicle emissions standards, phasing in through the first decade of the twenty-first century, are complex and allow vehicle manufacturers some flexibility in emissions among their different vehicles.

So how have we done? Have these complex regulations and standards improved air quality? The answer is an unequivocal yes. Figure 6.21 shows that emissions of the criteria pollutants have declined substantially since the 1970s, despite rising energy use, vehicle miles traveled, and GDP. However, most of this reduction occurred before 1995; wringing further emissions reductions from technological advances in the face of rising energy consumption and vehicle use is proving more difficult. And remember that emissions reductions are only a means to the goal of improving ambient air quality. There,

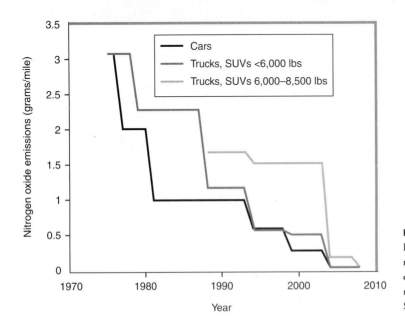

Figure 6.20
EPA standards for NO$_x$ emissions from motor vehicles. The tightening of NO$_x$ and other emissions standards over time is largely responsible for cleaner air in the United States.

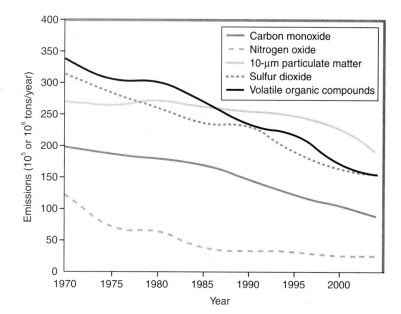

Figure 6.21
Pollutant emissions have declined significantly under the U.S. Clean Air Act. The curves show nationwide total emissions, in millions of tons per year for CO and in 100,000-ton units for other pollutants.

Figure 6.22
U.S. counties not meeting EPA air quality standards for 8-hour average levels of ozone, as of April 2005. Although the area covered is small, these counties are home to more than half of the U.S. population.

too, we've seen substantial improvement, but as Fig. 6.22 suggests, much of the U.S. population still breathes air that fails to meet national standards.

From health effects of air pollution to oil spills, from strip mining to coal-train accidents, the impacts of fossil fuels are vast and often environmentally deleterious. Yet the world will continue to rely on fossil fuels for the major share of its energy for at least several decades to come. And the biggest single emission from fossil fuel consumption—CO_2—has consequences we haven't even begun to discuss. Before we go there, we'll spend the next five chapters exploring alternatives to fossil fuels.

Chapter 6 Chapter Review

BIG IDEAS

6.1   Pollutants are toxic by-products of combustion, removable in principle by engineering the fuel, the combustion process, or the treatment of combustion products. Carbon dioxide, in contrast, is an essential product of fossil fuel combustion.

6.2   Air **pollution** is the most serious traditional impact of fossil fuel combustion. Air pollutants include particulate matter, sulfur compounds, carbon monoxide, nitrogen oxides, mercury and other heavy metals, and even radiation. Air pollutants under the influence of sunlight produce **photochemical smog**. Technologies exist to reduce air pollution significantly, but they're expensive and they consume energy. Air pollution from fossil fuels continues to have a major impact on human health.

6.3   Other impacts of fossil fuels include environmental degradation during extraction, refining, and transportation of fuels. Discharge of waste heat from power plants constitutes thermal pollution.

6.4   In the United States, the **Clean Air Act** and its amendments have improved air quality substantially. Nevertheless, pollution from fossil fuels continues to be a major problem both in the United States and globally.

TERMS TO KNOW

acid mine drainage (p. 161)	criteria pollutant (p. 169)
acid rain (p. 150)	cyclone (p. 147)
baghouse (p. 147)	dry cooling tower (p. 168)
catalytic converter (p. 154)	electrostatic precipitator (p. 147)
Clean Air Act (p. 152)	filter (p. 146)
Clean Air Act Amendments (p. 168)	flue gas desulfurization (p. 150)

fly ash (p. 148)
inversion (p. 155)
lapse rate (p. 156)
mountaintop removal (p. 160)
National Ambient Air-Quality Standards
   (p. 169)
ozone (p. 157)
particulate pollution (p. 145)
photochemical reaction (p. 155)

photochemical smog (p. 155)
pH scale (p. 150)
pollution (p. 144)
primary standards (p. 169)
scrubbing (p. 150)
secondary standards (p. 169)
strip mining (p. 160)
sulfate aerosol (p. 150)
wet cooling tower (p. 167)

GETTING QUANTITATIVE

Particulate size that has an impact on human health: <10 μm

Premature deaths caused by coal pollution in the United States: $\sim$24,000 per year

Acid rain formation: $2SO_2 + O_2 \rightarrow 2SO_3$, $SO_3 + H_2O \rightarrow H_2SO_4$

Neutral pH: 7

QUESTIONS

1. A gallon of gasoline weighs about 6 pounds. Yet combustion of that gallon yields nearly 20 pounds of CO_2. How is that possible?

2. Strip mining obviously presents serious environmental challenges. Yet strip-mined coal from the American West is in demand partly because it's less environmentally damaging in one aspect than is eastern coal. What aspect is that?

3. Why is CO harmful to human health?

4. Carbon dioxide is heavier than air because its triatomic molecules are more massive than the O_2 and N_2 molecules that dominate the atmosphere. Although CO_2 is nontoxic, people have been killed in accidents or natural disasters that release large quantities of CO_2. How is this possible?

5. Explain the difference between primary and secondary air quality standards.

6. Acid rain is mostly a problem in the northeastern United States and northern Europe. Why?

7. What is the purpose of an automobile's catalytic converter?

8. What coal pollutant is generally removed by scrubbing? Why won't filters work on this pollutant?

9. Most constituents of smog aren't emitted directly in vehicle exhaust, and yet vehicle emissions are largely responsible for smog. Explain.

10. Why do coal-burning power plants emit radiation?

11. Switching from coal to natural gas for electrical energy production lowers CO_2 emissions per unit of electrical energy for what two distinct reasons?

12. Figure 6.22 shows that only a small area of the United States fails to meet the National Ambient Air-Quality Standards for ozone. Why is the impact on population more significant than a quick glance at the figure might suggest?

EXERCISES

1. Estimate the amount of CO_2 released in burning 1 gallon of gasoline. Give your answer in both kilograms and pounds. The density of gasoline is about 730 kg/m^3, and gasoline is about 84 percent carbon by weight.

2. Choose the automatic-transmission gasoline Volkswagen Beetle discussed in Box 6.1 and your fuel mileage drops to 26 miles per gallon. Suppose you buy the manual-transmission diesel and your friend buys the automatic-transmission gasoline model. If you each drive 10,000 miles per year, how do (a) your annual fuel consumption and (b) your annual CO_2 emissions compare? You'll find the answer to Exercise 1 useful.

3. In the section of this chapter about the transportation of fossil fuels, I pointed out that humankind uses about 400 EJ of energy each year, nearly all of it from fossil fuels, and that the energy content of fossil fuels is around 30 to 60 MJ/kg. (a) Using an approximate figure of 40 MJ/kg, estimate the total weight of fossil fuels we burn each year. (b) Assume that each kilogram of fuel has to be moved 1,000 km from its source to where it's burned (more for imported oil, less for some coal), at an average speed of 20 km per hour. Estimate the total weight of fuel that must be in transit at a given time.

4. A large LNG tanker carries 138,000 m^3 of LNG. (a) Use Table 3.3 and the fact that LNG is 600 times denser than the gaseous state to estimate the total energy content of the LNG tanker's cargo. (b) How long would all that natural gas last if it were used to satisfy the total U.S. energy consumption rate of about 10^{12} W?

5. If all the CO in car exhaust were converted to CO_2, what would be the mass of CO_2 emitted for every kilogram of CO?

6. How much uranium is in each tonne of North Dakota lignite coal, which contains a whopping 1,800 parts per million of uranium?

7. Manual-transmission cars typically average 1 to 2 miles per gallon better than their automatic-transmission counterparts (although the gap has narrowed considerably thanks to computer-controlled transmissions and engines). Assume there are about 150 million cars in the United States, each driven about 12,000 miles per year, and that 90 percent are automatics. How much gasoline would we save each year if all drivers of automatics switched to comparable manual-transmission cars? Assume an average mileage of 25 miles

per gallon for manual transmissions and 23 miles per gallon for automatics. Express your answer in (a) gallons and (b) oil tanker capacity of 1.48 million barrels, which was the capacity of the Exxon *Valdez*.

8. Repeat the calculation used in Example 6.1 for a scrubber using magnesium carbonate ($MgCO_3$) and forming magnesium sulfate ($MgSO_4$); your answer should be the daily production of $MgSO_4$.

9. If a coal-burning power plant with an electric power output of 1 GWe and an efficiency of 33 percent were to emit the maximum amount of SO_2 allowed as listed in Table 6.2, how much SO_2 would it emit in a year? Note that the energy listed in Table 6.2 refers to the primary energy in the coal, not the electrical output.

10. A 500-MWe gas-fired power plant with 48 percent efficiency emits 2.4 tonnes of particulate matter each day. Is it in compliance with the particulate-matter standard listed in Table 6.2? Justify your answer quantitatively.

11. If you drive your car 12,000 miles each year, what's the maximum amount of particulate matter you could emit over the year and still meet the standards of Table 6.2? Answer separately for the Tier 1 and Tier 2 standards.

12. If you bought an SUV subject to Tier-1 standards instead of a Tier-1 car, how much additional NO_x would you be responsible for annually if each vehicle emitted the maximum allowed in Table 6.2? Assume you drive 12,000 miles per year.

13. By what factor does the concentration of hydrogen ions in rainwater change when the pH drops from 5 to 4.5?

14. A city occupies a circular area 15 km in diameter. If the city's air just barely meets Table 6.1's 24-hour standards for particulate matter of all sizes, what is the total mass of particulate matter suspended in the air above the city to an altitude of 300 m, about the height of the tallest buildings? Assume the particulate matter is uniformly distributed over this volume.

15. During winter, the Vermont Yankee nuclear power plant draws cooling water from the Connecticut River at the rate of 360,000 gallons per minute. The plant produces 650 MW of electric power and has an efficiency of 34 percent. (a) What is the rate at which the plant discharges waste heat? (b) By how much does the temperature of the cooling water rise as a result of the waste heat it absorbs? You may want to review Section 4.4 on specific heat before tackling this exercise.

16. Ship transportation requires approximately 300 kJ of fuel energy per kilometer for each tonne transported. Consider a supertanker so big it can't fit through the Suez Canal; it carries Middle Eastern oil to the United States on a 21,000-km route around the Cape of Good Hope. Calculate the total fuel energy required for each tonne of oil transported, and compare your result with the energy contained in a tonne of oil.

RESEARCH PROBLEMS

1. Check the EPA web site to find the fuel economy (city and highway), air pollution score, and greenhouse gas score for your or your family's car, SUV, or light truck.

2. Find a source of daily air-quality reports for your community (or the nearest big city for which such reports are available). Keep a week-long record of either overall air quality or the concentration of a particular pollutant, if quantitative data on the latter are available.

3. Look up the most recent large oil spill anywhere in the world. What were its environmental and economic consequences?

4. Determine the five most mercury-contaminated species of fish sold commercially in your country, and rank them in order by degree of contamination.

5. Describe any clean-air legislation currently before the U.S. Congress, your state legislature, or, if you're not from the U.S., before your national or regional government.

6. Has the number of air-quality alerts for your community (or the nearest big city) increased or decreased over the past decade? Back up your answer with a graph or table.

7. The U.S. Clean Air Act requires a New Source Review of all new stationary air pollution sources in the United States before they begin operating. (a) Find out what sources are exempt from the New Source Review. (b) Find out and describe how you, as an ordinary citizen, can have input into the review process.

NUCLEAR ENERGY

Fossil fuels supply nearly 90 percent of humankind's energy. In distant second place is nuclear energy, used almost exclusively to generate electricity (one exception being nuclear-powered ships and submarines, which are strategically important but energetically insignificant). Today, nuclear energy accounts for some 6 percent of the world's primary energy and supplies about 16 percent of its electrical energy. Those figures vary dramatically from country to country; France, for example, gets nearly 80 percent of its electricity from nuclear power plants due to a national policy to pursue energy independence. Although the figure in the United States is much lower, with about 20 percent of total electricity generated from nuclear energy, some regions of the United States depend heavily on nuclear power. Figure 7.1 shows the fraction of nuclear-generated electricity in a number of countries and individual U.S. states. Although the overall percentage of U.S. electricity generated from nuclear energy is relatively low, Figure 7.2 shows that the United States is nevertheless the leader in nuclear power generation capability.

Nuclear energy is one of only two technologically proven nonfossil energy sources capable of making a significant contribution to the world's energy supply today (the other is hydropower). This is not to denigrate other possible alternatives, about which I'll have much to say in subsequent chapters, but it is to be realistic about technologies that are available to us right now and that are known to work at the scales needed to meet humankind's prodigious energy demands. Despite continuing controversy over nuclear energy, its demonstrated technological feasibility means that no serious look at the world's energy prospects should ignore the nuclear option.

Understanding nuclear energy means, first and foremost, understanding the atomic nucleus. Understanding humankind's use of nuclear energy means understanding the technologies that allow us to control reactions within and among nuclei. This chapter deals with both topics, but before we start, I want to make a point about terminology: What I'm talking about in this chapter is *nuclear energy*. You won't see the phrase *atomic*

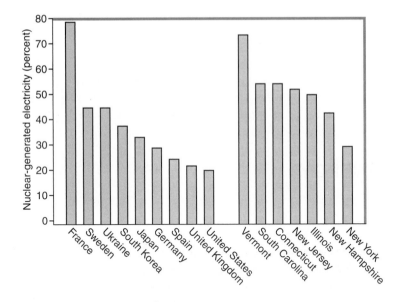

Figure 7.1
Percentage of electricity generated from nuclear power plants in selected countries and in the most nuclear-dependent U.S. states. Because of interstate electricity sales, the U.S. state figures don't necessarily reflect the fraction of nuclear electricity consumed within the state.

energy, because that term, although widely associated with things nuclear, is ambiguous. The chemical energy that comes from rearranging *atoms* to form new molecules—for example, by burning fossil fuels—has just as much claim to the term "atomic energy" as does the energy we get from splitting atomic nuclei. So I'll be unambiguous and speak of *nuclear* energy, *nuclear* power plants, *nuclear* waste, *nuclear* weapons, and *nuclear* policy.

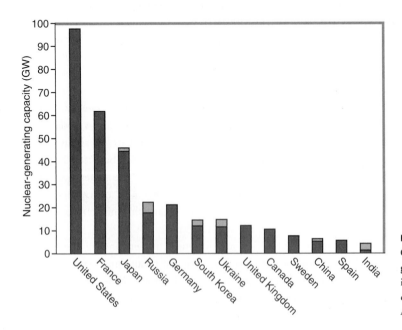

Figure 7.2
Countries with the largest nuclear power generation capability. Gray portion of bars indicates nuclear facilities currently under construction; note that none are in the Americas or Western Europe.

7.1   The Atomic Nucleus

Back in Chapter 3, I introduced the fundamental forces that govern all interactions in the universe—gravity, electromagnetism, and the nuclear force. The chemical energy of fossil fuels is a manifestation of electromagnetism alone, in its incarnation as electric fields and the electric force. As you might guess, nuclear energy involves the nuclear force. But that's not all: Nuclear energy actually depends on a struggle between the nuclear force and the electric force.

Discovery of the Nucleus

By the turn of the twentieth century, scientists understood that atoms—long thought indivisible—actually consisted of smaller components, including the negatively charged electron. In 1909 Ernest Rutherford and his colleagues bombarded a thin foil with high-energy subatomic particles from a radioactive source. Most passed right through the foil, but on rare occasion one bounced back, indicating that the particle had hit something tiny but massive within the foil. Rutherford's group had discovered the **atomic nucleus**, and from their work came the familiar picture of the atom as an electrically positive nucleus surrounded by orbiting electrons, much like a miniature solar system (Fig. 7.3). Although this picture is not fully consistent with modern quantum physics, it's adequate for our purposes here.

By the early 1930s, scientists had established that the nucleus itself consists of two kinds of particles: electrically positive **protons** and electrically neutral **neutrons**. Collectively, protons and neutrons are called **nucleons**. Being positively charged, nuclei attract negatively charged electrons, forming atoms. Atoms are normally neutral, meaning they have the same number of electrons as protons. The outermost electrons determine the chemical properties of an atom, affecting how it bonds with other atoms to make molecules. Since the number of electrons equals the number of protons, the protons ultimately determine an atom's chemical species. Every chemical element has a distinct num-

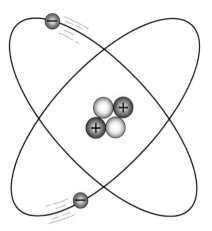

Figure 7.3
"Solar system" model of the helium atom, with a nucleus consisting of two protons and two neutrons. The figure is not to scale; the electrons are typically some 100,000 times more distant than suggested here.

ber of protons in its nucleus, a quantity called the **atomic number** and given the symbol Z. Hydrogen, for example, is atomic number 1 ($Z = 1$); helium is 2; carbon is 6, oxygen is 8, iron is 26, and uranium is 92. The atomic number uniquely identifies the element; for example, the element name *nitrogen* is synonymous with atomic number 7.

Keeping Nuclei Together

What holds the nucleus together? Every nucleus except hydrogen contains more than one proton, and those positively charged particles repel one another because of the electric force. Therefore there must be another, stronger force to hold the nucleus together. This is the nuclear force that I introduced in Chapter 3, and it acts between nucleons without regard for their electric charge. Thus the nuclear force attracts protons and neutrons, neutrons and neutrons, and even protons and protons. Without it there would be no nuclei, no atoms, no you or me.

Incidentally, if you know something about particle physics, you've probably heard of quarks, the sub-subatomic particles that make up protons and neutrons. What I'm calling the *nuclear force* is really a residual effect of the so-called strong force that binds quarks together.

The nuclear force is strong, but its range is very short. Its strength drops exponentially with increasing distance, so the nuclear force becomes negligible over distances a few times the diameter of a single nucleon—about 1 femtometer (fm, or 10^{-15} meter). The electric force decreases with distance, too, but it does so as the inverse square of the distance, which is a lot slower than exponential decrease. This means that protons feel unambiguously bound to nearby neutrons and even protons, but they feel predominantly repulsion from more distant protons. For the nucleus to stick together, nuclear attraction has to overcome electrical repulsion. The neutrons provide the "glue" that makes this possible. Since they offer nuclear attraction without electrical repulsion, the presence of neutrons stabilizes the nucleus against the tendency of electrical repulsion to tear it apart.

Do you know how many neutrons and protons are in the nucleus of carbon? There are six of each, at least in most carbon (more about this caveat in the discussion of isotopes below). How about oxygen? Eight of each. Helium has two protons and two neutrons, and most nitrogen has seven of each. But iron, with 26 protons, has 30 neutrons in most of its nuclei. Iodine has 53 protons and 74 neutrons, whereas uranium has 92 protons and, usually, 146 neutrons. Do you see a pattern here? The larger the nucleus, the greater the proportion of neutrons. There's a good reason for this fact: Larger nuclei have protons that are relatively far apart. They experience electrical repulsion but, because of the short range of the nuclear force, essentially no nuclear attraction. To compensate for this excess repulsion, larger nuclei need more nuclear glue (neutrons) to stick together. By the way, this compensation works only up to a point; for elements with 84 or more protons, electrical repulsion eventually wins out and the nucleus is radioactive, spontaneously decaying into smaller particles.

Isotopes

Most carbon has 6 protons and 6 neutrons; *most* iron has 26 protons and 30 neutrons; *most* uranium has 92 protons and 146 neutrons. Why *most?* First, absolutely *all* carbon nuclei have six protons; this is what determines the number of electrons and hence the chemistry of the element—in other words, this is what it means to be carbon. To be iron is to have a nucleus with 26 protons, and to be uranium is to have a nucleus with 92 protons—period. But the neutrons play no electrical role; adding or subtracting a neutron doesn't affect the number of electrons in the atom and thus doesn't change its chemistry. Carbon with seven neutrons instead of six is still carbon. Oxygen with 10 neutrons is still oxygen. Iron with 28 instead of the usual 30 neutrons is still iron. And uranium with 143 neutrons is still uranium. What I'm describing here are different **isotopes** of the same element—versions of the element with different numbers of neutrons. They're *chemically* identical (except for very subtle changes resulting from their slightly different masses), but their nuclear properties can vary dramatically.

We designate isotopes with a shorthand made up of the chemical symbol for a given element preceded by a subscript number indicating the atomic number and a superscript number giving the total number of nucleons. This superscript number is also called the **mass number,** and it's approximately equal to the nuclear mass in unified atomic mass units (u), being approximately the mass of one proton, or 1.67×10^{-27} kg. For example, the most common form of carbon, which has six neutrons, is written $^{12}_{6}$C. Figure 7.4 shows some isotopes of well-known elements, along with their symbols.

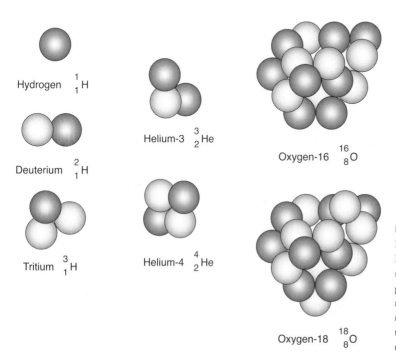

Hydrogen $^{1}_{1}$H

Deuterium $^{2}_{1}$H

Tritium $^{3}_{1}$H

Helium-3 $^{3}_{2}$He

Helium-4 $^{4}_{2}$He

Oxygen-16 $^{16}_{8}$O

Oxygen-18 $^{18}_{8}$O

Figure 7.4
Isotopes of hydrogen, helium, and oxygen. Each isotope is designated by its element symbol (H, He, O), preceded by a subscript giving the number of protons (atomic number) and a superscript giving the total number of nucleons (mass number). Only the hydrogen isotopes are given unique names.

Nature provides a variety of isotopes for each element. Some combinations of neutrons and protons just can't stick together and don't occur. Others do stick together for a while, but are unstable and undergo radioactive decay. In most cases, however, there are a few stable isotopes for each element. Carbon-13, also designated as C-13 or $^{13}_{6}$C, is a stable isotope that comprises 1.11 percent of natural carbon; the rest is C-12. The stable isotope iron-54 ($^{54}_{26}$Fe), with 28 neutrons, comprises about 6 percent of natural iron; most of the rest is Fe-56, with 30 neutrons, but there's also the more rare Fe-57 at about 2 percent and Fe-58 at 0.3 percent. Even the hydrogen nucleus doesn't always have a single proton; one in about every 6,500 hydrogen nuclei is H-2 ($^{2}_{1}$H), also called deuterium (and sometimes given its own symbol, D), with a nucleus consisting of one proton and one neutron. And two uranium isotopes, U-235 and U-238—neither completely stable but both usually holding together over geological time scales—occur with natural abundances of 0.7 percent and 99.3 percent, respectively. That particular difference has enormous consequences for our security in the nuclear age.

Example 7.1   Isotopes and Atomic Mass

The two stable isotopes of chlorine are Cl-35 and Cl-37; they occur naturally in abundances of 75.8 percent and 24.2 percent, respectively. Assuming their atomic masses are very nearly equal to their mass numbers (which is accurate to 0.1 percent), estimate the average atomic mass of natural chlorine.

Solution

We find the average atomic mass by weighting each isotope's mass by its proportion in natural chlorine. Thus the average atomic mass of chlorine is

$$(35)(0.758) + (37)(0.242) = 35.5$$

To three significant figures, this is the value listed for chlorine in the periodic table of the elements. By the way, you'll find many elements in the periodic table whose atomic masses are very close to whole numbers; these typically have one dominant isotope in which the atomic mass is close to but not exactly equal to its mass number. Chlorine is an exception, with two stable isotopes that contribute substantially to the mix of natural chlorine.

7.2 **Energy from the Nucleus**

Drop a heavy rock on hard ground, and you'll make a loud noise and also, less obviously, generate some heat. You've given up gravitational potential energy in favor of kinetic energy as the rock falls, and then the kinetic energy turns into the energy of sound and heat as the rock hits the ground. You'd have to do some work—supply some energy, that is—to lift the rock once again. To put this more abstractly, the formation of (rock + Earth) from separated rock and Earth is a process that releases energy. Once (rock + Earth) forms, you would have to supply that much energy to separate them again. The energy released in forming (rock + Earth) and, equivalently, the energy needed to separate them again, is called **binding energy**, because it describes quantitatively the energy associated with the binding together of different objects, here rock and Earth.

The Curve of Binding Energy

Binding energy works the same way with nuclei. When separate protons and neutrons bind together to make a nucleus, energy is released. For very small nuclei, the individual nucleons don't have many neighbors to exert attractive nuclear forces, and the binding energy isn't particularly large. But as nuclei get larger, the force on any one nucleon grows, and with it the binding energy. In even larger nuclei, the binding energy per nucleon begins to level off because additional nucleons are so far away that the short-range nuclear force doesn't contribute much more to the binding energy. Then at a certain point—at the size of the iron nucleus, in fact—the repulsive effect of the protons begins to tip the balance. As a result, binding energy per nucleon has the greatest value for iron and decreases for heavier elements. This trend is evident in the **curve of binding energy**, a plot of the binding energy per nucleon versus mass number (Fig. 7.5).

The curve of binding energy shows that there are two fundamentally different strategies for extracting energy from atomic nuclei. Just as dropping a rock releases energy,

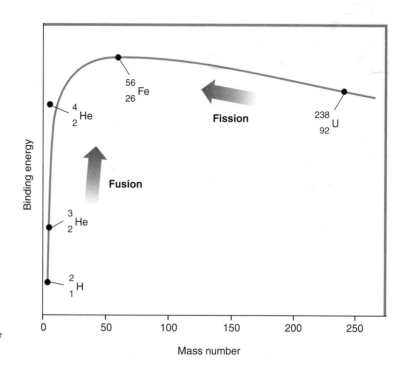

Figure 7.5
The curve of binding energy, a plot of binding energy per nucleon versus mass number. Individual isotopes may lie slightly off the general curve, as with $^4_2$He. Arrows indicate the energy-releasing paths of nuclear fusion and fission.

so too does "dropping" two lighter nuclei together to make a heavier one with greater binding energy. The energy released is equal to the difference in binding energies. I've marked this path to nuclear energy with an arrow on the left-hand side of Figure 7.5. This process of combining nuclei is called **nuclear fusion**, and it's what powers the Sun and other stars. The Sun, specifically, fuses hydrogen to form helium; in more massive stars, fusion makes heavier elements as well. Thus the energy that comes to Earth in the steady flow of sunlight, as well as the energy locked away in fossil fuels, is energy that originated from nuclear fusion deep in the Sun. Fusion is also the dominant energy source in our thermonuclear weapons, or "hydrogen bombs." Unfortunately, we haven't yet learned to control fusion for the steady generation of energy. I'll describe the techno-logical challenges of fusion in Chapter 11.

The curve of binding energy heads downward beyond iron, showing that nuclei much heavier than iron have less binding energy per nucleon than those only a little heavier. Therefore the process of splitting a very massive nucleus into two medium-weight nuclei is also an energy-releasing process. This is **nuclear fission**. Unlike fusion, fission plays almost no role in the natural universe, but it's fission that drives our nuclear power plants and nuclear-powered ships. (A natural fission reaction did occur 2 billion years ago in an area that is now a uranium mine in Africa; it has provided us with valuable infor-mation about the migration of nuclear wastes through the environment.) I've marked an arrow symbolizing fission energy release on the right-hand side of Figure 7.5.

The Nuclear Difference

Chemical reactions involve the rearrangement of atoms into new molecular configura-tions, but the nuclei of the atoms remain unchanged. The energy released in such reac-tions is modest; for example, formation of one CO_2 molecule from carbon and oxygen releases about 6.5×10^{-19} J, or somewhat under 1 attojoule (aJ). In contrast, the energy released in forming a helium nucleus is about 4 million aJ, and splitting a single ura-nium nucleus gives about 30 million aJ. So nuclear reactions provide, in rough terms, something like 10 million times the energy of chemical reactions. I call this the **nuclear difference**, and it's the reason nuclear fuel is such a concentrated energy source. The nuclear difference has a great many practical, and sometimes ominous, implications. For example, it's why we need many 100-car trainloads of coal each week to fuel a coal-burning power plant (recall Example 5-2), while a nuclear plant refuels maybe once a year or so with a single truckload of uranium fuel (Fig. 7.6). It's also the reason why ter-rorists could, in principle, fit a city-destroying bomb in a suitcase. And it's why the vol-ume of waste produced by the nuclear power industry is infinitesimal compared with the tons and tons of carbon dioxide, fly ash, and sulfurous waste produced in burning coal. This is a comment about quantity, not danger; nuclear waste is, on a pound-for-pound basis, a lot more dangerous than the products of fossil fuel combustion. But there are a lot fewer pounds of it—by, again, that factor of roughly 10 million.

Physically, the nuclear difference reflects the relative strengths of the forces binding nuclei and molecules. For nuclei, it's the very strong, short-range nuclear force that holds

Figure 7.6
Unloading fuel at a nuclear power plant. Refueling takes place typically once a year or less and involves only one or two truckloads of fuel. Compare with Figure 5.10, which shows a 110-car coal train arriving at a fossil-fueled power plant, one of the fourteen such fuel deliveries per week. That's the nuclear difference!

nucleons together. Because this force is so strong, rearranging nuclei necessarily involves a lot of energy. Chemical reactions, in contrast, involve the much weaker electrical forces on the outermost atomic electrons. Consequently the energies involved in atomic—that is, chemical—rearrangements are much smaller.

Matter and Energy: $E = mc^2$

Most people think of Einstein's famous equation $E = mc^2$ as the basis of nuclear energy. In a sense that's true, but only in the sense that Einstein's equation applies to *all* energy-releasing processes, from the burning of fossil fuels to the metabolism of food in your body to the snapping of a rubber band to the fissioning of uranium. What $E = mc^2$ expresses is a *universal* equivalence—or, better, interchangeability—between matter (quantified by its mass, m) and energy (E). The proportionality between the two is the square of the speed of light, c. In SI units c is a big number, very nearly 3×10^8 meters per second.

As I pointed out at the beginning of Chapter 3, matter and energy aren't separately conserved; instead, we think of a single universal "stuff," mass-energy, that can manifest itself as either matter or energy. In any process that releases energy, the total mass of the matter involved decreases by an amount given by

$$\Delta m = \frac{E}{c^2} \tag{7.1}$$

This equation follows, obviously, from $E = mc^2$, but I've used Δm instead of m to emphasize that I'm talking about a *change* in mass occurring in a process that releases energy. If you add energy to a system instead of releasing energy, then the mass goes *up* by the amount given in Equation 7.1. Stretch a rubber band, for example, and it weighs a bit more—by the amount E/c^2, where E is the energy you supplied in stretching it. Weigh a gallon of gasoline and also all the oxygen that goes into your car to burn it, and then weigh all the products of combustion, and you'll find—in principle—that the latter weigh less, by the amount given in Equation 7.1. Combine an atom of carbon and a molecule of oxygen to make a CO_2 molecule, and you release some energy in the process; if you weigh the CO_2 molecule, you'll find it weighs less than the carbon and oxygen that went into making it. Weigh two protons and two neutrons and then fuse them together to form a helium nucleus ($^4_2$He); you'll find the helium weighs less than its constituent particles. (This, by the way, is the main reason why the mass of a nucleus isn't given exactly by its mass number.) And if you could weigh the Sun and then weigh it again 1 second later, you'd find its mass has decreased by some 4 million tons as a result of the energy released by nuclear fusion in its core (see Exercise 2). It doesn't matter whether these processes are chemical or nuclear. *Any* process that releases energy results in an equivalent decrease in mass. Here's one more example: Suppose two electric power plants, one nuclear and one fossil-fueled, have exactly the same efficiency and produce the same amount of electric power. They therefore convert matter to energy at exactly the same rate—the fossil-fueled plant no more and no less than the nuclear plant. If you weigh all the fuel that goes into the nuclear plant in a year and all the nuclear waste that comes out, the latter will weigh less. If you weigh all the fuel and oxygen going into the fossil-fueled plant and then all the products of combustion, the latter will weigh less—and the difference will be the same as in the nuclear case.

In the previous paragraph I added the caveat "in principle" when I talked about burning gasoline and weighing its combustion products to find the mass change. I should have added this same caveat to all but the nuclear interactions as well, because chemical reactions convert only a tiny, essentially immeasurable, fraction of the matter involved to energy. Nuclear reactions convert much more—about 10 million times more. This is another way to describe the nuclear difference I introduced in the preceding section. Although those two power plants in the previous paragraph convert matter to energy at the same rate, the fraction of the fuel mass that's converted is much larger for the nuclear plant, which is another way of understanding why the fossil-fueled plant uses so much more fuel. Even so, the energy-producing reactions in our nuclear reactors and in the Sun still convert less than 1 percent of the available matter to energy (see Exercise 3). That's measurable, though, and thus $E = mc^2$ provides us with a way to measure the energy output of nuclear reactions. Examples 7.2 and 7.3 explore this idea.

Example 7.2 Burning Gasoline

Determine the mass converted to energy in burning 1 kg of gasoline.

Solution

Table 3.3 shows that the energy content of gasoline is about 44 MJ/kg. So the mass converted to energy in burning this amount of gas is

$$\Delta m = \frac{E}{c^2} = \frac{44 \times 10^6 \text{ J}}{(3 \times 10^8 \text{ m/s})^2} = 4.9 \times 10^{-10} \text{ kg}$$

That's less than a billionth of the original mass of the gasoline, and it's an even smaller fraction of the total mass of gasoline plus oxygen that went into this chemical reaction.

Example 7.3   Making Helium

The masses of the proton and neutron are both approximately 1.67×10^{-27} kg. The mass of a He-4 nucleus ($^4_2$He) is about 6.64×10^{-27} kg. Estimate the energy released in the formation of a helium nucleus from two protons and two neutrons.

Solution

The mass lost in forming the helium nucleus is the difference between the masses of the constituent particles (two protons, two neutrons) and the helium itself:

$$\Delta m = (2m_p + 2m_n) - m_{He} = (4)(1.67 \times 10^{-27} \text{ kg}) - 6.64 \times 10^{-27} \text{ kg} = 4 \times 10^{-29} \text{ kg}$$

We then convert this to energy using Einstein's equation, here with Δm in place of m:

$$E = \Delta m \, c^2 = (4 \times 10^{-29} \text{ kg})(3 \times 10^8 \text{ m/s})^2 = 4 \times 10^{-12} \text{ J}$$

(I've rounded this result to one significant figure because my rough values for the particle masses left only one significant figure after the subtraction to get Δm.) With 1 aJ = 10^{-18} J, this value is indeed the 4 million aJ I cited in the previous section as being about 10 million times the energy released in forming a CO_2 molecule. By the way, you can easily show that the mass change here is a little less than 1 percent of the total mass involved in this reaction.

7.3   Nuclear Fission

In principle, splitting any massive nucleus should result in energy release. But for most nuclei, splitting doesn't happen readily. The exceptions, which occur among the heavier nuclei, are designated **fissionable**. Some nuclei undergo spontaneous fission, but this process occurs so rarely that the total energy release is insignificant on a macroscopic scale. More productive is **neutron-induced fission**, in which a nucleus captures a neutron and becomes a new isotope. For most nuclei, that's the end of the neutron-capture process (although the new isotope may be unstable and eventually undergo radioactive decay). But with fissionable nuclei, neutron capture can cause the nucleus to split into two middleweight nuclei (Fig. 7.7). These **fission products** fly apart with a great deal of kinetic energy; in fact, most of the approximately 30 picojoules (pJ, or 30×10^{-12} J) released in uranium fission goes into the kinetic energy of the fission fragments.

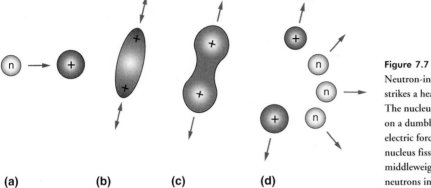

(a)　　　　**(b)**　　　**(c)**　　　　**(d)**

Figure 7.7
Neutron-induced fission. (a) A neutron strikes a heavy nucleus and is absorbed. (b) The nucleus begins to oscillate. (c) It takes on a dumbbell shape, and the repulsive electric force begins to dominate. (d) The nucleus fissions into two unequal middleweight nuclei, emitting several neutrons in the process.

Many heavier nuclei are fissionable, including the common uranium isotope U-238. In most such cases, though, the bombarding neutron must itself deliver a large amount of energy, making fission difficult to achieve. A much smaller group of nuclei undergo fission with neutrons of arbitrarily low energy. Principal among these **fissile isotopes** are U-233, U-235, and plutonium-239 (Pu-239). Of these, only U-235 occurs in significant abundance in nature, and it comprises just 0.7 percent of natural uranium (the rest is U-238). Uranium-235 is therefore the important ingredient in fuel for nuclear power plants, although U-238 and Pu-239 play significant roles in nuclear energy generation as well.

The Discovery of Fission

The discovery of nuclear fission marks an important moment in human history. In the 1930s, the Italian physicist Enrico Fermi and his colleagues explored nuclear interactions by bombarding uranium with neutrons, in the process creating a host of new radioactive materials. Most were so-called **transuranic** elements, which are heavier than the heaviest naturally occurring element, namely uranium. The German chemist Ida Noddack was the first to suggest that such experiments might also lead to what we now call fission—the splitting of uranium into large chunks. Noddack's suggestion went largely unnoticed, but the neutron-uranium experiments continued.

In 1938, the German chemists Otto Hahn and Fritz Strassmann analyzed the results of their experiments on neutron bombardment of uranium. To their surprise, they found the element barium among the products. Barium's atomic number is only 56, just over half that of uranium's 92. So where did the barium come from? Hahn and Strassmann's former colleague, the physicist Lise Meitner, found the answer in December 1938. Now in Sweden after fleeing from Nazi Germany, Meitner and her nephew, Otto Frisch, discussed the uranium experiments during a walk in the Swedish countryside. Meitner drew a sketch similar to Figure 7.7 that suggested how uranium might split into two middleweight fragments, one of which could be Hahn and Strassmann's barium. Meitner and Frisch calculated that the energy released in the process would be tens of millions of times that of chemical reactions—a recognition of the nuclear difference. In a paper describing their work, Meitner and Frisch were the first to use the word *fission* in the nuclear context.

Fission's discovery came on the eve of World War II and involved scientists on both sides of Europe's great divide. The possible wartime use of fission was obvious, and within a few years both sides had active and highly secret fission research programs. In the United States, where a stream of scientists fleeing fascism contributed much to the effort, early success came when a group under Fermi achieved the first sustained fission reaction in a reactor built under the stadium stands at the University of Chicago. Detonation of the first fission bombs came in 1945, with the test at the Trinity site in New Mexico, followed only weeks later by the bombing of the Japanese cities Hiroshima and Nagasaki. Since then the destructive potential of fission bombs and even more advanced nuclear weapons has been a guiding force in international affairs and an ongoing threat to humankind's very existence.

The Chain Reaction

Fission of a U-235 nucleus produces more than fission products and energy. As Figure 7.7 shows, several neutrons also emerge in the process, typically two or three. And neutrons induce U-235 fission in the first place, so each fission event in U-235 releases neutrons that can go on to cause more fission. The result is a nuclear **chain reaction**, a self-sustaining process that may continue until there's no U-235 left to fission (Fig. 7.8).

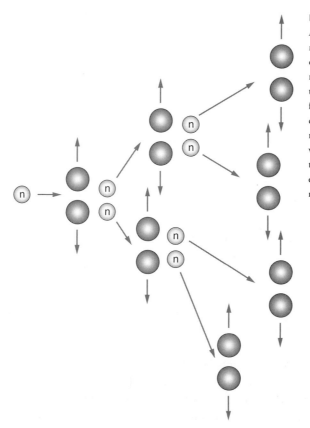

Figure 7.8
A nuclear chain reaction. At left, a neutron strikes a U-235 nucleus, causing it to fission and, in this case, release two neutrons that go on to cause two additional fissions. Each of those fission events releases neutrons that cause more fission, and the chain reaction grows exponentially. This is what happens in a bomb. In a reactor, the neutrons are carefully controlled to ensure that, on average, each fission results in only one additional fission.

Imagine starting with a lump of uranium containing at least some of the fissile isotope U-235. Sooner or later, a U-235 nucleus fissions spontaneously, releasing two or three neutrons. If at least one of those neutrons strikes another U-235 nucleus, the reaction continues. But that's a big "if," for several reasons. First of all, most of the nuclei in natural uranium—99.3 percent—are nonfissile U-238. So a neutron striking U-238 is "lost" to the chain reaction (although not necessarily forever; more on this in the section "Breeder Reactors" on p. 195). Other substances in or around the nuclear fuel may also absorb neutrons and thus squelch the chain reaction. And if the volume of nuclear fuel is small, some neutrons escape altogether rather than encounter a nucleus. For this last reason there's a **critical mass** of fissile material required to sustain a chain reaction. The exact value of the critical mass depends on the geometrical shape of the material and on the concentration of fissile isotopes. For pure U-235, the critical mass is about 30 pounds; for Pu-239 it's 5 pounds or so.

Those numbers are frighteningly small. Consider a chain reaction like the one shown in Figure 7.8 in which, on average, two fission-released neutrons go on to cause additional fission. Soon after the initial fission event, two fissions occur; soon after that 4, then 8, then 16—the fission rate grows exponentially. In a mass of pure U-235, the time between successive fissions is about 10 nanoseconds (10 billionths of a second). At that rate, a 30-pound mass would fission in a microsecond, releasing about 10^{15} J of energy in the process. This is the energy contained in 30,000 tonnes of coal (three of those 100-car trains; see Exercise 5). What I've described, of course, is a nuclear fission bomb, and you can see why the relatively small size of a critical mass is so frightening in today's world.

Example 7.4　Critical Mass!

Using the value of 30 pJ per fission event in U-235, confirm that fissioning of a 30-pound critical mass releases about 10^{15} J.

Solution

A mass of 30 pounds corresponds to (30 lb)/(2.2 kg/lb) = 13.6 kg. The mass of a U-235 nucleus is approximately 235 u, or (235 u)(1.67 × 10^{-27} kg/u) = 3.92 × 10^{-25} kg. So our critical mass contains

$$\frac{13.6 \text{ kg}}{3.92 \times 10^{-25} \text{ kg}} = 3.47 \times 10^{25} \text{ U-235 nuclei}$$

At 30 pJ (30 × 10^{-12} J) per fission, the total energy release is

$$(3.47 \times 10^{25} \text{ nuclei})(30 \times 10^{-12} \text{ J/nucleus}) = 1.04 \times 10^{15} \text{ J}$$

By the way, that's just about the equivalent of 250,000 tons of chemical explosive, making this a 250-kiloton bomb—some twenty-five times more powerful than the Hiroshima bomb, which actually fissioned only about 1 kg of its 100 kg of uranium.

Controlling the Chain Reaction

To produce useful energy from nuclear fission, we obviously need to control the chain reaction. Control here has a very specific meaning: On average, exactly one of those neutrons released by each fissioning nucleus should go on to induce another fission. Less than one, and the reaction fizzles out. More than one, and the reaction grows explosively, although perhaps not as fast as in a bomb. The all-important quantity here is the average number of neutrons from each fission event that cause additional fission; this is called the **multiplication factor**.

Nuclear reactors for power generation use a variety of approaches to controlling the chain reaction. We'll explore the technological details in the next section; here I present the basic principles behind reactor control. In designing a reactor, a nuclear engineer has two goals: first, to ensure that there are enough neutrons to sustain a chain reaction, and second, to ensure that the reaction rate doesn't start growing exponentially. In other words, the goal is to keep the multiplication factor at 1, not more and not less. Because the reaction rate varies with factors such as temperature, the age of the nuclear fuel, and the presence of contaminants, a nuclear reactor needs active systems that can respond to such changes by keeping the multiplication factor at exactly 1.

It turns out that low-energy, slow-moving neutrons are far more likely to induce fission in U-235 than the fast neutrons that emerge from fissioning uranium. This is not an issue in a bomb, where fissile uranium is so tightly packed that fast neutrons can still do the job. But it does mean that, in most nuclear reactor designs, it's necessary to slow the neutrons so they can be effective in causing more fission. The substance that slows neutrons is called the **moderator**. This term can be a bit misleading; you might think it means something that "moderates" or limits the chain reaction, but the opposite is true: The moderator slows down the neutrons, helping to sustain the chain reaction. Without a moderator, the chain reaction in most nuclear reactors would stop.

What to use for a moderator? The most effective way to slow something down is to have it collide with an object of similar mass. (You've learned this if you've had introductory physics or if you've played pool: If one ball hits another head-on, the first ball stops and transfers all its energy to the second ball.) Something with very nearly the same mass as the neutron is the proton. There are protons galore in water (H_2O), since the nucleus of ordinary hydrogen (^1_1H) is just a single proton. So water makes an excellent moderator. However, ordinary water has an undesirable property, namely that hydrogen absorbs neutrons readily, turning into the heavier hydrogen isotope deuterium (^2_1H, or D). A neutron absorbed is a neutron that can't cause fission. This problem, coupled with the absorption of neutrons by the common but nonfissile uranium isotope U-238, makes it impossible to sustain a chain reaction in natural uranium with ordinary water as the moderator. Fuel for such reactors must be enriched in U-235, from that isotope's natural abundance of 0.7 percent up to about 4 percent. Such **enrichment** is expensive, and more important, it's dangerous: Once a nation or other group has uranium enrichment technology, there's nothing to prevent enrichment to 80 percent or more U-235, which is bomb-grade material.

An alternative to ordinary, so-called light water is **heavy water** ($^2_1H_2O$, or D_2O) containing the hydrogen isotope ^2_1H (D). Deuterium absorbs fewer neutrons, so the reactor fuel can be unenriched natural uranium. But heavy water is rare, and extracting it from ordinary water is an expensive process. A third possible moderator is carbon, in the form of graphite. It has the advantage of being a solid, although carbon's greater mass means it's less effective in slowing neutrons. And graphite is flammable—a significant safety consideration. Overall, there's no obvious choice for the ideal moderator. In the next section I explain how the choice of moderator figures centrally in the different nuclear reactor designs in use today.

Precise control over the chain reaction is achieved by inserting neutron-absorbing substances into a nuclear reactor. Usually these take the form of **control rods** that drop into the space between uranium fuel elements, thereby absorbing more neutrons and decreasing the reaction rate. Pulling the rods out increases the reaction rate. The whole system can be automated, using feedback, to maintain the multiplication factor at exactly 1 and therefore keep the chain reaction going at a steady rate. In an emergency shutdown, the control rods are inserted fully and the chain reaction quickly stops.

All those fission reactions produce heat, which arises as the high-energy fission products and neutrons collide with the surrounding particles, giving up their energy in what becomes random thermal motion. If that energy isn't removed, the nuclear fuel would soon melt—a disaster of major proportions. More practically, we want to use that energy, usually to generate electricity. So every nuclear power reactor needs a **coolant** to carry away the energy generated by nuclear fission.

The moderator and control rods are the essential elements that control the nuclear chain reaction; the coolant extracts the useful energy. Together, the choice of these three materials largely dictates the design of a nuclear reactor.

7.4 Nuclear Reactors

A nuclear power reactor is a system designed to sustain a fission chain reaction and extract useful energy. In most of today's reactors, the fuel takes the form of long rods containing uranium, usually as uranium oxide (UO_2; Fig. 7.9). The moderator, control rods, and coolant occupy the space between the fuel rods. But there the similarity ends, as the choice of these three components, especially of the moderator, leads to very different reactor designs.

Light-Water Reactors

All power plant reactors in the United States today are **light-water reactors** (LWRs), meaning they use ordinary water as the moderator. The very same water also serves as the coolant. A thick-walled **pressure vessel** contains the nuclear fuel, and the moderator/coolant water circulates among the fuel rods to slow the neutrons and carry away the thermal energy. The simplest LWR design is the **boiling-water reactor** (BWR), in which water in the reactor vessel boils to make steam that then drives a turbine connected to

Figure 7.9
Bundles of nuclear fuel rods being lowered into a reactor.

an electric generator, just as in a fossil-fueled power plant. About one-third of the approximately one hundred U.S. power reactors are BWRs. The rest are **pressurized-water reactors** (PWRs), in which the water is kept under such high pressure that it can't boil. Heated water from the reactor flows through a **steam generator**, where it contacts pipes containing water at lower pressure that's free to boil. Steam in this secondary water system then drives a turbine. One advantage of the PWR is that water in the secondary loop never contacts the nuclear fuel, so it doesn't become radioactive. In the BWR design, in contrast, the steam that drives the turbine is mildly radioactive. Figure 7.10 shows what's inside the pressure vessel of a typical LWR.

Light-water reactors are based on designs first used to power submarines in the 1950s. They're fairly straightforward and have one important safety feature: A loss of coolant—among the most serious of possible reactor accidents—also deprives the reactor of its moderator, which causes an immediate shutdown of the chain reaction. This is not quite as promising as it sounds, though, because radioactive decay immediately following the shutdown continues to produce energy at nearly one-tenth the rate of the full-scale fission reaction. But it does mean that the nuclear chain reaction itself stops automatically in the event of coolant loss.

Refueling an LWR is difficult, because it requires shutting down the entire operation and opening the pressure vessel. For that reason, refueling takes place rarely—at

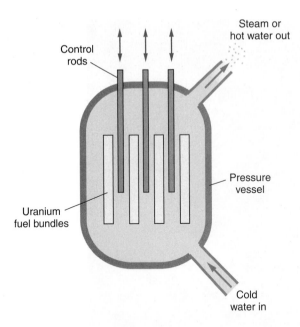

Figure 7.10

Inside the pressure vessel of a nuclear reactor. Fuel bundles are assemblies of individual fuel rods as shown in Figure 7.9. Control rods made of a neutron-absorbing material move up and down to control the chain reaction.

intervals of a year or more—and usually entails weeks of down time (Fig. 7.11). There's a security advantage here, though, because it's difficult to remove fuel from an operating reactor to extract plutonium for weapons.

Other Contemporary Reactor Designs

Canada has chosen its own route to nuclear power. In the current CANDU (for Canadian-deuterium-uranium) reactor design, moderator and coolant are both heavy water, although the two functions are kept separate. The deuterium in heavy water absorbs fewer neutrons, so CANDU reactors use natural, unenriched uranium. That's an advantage as far as weapons-proliferation potential is concerned, but there's a disadvantage: The CANDU design replaces the pressure vessel with coolant-carrying pipes in proximity to the nuclear fuel. This makes for easy refueling, but it could facilitate the illicit removal of nuclear materials. CANDU reactors are used throughout the world and comprise somewhat less than 10 percent of operating reactors. Their continuous refueling capability minimizes down time and makes the modern CANDU the highest-performing commercial reactor design.

Still other reactors use solid graphite as the moderator, with fuel rods and coolant channels interspersed among graphite blocks. This design used to be popular in Russia and other states of the former Soviet Union. Like the CANDU reactor, the Russian design can be refueled continually, and in fact some Soviet power reactors once served the dual purpose of producing both civilian electric power and plutonium for nuclear weapons. Graphite reactors lack the inherent safety of the U.S. light-water design, because loss of coolant doesn't stop the nuclear reaction. In fact, because of the neutron-absorbing prop-

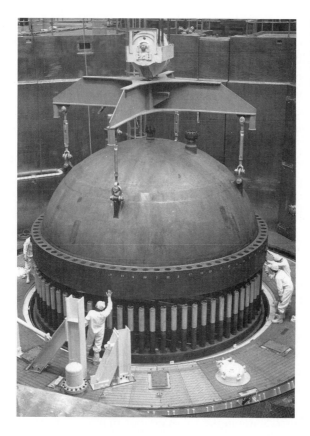

Figure 7.11
Refueling a light-water reactor is a major operation, undertaken once a year or even less frequently. Here workers remove the top of the pressure vessel.

erties of the hydrogen in light water, loss of coolant can actually accelerate the chain reaction. That phenomenon played a significant role in the 1986 Chernobyl accident. So did combustion of the graphite moderator, which burned fiercely and helped spew radioactive materials into the environment.

A variant on the graphite-moderated design is the high-temperature gas-cooled reactor. Here, helium gas flows through channels in a solid graphite moderator to carry away the heat from fissioning uranium. In the United States, a gas-cooled power reactor was in operation in Colorado in the 1970s and 1980s, but its performance was poor and the plant was eventually converted to a natural gas–fired facility.

Breeder Reactors

I hinted earlier (see p. 190) that neutrons striking nonfissile U-238 may not be lost forever. When U-238 absorbs a neutron, it becomes U-239. Through the process of beta decay (described in more detail in the section "Radioactivity" below), a neutron in the U-239 nucleus decays into one proton and one electron. The electron is emitted with great energy, and the appearance of another proton in the nucleus makes a new element, neptunium-239. The Np-239 nucleus undergoes an additional beta decay, becoming

Pu-239. Plutonium-239 is fissile, just like U-235, except with a lower critical mass and other properties that make it more desirable and dangerous as a bomb fuel, although one that's more difficult than uranium to engineer into a bomb.

Conversion of nonfissile U-238 into fissile Pu-239 occurs gradually in a normal nuclear reactor. Plutonium then joins in the fission chain reaction. In a typical reactor, there are an average of about two Pu-238 nuclei formed for every three U-235 fission events. By the time nuclear fuel approaches the end of its useful life in a reactor, most of the energy it generates actually comes from plutonium fission.

It is possible, though, to make a lot more plutonium. A **breeder reactor** is designed to increase plutonium production, actually making more plutonium than the uranium-235 it consumes. Breeder reactors could, in principle, utilize much of the 99.3 percent of natural uranium that is nonfissile U-238. But this "breeding" of new fuel comes at a cost: Breeding works best with fast neutrons and hence no moderator, which means breeders need to be more compact, sustain a more vigorous chain reaction, and operate at higher temperatures than ordinary slow-neutron reactors. In other words, a breeder is a little more like a bomb. No slow-neutron reactor could ever blow up like a bomb (although plenty of other things can still go wrong). That's probably true of breeders as well, but the breeder design entails the risk of more serious nuclear accidents. And the breeder is chemically more dangerous because it uses liquid sodium as its coolant, a substance that burns spontaneously on contact with air.

More significant, an energy economy dependent on breeder reactors is an economy that necessarily traffics in plutonium, which is a serious threat to world security. For that reason, the United States in the 1970s decided to forgo work on breeder reactors. Other nations have gone forward, especially France, although the results have not been entirely satisfactory. Today, only one of France's fifty-nine commercial power reactors is a breeder, and a modest 230-MW one at that; a much more ambitious 1,200-MW unit was shut down in 1998, ostensibly for economic reasons but with political opposition probably playing a role as well.

Example 7.5   Clean Clothes and Bombs

You can either toss your clothes into a 5-kW electric clothes dryer for an hour, or you can hang them on the clothesline to dry. If your electricity comes from a nuclear plant with 33 percent efficiency, how much Pu-239 are you responsible for making if you choose to use the electric dryer?

Solution

That 5-kW dryer uses energy at the rate 5 kJ/s; in an hour it uses (5 kJ/s)(3,600 s) = 18 MJ. This energy comes from a 33 percent efficient nuclear power plant, so back at the plant a total of 3 × 18 MJ = 54 MJ of energy is released from fissioning uranium.

As I noted earlier, each uranium fission releases 30 pJ, so the total number of U-235 nuclei that fission to dry your clothes is

$$\frac{54 \times 10^6 \text{ J}}{30 \times 10^{-12} \text{ J}} = 1.8 \times 10^{18} \text{ U-235 nuclei}$$

Since two Pu-239 nuclei are produced for every three U-235 fission events, your choice to use the electric clothes dryer results in the production of some 1.2×10^{18} Pu-239 nuclei. That's a million trillion plutonium nuclei, and as you can show in Exercise 4, it amounts to about 0.5 mg of Pu-239. Plutonium has a critical mass of about 5 kg, so it would take 10 million households running their dryers for an hour to produce one bomb's worth of plutonium. In the United States, this plutonium ends up as nuclear waste, not bombs, but the example illustrates quantitatively a potential link between nuclear power and nuclear weapons.

Advanced Reactor Designs

Today's reactor designs are hardly the ultimate in efficiency and safety. Already, newer LWRs operating in Japan, Korea, and other countries are probably safer than any of the existing U.S. reactors. That's because these newer designs have had the benefit of several more decades' experience with nuclear power in an era of acute concern for safety and environmental impact. But these new reactors are still only modifications of the basic light-water designs that trace their roots to the nuclear submarine programs of the 1950s. Can we do better?

The U.S. Nuclear Regulatory Commission and its overseas equivalents have recently certified several advanced LWRs. These incorporate a number of passive safety features to make them much less dependent on active systems such as electrically driven pumps that handle accidents involving loss of coolant. The more innovative of these new designs—the 600-MWe Westinghouse AP-600—has an estimated probability of a core-damaging accident that is a factor of 1,000 lower than the Nuclear Regulatory Commission requires. And the new designs are standardized so that essentially identical reactors can be built anywhere. This standardization gives prospective buyers more confidence in construction costs and schedules, and it greatly enhances the economic viability of the new reactor designs.

Advanced light-water designs have been developed elsewhere in the world as well. A new European design will boast the highest efficiency of any commercial nuclear power plant, some 36 percent. Russia has developed enhanced-safety reactors to replace Chernobyl-style plants. And Canada continues to advance its heavy-water-moderated CANDU design with new reactors that are simpler, safer, cheaper to operate, and more efficient. India, which has nearly a third of the world's thorium, is advancing the heavy-water design with a reactor that breeds fissile U-233 from Th-232.

Another innovation is the high-temperature gas-cooled reactor (HTGCR). Although this basic idea was used in early reactors for plutonium production and in a handful of commercial reactor applications, the new HTGCR designs are radically different. South Africa is working on a so-called pebble-bed reactor, in which tiny pellets of uranium fuel are encased in ceramic materials capable of withstanding high temperature. The ceramic serves as a containment system for radioactive fission products. Hundreds of thousands of such "pebbles" fuel a reactor. Gaseous helium carries off the heat, which is transferred to water that drives a conventional steam turbine and generator. Graphite serves as the moderator. A prototype reactor of this sort was operated at maximum power and with no coolant flow—conditions that would have destroyed a LWR—yet no damage occurred. Another gas-cooled reactor, under design in the United States and Russia, would use the heated gas to drive a gas turbine; because of the high temperature, this design's thermal efficiency is close to 50 percent. The first such reactor will literally beat swords into plowshares by "burning" plutonium taken from retired nuclear weapons.

Finally, some innovative proposals combine conventional nuclear fission with high-energy particle accelerators to "burn" a variety of fissionable but not necessarily fissile isotopes. Such reactors could operate with less than a critical mass, since the accelerator provides the extra particles needed to induce fission. This concept has a big safety advantage, since it means that turning off the accelerator immediately stops the reaction. But accelerator-based designs have a long way to go before they prove themselves either technologically or economically viable.

There is no obvious best design for a nuclear power reactor. Some offer greater safety against nuclear accidents, others against different aspects of weapons proliferation. Some are more economical, others offer the promise of nearly unlimited fuel through the breeding of fissile isotopes. All reactor designs have shortcomings, and none is ideal. But neither, of course, is any other energy technology.

Nuclear Power Plants

The reactor is the heart of any nuclear power plant. Its job is to produce thermal energy from fissioning uranium. From there the technology is similar to that of fossil-fueled plants: Thermal energy generates steam, either right in the reactor or in a separate steam loop. High-pressure steam drives a turbine connected to an electric generator. Spent steam is condensed, typically using water from a nearby river, lake, or bay, and the cycle continues. Cooling towers usually dump some of the waste heat to the atmosphere to avoid thermal pollution of surface waters. As always, it's the second law of thermodynamics that limits the overall efficiency. The practical second-law limit for nuclear plants is similar to that for fossil plants, although in U.S. LWRs the need for a single, large, thick-walled pressure vessel has the effect of limiting maximum steam temperature and thus reducing efficiency somewhat from that of a comparable fossil plant. But the similarities between nuclear and fossil fuel plants are striking—so much so that in at least once instance a nuclear power plant under construction was converted into a fossil-fueled plant. Figure 7.12 diagrams a typical nuclear power plant.

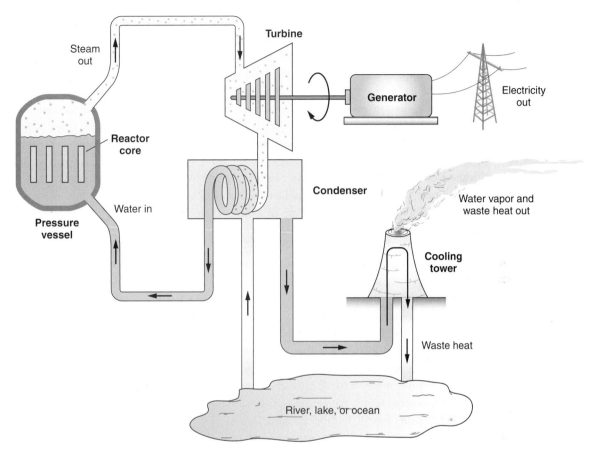

Figure 7.12
A nuclear power plant using a boiling-water reactor. Compare with
the fossil-fueled plant shown in Figure 5.9; the reactor replaces the
boiler, but otherwise they're essentially the same. The reactor core
comprises the nuclear fuel and other components immersed in the
cooling water.

7.5   Nuclear Radiation

The dominant environmental concern with almost every aspect of nuclear power—from
mining to power plant operation to waste disposal—is nuclear radiation. Here I describe
how nuclear radiation originates and explore its properties. The next section deals explicitly with the health and environmental impacts of radiation.

Radioactivity

I've alluded earlier to the fact that some nuclei can't stick together indefinitely, but
undergo **radioactive decay** by spewing out subatomic particles. Scientists making the
first studies of radioactivity, around the turn of the twentieth century, quickly identi-

fied three types of radioactive emissions. They called these **alpha, beta,** and **gamma radiation**. Today we know that alpha radiation consists of ordinary helium nuclei (two protons and two neutrons, ^4_2H) ejected with high energy from larger nuclei, that beta radiation consists of electrons (or their antimatter opposite, called *positrons*), and that gamma radiation is just a high-energy form of electromagnetic radiation.

Alpha radiation results from the mutual repulsion of protons within the nucleus, although a quantum-mechanical effect called *tunneling* helps the alpha particles escape from the nucleus. Most naturally occurring heavier nuclei decay by alpha emission. Alpha particles tend to have the lowest energy of the three types of radioactive emissions, and a sheet of paper, a layer of clothing, or a few centimeters of air readily stop them. A typical alpha-emitting process is the radioactive decay of U-238, yielding Th-234. We write this reaction similarly to the way we describe a chemical reaction:

$$^{238}_{92}U \rightarrow \, ^{234}_{90}Th + \, ^4_2He \text{ (alpha decay)} \qquad (7.2)$$

Notice how this equation balances: The number of positive charges (atomic number, at the bottom of each nuclear symbol) is 92 on the left and $90 + 2 = 92$ on the right. The mass numbers at the top also agree, with 238 on the left and $234 + 4 = 238$ on the right. Remember that the mass number is actually the total number of nucleons; it's only approximately the nuclear mass in unified atomic mass units.

Beta radiation generally occurs when a nucleus has too many neutrons. In a typical beta decay, a neutron in the nucleus decays into a positive proton, a negative electron, and an elusive, nearly massless, neutral particle called a *neutrino*. The electron emerges with high energy (that's the beta radiation), leaving a nucleus with one more proton than before. Thus its atomic number actually increases, although given the small mass of the electron, its mass number is unchanged. A typical beta decay is that of C-14, used in the age-determining procedure called *radiocarbon dating*:

$$^{14}_6C \rightarrow \, ^{14}_7N + \, ^{0}_{-1}e + \bar{\nu} \text{ (beta decay)} \qquad (7.3)$$

Here e is the electron, with the subscript -1 indicating that it carries 1 unit of negative electric charge and the superscript 0 showing that its mass is negligible in comparison with that of the nucleons. The final symbol, $\bar{\nu}$, is the neutrino (actually an antineutrino, which is what the bar indicates). Note again how the numbers add up: The subscripts show 6 units of charge on the left and $7 + (-1) = 6$ units on the right. The masses agree, too, with 14 on each side of the equation. By the way, the nuclear end product of this beta decay is ordinary nitrogen-7, the dominant nitrogen isotope. Beta radiation is somewhat more penetrating than alpha radiation, capable of traversing several feet of air or a fraction of an inch in solids, liquids, and body tissues.

Example 7.6 Making Plutonium

In a nuclear reactor, U-238 absorbs a neutron (^1_0n; no charge but 1 mass unit) to make U-239: $^{238}_{92}U + \, ^1_0n \rightarrow \, ^{239}_{92}U$. The U-239 then undergoes two successive beta decays, like that of C-14 given in Equation 7.3. Write equations for these two decays, and identify the final products.

Solution

Each beta decay increases the atomic number by 1, so we're going to end up with elements 93 and 94. A look at the periodic table shows that these are, respectively, neptunium and plutonium. So the reactions are:

$$^{239}_{92}U \rightarrow \, ^{239}_{93}Np + \, ^{\ 0}_{-1}e + \bar{\nu}$$

and

$$^{239}_{93}Np \rightarrow \, ^{239}_{94}Pu + \, ^{\ 0}_{-1}e + \bar{\nu}$$

The end product is Pu-239, the fissile isotope and potential bomb fuel.

Gamma radiation consists of "bundles" of high-energy electromagnetic radiation. Because it carries no electric charge, gamma radiation can penetrate several feet of concrete or even lead. Gamma radiation occurs when an atomic nucleus is excited to a higher energy state and then drops back down to its so-called ground state. The gamma radiation carries off the excess energy. Gamma radiation has neither mass nor charge so, unlike the alpha and beta processes, gamma decay doesn't change the nucleus.

Half-Life

All radioactive nuclei eventually decay, but when? For any single nucleus, quantum physics rules out giving a definite answer. But we can answer the question statistically: Given any number of identical radioactive nuclei, after a certain time very nearly half of them will have decayed. This time is the **half-life**. Half-lives of radioactive isotopes vary dramatically, from tiny fractions of a second to billions of years. The radioactive species found in nature are either those with very long half-lives, so long they haven't had time to decay away since their creation billions of years ago, or shorter-lived nuclei that result from the continual decay of longer-lived species. For example, U-238 has a half-life of 4.5 billion years, and among its decay products is the radioactive gas radon-222. Radon-222 has a 3.8-day half-life, but it's continually being formed from the decay of radium-226. The latter, in turn, arises after U-238 undergoes first an alpha decay to Th-234, two beta decays to U-234, and then two more alpha decays (see Exercise 7). By the way, natural radon emanating from rocks and soils and leaking into basements can be a significant health hazard that is much greater than anything we face from the ordinary operation of nuclear power plants. Another source of short-lived isotopes in nature are natural nuclear reactions, which on Earth involve cosmic rays—themselves high-energy particles of cosmic origin—impinging on the upper atmosphere. Carbon-14, the radiocarbon dating isotope, forms in this way.

So given a radioactive substance, half of its nuclei will have decayed after 1 half-life. Wait another half-life and half of those remaining nuclei are gone, too, leaving only one-

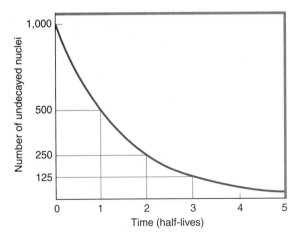

Figure 7.13

Decay of a radioactive sample initially containing 1,000 nuclei.

fourth of the original number. After 3 half-lives, only one-eighth of the original nuclei remain undecayed. Wait 10 half-lives, and only about one-thousandth remain (actually $1/2^{10}$, or $1/1,024$). Wait 20 half-lives, and you're down to about a millionth of the original sample. Figure 7.13 depicts this exponential decay, which has the same mathematics as the exponential growth described in Box 5.2, but with a minus sign.

Table 7.1 shows some important radioactive isotopes, their half-lives, and their scientific, environmental, or health significance. Note that the two naturally occurring uranium isotopes have long but considerably different half-lives. The much shorter half-life of U-235 (700 million years, versus 4.5 billion years for U-238) means that U-235 used to be more abundant. That's why a natural fission reactor was possible 2 billion years ago, but not today. Several of the biologically significant isotopes listed in Table 7.1 are fission products. You can see why these are necessarily radioactive if you remember that more massive nuclei have a greater ratio of neutrons to protons, in order to overcome electrical repulsion. Uranium is one such nucleus, and when it splits, the resulting middleweight nuclei still have uranium's ratio of neutrons to protons. This is too many neutrons for a stable middleweight isotope, so these fission products are necessarily radioactive and, given their relatively short half-lives, much more intensely so than the original uranium.

Example 7.7   Recovering from Nuclear Terrorism

A serious concern in this age of international terrorism is the prospect of a so-called dirty bomb that would use conventional chemical explosives to disperse radioactive material. Fission products taken from spent nuclear reactor fuel would make a particularly dirty weapon. Suppose such a device were made with strontium-90 and ended up spreading Sr-90 into the environment at 100,000 times the level considered to be safe. If the Sr-90 could not be removed physically, how long would the contaminated region remain unsafe?

Table 7.1 Some Important Radioactive Isotopes

Isotope	Half-life (approximate)	Significance
Carbon-14	5,730 years	Used for radiocarbon dating objects up to about 60,000 years old.
Iodine-131	8 days	Fission product released into the environment from nuclear weapons tests and reactor accidents. Lodges in the thyroid gland where it can cause damage or cancer.
Potassium-40	1.25 billion years	Natural isotope comprising 0.012 percent of natural potassium. The dominant radiation source within the human body. Used for dating ancient rocks and to establish Earth's age.
Plutonium-239	24,000 years	Fissile isotope produced in fission reactors from U-238. A nuclear weapons fuel with low critical mass.
Radon-222	3.8 days	Gas formed in the decay of natural radium in rocks and soils, ultimately from uranium-238. A health hazard when it seeps into buildings.
Strontium-90	29 years	Fission product that chemically mimics calcium, so it is absorbed into bone. Still at measurable levels in the environment following aboveground nuclear bomb tests of the mid-twentieth century.
Tritium (hydrogen-3)	12 years	Radioactive hydrogen isotope used to "tag" water and other molecules for biological studies. Strategically important because it can boost the explosive yield of fission weapons.
Uranium-235	704 million years	Fissile isotope comprising 0.7 percent of natural uranium.
Uranium-238	4.5 billion years	Nonfissile isotope incapable of sustaining a chain reaction. Fissions on bombardment with neutrons of very high energy; used to boost yield of thermonuclear weapons. Depleted uranium (after removal of most U-235) is used for armor-penetrating conventional weapons because of its high density. Causes modest radioactive contamination and is chemically toxic.

Solution

In 1 half-life, the level of radioactive material drops to half its original value; in 2 half-lives to one-fourth, and so on—dropping to $1/2^n$ after n half-lives. Here we need the level to drop to 1/100,000 of its original value, so we want n such that $1/2^n = 1/100,000$, or inverting it, $2^n = 100,000$. We can isolate the n in this equation by taking logarithms of both sides:

$$\log(2^n) = \log(100,000)$$

But $(2^n) = n \log(2)$, and $\log(100,000) = \log(10^5) = 5$, using base 10 logarithms. So we have

$$n \log(2) = 5$$

or

$$n = \frac{5}{\log(2)} = \frac{5}{0.301} = 16.6$$

This result (16.6) is the number of half-lives. Table 7.1 shows that Sr-90's half-life is 29 years, so this answer amounts to (29 years/half-life) (16.6 half-lives), or just under 500 years. That's a long time!

By the way, if you're not fluent with logarithms, you might recall that 20 half-lives drop the level of radioactivity by a factor of a million, which is more than we need here. Working backward, 19 half-lives drops the level by a factor of half a million, 18 by a quarter-million, 17 by one-eighth of a million or 125,000—just over what we need. So the answer must be a little under 17 half-lives, as our more rigorous calculation shows.

Measuring Radiation

If you look seriously at the health and environmental impacts of nuclear power, you'll encounter a somewhat bewildering array of units for quantifying nuclear radiation. The most basic is the **becquerel** (Bq), named after Henri Becquerel, who discovered natural radioactivity in 1896. The becquerel simply measures the rate at which a given sample of radioactive material decays, without regard to the type or energy of the radiation. One becquerel is 1 decay per second. An older unit, the **curie**, is defined as 37 billion decays per second and is approximately the radioactivity of 1 gram of radium-226. Activity in becquerels or curies is closely related to half-life. Given equal numbers of atoms of two different radioactive materials, the one with the shorter half-life must have the greater activity because it decays faster.

We're concerned less with radiation itself than with its effect on materials, especially biological tissue. Two additional units address these effects. The **gray** (Gy) is a purely physical unit that describes the energy absorbed when an object is exposed to radiation. One gray corresponds to the absorption of 1 J of radiation energy per kilogram of the exposed object. An older unit still in widespread use is the **rad**, equal to 0.01 Gy.

The biological effects of radiation depend not only on the total energy absorbed but also on the type and energy of the particles comprising the radiation. Yet another radiation unit, the **sievert** (Sv), measures absorbed energy per kilogram, but it's adjusted to account for the biological impacts of different radiation types and energies. Again, you'll often encounter an older unit, the **rem**, equal to 0.01 Sv.

7.6 Environmental and Health Impacts of Nuclear Radiation

Throughout this book I've been including the effects on human health among the environmental impacts of human energy use. In the case of nuclear energy, health effects loom larger than other environmental impacts. This is in part a consequence of the nuclear difference—the 10 million times greater energy content of nuclear over chemical fuels that decreases the sheer volume of material going into and out of nuclear power plants. As a result, nuclear energy doesn't usually result in the more blatant and large-scale forms of environmental degradation that are associated with strip mining of coal, or acid precipitation, or visible diminution of air quality over thousands of square miles. Instead, the high energy associated with nuclear radiation makes for impacts that are most noticeable at the level of biological cells.

Biological Effects of Radiation

The high-energy particles that make up nuclear radiation affect biological systems by breaking, ionizing, or otherwise disrupting the biochemical molecules within living cells. The results include mutations, cancer, and outright cell death. The likelihood of these outcomes, and the overall impact on a living organism, depends on the total dose of radiation.

A handful of nuclear accidents and the nuclear bombings of Hiroshima and Nagasaki at the end of World War II have shown us the effects that large radiation doses have on human beings. A dose of 4 Sv delivered in a relatively short period, for example, kills approximately half the people exposed. Death rates go up with higher exposures, and severe radiation sickness persists after somewhat lower doses.

However, very few people are exposed to such high radiation levels. Neighbors of Pennsylvania's Three Mile Island (TMI) reactor experienced average individual exposures of about 10 microsieverts (μSv), or 1 millirem, during TMI's partial meltdown in 1979, which is the worst commercial nuclear accident to date in the United States. In contrast, the average American is exposed to about 500 μSv each year through X rays and other radiation-based medical procedures. At such low doses it's impossible to ascertain with certainty a direct relationship between radiation dose and health effects. Many radiation standards are based on the idea that the effects of a large radiation exposure extrapolate linearly to low doses. For example, a large dose of 1 Sv to an individual human being is known to double the spontaneous mutation rate, causing twelve extra radiation-induced mutations for every 1,000 births. Using linear extrapolation, this means that TMI's 10-μSv (1/100,000 Sv) exposure would cause twelve mutations for every 1,000 × 100,000 = 100 million births, or about one mutation in every 10 million births. Obviously, that's vastly more births than took place in the TMI neighborhood, so based on the linear assumption, we really shouldn't expect any mutations from the TMI accident. Others estimate the effect of low radiation doses as even smaller, arguing that so-called **repair mechanisms** in cells can repair damage caused by the occasional high-energy particle that might strike a cell during low-dose radiation exposure. On the other hand, radiation preferentially damages fast-growing cells, which means that children and the unborn are especially susceptible. This preferential effect on fast-growing cells also means that radiation—itself a carcinogen—is useful for fighting cancer.

The U.S. National Academy of Sciences, in the latest of its studies entitled *Biologic Effects of Ionizing Radiation* (BEIR VII, 2005), concludes that exposure to 100 mSv gives an individual 1 chance in 100 of developing radiation-induced cancer in his or her lifetime and roughly half that chance of developing a fatal cancer. The same individual has 42 chances in 100 of developing cancer from other causes. By comparison, a typical chest X ray gives a radiation dose of 0.1 mSv, and a full CAT scan gives 10 mSv. Cancer risk is assumed to scale linearly with dose, meaning that a 10-mSv CAT scan results in 1 chance in 1,000 of developing cancer. Because the effect is assumed to be linear, it makes no difference whether that 100-mSv dose is delivered to a single individual, who then has a 1 percent chance of developing cancer, or is spread over an entire population of, say, 1 million people. In that case, each individual has only 0.01 chance in a million—10 chances in a billion—of developing cancer. It's impossible to verify this estimate empirically, and credible scientists have come up with other estimates that vary by as much as a factor of 10 in either direction. It's also impossible to ascribe any given case of cancer to low-level radiation; the most we can do is to look for statistically significant increases in cancer among radiation-exposed populations. Given that cancer is common and a leading cause of death, it takes a large radiation dose for radiation-induced cancers to stand out against the background cancer rate.

Example 7.8   Cancers from Three Mile Island?

The 1979 TMI nuclear accident resulted in a total radiation dose on the order of 30 Sv. How many cancers would be expected from this dose?

Solution

At 0.01 cancer per 100 mSv, or 0.1 cancer per 1 Sv, the TMI radiation should have resulted in

$$(30 \text{ Sv})(0.1 \text{ cancer/Sv}) = 3 \text{ cancers}$$

about half of them fatal. An increase that small would be statistically undetectable against the normal rates of cancer and cancer deaths in the population living near the TMI nuclear plant.

Background Radiation

Any discussion of radiation from nuclear power needs to take place in the context of other exposures to natural and artificial radiation, called **background radiation**. On average, individual radiation exposure in the United States amounts to 3.6 mSv (360 millirem) per year; worldwide it's about two-thirds of that figure. Exposure varies substantially with such factors as where you live, how much you travel, and even the construction of your home. Move to high-altitude Denver, or fly across the Atlantic, and your exposure to cosmic rays goes up. Build a house with a cinder-block basement over bedrock that has a high natural uranium content, and you can be exposed to significant levels of radioactive radon gas right in your own home. In fact, radon is the dominant source of radiation exposure for most people, at just over half the total annual radiation dose. Rocks and soils contribute just under 10 percent, as do cosmic rays. Your own body provides more than 10 percent of your radiation dose, mostly from natural potassium-40. Medical procedures amount to another 15 percent of the average American's yearly radiation dose, and consumer products such as smoke detectors, exit signs, and TVs contribute some 3 percent. Radiation from nuclear power and nuclear weapons activity contributes less than 1 percent of the average American's radiation exposure (Fig. 7.14).

Radiation Doses from Nuclear Power

In normal operation, nuclear power plants emit very small amounts of radioactive materials, mostly to the atmosphere. These include the chemically inert gases krypton and xenon, and small amounts of iodine-131. Of these, only iodine is biologically significant. In the United States, the Nuclear Regulatory Commission limits the exposure of the general public to a maximum yearly dose of 0.08 mSv from nuclear power plant emissions— about 2 percent of the normal background rate. Actual exposures are far less; for example, the immediate neighbors of nuclear plants typically receive about 0.01 mSv. If the entire U.S. population were exposed to this dose, the result would be about 150 fatal cancers (see Exercise 10). Even the workers in the nuclear power industry are allowed a maximum yearly exposure of only 0.05 Sv, about fifteen times the average background dose. Aircraft flight crews on northern intercontinental routes routinely experience considerably higher doses from cosmic radiation.

Accidents and Terrorism

The inventory of radioactive material in the core of a nuclear power reactor is huge— enough to cause large-scale environmental contamination and widespread health effects in the event of an accident or intentional release. The TMI accident involved a negligible release of radiation, probably resulting in no radiation-induced mutations and perhaps one or two additional cancer deaths. Some have argued that the greatest loss of life caused by the TMI incident may have come from the negative health effects of having to burn additional coal in fossil-fueled plants to make up for the power lost when the TMI reactor went down.

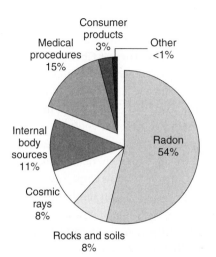

Figure 7.14
Sources of background radiation for the average U.S. resident. The "other" category includes nuclear power, radioactive waste, and fallout from weapons tests. Most background radiation is natural; artificial sources are in the separate wedge at top left.

The 1986 Chernobyl reactor accident is a different story. This event resulted, ironically, from a runaway nuclear reaction that occurred during a test of the reactor's safety systems. The accident spewed measurable radiation across the entire northern hemisphere, prompting short-term bans on milk and other contaminated products in many parts of Europe. Several hundred thousand people were evacuated from the area around the reactor, and even today thousands of square miles remain officially unlivable, although radiation levels have dropped substantially since the accident. Nevertheless, fewer than fifty deaths are definitively attributable to Chernobyl's radiation, while hundreds of people—nearly all emergency workers—have recovered from radiation sickness. An authoritative study published almost 20 years after the accident suggests that some four thousand deaths worldwide will result from Chernobyl in the 50 years following the accident. Any serious student of energy alternatives should compare those numbers with the 24,000 premature deaths that probably occur in the United States alone each year as a result of pollution from coal-burning power plants (recall Fig. 6.2).

Nuclear accidents on the scale of Chernobyl or even the much lesser TMI are rare, but not impossible. And the Chernobyl accident released only 3 percent of the reactor's radioactive solids. In an age of terrorism, commercial power reactors are tempting targets. How are we to assess the potential health and environmental effects of catastrophic reactor incidents, accidental or intentional?

As far as accidents are concerned, there is broad consensus that a Chernobyl-type nuclear accident is impossible in an LWR of the sort used in the United States. That's because, as I described earlier, the coolant and moderator in a U.S. LWR are one and the same substance, so a loss of coolant shuts down the nuclear reaction. At Chernobyl, the water-cooled graphite-moderated reactor became unstable and, with loss of water, the reaction rate rose to some five hundred times its design value in a mere five seconds. A steam explosion ensued, blowing the top off the reactor building. The graphite moderator caught fire, helping to disperse radioactive materials into the atmosphere. None of that could occur in an LWR. However, there might be other surprises. During the TMI accident, for example, an unexpected accumulation of hydrogen gas developed in the reactor containment building, and for a while there seemed a very real threat of a chemical explosion that would disperse radioactive material.

Although nothing approaching the severity of Chernobyl has occurred in the intervening decades, a 2007 earthquake shut down the world's largest nuclear power complex at Niigata, Japan. Although damage and radiation leakage were minimal, the incident took 8 GW of generating capacity offline and did little to inspire confidence in the nuclear option, especially in seismically active zones.

The problem with assessing the environmental and health impacts of large-scale nuclear power accidents is that these are extremely rare events, but they have potentially large consequences. In this sense, nuclear power's impacts are very different from those of fossil fuels, where a steady output of pollution results in significant but—for better or worse—politically acceptable rates of death, illness, and environmental degradation. A nuclear incident with even a small fraction of the impact we experience from routine fossil fuel pollution would be totally unacceptable.

A number of studies have attempted to assess the probability of severe nuclear accidents at U.S. power reactors. A report by the Nuclear Regulatory Commission (NUREG-1150) looked at five specific U.S. power plants and concluded that a large-scale radiation release to the environment beyond a plant's boundaries might be expected to occur, on average, about once every 250,000 "reactor-years." A reactor-year represents one nuclear reactor operating for one year. With approximately one hundred reactors operating in the United States, this means we could expect a large-scale radiation release once every 2,500 years. Over the next 25 years, which takes us beyond the operating lifetime of nearly all U.S. nuclear plants, there's a 1 percent chance of such an accident.

The Nuclear Regulatory Commission study also looked statistically at potential fatalities from nuclear reactor accidents. For the five reactors studied, fatality rates ranged from one per 1,000 reactor-years to one per 50 reactor-years. Again, those numbers compare with some 24,000 deaths per year from air pollution due to coal-fired power plants, or around one hundred deaths per coal-plant-year.

The terrorist attacks of September 11, 2001, raised the frightening prospect of an attack on a nuclear power plant. In the United States, thick concrete containment structures surround every commercial power reactor. (Chernobyl, by the way, had no such containment.) License hearings for every U.S. nuclear plant include discussion of accidental aircraft impacts. But given the relatively small size of nuclear facilities, the probability of an accidental airplane crash into a reactor containment building is minuscule. Clearly that's not the case with intentional terrorism. The Nuclear Regulatory Commission admits that existing plants are not designed explicitly to withstand impacts by large airliners. However, a 2002 study by the Electric Power Research Institute done for the nuclear industry's Nuclear Energy Institute used computer modeling to suggest that an intentional crash of a Boeing 767 would not breach a reactor containment structure. The study looked at a variety of impact scenarios, including the effect of the entire aircraft and the specific impact of a heavy, compact jet engine. In no case did any part of the simulated aircraft enter the simulated containment building. The study also claimed that on-site nuclear waste storage facilities, both water-filled spent-fuel pools and so-called dry-cask systems, would survive an aircraft attack. A 2005 report by the U.S. National Academy of Sciences was less optimistic, suggesting particularly that spent fuel stored in water-filled pools at nuclear power plants is vulnerable to terrorist attack. The National Academy of Sciences did conclude that such an attack would be difficult, and it proposed strategies for minimizing radiation release. In any event, no real-life experiment has been performed, so the vulnerability of nuclear plants to terrorist attack remains uncertain.

Comparative Risks

Nuclear energy and its perceived dangers stir worries and passions far beyond the emotions associated with fossil fuels. The link—part real, part imagined—between nuclear energy and the horrific destructive power of nuclear weapons is one reason for public concern with all things nuclear. The invisible, odorless, tasteless nature of nuclear radiation is another. The sheer concentration of nuclear energy—that 10-million-fold nuclear difference—adds to the sense of nuclear danger. And the long-term hazard of nuclear waste makes nuclear energy look like a burden that humankind might not want to take on.

No energy source is without risk, however, and a serious examination of energy alternatives should compare those risks. As I've suggested here and in Chapter 6, objective analyses suggest that the risks of nuclear energy are far below those associated with fossil fuels. There's simply nothing close to 24,000 premature deaths each year in the United States alone from nuclear power; if there were, nuclear power would be banned immediately. Yet we tolerate those deaths when they're from coal energy. Table 7.2 compares impacts from coal and nuclear power plants. The estimates of total deaths from a nuclear plant ranges from the most conservative to those of the most virulently antinuclear groups; note that even the latter is an order of magnitude below the death rate from coal. The impacts in Table 7.2 represent power plant operation only and don't include such things as fuel processing and transportation. It takes a considerable amount of fossil energy to prepare nuclear fuel, and this fossil combustion is responsible for some traditional pollutants and greenhouse gases. A German study on this issue concluded that, for example, greenhouse emissions associated with nuclear power are a few percent those of coal.

The Weapons Connection

There is one scenario in which, I believe, there could be serious environmental and health impacts resulting indirectly from the use of nuclear power—namely, the development of nuclear weapons as an offshoot of a nuclear power program and the subsequent wartime use of those weapons. Although nuclear weapons and power reactors are very different things, they share some connections. Enrichment of uranium to reactor-fuel levels is

Table 7.2 Impacts of Typical 1-GWe Coal and Nuclear Power Plants

Impact	Coal	Nuclear
Fuel consumption	360 tons coal per hour	30 tons uranium per year
Air pollutants	400,000 tons per year	6,000 tons per year
Carbon dioxide	1,000 tons per hour	0
Solid waste	30 tons ash per hour	20 tons high-level radioactive waste per year
Land use (includes mining)	17,000 acres	1,900 acres
Radiation	1 MBq per minute from uranium, thorium, and their decay products, but varies and can exceed nuclear plant radiation	50 MBq per minute from H-3, C-14, noble gases, and iodine-131
Mining deaths	1.5 per year from accidents; 4 per year from black lung disease	0.1 per year from radon-induced lung cancer
Deaths among general public	100 premature deaths per year from air pollution	0.1–10 deaths per year from radiation-induced cancer

harmless enough, but the same enrichment capability can be taken further to make weapons-grade uranium. Nuclear power plants produce Pu-239, which is a potent weapons fuel. Extracting and purifying plutonium is a difficult and hazardous procedure, but it can be done. More generally, the technical personnel and infrastructure needed to maintain a civilian nuclear power program can easily be turned to weapons activity. Nuclear weapons, especially in the hands of rogue nations or terrorist groups, are a potentially grave threat to world security. Discouraging weapons proliferation will require that any global expansion of nuclear power be accompanied by strict international controls on nuclear materials—especially uranium enrichment and spent-fuel reprocessing.

7.7 The Nuclear Fuel Cycle and Uranium Reserves

Coal and oil are extracted from the ground, are processed and refined, and are burned in power plants. Electrical energy is produced, along with waste heat, CO_2, fly ash, sulfurous waste, and a host of toxic and environmentally degrading air pollutants. An analogous sequence happens with nuclear energy: Uranium is mined from the ground, processed into nuclear fuel, and fissioned inside a reactor. Electrical energy is produced, along with waste heat and radioactive waste. A little of that radioactive waste is released to the environment on a regular basis, but the vast majority of it stays locked physically in the nuclear fuel rods. Eventually the rods' fission energy release declines, and they're replaced with new uranium fuel. Those spent fuel rods constitute the major waste product of nuclear power plants, and they have to be disposed of somehow.

Uranium Mining

What I've just described is a "once through" version of the **nuclear fuel cycle**—the entire sequence from uranium mining to processing, to fission in a reactor, to removal and disposal of nuclear waste (Fig. 7.15). As with the fossil fuels, each step in the fuel cycle has environmental consequences. Although uranium itself is only mildly radioactive, its natural decay products are more so and include the gas radon. Early in the nuclear age, radon-induced lung cancer was a serious health problem for uranium miners; however, advances in mining technology and radiation protection have dropped the exposure levels by factors of 100 to 1,000 in recent years. Another mining issue is the disposal of tailings, the volumes of rock that are pulverized to extract uranium. Uranium mine tailings are mildly radioactive, and wind can easily spread the radioactive material from tailings piles. In the past, uranium tailings were actually incorporated into construction materials or used for fill; in the 1950s and 1960s, some four thousand buildings in Colorado were contaminated with uranium tailings. Today, tailings piles in the United States must be covered, and this particular source of radioactive pollution has been substantially reduced.

It's important here to remember the nuclear difference. When you think of old-time uranium miners digging away in radon-contaminated air, remember that we would need only about one ten-millionth as much uranium as coal, even if nuclear power and coal

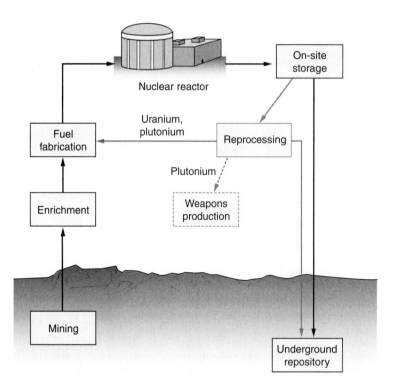

Nuclear reactor

On-site storage

Uranium, plutonium

Fuel fabrication

Reprocessing

Plutonium

Enrichment

Weapons production

Mining

Underground repository

Figure 7.15
Nuclear fuel cycles. The once-through cycle is shown in black; additional steps in a reprocessing cycle are in gray. On-site storage involves pools of water for short-term storage of fresh, highly radioactive waste. Power plants are increasingly using longer-term dry-cask storage in the absence of an established underground repository.

provided equal amounts of energy, which of course they don't. Consequently, there are far fewer uranium miners at work than there are coal miners. And the latter are subject to their own health hazards, especially mining accidents and black lung disease. In the early 2000s there were about 100,000 coal miners in the United States, compared with only a few hundred uranium miners.

Uranium Enrichment

Uranium comes from its ore in the form of "yellowcake," the oxide U_3O_8. But only 0.7 percent of that uranium is the fissile isotope U-235. For all reactor designs except the Canadian CANDU, uranium fuel must be enriched in U-235. Because the uranium isotopes are chemically similar, enrichment schemes generally make use of the small mass difference between the two isotopes. In the most common approach today, uranium first reacts chemically to produce the gas uranium hexafluoride (UF_6). This is injected into centrifuges (spinning cylinders), where its slightly greater inertia causes U-238 to concentrate more toward the outside of the cylinders, leaving enriched UF_6 preferentially toward the interior. The enriched gas is fed to additional centrifuges, eventually producing reactor fuel at around 4 percent U-235, or weapons-grade uranium at 90 percent or more U-235.

Uranium enrichment itself has little direct environmental impact, although it is energy intensive. But its waste product, so-called depleted uranium, with less U-235, is still mildly radioactive. Because of its high density and other physical properties, depleted uranium is made into armor-piercing ammunition, leaving some modern-day battlefields strewn with this weakly radioactive material. Soldiers injured with depleted uranium munitions have detectable radiation levels in their urine. This environmental impact pales in comparison to any single effect of fossil fuels, but it shows once again that our energy use can affect the environment in surprising ways.

Once again I'll reiterate that, although uranium enrichment itself is not of great environmental significance, enrichment is one of the two main routes to nuclear weapons (the other being the extraction of plutonium from spent reactor fuel). The environmental impact of nuclear war would be horrendous, and if the spread of enrichment technology enabled such a war, then my statement here about the minimal environmental impact of uranium enrichment would prove horribly false.

Nuclear Waste

Following enrichment, uranium is fabricated into the fuel rods that go into nuclear reactors. After about three years, most of the U-235 has fissioned; plutonium-239, the fissile isotope produced when nonfissile U-238 absorbs a neutron, has reached a level where it's being fissioned as fast as it's created; and a host of fission products "poisons" the fuel by absorbing neutrons and reducing the efficiency of the chain reaction. At this point, it's time to refuel (Fig. 7.16).

In a typical U.S. LWR, about one-third of the fuel is replaced each year, with each batch allowed to spend about three years in the reactor. The incoming uranium is so mildly radioactive that you could handle it without harm. But spent fuel is nasty stuff because of all those highly radioactive fission products. A few minutes standing next to a spent-fuel bundle would kill you. Consequently, removal of spent fuel takes place under water, using remote-controlled equipment (Fig. 7.17). Energy released from radioactive

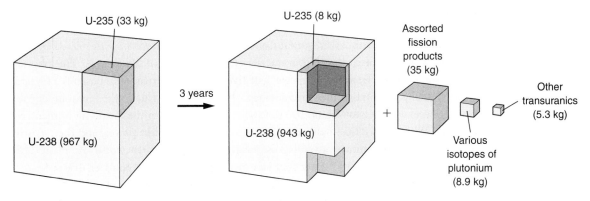

Figure 7.16
Evolution of 1,000 kg of 3.3 percent enriched uranium in a nuclear power reactor. After three years the fission products interfere with the chain reaction, and it's time to refuel.

Figure 7.17
Spent fuel bundle being moved underwater during reactor refueling. The glow around the fuel bundle results from high-speed electrons (beta radiation) interacting with the water. The bundle is shown in the spent-fuel pool; at the back is a water-filled channel that connects to the reactor vessel.

decay also makes the spent fuel so hot (in the old-fashioned thermal sense) that for this reason, too, it has to be kept under water.

After a few years, the shortest-lived and therefore most intensely radioactive isotopes have decayed away and both the radioactivity and thermal energy have decreased enough that spent fuel can be removed from the reactor site for disposal. But where? The longest-lived isotopes, the transuranics, have half-lives as long as tens of thousands of years. Therefore any disposal site must keep the radioactive waste out of the environment for at least several half-lives—some 100,000 years, far longer than the history of human civilization. We're more than half a century into the age of nuclear power, and there is still no operational long-term waste disposal facility anywhere in the world. At many U.S. nuclear plants, decades-old spent fuel sits in temporary cooling pools or dry-cask storage because there's no other place for it to go.

Scientists have considered a range of techniques for nuclear waste management (Table 7.3). Today many experts believe that underground storage of nuclear waste is the best long-term solution and that well-engineered underground repositories in geologically appropriate settings can keep radioactive materials out of the environment for the tens of thousands of years it takes the longest-lived isotopes to decay. Others argue that long-

Table 7.3 Some Nuclear Waste Options

Option	Advantages	Disadvantages
Dry-cask storage	Available short-term option Waste kept at nuclear plant site, so no need to transport	Most nuclear plants are near population centers and waterways
Shallow burial (~1 km)	Relatively easy construction and access Waste is recoverable and easily monitored	Proximity to groundwater poses contamination risk Subject to geologic disturbance
Sub-seabed burial		Probably not recoverable Requires international regulatory structure Currently banned by treaty
Deep-hole burial (~10 km)	Keeps waste well below groundwater	Not recoverable Behavior of waste at high temperature and pressure not well understood Subject to geologic disturbance
Space disposal (dropped into Sun or stored on Moon)	Permanent removal from Earth's environment	Impractical and economically prohibitive Risk of launch accidents too great
Ice sheet disposal	Keeps waste far from population centers	Expensive due to remoteness and weather Recovery difficult Global climate change is diminishing ice sheets Treaty bans radioactive waste from Antarctica
Island disposal	Burial under remote islands keeps waste away from population centers	Ocean transport of waste poses safety issues Seawater leakage into waste repository possible Seismic and volcanic activity common at island sites
Liquid-waste injection	Waste locked in porous rock below impermeable rock	Requires processing waste to liquid form Movement of liquid waste might result in radiation release
Transmutation	High-energy particle accelerators or "fusion torches" induce nuclear reactions that render waste nonradioactive or very short-lived	Technology not proven or available Requires a recoverable storage option until the technology is operational

term geological stability cannot be guaranteed, regardless of a site's history, and that the effect of radiation-generated heat can lead to the migration of water to the waste site, or the fracturing of rock, or other processes that might provide pathways for the escape of radioactive waste. Some would have us bury nuclear waste in holes miles deep, where it could never be recovered. Others claim that wastes may someday be valuable or that we'll develop technologies to render nuclear waste harmless, making shallow burial (typically a half mile down or so) more appropriate. As these arguments proceed, nuclear waste continues to pile up. Again, the nuclear difference means that the sheer volume of waste is relatively small; the spent-fuel waste from one year's operation of a large (1-GWe) nuclear plant might fit under your dining-room table. But this is nasty stuff, and no one thinks that on-site storage at nuclear power plants is a viable long-term solution.

In the United States, the hope for a permanent nuclear waste solution has been focused on a single site, inside Yucca Mountain in arid Nevada. Tunnels and storage areas are in place, but technical, geological, and political issues keep postponing the repository's opening. One thing that's not in question is who will pay for it: An assessment of one-tenth of a cent on every kilowatt-hour of nuclear-generated electricity has been accumulating since 1982 to pay for nuclear waste disposal; the fund now contains some $20 billion.

Reprocessing

The United States has chosen the once-through nuclear fuel cycle, disposing of nuclear fuel rods without further processing. This makes for an inexpensive and proliferation-resistant nuclear fuel cycle, but it wastes valuable energy. Europe and Japan, on the other hand, have opted for **reprocessing** spent fuel to recover plutonium and "unburned" uranium (see Figure 7.15). Reprocessing is expensive and technically difficult. It also introduces dangers associated with increased transport of radioactive materials and with radiation releases from reprocessing facilities, and it increases the possibility that terrorists or rogue nations could steal plutonium for use in illicit weapons programs. On the other hand, plutonium extracted in reprocessing is mixed with uranium and recycled into reactor fuel, reducing the need for raw uranium and extending our uranium reserves. And reprocessing significantly reduces both the level and duration of the radiation danger from nuclear waste (Fig. 7.18). At this point it is unclear from either a safety or economic standpoint whether the U.S. once-through fuel cycle or the European/Japanese reprocessing model is preferable.

The Uranium Resource

Nuclear fission looks like a serious alternative to fossil fuels—at least for generation of electric power—although questions of safety, routine radiation releases, and weapons proliferation all cloud the picture. But if we were to choose a nuclear route to replace fossil fuels, would we have enough uranium?

Surprisingly, a superficial check of the proven uranium resource—some 3 million tons, containing some 1,500 EJ of energy—suggests that the time until we face uranium shortages could, as for oil, be only a matter of decades. Currently there are some 440 reactors operating worldwide, producing 16 percent of humanity's electrical energy. If we quintupled those numbers to 2,200 reactors producing 64 percent of our electricity, identified reserves of high-grade uranium ore would last only one decade!

One solution to this apparent uranium shortage is the breeder reactor, which could convert the U-238 that makes up 99.3 percent of natural uranium into fissile Pu-239. This alone would increase our nuclear fuel reserve by a factor of 99.3/0.7, or nearly 150-fold. Even European-style reprocessing of fuel from ordinary reactors would stretch the uranium supply somewhat.

But we don't need breeders or other exotic technologies. Nuclear energy hasn't seen the growth once expected, and as a result the world is awash in cheap uranium. At

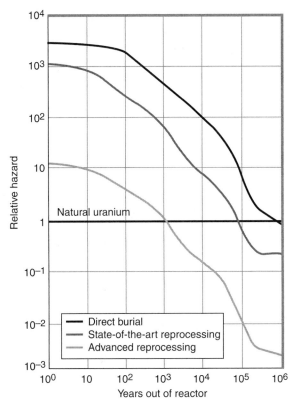

Figure 7.18
Decline in hazardousness of nuclear waste with direct burial, today's state-of-the-art reprocessing, and future advanced reprocessing that would remove 99.9 percent of transuranic isotopes. The hazard is measured relative to natural uranium. Even without reprocessing, the waste hazard drops by an order of magnitude in the first thousand years. With advanced reprocessing, waste becomes no more hazardous than natural uranium after the same amount of time.

present there's little incentive to explore for more, even though such exploration would surely increase the known reserves substantially. There's also plenty of lower-grade ore—probably 100 to 300 million tons—that could be exploited, although it would cost more to extract and prepare. However, fuel plays a much smaller part in the cost of operating a nuclear plant than it does in a fossil plant (again, the nuclear difference), so the substantial cost of extracting uranium from low-grade ore wouldn't have much impact on the cost of nuclear-generated electricity.

Uranium is not a particularly rare substance. It's dispersed throughout much of the Earth's crust at an average concentration of about 4 parts per million. And it's soluble in water to the extent that uranium is more abundant in the oceans than some common metals such as iron. In fact, the world's oceans contain some 4.5 billion tons of uranium—enough to last for thousands of years. At present it's much too expensive to extract, but this uranium energy resource is certainly available and could become economically viable as energy costs rise and technological advances bring down extraction costs.

The bottom line is that high-grade uranium ore could become scarce in a fission-powered world, but the overall uranium resource is sufficient to allow fission to become a serious alternative to fossil fuels, at least for stationary power sources such as electric power plants.

7.8 Policy Issue: Economics, Politics, and the Future of Nuclear Power

The number of nuclear power plants—some 440 worldwide today—and their total power output has grown substantially in the decades since the first commercial reactors went online in the mid-1950s. But growth has been far less than nuclear advocates had anticipated in those early years, when it was thought nuclear electricity would be "too cheap to meter" and peaceful uses of nuclear energy would include not only electric power but also nuclear explosives for excavation and even nuclear-powered spaceflight. Projections for the future suggest that, absent a major nuclear revival, nuclear energy's share of world electrical energy production is likely to shrink in the next few decades. Figure 7.19 shows historical and projected future growth in nuclear power.

Today, the use of nuclear power is growing slowly, but that growth is hardly uniform across the planet. The most rapid growth is in Asia, where burgeoning economies fuel increasing demand for energy. European nuclear energy is fairly stagnant, with some modest growth in Eastern Europe. In the United States the number of operational reactors has been declining as old plants are retired, and no replacements were ordered for some 30 years. In 2007 a combination of government incentives, high fossil fuel prices, and concern over climate-changing fossil carbon emissions led to the first nuclear license applications in decades. Even if these are approved, it will be years before any new plants come online.

What happened to dampen the nuclear power industry in the late twentieth century? In the United States, and to a lesser extent the world, the 1973 oil shortage was a wake-up call to the need for greater energy efficiency. Growth in demand for electrical energy slowed, and with it the need for more nuclear power. Then in 1979 the TMI accident seriously damaged public confidence in nuclear power. Economically, TMI also made nuclear power less competitive because of expensive safety requirements promulgated in

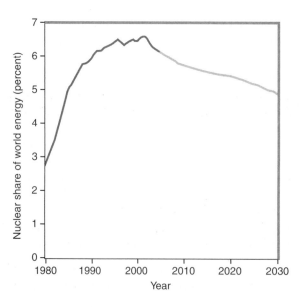

Figure 7.19

Historical trends and projections of nuclear energy as a percentage of total world energy consumption. Barring a turnaround in nuclear power development, the nuclear percentage may have peaked in the early 2000s.

the wake of the accident. For example, the Vermont Yankee nuclear plant began generating power in 1972 after construction costing $238 million. The Seabrook plant in neighboring New Hampshire, among the last U.S. nuclear plants to go online, was completed in 1986 at a cost of $4.5 billion. This was the year of the Chernobyl accident, which further eroded the political and economic viability of nuclear power. In the wake of Chernobyl, opponents kept Seabrook off-line until 1990, by which time the plant's total cost approached $7 billion. Seabrook fared better than its contemporary, the Shoreham plant on New York's Long Island. Completed in 1985 at a cost of $5 billion—fifty times that of an earlier but nearly identical plant in Connecticut—Shoreham never saw commercial operation. Instead, the plant was sold to the State of New York for $1 so it could be shut down and decommissioned.

A Nuclear Renaissance?

Despite its economic woes, nuclear power remains potentially a strong candidate to replace fossil fuels for electric power generation. As concern over fossil greenhouse emissions and climate change continues to mount, some predict a major renaissance for nuclear power. Although few technologies inspire as much fear and controversy, nuclear power is arguably less harmful to the environment than continued use of fossil fuels, and it's clearly superior from a climate-change standpoint. However, the small but nonzero chance of catastrophic accidents, uncertainties over the disposal of nuclear waste, and especially the connection between nuclear power and the nuclear weapons capable of destroying civilization cloud the future of nuclear power. Even if technological advances and international agreements mitigate these concerns, there's the practical question of whether we can build new reactors sufficiently quickly to displace enough fossil fuel power plants to make a significant difference.

So should we embrace nuclear power to satisfy, temporarily or permanently, our prodigious and growing energy appetite? Or are we better off with fossil fuels, despite the limited resources, the death-dealing air pollution, and the continued buildup of climate-changing CO_2 in the atmosphere? Or are there alternatives that might reduce our need for both nuclear power and fossil fuels?

Chapter 7 Chapter Review

BIG IDEAS

Nuclear energy is a distant second to fossil fuels in its share of the world's energy supply.

7.1   The **atomic nucleus** consists of **neutrons** and **protons** bound by the strong nuclear force. The **atomic number** is the number of protons, which determines the chemical element. Different **isotopes** of an element have different numbers of neutrons.

7.2   **Binding energy** is the energy released when a nucleus forms; the binding energy curve peaks at the iron nucleus, which has 26 protons. It's possible to extract nuclear energy by

fission of elements heavier than iron or by **fusion** of elements lighter than iron. This chapter deals primarily with fission; Chapter 11 discusses fusion. Nuclear reactions release some 10 million times more energy than chemical reactions do.

7.3 **Fissile isotopes** fission when struck by low-energy neutrons; they include U-235 and Pu-239. Fission produces intermediate-size nuclei called **fission products**; it also releases energy and typically two to three neutrons. The neutrons cause additional fission, resulting in a **chain reaction**. To generate energy safely from nuclear fission, it's necessary to control the chain reaction. In most cases it's also necessary to slow the neutrons with a **moderator** so they're effective in inducing fission.

7.4 Nuclear power reactors come in many different designs. Most common are the **light-water reactors** that use ordinary water as a moderator and coolant. Other designs include heavy-water reactors and gas-cooled reactors. **Breeder reactors** convert nonfissile U-235 into fissile Pu-239. New reactor designs promise greater intrinsic safety. Whatever the reactor type, nuclear reactors are used to boil water and drive a steam cycle as in fossil-fueled power plants.

7.5 Nuclear reactions produce **radiation**, consisting of high-energy particles. The **half-life** is the amount of time it takes half the nuclei in a radioactive material to decay. Radioactivity is quantified by the number of decays per unit time, and by the energy delivered and its effects on biological tissue.

7.6 The dominant environmental impact of nuclear power is the health effects of radiation, through mutations, cancer, or cell death. The effects of high radiation doses are well known, but it's hard to quantify the effects of low doses against the background of mutations and cancers with other causes. Radiation from nuclear power constitutes a tiny fraction of the radiation humans receive from natural and anthropogenic sources. Realistic estimates of nuclear energy's impact suggest that it's far lower than that of fossil fuels. However, nuclear power has the potential for large-scale accidents and is vulnerable to terrorism, and the technology and expertise that enable nuclear power programs can be used to produce nuclear weapons.

7.7 The **nuclear fuel cycle** includes uranium mining, enrichment to increase the U-235 content, fuel fabrication, "burning" in a reactor, and waste storage or reprocessing. Nuclear waste includes isotopes with half-lives as long as tens of thousands of years, and isolation of these wastes from the environment is one of nuclear power's thornier issues. **Reprocessing** removes fissile plutonium from spent nuclear fuel rods for use in new reactor fuel, and at the same time renders the remaining waste less radioactive. But reprocessing is difficult and dangerous, and it makes plutonium a commercial substance. Reprocessing, especially when combined with breeder reactors, could extend the life of Earth's uranium reserves to many centuries.

7.8 Nuclear power has grown more slowly than anticipated, because of safety and economic issues. The need for carbon-free energy coupled with a new generation of simpler, less expensive, and intrinsically safer nuclear reactors could bring about a resurgence of nuclear power. Such a resurgence must confront the link between nuclear power and nuclear weapons.

TERMS TO KNOW

alpha radiation (p. 200)
atomic nucleus (p. 179)
atomic number (p. 180)
background radiation (p. 206)
becquerel (p. 204)
beta radiation (p. 200)
binding energy (p. 183)
boiling-water reactor (p. 192)
breeder reactor (p. 196)
chain reaction (p. 189)
control rods (p. 192)
coolant (p. 192)
critical mass (p. 190)
curie (p. 204)
curve of binding energy (p. 183)
enrichment (p. 191)
fissile isotope (p. 188)
fissionable isotope (p. 187)
fission product (p. 187)
gamma radiation (p. 200)
gray (p. 204)
half-life (p. 201)
heavy water (p. 192)

isotope (p. 181)
light-water reactor (p. 192)
mass number (p. 181)
moderator (p. 191)
multiplication factor (p. 191)
neutron (p. 179)
neutron-induced fission (p. 187)
nuclear difference (p. 184)
nuclear fission (p. 184)
nuclear fuel cycle (p. 211)
nuclear fusion (p. 184)
nucleon (p. 179)
pressure vessel (p. 192)
pressurized-water reactor (p. 193)
proton (p. 179)
rad (p. 204)
radioactive decay (p. 199)
rem (p. 204)
repair mechanism (p. 205)
reprocessing (p. 216)
sievert (p. 204)
steam generator (p. 193)
transuranic (p. 188)

GETTING QUANTITATIVE

Isotope symbol example: $^{4}_{2}\text{He}$

U-235 as a fraction of natural uranium: 0.7 percent

Mass-energy equivalence: $\Delta m = E/c^2$ (Equation 7.1; p. 185)

Nuclear difference: $\dfrac{\text{Energy released in typical nuclear reaction}}{\text{Energy released in typical chemical reaction}} \sim 10^7$

Alpha decay example: $^{238}_{92}\text{U} \rightarrow {}^{234}_{90}\text{Th} + {}^{4}_{2}\text{He}$ (Equation 7.2; p. 200)

Beta decay example: $^{14}_{6}\text{C} \rightarrow {}^{14}_{7}\text{N} + {}^{\ 0}_{-1}e + \overline{\nu}$ (Equation 7.3; p. 200)

Cancer incidence from radiation: 1 percent lifetime chance of radiation-induced cancer per 100 mSv radiation dose; ~50 percent of cancers fatal

Background radiation dose, U.S. average: 3.6 mSv per year

Waste from 1-GWe nuclear power plant: 20 tons high-level radioactive waste per year

Nuclear power reactors worldwide, early twenty-first century: ~440

QUESTIONS

1. Your state needs new sources of electricity, and the local utility is considering either a coal-burning power plant or a nuclear plant. You live downwind of the proposed plant site. Which type of plant would you rather have, and why?

2. Explain the role of the moderator in a nuclear reactor, and describe why the use of the same light water for both moderator and coolant is an inherent safety feature of LWRs.

3. In what sense does a breeder reactor produce more fuel than is put into it? Do breeders violate the principle of energy conservation?

4. What are some of the possible connections between nuclear power and nuclear weapons?

5. In Chapter 6 I estimated that approximately one hundred people die each year in the United States in collisions between automobiles and coal trains. Why would you expect fewer deaths in accidents with vehicles transporting uranium? Explain in terms of basic science.

EXERCISES

1. Copper's atomic mass is 63.6 u. Natural copper consists of just two stable isotopes, and 69.1 percent is Cu-63. What is the other isotope?

2. The Sun generates energy at the rate of about 3.9×10^{26} W. Use this value to confirm Section 7.2's assertion that the Sun loses mass at the rate of about 4 million tons per second.

3. Fission of a U-235 nucleus releases about 30 pJ (30×10^{-12} J). If you weighed all the particles resulting from a U-235 fission, by what percentage would their total mass differ from that of the original uranium nucleus?

4. Find the mass of the Pu-239 produced by your decision to dry your clothes in an electric dryer, as described in Example 7.5.

5. Confirm the assertion in Section 7.3 that the 10^{15} J of energy released in fissioning 30 pounds of U-235 is approximately equal to that contained in three 100-car trainloads of coal, with each car carrying 100 tonnes.

6. For every three U-235 nuclei that fission in a reactor, approximately two nuclei of Pu-239 are produced. Suppose you walk out of your room and leave your 100-W light on for one hour. If your electricity comes from a 33%-efficient nuclear plant, how many Pu-239 nuclei are produced during that time to keep the lightbulb burning?

7. Radium results from a sequence of reactions, beginning with U-238 undergoing the alpha emission described in Equation 7.2. Two beta decays follow, and then two more alpha decays. Write equations for these four additional reactions, correctly identifying the intermediate isotopes.

8. After the Chernobyl accident, Austria reported an iodine-131 radioactivity level of 1,500 Bq per liter of milk. Austria's milk safety guidelines called for a maximum level of 370 Bq/L. Given iodine-131's half-life of 8.04 days, how long did Austrians have to wait for their milk to become safe?

9. (a) In the U.S. population of approximately 300 million, how many cancer cases should result each year from the average background radiation dose of 3.6 mSv per person per year? (b) Consult Figure 7.14 to determine the number of these deaths attributable to natural and artificial radiation sources.

10. Confirm the text's estimate of 150 fatal cancers per year if everyone in the United States was exposed to radiation at the 0.01-mSv dose received in the neighborhood of nuclear power plants (recall that roughly half of the radiation-induced cancers are fatal).

RESEARCH PROBLEMS

1. Determine the fraction of electricity that is generated by nuclear power plants in your state, country, or local electric utility's region.

2. Locate the nearest nuclear plant to your home and determine as many of the following statistics about it as you can: (a) type of reactor; (b) maximum rated power output in MWe or GWe; (c) year of initial operation; (d) capacity factor (actual annual energy production rate as a fraction of the production if the plant produced energy at its maximum rate all the time); (e) status of its high-level nuclear waste (spent fuel).

3. If you're a resident of the United States, find your average annual radiation dose using the EPA's calculator at www.epa.gov/radiation/students/calculate.html.

ENERGY FROM EARTH AND MOON

Now that we've covered fossil and nuclear energy sources, we've reviewed all the Earth-stored fuels that we humans currently use to produce most of our energy. We'll look at one futuristic fuel in Chapter 11, but for the technologies available today, the fossil fuels and uranium are all we have. Fuels—substances that store energy over the long term—aren't the only assets in Earth's energy endowment, however. As we saw in Chapter 1, there are also steady flows of energy that we can tap into. This chapter deals with two of those flows, namely geothermal and tidal energy, and Chapters 9 and 10 will cover the much more significant solar energy flow.

8.1 The Geothermal Resource

Earth's interior is hot, a result of both naturally occurring radioactive materials and primordial energy released in the accretion of material that built our planet. The temperature difference between Earth's interior and its surface drives an upward flow of heat. The rate of that heat flow determines the size of the sustainable geothermal energy resource, although there's nothing that keeps us from using that resource, at least temporarily, at an unsustainable rate.

Geothermal energy flow supplies only about 0.025 percent of the energy coming to Earth's surface; nearly all the rest is from the Sun (recall Fig. 1.8). Averaged over the entire planet, this 0.025 percent amounts to only about 0.087 W/m^2 (or 87 mW/m^2). That's pretty feeble compared with the average solar flow, in direct sunlight, of about 1,000 W/m^2 (or 1 kW/m^2). As you can show in Exercise 1, the total geothermal flow amounts to some 40 TW, about three times humankind's current energy-consumption rate. But the low temperature associated with near-surface geothermal energy means this is energy of very low quality. That, along with the diffuse nature of the geothermal flow, tells us that we'll never be able to use more than a tiny fraction of the geothermal energy flow for high-quality energy needs (see Exercise

2 for a quantitative look at this point). So we can already conclude that sustained use of geothermal energy will never go very far toward meeting humankind's total energy needs. Nevertheless, geothermal energy today ranks third among so-called renewable energy sources (although geothermal energy is not always used in a renewable way), behind waterpower and biomass but well ahead of direct solar and wind power. And unlike solar and wind, geothermal resources provide a steady, uninterruptible energy flow that can be used for baseload power generation.

Geothermal energy flows to Earth's surface because of the greater temperature at depth. On average, the temperature in Earth's crust increases by about 25°C for every kilometer of depth—a figure known as the **geothermal gradient**. In Chapter 4 we looked at examples of temperature-driven energy flows in connection with building insulation, in which the relevant material property is the thermal conductivity (k). Using the k value for rock given in Table 4.1, you can show in Exercise 3 that the geothermal energy flow and geothermal gradient are indeed mutually consistent with a temperature-difference-driven flow through rock.

On the other hand, there's nothing to keep us from extracting geothermal energy at a greater rate than it's being replenished by the geothermal flow from below, although such energy use doesn't count as renewable or sustainable. With a geothermal gradient of 25°C/km, the average temperature in the top 5 km of Earth's crust is 75°C, which is about 60°C higher than the average surface temperature. As Example 8.1 shows, this amounts to a substantial value for the associated thermal energy—somewhere around 300 million EJ, far more than the known reserves of all fossil fuels. However, the relatively low temperature difference makes for low thermodynamic efficiency, leaving only a small fraction of the energy available for most uses. Exercise 4 explores this use of the geothermal resource in such a nonsustainable way, and it suggests that large-scale extraction of geothermal energy from the average near-surface crust is simply not practicable. So again, geothermal energy is not likely to make a substantial contribution to humankind's total energy needs.

Example 8.1   Hot Rocks!

Estimate the thermal energy extractable from the rock that constitutes the top 5 km of Earth's crust, assuming it's granite (density 2,700 kg/m^3) and that the average temperature is 60°C higher than the average surface temperature. Here "extractable" means the energy that would become available in cooling this entire layer to the surface temperature. Compare this with Earth's total known reserves of fossil energy.

Solution

The total thermal energy content is the energy that would be needed to increase the temperature of the 5-km layer by 60°C. That energy, Q, depends on the mass (m) of material being heated and on its specific heat (c), as described in Equation 4.3: $Q = mc\Delta T$.

We can calculate the mass of rock involved from the given density and the volume, which is the Earth's surface area $4\pi R_E^2$ multiplied by the 5-km layer thickness:

$$m = \text{density} \times \text{volume} = (2{,}700 \text{ kg/m}^3)(4\pi)(6.37 \times 10^6 \text{ m})^2(5{,}000 \text{ m}) = 6.88 \times 10^{21} \text{ kg}$$

Here we found Earth's radius (R_E) in the physical constants list inside the back cover. For the next step we need the specific heat of granite, which is given in Table 4.3 as 840 J/kg/K but can also be written as 840 J/kg/°C because kelvins and Celsius degrees are equivalent as far as temperature changes are concerned. Then Equation 4.3 gives

$$Q = mc\Delta T = (6.88 \times 10^{21} \text{ kg})(840 \text{ J/kg/°C})(60°C) = 3 \times 10^{26} \text{ J}$$

rounded to one significant figure. That's 300 million EJ. Adding together the known reserves of oil (about 6,000 EJ), natural gas (another 6,000 EJ), and coal (roughly 25,000 EJ) shows that this geothermal energy resource is nearly ten thousand times the energy content of the known fossil fuels combined. But again, the low energy quality and the diffuse nature of the resource—spread over the entire globe, with some 70 percent beneath the oceans—makes this comparison rather meaningless.

Nevertheless, geothermal energy can be important in limited geographical regions. That's because the geothermal gradient, and hence the rate of geothermal energy flow, varies substantially with location. In some regions, hot magma lies much closer to Earth's surface, resulting in a more rapid increase in temperature with depth—that is, a larger geothermal gradient. This helps in two ways: First, it means the sustained flow of geothermal energy is greater. Second, it puts high temperatures and hence higher-quality energy closer to the surface, making it technically and economically more viable to reach. This, in turn, leads to higher thermodynamic efficiency should we wish to produce high-quality energy—for example, by running a heat engine to drive an electric generator.

Not surprisingly, regions of strong geothermal energy potential are often associated with volcanic activity in the present or the geologically recent past. Obvious manifestations of geothermal potential include not only volcanoes themselves but also thermal features such as hot springs and geysers. Yellowstone National Park in the United States, for example, lies atop a huge "supervolcano" that last erupted some 640,000 years ago. The geothermal energy flow at Yellowstone is some thirty to forty times the global average. Geothermal features found there today are reminders of the near-surface magma that still powers geothermal activity and is likely one day to result in another colossal eruption.

Geothermal sites with the highest energy-producing potential generally lie near the boundaries of the Earth's tectonic plates, where earthquakes and volcanic eruptions are most likely. Other sites, such as the Hawaiian Islands, are located over local "hot spots" associated with individual plumes of magma that extend close to the surface. Although it's one of the most volcanically active places on Earth, Hawaii lies in the middle of the Pacific plate, about as far from plate boundaries as one can get. Figure 8.1 shows Earth's system of tectonic plates, along with areas of strong volcanism. Note in particular the

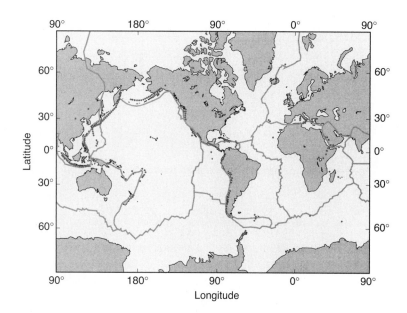

Figure 8.1
Regions with some of the greatest geothermal energy potential are located along geologic plate boundaries (gray lines) and close to active volcanoes (dots).

"ring of fire" surrounding the Pacific Ocean. This region is home to some of the most active volcanoes, some of the most damaging earthquakes, and some of the most promising sites for geothermal energy production.

A closer look at geothermal resources in the United States reveals that the sites with the greatest potential are located, not surprisingly, in the western states. Figure 8.2 maps the geothermal gradient throughout the continental United States, and you can see that the highest gradients average some three to four times the lowest values; more detailed data show measured gradients ranging from less than 10°C/km to more than 60°C/km.

As I indicated earlier, geothermal energy isn't necessarily extracted sustainably. When it's not, the local geothermal resource is gradually depleted, and we really shouldn't consider geothermal energy to be renewable under such circumstances. On the other hand, the geothermal resource may eventually recover. How long that takes depends on the size of the underlying magma body and whether heat transfer is by conduction or, more rapidly, by the motion of water or magma itself. Exercise 4 explores this point.

Wet or Dry? Types of Geothermal Environments

Temperature and quantity of thermal energy aren't the only factors that determine the potential of a given geothermal energy resource. Also important is the associated geological structure and the presence or absence of water. Geothermal resources come in two broad categories, wet and dry. Wet systems are far and away the most valuable for energy production. In these systems, fractures in the rock allow water to penetrate to depths where the temperature is high. Systems in which water or steam can circulate freely are called **hydrothermal systems**. In the most valuable hydrothermal systems, termed **vapor-dominated systems**, fractures and pores in the rock are saturated with steam at high

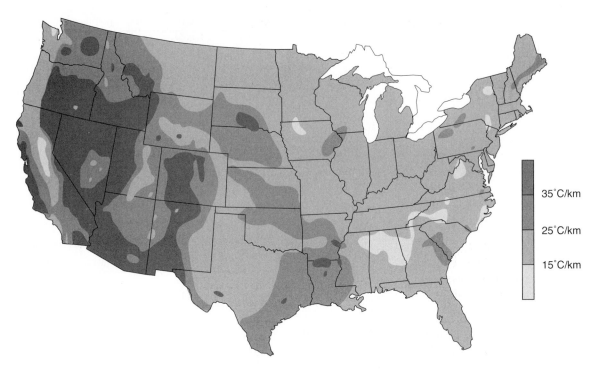

Figure 8.2
Geothermal gradient in the continental United States. Note the
preponderance of geothermal resources in the western states.

temperature and pressure. As we'll see shortly, this steam can be used directly to drive
a turbine-generator. Others systems contain hot, liquid water under pressure; when it's
brought to the surface, some of the water boils to steam and can again drive a turbine.
But liquid-based systems are harder to exploit because of the need to separate liquid
water from steam. Finally, many systems contain water at only moderate temperatures.
These cannot produce steam and they require more complex technologies to extract high-
quality energy to generate electricity, although they may be used directly to supply low-
quality energy for heating. We'll look at geothermal technologies in the next section.

In addition to thermal energy, some hydrothermal systems contain considerable
mechanical energy associated with high water pressure. Water pressure naturally increases
with depth due to the weight of the overlying material, but in so-called **geopressured
systems** the water-saturated zone is effectively cut off from the surface, usually by layers of sedimentary rock forming at rates that are rapid by geological standards. Water
trapped in rock pores is literally squeezed from the rocks, and excess pressure builds. In
principle, both thermal and mechanical energy could be extracted simultaneously from
such systems, although the technology to accomplish this extraction has not proved economical. Geopressured systems often contain dissolved methane, which also makes them
sources of chemical energy.

Dry geothermal environments lack either water or the fractures and pores necessary for groundwater to contact hot rock. Drill deep enough, though, and you'll find hot, dry rock anywhere on Earth; drill in a region with a larger-than-average geothermal gradient and you'll strike hot rock sooner. At present, however, there's no economical technology for dry-rock energy extraction. The same goes for magma itself, which, because it's found at temperatures of up to 1,300°C, is the highest-quality source of geothermal energy. Magma, of course, is available near the surface predominantly in regions of active or recent volcanism.

8.2 Geothermal Energy Technology

Humankind's use of geothermal energy goes back thousands of years, if not longer. Surface geothermal activity such as geysers, steam vents, and hot springs would have been obvious to our prehistoric ancestors, and it's likely that gentler geothermal features were used for bathing and possibly for food preparation. Ancient Japanese, Greeks, and Romans constructed spas at geothermal sites, and the use of geothermal hot springs for bathing, relaxation, and physical therapy continues to this day (Fig. 8.3).

Geothermal Heating

Geothermally heated water at modest temperatures can be used directly for purposes requiring low-quality thermal energy. Temperatures in such direct-use geothermal applications range from as low as about 40°C to about 150°C, with most below the boiling point of 100°C at atmospheric pressure. Geothermal resource temperatures below 90°C are classified as low-temperature and are generally suitable only for direct-use applications. Temperatures between 90°C and 150°C are considered moderate and may be used directly as thermal energy or, with some difficulty and low efficiency,

Figure 8.3
Bathers enjoy geothermally heated waters in Iceland's Blue Lagoon.

Figure 8.4
Geothermal hot water carried through tubing installed in city sidewalks melts snow from walkways in Klamath Falls, Oregon.

to produce high-quality energy. Resources above 150°C are most valuable for producing high-quality energy, particularly electricity (more on this in the next section).

In heating applications, hot water from surface features or shallow wells is circulated through individual buildings or through community-wide hot-water heating systems. Several cities, including Reykjavik, Iceland; Copenhagen, Denmark; Budapest, Hungary; Paris, France; and Boise, Idaho, are heated in part with geothermal energy. A novel community use of direct geothermal heat is for melting snow on the sidewalks of Klamath Falls, Oregon (Fig. 8.4).

Many industrial and agricultural processes require low-quality energy associated with modestly heated water. Direct use of geothermal energy can substitute for fossil fuels in such applications. Geothermally heated greenhouses provide fresh produce year-round in cold regions with abundant geothermal resources. Dairies, fish farms, mushroom growers, and other light industries also benefit from the use of low-grade geothermal heat. Figure 8.5 shows the distribution of direct-use geothermal energy applications in the United States. Overall, systems making direct use of geothermal energy in the United States have a modest capacity of about 600 MW, but their average output is only about 50 percent of this capacity.

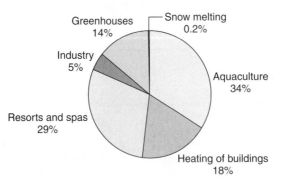

Figure 8.5
Direct uses of geothermal energy in the United States.

Electric Power Generation

Higher-temperature geothermal sources can generate electricity, albeit with only modest thermodynamic efficiency. Development of a method to tap the energy of high-temperature magma would greatly increase efficiency (see Exercise 5).

The most valuable geothermal resources for electric power generation are the vapor-dominated systems in which a hydrothermal environment is saturated with steam. Not only is the steam at high temperature and pressure, but also the absence of liquid water simplifies power plant design. Drill a well into such a system, and up comes the steam, ready to drive a turbine-generator directly. Such a geothermal power plant is like the fossil and nuclear plants we saw in earlier chapters, with the geothermal steam source replacing a fossil-fueled boiler or fission reactor. The rest of the system is the same: High-pressure steam turns a turbine connected mechanically to an electric generator. The spent steam is sent through a condenser, which extracts the waste heat required by the second law of thermodynamics, turning the steam into liquid water in the process. The water is reinjected into the geothermal heat source to replace the extracted steam; if this were not done, the plant's power output would decline not because of the exhaustion of energy but of water. Additional makeup water is sometimes needed, as some steam is inevitably lost to the atmosphere. In an environmental win–win situation, The Geysers geothermal power plant complex in California managed to stem a water-limited decline in power output by injecting wastewater from Santa Rosa and other nearby communities (see Box 8.1). Figure 8.6 shows the essential features of a typical steam-based geothermal power plant.

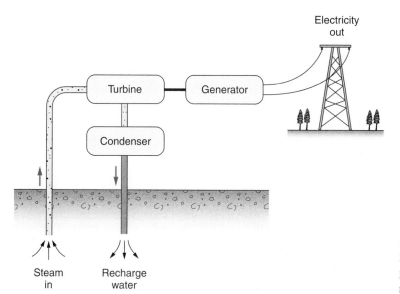

Figure 8.6

Essential features of a geothermal power plant using steam from a vapor-dominated geothermal resource.

Box 8.1   The Geysers Geothermal Power Plants

The Geysers, a region in northern California about 120 km north of San Francisco, has long been known for its geothermal features, including steam vents, spouting hot springs, and mudpots, although it has no true geysers. These relatively modest features belie the huge geothermal resource that lies beneath the mountainous region, which is adjacent to a volcanic area that erupted as recently as a few thousand years ago. In fact, The Geysers region sits atop a large vapor-dominated hydrothermal system capable of supplying steam at high temperature and pressure.

Wells first drilled in 1924 used this geothermal resource to drive turbine-generators capable of a modest few kilowatts of power output. By the 1950s, wells several hundred meters deep tapped into the main hydrothermal system, and by 1960 a single power plant at The Geysers was producing 12 MW of electric power (12 MWe; as with any heat engine, the thermal power extracted from the ground was greater, thanks to the second law of thermodynamics). Development at The Geysers continued to accelerate until about 1990, at which time total generating capacity was more than 2 GWe—the equivalent of two large coal-fired or nuclear power plants, although at The Geysers the power production is distributed among some twenty-six individual plants with power outputs up to about 120 MWe (Fig. 8.7).

Given its high rates of energy extraction, however, The Geysers definitely does not qualify as a sustainable or renewable energy source, and in fact power production began declining after 1988 (Fig. 8.8). Much of this decline was due to loss of steam, which was not fully compensated for by the injection of condensate water into the ground. In 1997, a 50-km pipeline began supplying treated municipal wastewater from nearby communities for ground injection; the resulting increase in geothermal steam immediately raised The Geysers' power output by 75 MWe while helping to dispose of wastewater in an environmentally sound way. In 2003 an additional pipeline began carrying wastewater from the city of Santa Rosa, resulting several years later in an 85-MWe increase in generating capacity. Nevertheless, The Geysers' total capacity of some 1.7 GW remains well below its late-1980s peak, and average power output is only about half of capacity, or about 850 MW. Indeed, some Geysers plants have been retired or dismantled, and one even has been designated a National Historical Mechanical Engineering Landmark.

In 2005, The Geysers' geothermally produced electricity was selling for about 3.5 cents per kilowatt-hour, which is comparable to the price of fossil-generated electrical energy. Energy prices for new geothermal installations at geothermal sites comparable to The Geysers' quality would be higher, but still very much competitive with conventional sources.

Figure 8.7
Units 3 and 4 at the Geysers geothermal complex in California. These two units have a combined electric power output of nearly 60 MW.

As with fossil and nuclear power plants, the waste heat from a geothermal steam cycle can itself be used to supply low-quality thermal energy. Iceland has pioneered such geothermal cogeneration plants, producing both electricity and heat for entire city districts. Although some geothermal sites are too remote for cogeneration to be practical, California is actively considering district heating and snow-melting using waste heat from geothermal plants at Mammoth Lakes.

High-quality vapor-dominated systems such as The Geysers are rare. More common and still valuable for power generation are liquid-water systems at temperatures above water's atmospheric-pressure boiling point of 100°C. High below-ground pressure keeps water liquid even at such high temperatures, but when the water reaches the surface in a geothermal well, some of it flashes (boils abruptly) to produce steam. The steam can

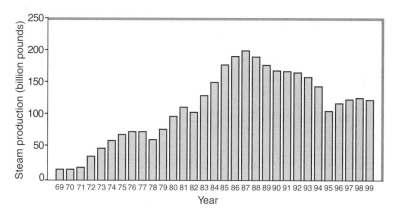

Figure 8.8
History of steam production at The Geysers geothermal field. The decline in output during the 1990s was halted by wastewater injection.

drive a turbine-generator, but first it's necessary to remove any liquid water present. Thus a hot-water-based geothermal power plant is more complicated than one using geothermal steam, because it needs a device to separate the steam and water. This drives up the cost, although hot-water geothermal systems can still compete with conventional sources of electrical energy. Figure 8.9a shows a typical water-based system; note that both the condensate and water separated from the steam are injected back into the ground.

Finally, even moderate-temperature hydrothermal systems—from about 90°C to 150°C—can produce electrical energy, but not by using steam turbines. Instead, geothermal hot water extracted from the ground passes through a **heat exchanger**, a device that transfers energy from one fluid to another. In this case the second fluid, called the *working fluid*, has a lower boiling point than water. The working fluid boils to vapor, which drives a turbine-generator. It then passes through a condenser, liquefies, and is recycled into the heat exchanger (Fig. 8.9b). Because it involves two fluids with separate pathways, such a power plant is called a **binary system**. Binary systems are more complex and more expensive than steam-based geothermal plants, and because of the lower temperatures involved, they have lower thermodynamic efficiency. But they do have some advantages. The geothermal water doesn't boil, which decreases the loss of both water and pressure in the geothermal resource and makes for more efficient heat transfer from the surrounding rocks. And binary technology minimizes some of the environmental

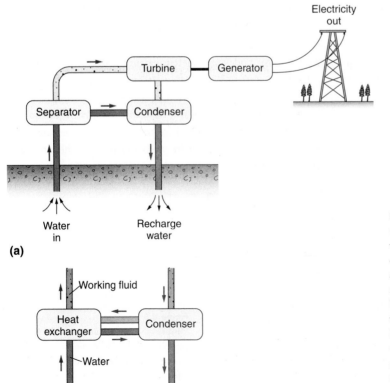

(a)

(b)

Figure 8.9
(a) In a hot-water geothermal power plant, water boils in the well pipe and a separator extracts the steam to drive a turbine. The separated water and condensate are then returned to the geothermal source. (b) A binary geothermal plant uses a closed loop with a separate working fluid, which has a boiling point lower than water's. The heat exchanger transfers energy from the warm geothermal water to the working fluid. Structures shown replace the separator and condenser shown in (a).

Figure 8.10
Binary-cycle geothermal power plant at Mammoth Lakes, California. The plant transfers heat from geothermal water at 170°C to its working fluid, isopentane, which boils at 28°C. Four separate units produce a total of 40 MW of electric power. The plant was designed to blend with its scenic surroundings, and the closed-loop binary cycle minimizes emissions.

problems associated with geothermal energy (more on this in Section 8.3). Figure 8.10 shows a 40-MW binary geothermal power facility at Mammoth Lakes, California.

Other geothermal resources that may prove useful in the future include hot, dry rock and magma itself. Energy extraction from dry rock entails drilling very deep holes into the hot rock layer, then pumping cold water down and back out as hot water. An experimental dry-rock program in New Mexico uses holes bored some 3 to 4 km deep into rock at temperatures of 175°C to 425°C. The results suggest that dry-rock energy extraction is probably not at present economically viable. Nevertheless, hot, dry rock is the most widely available geothermal energy resource, occurring at kilometer-scale depths everywhere on the globe.

Near-surface magma is generally available in regions of active volcanism, but extreme temperatures, corrosive fluids, physical danger, and economic uncertainties associated with the long-term survival of installations near active volcanoes all suggest that we'll not soon be extracting energy directly from magma. Nevertheless, an experiment in Hawaii in the 1980s demonstrated the technical feasibility of tapping energy from liquid magma beneath the solid surface of a recently formed lava lake. In this case, however, there's a vast gap between technical feasibility and practical, economic viability.

Geothermal Energy Production: A Worldwide Look

The United States leads the world in total geothermal energy production, which is not surprising, given that it has the third-largest population and is among the most technologically advanced nations. But geothermal sources supply only about 0.4 percent of U.S. electrical energy. Other countries have been much more aggressive in developing geothermal energy, usually because they have substantial geothermal resources. Table 8.1 ranks the leading producers of geothermal electricity by the percentage of their electricity generation capacity that comes from geothermal sources. I've included California in the list because the state produces one-third of the world's geothermal electricity and its

Table 8.1 Geothermal Electricity Generation: The Top Nine Countries plus California, the United States, and the World

Country	Portion of electricity generation capacity from geothermal sources (percent)	Total geothermal power generation capacity (MWe)
Nicaragua	17	78
Philippines	16	1,931
El Salvador	15	105
Kenya	15	127
Iceland	13	200
Costa Rica	7.8	162
New Zealand	5.1	453
California	4.9	2,156
Indonesia	3.0	807
Mexico	2.4	953
United States	0.37	2,395
World	0.25	8,700

population exceeds that of some countries. Do you notice anything that the countries with the highest percentage of geothermal electricity have in common? All are located in regions of high volcanic activity, as you can see with a look back at Figure 8.1.

The data in Table 8.1 are just for electrical energy. The geothermal energy extracted worldwide for direct heating is more difficult to quantify, but it's roughly comparable to what's used for electrical energy. And some countries get substantially more of their *total* energy from geothermal sources than Table 8.1 might suggest. Iceland, for instance, uses geothermal energy extensively for community heating, with some 87 percent of Iceland's homes relying on geothermal heat. Coupled with its significant geothermal electricity production (nearly all the rest is hydropower), this makes Iceland the world leader with more than 50 percent of its total primary energy coming from geothermal sources.

8.3   Environmental Impacts of Geothermal Energy

Geothermal energy is unquestionably more "green" than fossil fuels, but no energy source is completely benign. We've already seen that geothermal energy isn't always renewable, in which case we need to worry about depleting the local geothermal resource. And geothermal energy extraction has a number of significant environmental impacts, although, again, they pale in comparison with the environmental consequences of fossil fuel combustion.

Most geothermal energy systems emit both pollutants and CO_2 to the atmosphere. That's because pressurized geothermal fluids contain dissolved gases, which escape to the atmosphere when fluids are brought to the surface. Typically, geothermal power plants produce about 13 percent as much CO_2 per unit of electricity generated as do coal plants, and more than one-third the CO_2 of gas-fired plants. Geothermal water also contains dissolved sulfur and nitrogen compounds, some of which escape to the atmosphere. But, again, emissions are small compared with fossil-fueled plants. The quantity of sulfur dioxide from a typical geothermal plant is only about 1 percent of what a comparable coal-fired plant emits, and NO_x emissions are even lower. Another gas that's particularly abundant in geothermal fluids is hydrogen sulfide (H_2S), which is toxic in high concentrations but is usually nothing more than a nuisance because of its characteristic rotten egg smell. You've smelled hydrogen sulfide if you've visited a natural geothermal area such as Yellowstone. Hydrogen sulfide reacts quickly with oxygen, so it's not a long-lasting pollutant.

Early geothermal plants simply dumped condensed steam and geothermal water to nearby rivers or ponds, but geothermal fluids are rich in salts and other contaminants—as much as 30 percent dissolved solids in water-based systems—and the result was significant water pollution. Again, you may have seen the natural analog in the streams that carry geothermal runoff through areas like Yellowstone. Today, geothermal fluid is reinjected into the ground to sustain the hydrothermal system and maintain geothermal energy output; incidentally, this process also minimizes water pollution. However, fallout from steam vented to the atmosphere can still cause contamination of land and surface waters.

Binary geothermal systems (recall Figs. 8.9b and 8.10) greatly reduce the potential for land and water pollution. That's because geothermal fluids in these systems never boil, and they flow in a strictly closed loop from the ground, through the heat exchanger, and back into the ground. Despite their greater complexity and lower efficiency, this environmental benefit is one reason these plants are becoming increasingly popular. Another is their ability to exploit moderate-temperature geothermal resources.

Geothermal energy also presents some unique environmental issues. Many geothermal sites are in scenic natural areas or in regions that are valued as attractions because of the surface geothermal features themselves. Siting industrial-scale power plants in such areas can be an obvious aesthetic affront, although, as Figure 8.10 shows, it's possible to reduce their visual impact. Geothermal plants are also noisy, particularly with the shrill sound of escaping steam. A more serious problem is that the removal of geothermal fluids from the ground can cause land subsidence—slumping of the ground level as the land drops to fill the void left by the extracted fluid and the pressure it exerted. Modern plants that use water reinjection have substantially reduced this problem. However, injection of geothermal fluid or additional wastewater poses another problem: The process lubricates seismic faults and increases the frequency of small earthquakes. At The Geysers, for example, studies show that the rate of "micro-earthquakes" correlates directly with steam extraction for power generation. This rate increased dramatically with the additional injection of Santa Rosa wastewater in 2003. Whether these quakes pose a hazard to geothermal power facilities or even to the public is an open question.

8.4 Heat Pumps

A heat pump is a refrigerator used to heat buildings. Your kitchen refrigerator extracts unwanted thermal energy from the food you put inside it, and transfers that energy to the refrigerator's surroundings. Consequently, the kitchen gets a bit warmer. A heat pump does exactly the same thing, except that it extracts thermal energy from the outdoor environment and transfers it into a building.

As I explained in Chapter 4, a refrigerator or heat pump is, conceptually, a heat engine run in reverse. A heat engine extracts energy from a hot source, delivers some high-quality mechanical energy, and transfers additional thermal energy to a lower-temperature environment. The smaller the temperature difference between the hot source and the environment, the lower the engine's efficiency—meaning that it converts a smaller fraction of the extracted thermal energy into mechanical energy and dumps more as waste heat. Equation 4.5, which we've used frequently, quantifies the thermodynamic efficiency of heat engines.

Reverse a heat engine and you have a refrigerator or heat pump—a device that moves thermal energy from a cool region to a hotter one. This isn't something that happens spontaneously, but only when there's an additional input of high-quality energy. You need a lot of high-quality energy to transfer thermal energy across a large temperature difference, and less energy if that difference is small. Figure 8.11 repeats Figures 4.13 and 4.16 in summarizing this relationship between heat engines and refrigerators or heat pumps. In practice, refrigerators and heat pumps aren't actually built like heat engines, but the conceptual similarity remains.

Heat pumps are valuable because they can transfer more thermal energy than their high-quality energy input; you can see this in Figure 8.11b, where the arrow indicating thermal energy delivered to the hot region includes both the thermal energy from the cool region and the high-quality energy supplied to the device. When it's used to heat a building, a heat pump is moving thermal energy from the outdoor environment—energy that's available for free—into the building. But the whole process isn't free, because we have to supply high-quality energy to effect the unnatural energy transfer from cool to hot. This high-quality energy is invariably in the form of electricity, and we'd like to minimize the amount used.

But a heat pump, being a heat engine operating in reverse, is subject to the same thermodynamic limitation expressed in Equation 4.5. In the case of a heat pump, the appropriate measure of efficiency is called the **coefficient of performance** (COP). For pumps used to heat buildings, the most useful definition of COP is the ratio of heat

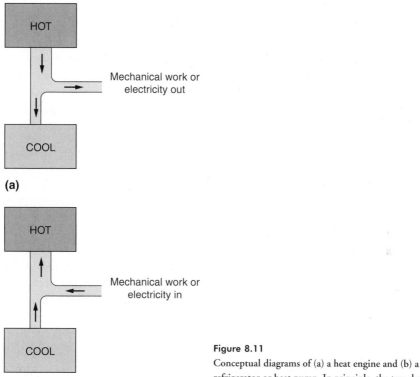

(a)

(b)

Figure 8.11
Conceptual diagrams of (a) a heat engine and (b) a refrigerator or heat pump. In principle, the two devices are similar, but operate in reverse.

delivered to the high-quality energy required to run the pump. The same physics analysis that leads to Equation 4.5 also gives an expression for the COP of a heat pump:

$$\mathrm{COP} = \frac{\text{Heat delivered}}{\text{High-quality energy required}} = \frac{T_b}{T_b - T_c} \tag{8.1}$$

Here T_b is the temperature inside the building and T_c the temperature of the outdoor environment from which heat is extracted. A look at the denominator of Equation 8.1 clearly shows that the COP increases as the temperature difference between interior and outside decreases. As with Equation 4.5, temperatures here must be given in absolute units (kelvins in the SI system). And both equations give theoretical maximum values; the COPs of real heat pumps are significantly lower than the limit given in Equation 8.1.

To understand the COP, consider a typical electrically driven heat pump with a COP of 4. This number tells us that for every one unit of electrical energy used, the pump delivers four units of energy into the building being heated. Where do the other three units come from? As Figure 8.11b shows, they're extracted from the cooler outside environment. So the heat pump gives us a good deal: four units of energy for the price of one! And those four units of heat energy delivered to the building carry the environmental impact associated with only one unit of electrical energy.

One caveat, though: If the electricity comes from a thermal power plant—fossil, nuclear, or even geothermal—with a typical efficiency around 33 percent, then that one unit of electrical energy means three units of fuel energy were consumed back at the power plant. Still, we get four units of thermal energy for those three units of fuel energy, which is better than if the fuel were burned directly to produce heat. Example 8.2 explores this idea further.

Example 8.2 Heat Pump versus Gas Furnace

A natural gas furnace operates at 85 percent efficiency, meaning it converts 85 percent of the fuel energy into useful heat. An electrically driven heat pump gets its electricity from a 45 percent efficient gas-fired power plant, and its COP is half the maximum possible value given in Equation 8.1. If the low-temperature source for the heat pump is at 8°C and it's used to produce water at 75°C for circulating through the house, which heating system uses the least fuel?

Solution

With 85 percent efficiency, the natural gas furnace uses $1/0.85 = 1.2$ units of fuel to supply one unit of thermal energy to the house. To find the comparable figure for the heat pump we need its COP, which we're told is half the value given by Equation 8.1. The Celsius temperatures we're given translate into kelvins by adding 273: $T_c = 281$ K, $T_h = 348$ K. Then we have

$$\text{COP} = (0.5) \frac{T_h}{T_h - T_c} = (0.5) \frac{348 \text{ K}}{348 \text{ K} - 281 \text{ K}} = 2.6$$

Note that COP is a dimensionless ratio; it has no units. Here our COP of 2.6 means that the pump delivers 2.6 units of heat for every unit of electrical energy. Equivalently, for every one unit of heat delivered to the house, the pump uses $1/2.6 = 0.38$ unit of electrical energy. But the electricity comes from a power plant that's only 45 percent efficient, so to produce this 0.38 unit of electrical energy requires $0.38/0.45 = 0.84$ unit of fuel energy at the power plant.

So here's our comparison between the two gas-fueled approaches to heating our house: Burning gas directly in the home furnace requires 1.2 units of fuel energy, whereas the heat pump uses considerably less, namely 0.84 unit of fuel energy. Both, by the way, are vastly more efficient than using electricity directly to heat the house; although this approach is essentially 100 percent efficient in converting high-quality electrical energy to heat, the power plant's inefficiency means that we would still need to burn $1/0.45 = 2.2$ units of gas to supply the house with one unit of heat.

Clearly, the heat pump in this example is energetically superior to the furnace. Whether it's economically superior depends on the relative costs of gas and electricity, tempered by the greater installation costs for the heat pump. But in the long run, the heat pump may well be the economic winner.

Ground-Source Heat Pumps

What does all of this have to do with geothermal energy? In warmer climates, so-called air-source heat pumps are often used to provide both winter heating and summer air conditioning. These units exchange thermal energy between a building's interior and the outside air. In winter, energy is pumped from the cooler outside air into the house—with, of course, the thermodynamically mandated input of high-quality electrical energy. In the summer, the heat flow is reversed and the pump removes thermal energy from inside and transfers it to the outside air, thus cooling the building. With the mild winter temperatures of warmer climates, the temperature difference is never great and the COP of Equation 8.1 is acceptable. But the COP of an air-source heat pump in regions with cold winter temperatures would be so low that, when coupled with power plant inefficiencies, the overall energy efficiency could be less than that of a heating system burning fuel directly.

Enter the **ground-source heat pump**. These units extract energy from the ground at depths of a meter or more—where the temperature stays essentially constant year round at what is the mean temperature for the region. In the northern United States, for example, the temperature a meter or so down is typically around 5°C to 10°C (41°F to 50°F). By avoiding low wintertime air temperatures, ground-source heat pumps maintain acceptable COPs of typically 3 to 4. The best available units have COPs exceeding 5.

Ground-source heat pumps come in several basic types. Some use tubing buried at shallow depths, with a heat-transfer fluid circulating in a closed loop (Fig. 8.12). Others draw water from surface ponds or from wells as much as several hundred meters deep. A heat exchanger within the pump transfers energy from the water to the pump's working

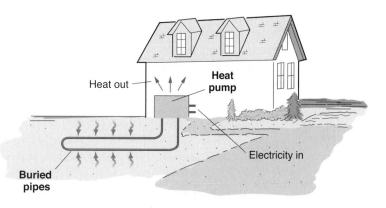

Figure 8.12

Heat-pump system using a closed loop with working fluid in buried pipes.

fluid, and the water is then returned to the environment. Ground-source heat pumps can be reversed to provide summer cooling. In fact, given favorable subsurface conditions—typically, a water-saturated region that's thermally isolated from its surroundings—energy pumped out of the building in the summer is stored as increased thermal energy beneath the building, which can be pumped back in the winter. Inevitable heat loss and the fundamental limitations of the second law of thermodynamics mean this process isn't perfect. Nevertheless, this is a rare case where some modest "energy recycling" takes place.

Ground-source heat pumps are often called *geothermal heat pumps,* and they are frequently listed among applications of geothermal energy, which is the reason they're included in this chapter. In fact, inventories of geothermal energy usage show that, worldwide, ground-source heat pumps comprise the dominant use of geothermal energy for direct heating. That's because ground-source heat pumps use the modest subsurface temperatures available anywhere, not the higher temperature resources available only in geologically active regions.

But do ground-source heat pumps really qualify as geothermal energy? In one sense, yes, because they extract energy from the ground. But the top layers of the ground are strongly influenced by surface conditions, and much of the thermal energy stored in those layers is actually solar in origin. One clue to this fact is that the near-surface geothermal gradient reverses in the summer, when the surface temperature is greater than the temperature a meter or so down. One has to go to significant depth, on the order of 100 m or more, before the surface influence disappears and the temperature starts to rise in response to Earth's high interior temperature. So although I'm including ground-based heat pumps in the chapter on geothermal energy, be aware that their energy source isn't unambiguously geothermal. Exercise 6 explores this point quantitatively.

8.5 Tidal and Ocean Energy

Tidal energy is unique in that it's the only one of Earth's energy flows that originates as high-quality mechanical energy. You're certainly aware that the Moon's gravity is in some way responsible for the tides. Actually, it's not so much gravity itself but the *variation* in gravity with position that causes tides. The Sun's *direct* gravitational effect on Earth is some two hundred times stronger than the Moon's, but the Moon's proximity makes the variation in its gravity more noticeable. As a result, the Moon's *tidal* effect is greater, although only about double the Sun's.

Origin of the Tides

Figure 8.13 shows how tides would occur in a simplified Earth covered entirely with water. The Moon's gravity decreases with distance, so the part of the ocean closest to the Moon experiences a stronger attractive force than the solid Earth itself. This results in a tidal bulge in the ocean facing the Moon. The solid Earth, in turn, experiences a stronger pull than the ocean water on the far side of the planet, and this effect leaves the water bulged outward on the far side as well. As Earth rotates beneath the Moon, the tidal bulges sweep by a given point twice a day. That's why there are (usually) two high tides each day. Because the Moon revolves around Earth every 27.3 days, it appears at a slightly

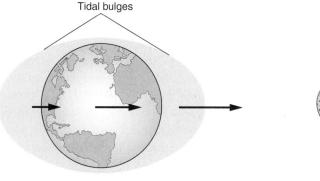

Tidal bulges

Earth

Moon

Figure 8.13
Tides result from the variation in the Moon's gravity (arrows). In an ideal ocean-covered Earth, there would be two tidal bulges on opposite sides of the planet. (The bulges shown are greatly exaggerated.)

different position in the sky at a given time on successive days, which is why the time of high tide shifts from day to day. The Sun's tidal contribution adds directly to the Moon's, making for especially high tides when the Sun, Moon, and Earth are all in a line—namely, at new Moon and full Moon.

But things aren't as simple as Figure 8.13 suggests. The tilt of the Moon's orbit introduces tidal variations with latitude. Tides affect the entire ocean from surface to bottom, and therefore water depth strongly affects tidal amplitude. In the open ocean the tidal bulge is just a fraction of a meter in height, but in shallow coastal water it can be much higher. Bays and inlets, particularly, affect tidal amplitude through their shape and bottom structure, and through resonance effects involving wave motion of tidal flows. The world's highest tides, at Nova Scotia's Bay of Fundy, range up to 17 m (Fig. 8.14).

Figure 8.14
Canada's Bay of Fundy has the world's highest tides. At high tide these rock pillars will be islands.

The Tidal Energy Resource

The energy associated with the tides comes ultimately from the motions of the Earth and Moon. The result is a minute slowing of Earth's rotation and an increase in the Moon's orbital radius. Our estimates of the tidal energy input to Earth come from sensitive measurements of the decrease in Earth's rotation. These yield a tidal energy rate of about 3 TW, approximately 30 percent of humankind's total 10^{13}-W (10 TW) energy-consumption rate. About two-thirds of this tidal energy, however, is dissipated in the open ocean. The remaining third appears in bays, estuaries, and coastal shallows, where it's ultimately lost to frictional heating. This is the energy that's available, before it dissipates, for conversion to useful forms.

Harnessing tidal energy is practical only where the tidal range is several meters or more, and where the tidal flow is concentrated in bays, estuaries, and narrow inlets. As a result, probably only about 10 percent of the 1 TW potential—about 100 GW, or the equivalent of one hundred large fossil or nuclear power plants—is actually available to us. Even that amount may not all be economically exploitable. Figure 8.15 shows some of the world's most promising tidal sites.

Harnessing Tidal Energy

Tidal power has been used for centuries to grind grain and for other tasks traditionally powered by running water. Impounding water at high tide and then letting it turn a water wheel as it flows out at low tide provides a reliable but intermittent power source. In the latter part of the twentieth century, this same idea saw large-scale application with the construction of a handful of tidal power plants. The first and largest, near the outlet of the La Rance River in France, was completed in 1967 (Fig. 8.16). The La Rance River plant has a peak power output of 240 MW and was designed to generate power from both the incom-

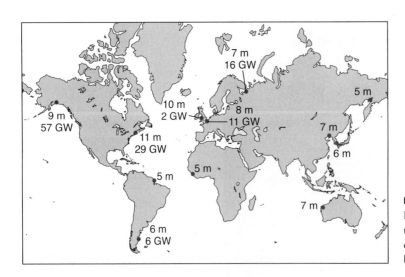

Figure 8.15
Locations of some of the world's highest tides, with tide amplitude shown and estimated tidal power generation capacity listed for some sites.

Figure 8.16
The world's first and largest tidal power plant, on France's La Rance River, has a peak electric power output of 240 MW.

ing and outgoing tides. The plant produces power near the times of highest and lowest tides, which translates into power generation equal to only about one-fourth of what would be obtained if the plant ran all the time. Furthermore, siltation problems at La Rance have made it difficult to operate with flows in both directions.

The twenty-first century saw a resurgence of interest in tidal power, especially in Europe. The new schemes use devices that are like underwater versions of the wind turbines we'll discuss in Chapter 10, capturing the energy in tidal currents without the need for impoundment dams like the one at La Rance. Like their wind-driven cousins, the tidal turbines rotate to harness energy, in this case operating on both inflow and outflow of the tides. Individual tidal turbines have a power output of around 300 kW, but plans call for large-scale turbine farms generating as much as 100 MW—significant, but still small compared with full-scale fossil and nuclear power plants.

Environmental Impacts of Tidal Energy

The traditional approach to tidal energy involves constructing large dams at the inlets to tidal bays or estuaries. The La Rance River dam, for example, is 750 m long and impounds some 17 km^2 of water. One effect of such dams is to reduce the overall tidal flow and the salinity of the water. Dams also impede the movement of larger marine organisms. Both effects can result in ecological changes. Shifting sediments and changes in water clarity may have additional adverse impacts. These changes are particularly significant because estuaries—nutrient-rich regions where fresh and salt water mix—harbor a rich diversity of marine life and act as the nurseries for many commercially important fish species.

The newer bottom-mounted turbines are more environmentally friendly, but in large numbers they, too, will reduce tidal flows and therefore alter the balance of fresh and salt water. To date there have been few comprehensive reviews of the environmental impact of tidal power.

Waves and Currents

Waves and currents, like the shallow "geothermal" heat pumps we discussed earlier, don't really belong in this chapter. That's because solar-induced temperature differences between tropical and higher-latitude waters drive the ocean currents, which are therefore mainly indirect manifestations of solar energy. Similarly, winds arise from atmospheric temperature differences, and they in turn drive the ocean waves. Because waves and currents are often considered "ocean energy" and thus related to tidal power, we'll deal with them here.

Today, wave energy is tapped mostly in small amounts to power remote ocean buoys. These devices use the compression of air as the buoy bobs up and down to drive a turbine that generates electricity. Because they operate best in the open ocean, wave-energy schemes present an obvious problem with transmission of power to land. Nevertheless, Portugal and Scotland have added several megawatts of wave power generation to their national grids (Figure 8.17), and the coming years should see modest increases in contributions from wave energy. A 2005 study by the Electric Power Research Institute and the National Renewable Energy Laboratory suggests the potential for 10 to 20 GW of wave power generation within U.S. territorial waters. But a lot of technology still needs to be developed before wave power makes a substantial contribution, and in any event 10 to 20 GW is but a small fraction of total U.S. electric power generation.

Even less has been done to harness ocean currents for energy production. Some estimates suggest that we may be able to capture as much as 450 GW of power worldwide

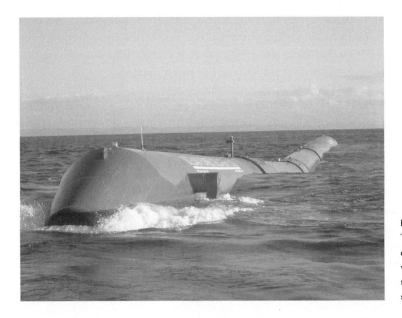

Figure 8.17
This snakelike device generates 750 kW of electrical power as it flexes in the ocean waves. "Wave farms" consisting of several of these units are linked to the shore with a single cable laid on the sea floor.

from these vast "rivers" that flow within the oceans. Although this is significant—about a quarter of the world's electric power consumption in the early twenty-first century—the high capital costs of producing equipment that can operate reliably for years on the ocean floor mean that it will be a long time, if ever, before ocean currents make even a modest contribution to our energy supply.

Prospects for Ocean Mechanical Energy

Tides, waves, and ocean currents all present intriguing possibilities for converting the mechanical energy of moving water into electricity. But the overall energy resources of all three are limited, and the technologies to take advantage of them have a long way to go. As with geothermal energy, there are local regions where ocean energy may become important, but realistically, these resources are unlikely to make a significant contribution to the global energy supply.

Chapter 8   Chapter Review

BIG IDEAS

8.1   Geothermal heat supplies only about 0.02 percent of the energy reaching Earth's surface, but in a few localized regions the geothermal resource is significant.

8.2   Technologies for extracting geothermal energy include direct use of geothermal heat as well as electric power generation.

8.3   Environmental impacts of geothermal energy include air and water pollution from geothermal fluids, CO_2 emissions, and land subsidence.

8.4   Geothermal heat pumps use electrical energy to transfer larger amounts of energy from the ground to heat buildings. Typical heat pumps move several times as much heat as they use in electrical energy.

8.5   Tidal energy originates from the Earth-Moon system and is a limited resource that can provide a practical energy supply only in a few localities. The energy available from ocean waves and currents is greater but technologically difficult to harness.

TERMS TO KNOW

binary system (p. 234)

coefficient of performance (p. 238)

geopressured system (p. 228)

geothermal gradient (p. 225)

ground-source heat pump (p. 241)

heat exchanger (p. 234)

hydrothermal system (p. 227)

vapor-dominated system (p. 227)

GETTING QUANTITATIVE

Geothermal gradient, average: ~25°C/km

Geothermal electricity, worldwide: ~9 GW, 0.25 percent of total electric power generation

Coefficient of performance: $\text{COP} = \dfrac{\text{Heat delivered}}{\text{High-quality energy required}} = \dfrac{T_h}{T_h - T_c}$

(Equation 8.1; p. 239)

QUESTIONS

1. What are the two sources of geothermal energy?

2. Why does a binary geothermal plant require a working fluid other than water?

3. In what sense are most "geothermal" heat pumps not actually a geothermal energy source?

4. Name some environmental impacts of geothermal electric power generation.

5. What are two factors that can lead to depletion of a geothermal resource?

6. Why does the confluence of tidal energy resources and estuaries increase the environmental impact of tidal power plants?

7. What is the ultimate source of the energy that powers a lightbulb whose electricity comes from a tidal power plant?

EXERCISES

1. Calculate the total global geothermal energy flow rate, assuming an average of 87 mW/m².

2. Consider a heat engine operating between Earth's surface, at a typical temperature of 295 K, and the bottom of a geothermal well 1 km deep. (a) Given a geothermal gradient of 25°C/km, what would be the maximum possible efficiency of such a heat engine? (b) If we could tap the geothermal flow over 10 percent of Earth's land area (exclude the oceans) with such heat engines, roughly what percentage of humankind's 14-TW energy use could geothermal energy supply?

3. Use granite's thermal conductivity from Table 4.1 to find the energy flow rate per square meter through granite when the geothermal gradient is 25°C/km.

4. An American city of 1 million people uses energy at the rate of about 10 GW (10⁴ W per person times 10⁶ people). Assume the city occupies a circular area 15 km in diameter. Make the approximation that the top kilometer of Earth's crust consists of granite (density 2,700 kg/m³ and specific heat 840 J/kg/K, and thermal conductivity 3.37 W/m·K), and that the average temperature in this layer is 60°C higher than the surface temperature. (a) Estimate the

total energy that could be extracted in cooling the 1-km-thick layer of crust beneath the city to the temperature of the surface. (b) If the efficiency in converting this low-quality thermal energy to useful energy is 5 percent, how long would the 1-km geothermal resource below the city last? (c) Show that the rate at which the geothermal energy is replenished by conduction is far less than the rate of extraction, implying that this use of geothermal energy is probably not renewable. (The assumption of conduction may or may not be valid; convection in water or magma itself can increase the rate at which geothermal energy is replenished.)

5. Compare the thermodynamic efficiencies of heat engines operating from (a) a 150°C geothermal source (considered about the minimum for electric power generation) and (b) magma at 1,000°C. Assume the lowest temperature accessible to the engine is the ambient environment at 300 K.

6. The average U.S. home consumes energy at the rate of about 3.5 kW. If the footprint of a typical house is 150 m^2, and if the average geothermal heat flow is 87 mW/m^2, what percentage of the home's energy needs could be supplied by the geothermal flow through the floor area?

7. (a) Find the maximum possible COP of a heat pump that pumps energy from the 10°C ground temperature to water at 60°C. (b) At what rate would such a pump consume electricity if it were used to heat a house requiring 15 kW of heating power?

8. The actual efficiency of a geothermal power plant using a 200°C geothermal resource is only about 7 percent. Take this value as typical of California's Geysers geothermal field, which produces 850 MW of electric power. (a) Find the actual rate of geothermal heat extraction at The Geysers. (b) It's estimated that 35 EJ of energy could be extracted from The Geysers before the temperature dropped from 250°C to the 150°C minimum for a vapor-dominated geothermal plant. Given your answer in (a), how long will The Geysers geothermal resource last before the temperature reaches 150°C?

9. The binary geothermal power plants at Casa Diablo, California, operate from a geothermal resource at 170°C. If the discharge temperature is 90°C, what is the maximum thermodynamic efficiency of these power plants?

10. Consider an ocean current flowing at 2.5 m/s. (a) How much kinetic energy is contained in a cubic volume of water that measures 1 m on each side? The density of seawater is about 1,030 kg/m^3. (b) If the flow is perpendicular to one of the cube faces, what is the rate at which the current carries kinetic energy across each square meter? Your answer gives an upper limit for the power that could be extracted from the flow, although it's an unrealistic limit because you'd have to stop the entire flow.

RESEARCH PROBLEMS

1. Use the latest list of "Existing Electric Generating Units in the United States," available from the U.S. Department of Energy's Energy Information Administration, to name five geothermal power plants along with their locations and rated power outputs.

2. Nearly half the states in the United States have at least one geothermal energy facility. Does yours? Find out by exploring the U.S. Geological Survey's Circular 1249, "Geothermal Energy—Clean Power from the Earth's Heat," available at http://pubs.usgs.gov/circ/2004/c1249. List the capacity, in megawatts, of both direct-use facilities and geothermal power plants in your state.

Chapter 9

DIRECT FROM THE SUN: Solar Energy

The first thing I want to emphasize in this chapter is that there's *plenty* of solar energy. Don't let anyone try to convince you that solar energy isn't practical because there isn't enough of it. In round numbers, the rate at which solar energy reaches Earth is some ten thousand times humankind's energy-consumption rate. And it's entirely renewable; it keeps on coming regardless of how much we might divert to human uses. That's not true of the fossil fuels, which, as we've seen, will last at most a few decades, or perhaps a little longer for coal. It's not true of nuclear fission fuels, because recoverable resources of U-235 are also measured in decades, and dangerous breeder technologies that convert U-238 to plutonium might buy us a few centuries to a millennium or so. Finally, as I made clear in Chapter 8, flows of geothermal and tidal energy just aren't sufficient to meet today's global energy demand. But the Sun will shine for another 5 billion years, bathing Earth in a nearly steady stream of energy into the unimaginable future.

Direct use of solar energy isn't going to displace fossil fuels as our dominant energy source in the next few decades, although there are plenty of near-term opportunities to develop significant solar contributions to our energy mix. For the long term, however, solar energy is the one proven source that can clearly and sustainably meet our energy needs. With the possible exception of technologically elusive nuclear fusion, which would essentially build miniature Suns here on Earth (as I'll discuss in Chapter 11), there's really no other viable long-term energy source for industrialized society.

9.1 The Solar Resource

Deep in the Sun's core, fusion of hydrogen into helium releases energy at the prodigious rate of 3.84×10^{26} W. At the core temperature of 15 million K, most of this power appears in the form of high-energy electromagnetic radiation—gamma rays and X rays. As it travels outward through cooler layers, the electromagnetic energy scatters off the dense solar material and its wavelength increases until, at the surface, the Sun radiates that 3.84×10^{26} W in the visible and near-visible portions of the electromagnetic spectrum. This energy radiates outward in all directions. The total remains essentially constant, but the intensity drops as the solar radiation spreads over ever-larger areas.

The Solar Constant

At Earth's average distance from the Sun (which is the average radius of the planet's orbit, $r = 150$ Gm), solar energy is spread over a sphere of area $4\pi r^2$; the intensity, or power per unit area, is then about 1,368 W/m^2 (Exercise 1 explores the calculation leading to this figure). There's some disagreement about the exact value at the level of a few watts per square meter, but I'll stick with 1,368 W/m^2 in this book. This number is the **solar constant**, S, and it measures the rate at which solar energy impinges on each Sun-facing square meter in space at Earth's orbital distance. The solar constant isn't quite constant; it varies by about 0.1 percent—around 1.4 W/m^2— over the 22-year solar cycle. There are also longer-term variations that may be climatologically significant, as we'll see in Chapters 13 and 14. But overall, the Sun is a steady and reliable star, and for the purposes of this chapter we can consider its energy output to be constant.

Insolation

If Earth were a flat disk of radius R_E facing the Sun, its area πR_E^2 would intercept a total solar power given by $P = \pi R_E^2 S$. With $R_E = 6.37$ Mm and $S = 1,368$ W/m^2, this amounts to 1.74×10^{17} W, or 174 PW (Exercise 2). In principle, this is the total rate of solar energy input that's available to us on Earth. Currently, humankind uses energy at the rate of about 15 TW (15×10^{12} W), so the total solar input is some 12,000 times greater than our energy needs. This is the basis of my assertion that there's plenty of solar energy, roughly ten thousand times as much as we currently use. Put another way, the Sun delivers to Earth every 40 minutes the amount of energy that humankind uses in a year (see Exercise 3). Earth isn't a flat disk, of course, but its sunward side nevertheless intercepts this same 174 PW (Fig. 9.1). However, our planet's curvature means that **insolation**—incoming solar energy per square meter of Earth's surface—isn't the same everywhere. Insolation is greater in the tropics, where the surface is nearly perpendicular to the incoming sunlight, than at high latitudes, where the surface is inclined away from the Sun. That's why it's generally cooler at high latitudes.

Figure 9.1 shows that the incident sunlight is actually spread over half the surface area of the spherical Earth—that is, over $2\pi R_E^2$ rather than the flat-disk area of πR_E^2. If all the power quantified by the solar constant S reached the surface, this would mean an average insolation of $S/2$ on the sunward-facing or daytime side. Averaging this $S/2$ energy over the area of the entire globe (both the nighttime and daytime sides) introduces another factor-of-2 decrease in average insolation, which would then be $S/4$ or 342 W/m^2. Reflection by clouds and other light surfaces returns some of the incident solar energy to space, leaving an average of about 235 W/m^2 actually available to the

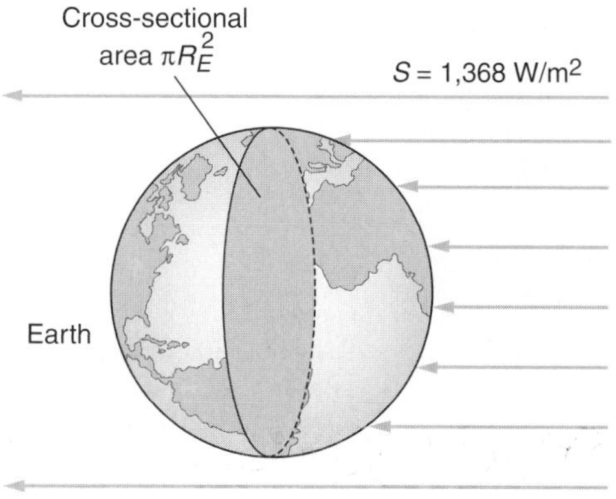

Cross-sectional area πR_E^2

$S = 1{,}368 \ \text{W/m}^2$

Earth

Figure 9.1

The round Earth intersects the same amount of solar energy as would a flat disk with an area equal to Earth's cross-sectional area (πR_E^2). The quantity S measures the power per unit area incident on an area oriented perpendicular to the incoming sunlight.

Earth-atmosphere system; of this, some is absorbed in the atmosphere before reaching Earth's surface. We'll take a closer look at these figures when we consider climate in Chapter 12. For now we'll consider that we have a global average of a little over $200 \ \text{W/m}^2$ of available solar energy, with more getting through to lower latitudes and regions with clearer-than-average skies. Another rough figure to keep in mind is that midday sunlight on a clear day carries about $1{,}000 \ \text{W/m}^2$, or $1 \ \text{kW/m}^2$, and that this energy is available on any surface oriented perpendicular to the incoming sunlight.

Example 9.1   Solar Houses versus Solar Cars

The roof area for a typical house is around $150 \ \text{m}^2$ and that of a car is around $7 \ \text{m}^2$. (a) Find the midday solar power incident on a typical house roof, and compare it with the average U.S. household energy-consumption rate of 3.5 kW. (b) Find the midday solar power incident on a car roof, and compare it with the 300-horsepower (hp) rating of a typical engine. Assess the practicality of solar-powered houses and solar-powered cars.

Solution

(a) At $1 \ \text{kW/m}^2$, the house roof is exposed to 150 kW in bright sunlight—far more than the household's average energy-consumption rate of 3.5 kW. Nonoptimal roof angles, conversion inefficiencies, clouds, seasonal variations, and nighttime periods all decrease the lopsidedness of this comparison, but nevertheless it's clearly not unrealistic to consider meeting all or most of a home's energy needs through the solar energy falling on its roof.

(b) The car is a different story. One kilowatt per square meter on its 7-m^2 area gives only 7 kW. Since 1 hp is about 0.75 kW, our 7 kW is about 10 hp, which conversion inefficiencies would reduce to perhaps 2 hp. So we'll not see cars powered by roof-mounted solar energy collectors, at least as long as we insist on having hundreds of horses under

Figure 9.2
Stanford University engineering students built this solar-powered car they named Solstice. It won the 2005 North American Solar Challenge.

our hoods. Pure solar cars have been built and raced across entire continents (Fig. 9.2), but they'll probably never make for practical everyday transportation. Cars powered indirectly by solar energy are another matter entirely, as we'll explore later in this chapter and in Chapter 11.

Distribution of Solar Energy

Figure 9.1 shows that insolation varies significantly with latitude; clearly it also varies with time of day, with season, and with local weather conditions. At sunrise the Sun's rays barely skim across the horizon, making for minimal insolation on a horizontal surface. As the day progresses, insolation increases through local noon, when the Sun is highest in the sky, and then decreases. On a perfectly clear day, the result would be a smooth curve of insolation versus time, but when passing clouds block the Sun, insolation fluctuates (Fig. 9.3). Note that even with cloud cover, the insolation in Figure 9.3 doesn't

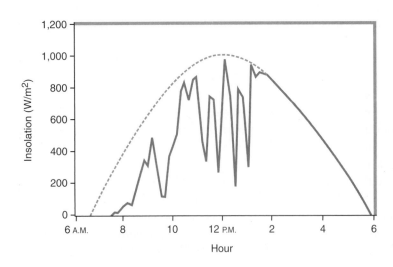

Figure 9.3
Insolation measured on a mid-October day in Middlebury, Vermont, at 44° north latitude. The dashed curve represents an ideal cloudless day.

drop to zero. That's partly because clouds aren't completely opaque, and also because Earth's atmosphere scatters sunlight. Scattering means that solar energy comes not only straight from the direction of the Sun—so-called **direct insolation**—but also from throughout the sky—called **diffuse insolation.** (Incidentally, atmospheric scattering is the reason why the sky appears blue.) Some types of solar energy collectors can use both direct and diffuse sunlight, so they continue to function, albeit less effectively, on cloudy days.

The total daily solar energy input on a horizontal surface depends on latitude, season, and length of day. As you move toward the equator, the peak insolation on a horizontal surface becomes higher, but in summer the length of day shortens as one nears the equator. The combined effect of the Sun's peak angle in the sky along with the length of day gives a surprising result for the maximum possible average daily summer insolation on a horizontal surface: It's actually lowest at the equator and greater toward the poles (Fig. 9.4). But the long polar night of the winter months overcompensates, giving lower latitudes a substantial edge in overall yearly insolation. Also significant is cloud cover, which tends to be greater in the equatorial zones and at higher latitudes; the result is that the actual average daily summer insolation peaks at around 30° latitude, in a zone containing most of the world's major deserts. Figure 9.4 shows actual insolation for selected cities in the Northern Hemisphere, with an approximate curve depicting this peaking effect at about 30° north latitude. Variations in climatic conditions with longitude also greatly influence availability of solar energy, as shown for the United States in Figures 9.5 and 9.6 and Table 9.1. Of course it doesn't make sense to put a solar collector flat on the ground, so Table 9.1 also lists insolation for surfaces tilted at a location's latitude—a common compromise orientation for simple solar collectors. A collector

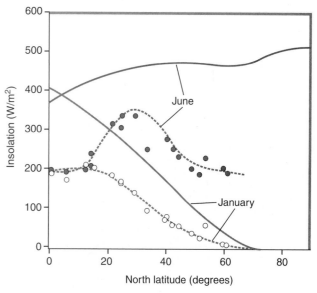

Figure 9.4

Theoretical maximum horizontal-surface insolation (solid curves) versus latitude for January and June. Also shown are actual data points for selected world cities in January (open circles) and June (solid circles). Dashed curves are trends in the insolation data, indicating the summer insolation peak at around 30° north latitude, above which the actual insolation drops because of increasing cloudiness. Cities, in order of latitude, are Singapore, Jakarta, Bangalore, Dakar, Manila, Honolulu, Abu Dhabi, Riyadh, Cairo, Atlanta, Ankara, Madrid, Detroit, Bucharest, Paris, London, Edmonton, Oslo, Anchorage.

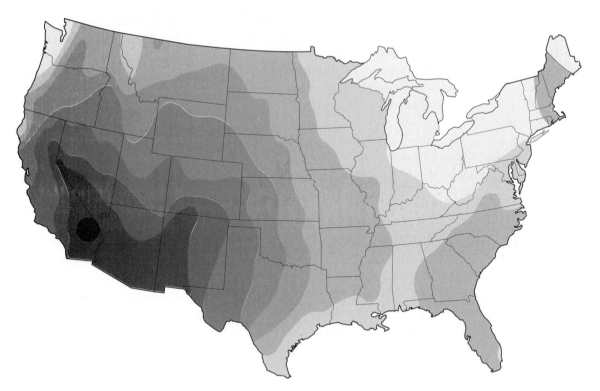

Figure 9.5
Annual average insolation in the continental United States, for surfaces oriented perpendicular to the incident sunlight. Values range from about 150 W/m² for the lightest shading to over 500 W/m² for the darkest.

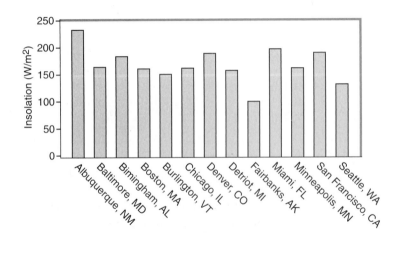

Figure 9.6
Annual average insolation on a horizontal surface for selected U.S. cities.

Table 9.1 Average Monthly Insolation (W/m²) and Temperature for Selected U.S. Cities

City	January Hori-zontal	January Tilted at latitude	January Average temperature (°C)	June Hori-zontal	June Tilted at latitude	June Average temperature (°C)	September Hori-zontal	September Tilted at latitude	September Average temperature (°C)
Albuquerque, NM	133	221	1.2	338	296	23.4	246	283	20.3
Baltimore, MD	88	146	−0.1	258	233	22.5	183	213	20.3
Birmingham, AL	104	154	5.3	258	233	24.6	200	221	23
Boston, MA	79	142	−1.9	254	229	19.8	179	213	18.2
Burlington, VT	67	121	−8.7	250	225	18.4	167	200	14.9
Chicago, IL	75	129	−6.2	263	238	20.3	175	204	18
Denver, CO	100	183	−1.3	288	254	19.4	208	250	16.8
Detroit, MI	67	113	−5.1	258	233	19.8	171	200	17.3
Fairbanks, AK	4	29	−23.4	233	196	15.4	96	142	7.5
Miami, FL	146	196	19.6	233	213	27.4	204	213	27.7
Minneapolis, MN	75	146	−11.2	263	233	20.1	171	208	15.8
San Francisco, CA	92	146	9.3	300	271	16.4	225	267	18.1

that tracks the Sun can maximize solar energy collection, and Figure 9.5 shows insolation for Sun-tracking surfaces. However, tracking mechanisms greatly increase the expense of solar collector systems and so aren't widely used (more in Section 9.4).

Figure 9.6 suggests that the solar resource varies substantially but not overwhelmingly with geographic location. Sunny Albuquerque, for example, gets less than twice the average insolation of cloudy Seattle, and only about 50 percent more than typical northern cities such as Chicago, Detroit, or Burlington, Vermont. So although some places are better endowed with solar energy, even the worst sites aren't orders of magnitude worse. It probably doesn't make sense to put expensive large-scale solar power stations in places like Seattle or Detroit, but some of the more individualized solar applications I discuss in this chapter could work effectively almost anywhere.

The Solar Spectrum

The amount of solar energy isn't the only factor important in describing the solar resource. Also significant is the distribution of wavelengths in sunlight. The Sun's spectrum is approximately the same as a hot, glowing object at about 5,800 K, a temperature at which most of the radiation lies in the visible and infrared, with a little in the ultraviolet. In the Sun's case, the shorter wavelengths—ultraviolet and deeper blue in the visible spectrum—are somewhat reduced from the theoretical distribution. The result, as shown in Figure 9.7, is that the solar output is nearly equally divided between infrared (47 percent) and visible light (46 percent), with the remaining 7 percent in the ultraviolet. These proportions are found at the top of the atmosphere, but most of the ultraviolet and some longer-wavelength infrared are absorbed in the atmosphere and don't reach Earth's surface.

You've probably seen a magnifying glass used to concentrate sunlight and set paper on fire. In principle, this approach can achieve temperatures up to that of the 5,800-K solar surface (Question 2 asks you to consider why we can't go higher). The high solar temperature makes sunlight a high-quality energy source. As you can show in Exercise 12, a heat engine operated between 5,800 K and Earth's average ambient temperature of around 300 K would have a thermodynamic efficiency of 95 percent. Practical limitations of solar energy concentration and heat engine design reduce this figure considerably, but nevertheless concentrated sunlight remains a high-quality energy source.

At the same time, the fact that just over half the solar energy comes to us as infrared radiation means that most solar photons have lower energy than those of visible light (recall Equation 3.1, which describes the quantization of electromagnetic wave energy). Later in this chapter we'll see how this limits the efficiency with which we can convert sunlight directly into electricity.

These energy-quality and photon-energy issues aren't relevant to the simplest of solar applications, namely solar heating, because solar heating systems absorb nearly all the incident solar energy and convert it into low-quality heat. We'll consider solar heating in the next two sections; after that we'll turn to the use of solar energy to generate electricity, an application for which energy quality and photon energy are important considerations.

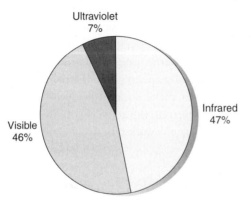

Ultraviolet
7%

Infrared
47%

Visible
46%

Figure 9.7

Distribution of solar energy among infrared, visible, and ultraviolet portions of the spectrum. Wavelengths between 380 nm and 760 nm make up the visible region.

9.2 Passive Solar Heating

A simple, straightforward, and effective use of solar energy is to heat buildings. In **passive solar heating** schemes, the building design itself maximizes the solar energy input and ensures that energy is distributed evenly throughout the building and over time. What distinguishes a passive system is the absence of active devices such as pumps and fans to move energy about.

Passive heating begins with building orientation. Siting a conventional house with the majority of its windows facing south (in the Northern Hemisphere) can provide considerable solar gain on clear winter days, as we found in Example 4.3. This, in turn, reduces the fuel consumption of a conventional heating system. But passive solar designs can do much better, with the best of them eliminating the need for conventional heating systems in all but the coldest, cloudiest climates.

Passive solar design entails four distinct elements. First is the collection of solar energy through large areas of south-facing glass. Sunlight passes through the glass, and its energy is absorbed in the building's interior, raising the temperature. There's no point in collecting solar energy only to have the solar-generated heat escape through walls and windows, so the second essential element is high-R insulation (recall Chapter 4's discussion of building insulation and R values). High-R insulation is important in *any* building, but it's especially crucial in a passive solar structure. Third, the solar design must promote the distribution of energy throughout the building—something that's done with pumps and fans in conventional heating. Finally, passive solar design must compensate for times when the Sun isn't shining, such as at night (predictable) and when it's cloudy (less predictable). So the fourth element in passive design is energy storage, which allows the building to accumulate and store excess energy when the Sun shines, then gradually release that energy as needed.

Figure 9.8 shows a simple house incorporating all of these design elements. The south wall is largely glass, to admit sunlight. The north-facing wall is windowless and heavily insulated. There's more insulation in the roof and under the floor. A thick, dark-colored concrete floor slab absorbs the incoming solar energy and warms. Concrete's relatively large thermal conductivity (1 W/m·K) ensures an even temperature throughout the concrete slab and therefore throughout the house. Concrete's substantial heat capacity (880 J/kg/K) means that it takes a lot of energy to change the concrete's temperature. Once heated, the concrete slab is therefore a repository of stored thermal energy, and it takes a long time to cool down. An added bonus is that this property of concrete helps to prevent overheating on sunny, warm days. Thus the concrete slab in Figure 9.8 functions as a **thermal mass**, which, because it's slow to heat and cool, helps to promote an even temperature over both time and space. Although the house in Figure 9.8 is designed for winter heating, it has one feature to combat summer heat: a roof overhang that reduces solar gain by blocking the more vertical summertime solar radiation.

A black concrete floor may not seem aesthetically appealing, but it really isn't essential. The dark color helps, but once sunlight gets into the building it reflects around and

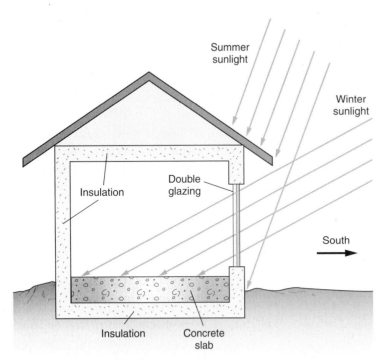

Figure 9.8
A passive solar house using direct gain and a concrete slab for thermal mass.

is eventually absorbed, whatever the color. A darker color absorber is more effective because it ensures immediate absorption before light has a chance to bounce right back out of the windows. And the thermal mass doesn't have to be concrete. As Table 4.3 shows, water's high specific heat makes it a particularly effective heat-storage medium. Although it's hard to use water as a building material, there are plenty of passive solar homes and greenhouses that use stacks of black-painted 55-gallon drums filled with water as their thermal mass. Even wood can function as a thermal mass; Table 4.3 shows that wood's specific heat is higher than that of concrete, but wood's much lower density means a much larger volume of wood is required for the same thermal storage.

Another approach to thermal energy storage is to use materials that change phase at appropriate temperatures. In Chapter 4 we discussed the considerable energy required to change ice to water, and water to steam. While such a phase change is occurring, the ice/water or water/steam mixture maintains a constant temperature of 0°C or 100°C, respectively. Changing back to the lower-energy phase then releases the energy that went into melting or boiling. Choosing a substance that melts at a temperature just a bit above what's comfortable in a house thus allows one to store large quantities of energy when the material is in the liquid phase. One such material is Glauber salt ($Na_2SO_4 \cdot 10H_2O$), which melts at 32°C (90°F). Glauber salt's heat of fusion is 241 kJ/kg, so a tonne of Glauber salt (1,000 kg) takes 241 MJ to melt. In a solar house this energy would come from sunlight, and it would then be released slowly during sunless times as the salt solidifies, all the while maintaining its 32°C melting-point temperature. Exercise 4 explores a Glauber-salt storage system.

The glass used in passive solar construction presents a dilemma. Ideally, it should admit the most solar energy while minimizing heat loss from within. The very presence of a solid material prevents the wholesale loss of heated air, which is the most overt form of convective heat loss. Double or triple glazing can further reduce the convective heat loss associated with air motions resulting from contact with cool glass. Glass naturally impedes the outflow of infrared radiation from the heated interior of the house, and special low-E coatings can reduce radiation losses from the glass itself. But all these measures also reduce the amount of solar energy passing into the house, so there's a tradeoff between maximizing solar energy collection and minimizing heat loss. Furthermore, there's no solar gain when the Sun isn't shining, but there's still heat loss. Even the best window—triple-glazed, argon-filled, with a low-E coating—has an R value of only around 5. This is well below a typical insulated wall, which might have an R value of 19 in conventional construction and 30 in a passive solar building. So it's important to provide extra insulation for the glass, especially at night or on very cloudy days when even diffuse solar radiation is at a minimum. This extra insulation can be as simple as heavy insulated curtains or slabs of foam that swing downward from the ceiling to cover the glass, preferably with a vapor barrier and edge seals to prevent air leakage and condensation. A more sophisticated approach is the **beadwall**, in which the space between the glass panes is filled with polystyrene beads to provide an R value equivalent to a well-insulated wall. Filling and emptying the beadwall is done with blowers and suction devices that can be triggered automatically based on sunlight intensity. Unfortunately, technical challenges involving static electricity, along with excessive cost, have limited the use of this elegant solution to the window dilemma.

Example 9.2   A Passive Solar House

A passive solar house like the one shown in Figure 9.8 has a level of insulation that gives a heat loss of 120 W/°C through the walls and roof. Its solar glazing measures 8 m wide by 3 m high, passes 82 percent of the incident sunlight, and has an R value of 4. On average, the admitted sunlight completely illuminates 48 m² of the black concrete floor. Use data from Table 9.1 to find the average interior temperature of the house in January if it was located in (a) Albuquerque and (b) Detroit. (Ignore the heat loss through the floor.)

Solution

We need to balance heat loss with solar gain to find the interior temperature. We're given a heat-loss rate of 120 W/°C through the walls and roof, which means the total loss through these surfaces is (120 W/°C)ΔT, where ΔT is the temperature difference between inside and outside. For the glazing we're given the R value and the area (3 m × 8 m = 24 m²), so we need to calculate the loss rate. Recall from Chapter 4 that R values are given in English units, namely $ft^2 \cdot °F \cdot h/Btu$. Here we're given all other values in SI units, so we need to convert the R value. Exercise 15 of Chapter 4 handles this calculation; the result is that an R value of 1 $ft^2 \cdot °F \cdot h/Btu$ is equivalent to 0.176 $m^2 \cdot °C \cdot s/J$, or 0.176 $m^2 \cdot °C/W$. The glazing has an R value of 4, or $0.176 \times 4 = 0.70 \ m^2 \cdot °C/W$, meaning that each square meter of glazing loses 1/0.70 W for each degree Celsius of temperature difference across the glazing. With 24 m^2 of glazing, the loss rate is then $24/0.70 = 34 \ W/°C$. Thus the glazing loses energy at the rate $(34 \ W/°C)\Delta T$. Adding the wall and roof loss gives a total loss of $120 + 34 = (154 \ W/°C)\Delta T$.

Now for the solar gain. The absorber here is the 48 m^2 of illuminated floor, so we use Table 9.1's entries for horizontal surfaces: 133 W/m^2 for Albuquerque and 67 W/m^2 for Detroit. Multiplying by the illuminated floor area and by 0.82 to compensate for the windows passing only 82 percent of the incident sunlight, we get average solar gains of $(133 \ W/m^2)(48 \ m^2)(0.82) = 5.23 \ kW$ for Albuquerque and, similarly, 2.64 kW for Detroit. On average the house is in energy balance, so we can equate loss and gain to get

$$(154 \ W/°C)\Delta T = 5.23 \ kW$$

or

$$\Delta T = \frac{5.23 \ kW}{0.154 \ kW/°C} = 34°C$$

for Albuquerque. Table 9.1 lists Albuquerque's average January temperature as 1.2°C, so the Albuquerque house is going to be uncomfortably warm at 35°C, or 95°F. No problem: Drawing some shades across the glass will reduce the solar input. Clearly, though, the Albuquerque house is over-designed and is easily capable of supplying 100 percent of its heat from solar energy.

A similar calculation for the Detroit house, with only 2.64 kW of solar input, gives $\Delta T = 17°C$. But Detroit averages $-5.1°C$ in January, so the interior temperature will be only 12°C, or 54°F. A backup heat source, more insulation, or a different solar design is necessary to keep the Detroit house comfortable.

The numbers in this example are averages; insolation and outside temperature fluctuate of course, and so therefore would the interior temperature. But a large thermal mass could minimize such fluctuations.

An alternative to the direct-gain passive solar house of Figure 9.8 and Example 9.2 is to use **indirect gain**, heating a Sun-facing thermal mass placed so that natural convection then carries energy throughout the house. One such indirect-gain design is the vented **Trombe wall** shown in Figure 9.9. Here sunlight strikes a massive vertical wall that serves as both absorber and thermal mass. Air confined between the wall and glass warms and rises, resulting in natural convection that circulates warmed air throughout the house and returns cold air to be heated at the wall. Radiation from the back of the wall provides additional heat transfer into the house.

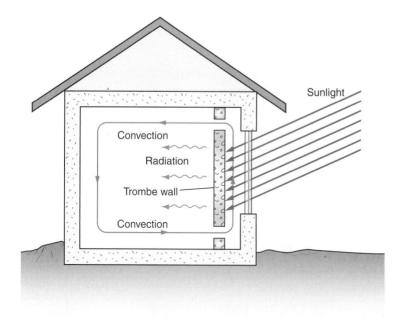

Figure 9.9
Indirect-gain solar house using a Trombe wall.

9.3 Active Solar Heating

Active solar heating systems use pumps or fans to move solar-heated fluids from **solar collectors** to an energy storage medium, and from there to where it's needed within a building. Thus active systems separate the individual functions that, in the passive houses we've just discussed, are combined and integral with the structure of the building. Active systems are more compact and allow for more flexible placement of solar components. As a result, active systems often fit better with conventional architecture. Because active solar collectors are just solar energy–catching devices that don't have to function as walls or floors, they're often more efficient than passive designs. Finally, active systems can achieve higher temperatures, making them particularly useful for domestic hot-water heating—usually the second-largest home energy use after space heating. The disadvantages of active systems are that they're more complex, with a host of moving parts from pumps to controls, and they require external power in the form of electricity. They also take more energy to manufacture, and cost more than most passive measures.

Flat-Plate Solar Collectors

The simplest and most economical solar collector is the **flat-plate collector**. The essential parts of a flat-plate collector are a black **absorber plate**, usually made of metal, and a glass cover. The back of the absorber plate is heavily insulated to prevent heat loss. A heat-transfer fluid flows through the collector, where it warms on contact with the absorber plate. With air-based systems, air simply passes over the absorber plate. With liquid-based systems, water or antifreeze solution flows through pipes either bonded to or built into the absorber plate. In liquid systems the absorber plate is usually copper or aluminum, which are chosen for their high thermal conductivity. Figure 9.10 shows a cross section of a typical liquid-based flat-plate collector.

Flat-plate collectors lose energy by radiation and conduction. As the absorber plate gets hotter, the losses increase and the collection efficiency goes down. At the so-called **stagnation temperature**, the loss equals the solar gain, and there's no net energy collected. The stagnation temperature depends on the ambient temperature and sky conditions, the insolation, and the characteristics of the collector. Again there are tradeoffs: Collectors with multiple glass covers have lower losses, but they admit less sunlight. They tend to perform better at high absorber-plate temperatures or low outdoor temperatures but not as well under the opposite conditions. A collector's efficiency can be increased by coating the absorber plate with a **selective surface**—one that is black in the visible spectrum, so it absorbs incident sunlight, but not in the infrared. Because absorptivity and emissivity are the same at a given wavelength (recall Section 4.3), this reduces radiation losses. Figure 9.11 shows efficiency curves for different flat-plate collector designs.

Space Heating

Active systems for space heating can be air or water based; the two are analogous to fossil-fueled heating systems using forced hot air and circulating water, respectively. In air-based systems, a fan delivers heated air from flat-plate collectors either directly to the building's interior or through an insulated bin of loose rocks or other heat-storage material that warms and thus serves as an energy-storage medium. Water-based systems circulate either water or an antifreeze mix through the solar collectors. In the former case, the heated water can go directly into a storage tank. Antifreeze, in contrast, circulates in a separate loop that includes a coil of pipe immersed in the storage tank. This arrangement constitutes a heat exchanger, and here the heated antifreeze transfers its energy to the water. Either way, the heated water is circulated through the house for heating, and it may also be withdrawn for domestic hot water. In addition, an auxiliary heat source can be used to boost the solar-heated water to the necessary temperatures when sunlight alone isn't sufficient to supply the building's heating needs.

Despite its potential, the use of solar energy as the primary source for space heating is not widespread. In the early twenty-first century, the U.S. Census Bureau's *American*

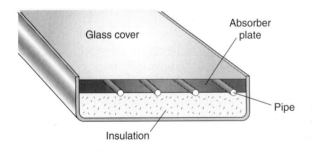

Figure 9.10

Cross section of a flat-plate solar collector using a liquid heat-transfer fluid.

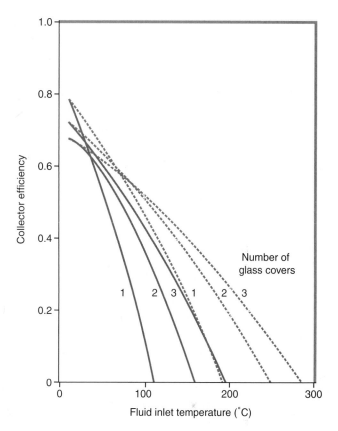

Figure 9.11
The efficiency of flat-plate solar collectors depends on the relative values of the insolation and energy loss; the latter depends on collector temperature and ambient conditions. Shown here are efficiency curves for collectors with one, two, or three glass covers, both with plain black absorbers (solid curves) and selective surfaces (dashed curves). Insolation is 1,000 W/m² and ambient temperature is 10°C. The horizontal axis is the temperature of the heat-transfer fluid as it enters the collector; temperatures where the curves intersect the horizontal axis are the stagnation temperatures under the given conditions. Multicover collectors would be best in conditions requiring high-temperature fluid or where ambient temperature is low. Single-cover collectors are better in warm climates where relatively low fluid temperatures are required, as in swimming-pool heaters.

Housing Survey lists only 28,000 solar-heated homes, about 0.02 percent of the total U.S. housing stock, or roughly 1 in 5,000 homes. Considerably more significant is the use of solar energy for domestic water heating.

Solar Domestic Hot Water

Hot-water heating accounts for some 20 percent of the energy use in a typical American home, and solar energy is well-suited for supplying this energy. An advantage of solar water heating is that it's easy to retrofit onto existing homes, in contrast to solar space-heating systems, which usually are incorporated into the initial design and construction. Nevertheless, only about 0.1 percent of U.S. homes have solar domestic hot-water systems, although a substantially higher percentage of swimming-pool heaters are solar powered. This percentage is a lot higher elsewhere: The Australian Bureau of Statistics reports that "only 3 percent" of households in South Australia have domestic hot-water systems, yet their "only" 3 percent describes a number thirty times greater than the U.S. fraction. In 2000 Barcelona, Spain, adopted its Solar Thermal Ordinance, requiring solar hot water on all new construction and renovations; and in Israel's sunny desert climate, solar hot-water heating is required for all residences.

In warm climates, a solar domestic hot-water system can be as simple as an insulated storage tank with glass on one side. Where freezing temperatures occur, a nontoxic antifreeze mixture circulates through flat-plate collectors and a heat exchanger at the bottom of the hot-water storage tank (Fig. 9.12). An alternative is the drain-back system, which pumps water through the collectors when solar energy is available, but drains it back into a storage tank when the Sun isn't shining, thus protecting the collectors and associated piping from freezing and bursting.

The system in Figure 9.12 shows a number of other essential details. The control unit compares temperatures from sensors at the collector and at the bottom of the storage tank; when the collector temperature exceeds the storage by some set amount—typically around 10°C—the control unit turns on the pump. Pumping continues as long as the temperature differential stays above a few degrees. In bright sunlight this strategy has the pump running continuously; under cloudy or intermittent sunlight it gives the collectors time to warm up before circulating the cool heat-transfer fluid for a short while until they cool down again.

It's crucial that the heat exchanger and tank sensor be at the bottom of the tank where the cold water enters. This is because hot water, being less dense, rises in the storage tank. The result is **stratification**, in which the tank temperature typically varies from cool at the bottom to hot at the top. Locating the heat exchanger near the bottom of the tank keeps the heat-transfer fluid cooler, and thus the collectors stay cooler. This, in turn, minimizes heat loss from the collectors and therefore increases the energy collection efficiency. On a very hot day the entire tank may reach a uniform temperature, but as soon as hot water is drawn off the top it's replaced by cold water at the bottom, and again

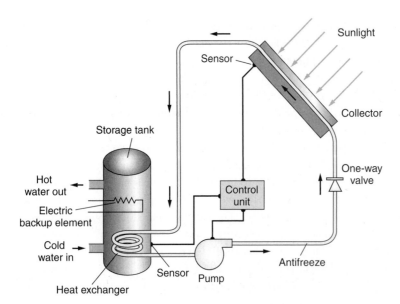

Figure 9.12

A solar domestic hot-water system.

the tank stratifies. Energy collection ceases if the heat-transfer fluid reaches the collectors' stagnation temperature, which, as Figure 9.11 suggests, varies with ambient temperature and insolation. In winter the stagnation temperature is lower and therefore heat loss limits the collection of solar energy.

Another feature in Figure 9.12 is the electric heating element, which provides backup energy for long cloudy periods. It's placed in the upper part of the tank and thus heats only the uppermost water; this leaves cool water at the bottom and allows solar collection at lower collector temperatures and therefore with less than full sunlight.

Finally, there's the one-way valve in the heat-transfer loop. Without that, heat flowing from the water into the heat-transfer fluid under cloudy or nighttime conditions could result in a density-driven flow of heat-transfer fluid backward through the collector, warming the collector and radiating the hard-earned energy back to the cool sky. However, if the tank were located *above* the collector, the density-driven flow would be in the right direction to collect energy rather than lose it. Such **thermosiphon** systems avoid the need for a pump but are architecturally difficult because the collectors themselves are usually on the rooftop.

Solar domestic hot-water systems are viable in almost any climate. Their technology is proven, simple, and straightforward, and they probably represent one of the easiest ways to displace a significant part of our fossil fuel use in homes and other buildings. Your author's house includes a solar hot-water system that's built into the roof and that supplies some 90 percent of the home's hot water from May to October in cloudy Vermont (Fig. 9.13).

Figure 9.13
The solar domestic hot-water system in your author's home, with collectors built into the roof structure. Inset photos show the basement storage tank (right) and control unit (left). Insulated pipes at the bottom of the tank carry heat-transfer antifreeze to and from the heat exchanger at the bottom of the tank. The control unit displays the collector temperature in degrees Fahrenheit, in this case on a hot, sunny summer day.

Example 9.3	Solar Water Heating

The solar collectors shown in Figure 9.13 each measure 2 feet by 8 feet, and there are six of them (the central glass is a skylight). Each has a single-pane glass cover that admits 84 percent of the incident sunlight. The loss rate for these collectors is 10 W/m²/°C, meaning that every square meter of collector loses energy at the rate of 10 W for every degree Celsius (or kelvin) of temperature difference (ΔT) between the absorber plate and the ambient air temperature. (Because radiation plays a significant role in the loss of heat from solar collectors, the correct measure of ambient temperature is a bit subtle, and it also depends on the clarity of the atmosphere. Here we'll stick with air temperature.)

(a) Find the maximum collector temperature on (i) a bright summer day with insolation $S = 950$ W/m² and ambient temperature of 30°C (86°F), (ii) a cloudy summer day with $S = 200$ W/m² and $T = 27$°C, and (iii) a sunny winter day with $S = 540$ W/m² and $T = -10$°C. (b) Find the rate of energy collection when the insolation is 900 W/m², the ambient temperature is 22°C, and the collector temperature—kept down by cool heat-transfer fluid—is 43°C.

Solution

(a) The maximum collector temperature—the stagnation temperature—occurs when collector loss equals the solar gain, or $0.84S = (10 \text{ W/m}^2/°C)(T_c - T_a)$, with T_c and T_a the collector and ambient temperatures, respectively, and where the factor 0.84 accounts for transmission through the glass. Solving for T_c gives

$$T_c = \frac{0.84S + (10 \text{ W/m}^2/°C)T_a}{10 \text{ W/m}^2/°C}$$

We have values of S and T_a for three cases; substituting and working the arithmetic gives (i) 110°C, (ii) 44°C, and (iii) 35°C. The first of these is higher than water's boiling point, and a well-designed solar heating system will have safety cutoffs to prevent the tank water from actually reaching this temperature. The other two temperatures, while well above ambient, are low for domestic hot water (44°C is 111°F, whereas typical water heaters would be set at around 120°F and a hot shower might be mixed with enough cold water to bring it down to around 105°F). So in both of these cases a backup heat source would be needed. Once the collectors reach these maximum temperatures, of course, they're no longer collecting energy, since at that point they lose as much energy as they gain. Note that these collectors do better on the cloudy summer day than on the sunny winter day; on the latter, the low Sun angle limits insolation and the low ambient temperature increases heat loss.

(b) In this scenario, the cool heat-transfer fluid keeps the collectors below their stagnation temperature, and there's a net energy gain equal to the difference between the solar input and the collector loss. We have six 2 ft × 8 ft collectors, for a total area of 96 ft², or 8.9 m². So the solar energy input is (0.84)(900 W/m²)(8.9 m²) = 6.73 kW.

Meanwhile the loss is $(10\ \mathrm{W/m^2/°C})(43°C - 22°C)(8.9\ \mathrm{m^2}) = 1.87\ \mathrm{kW}$. This leaves a net gain of 4.86 kW, close to the power of a typical 5-kW electric water heater. Incidentally, the collector efficiency under these conditions is 4.86 kW/6.73 kW = 72 percent.

9.4 Solar Thermal Power Systems

In describing the solar resource, I pointed out that the Sun's high temperature makes solar energy a high-quality energy source. By concentrating sunlight to achieve high temperatures, we have the potential to run heat engines with substantial thermodynamic efficiency. Such a heat engine can then drive an electric generator, supplying electricity to the power grid. The combination of concentrator, heat engine, and generator constitutes a **solar thermal power system**. Quantitatively, such systems are characterized by their **concentration ratio**—the ratio of concentrated sunlight to direct sunlight intensity—and the temperature they achieve. For the flat-plate collectors discussed in Section 9.3, the concentration ratio is 1, and the highest temperatures achievable are somewhat over 100°C.

Concentrating Solar Energy

Both lenses and mirrors can serve as light-concentrating devices. But large-scale lenses are expensive, fragile, and hard to fabricate, so mirrors are the concentrators of choice in solar power systems. The most effective concentrator is a **parabolic reflector** (Fig. 9.14).

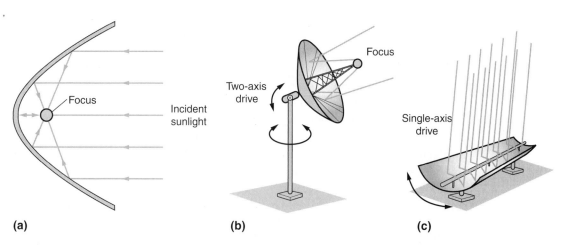

(a)   **(b)**   **(c)**

Figure 9.14
Solar concentrators. (a) A parabola reflects parallel rays of sunlight to a common focus. (b) A parabolic dish focuses sunlight to a point and requires two-axis tracking. (c) A trough concentrator focuses sunlight to a line and requires only one-axis tracking.

A parabola is a curve with the property that parallel rays reflect to a common point called the **focus**, as shown in Figure 9.14a. The Sun is so distant that its rays at Earth are very nearly parallel, so a three-dimensional **parabolic dish** (Fig. 9.14b) brings sunlight very nearly to a point (although not quite, otherwise we could achieve an infinite temperature, in violation of the second law of thermodynamics; see Question 2). You're familiar with parabolic dish reflectors in another application: They're the satellite dishes used to focus TV signals from communications satellites. And parabolic reflectors—or their spherical approximations—are widely used as the light-gathering elements in astronomical telescopes. Parabolic dish concentrators can have very high concentration ratios, and they reach temperatures measured in thousands of kelvins. Such temperatures are too high to handle in a practical heat engine, but they're useful in materials research. A "solar furnace" at Odeillo, France, has a concentration ratio around 10,000 and focuses 1 MW of solar power to achieve maximum temperatures of 3,800 K.

For maximum energy collection, a parabolic dish concentrator needs to point toward the Sun, which requires a complex and expensive mechanism to steer the entire dish and the energy-absorbing component at its focus in two independent directions, as illustrated in Figure 9.14b. This makes the parabolic dish a **two-axis** concentrator. A simpler and lower-cost alternative is the **parabolic trough concentrator** shown in Figure. 9.14c, which focuses sunlight to a line rather than a point. Trough concentrators track the Sun in only one direction, so they're **single-axis** devices, with a correspondingly less complex steering mechanism. The disadvantage is that the concentrated sunlight is spread out along a line, rather than being concentrated into a point, and therefore trough concentrators achieve lower temperatures than dishes.

Another concentration strategy is to use multiple flat mirrors to approximate a parabolic shape. Individual mirrors can be steered to keep sunlight concentrated on a central absorber that itself doesn't need to move. This approach—using individual steerable mirror segments—is also used in some of the largest astronomical telescopes.

The concentrating schemes I've just introduced are all capable of forming actual images. In fact, nighttime astronomers have actually used a mirror-field solar power system to image astronomical objects. But solar power systems don't need imaging, just energy concentration. **Compound parabolic concentrators** achieve such concentration using two parabolic surfaces that funnel light incident on a broad opening to a narrower outlet (Fig. 9.15). Although they achieve lower concentration ratios than imaging concentrators, these devices have the advantage that they concentrate light from a wide range of incidence angles. They can therefore work without tracking the Sun or, for better performance, with modest tracking mechanisms.

Parabolic Dish Systems

It's expensive to build large, steerable parabolic dish reflectors, and it's difficult to pipe heat-transfer fluids from the focus of a movable dish into a stationary turbine. Therefore parabolic dish technology for solar thermal power generation presents significant challenges. Nevertheless, a number of experimental systems have been constructed. In the

Incident light

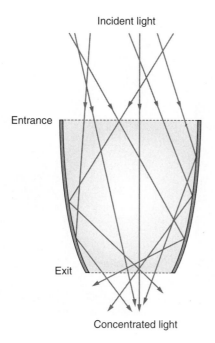

Entrance

Exit

Concentrated light

Figure 9.15
A compound parabolic concentrator funnels incident light
from a range of angles, providing concentration without
imaging.

1980s an array of fourteen parabolic dishes built by the Australian National University (ANU) supplied some 25 kW of electric power to the town of White Cliffs, Australia. Australia continues to do pioneering work in parabolic dish solar systems, with an ANU-designed 400-m^2 reflector seeing its first commercial development agreement in 2005 (Fig. 9.16a).

An approach that avoids the difficulties of piping hot fluids from the dish focus to a turbine is to place a heat engine right at the focus. The so-called Stirling engine is well-suited to this purpose, because it's an external-combustion engine that can use concentrated solar energy in place of heat released through combustion. Invented in 1816 by a Scottish minister but never widely used, the Stirling engine's working fluid is a gas that's permanently sealed into a piston-cylinder system. The American company Stirling Energy Systems, with funding from the U.S. Department of Energy, has developed a 25-kW parabolic dish and Stirling engine combination (Fig. 9.16b). The company has contracts with California utilities for a 20,000-dish solar electric power plant. Its 500-MW power output would be comparable to a medium-sized fossil or nuclear plant.

Power Towers

Power tower solar thermal power plants use a field of **heliostats**—flat, Sun-tracking mirrors—to focus sunlight on an absorber at the top of a central tower. Steam is generated either directly at the tower or indirectly via a heat-transfer fluid. A demonstration unit, Solar One in the Mojave Desert near Barstow, California, operated from 1982 to 1988 and produced 10 MW of electric power by boiling water atop a 100-m tower (Fig. 9.17). In 1995

(a)

(b)

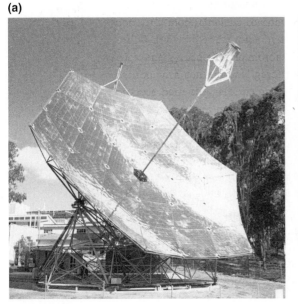
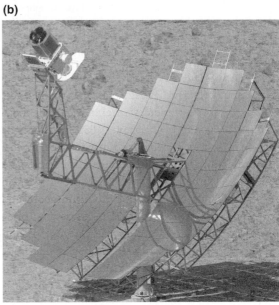

Figure 9.16

Two approaches to solar thermal electric power generation using parabolic dish concentrators. (a) Australian National University's "Big Dish" solar concentrator produces steam at 500°C to drive an external turbine and a 100-kW electric generator. (b) Stirling Energy Systems places a Stirling engine and generator at the dish focus, producing 25 kW of electric power.

Figure 9.17

Solar Two, a demonstration solar thermal electric power plant, generated 10 MW of electric power by concentrating sunlight from a field of nearly two thousand 39.1-m^2 heliostat mirrors onto a central absorber, producing temperatures of 566°C in a liquid salt heat-transfer fluid.

Solar One was retrofitted with an additional ring of heliostats and was modified to use molten salt as a heat-transfer fluid; renamed Solar Two, it operated until 1999. The use of molten salt allowed an operating temperature of 566°C, and latent heat of the liquid phase provided a measure of thermal energy storage for cloudy conditions. The salt transferred its energy to water through a heat exchanger, boiling the water to steam that drove a conventional turbine and generator. As with fossil and nuclear thermal power plants, second-law and other inefficiencies make it necessary to distinguish thermal power from electric power. Solar Two's 10-MWe output corresponds to 42 MW of solar thermal power at the tower. Exercise 11 explores the implications for Solar Two's efficiency.

Although the Solar One/Solar Two demonstration project has ended, a successor, the Solar Tres power tower in Spain, will produce 15 MW of electric power when operational. With 16-hour thermal storage capacity in its molten salt system, Solar Tres should generate power nearly 24 hours a day.

Solar Chimneys

A very different power tower technology is the **solar chimney**, which consists of a vast glass or transparent plastic roof with a tall, hollow central tower—the chimney. Sunlight enters the roof, heating the air inside and driving a hot wind up the chimney. Turbines in the air stream extract kinetic energy from the moving air and drive electric generators. A prototype of this design operated successfully in Spain during the 1980s; its chimney was 195 m high and it generated 50 kW of peak power. The Australian company EnviroMission is developing a colossal solar chimney with a roof some 5 km in diameter and a 1-km-high chimney—the tallest engineered structure on Earth. The system will incorporate thirty-two 6.5-MW turbines, for a total electric power output of 200 MW. Thermal storage materials under the roof will keep the wind—and the electric power—flowing 24 hours a day. Figure 9.18 shows the essential features of a solar chimney.

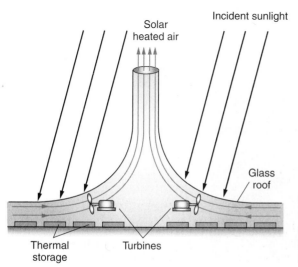

Figure 9.18
Cross section of a solar chimney; the actual structure would be circular. Thermal storage allows the system to generate power 24 hours a day. The chimney for the proposed 200-MW Australian Solar Tower project will be 1 km high.

Although the solar chimney may seem unrelated to the power towers I described in the preceding section, it, too, is basically a solar-driven heat engine with the heated air for its working fluid. Because it doesn't concentrate solar energy and therefore operates at much lower temperatures, its efficiency is a lot lower—a few percent at best (see Exercise 13). However, its few moving parts and lack of complex fluid plumbing may help keep it economically competitive with other solar tower concepts.

Trough Concentrators

Parabolic troughs are simpler than dishes because they need to rotate around only one axis to track the Sun. The downside, a lower operating temperature and therefore lower efficiency, may be acceptable because of lower construction costs. In the 1980s an American company, Luz International, built a series of nine solar power plants using trough concentrators; Figure 9.19 shows one of the Luz plants. Luz's nine-plant experience allowed the company to lower costs, raise operating temperatures, and increase efficiency and total output of successive plants. By 1990 they could produce electricity at a cost roughly double that of conventional methods. That wasn't good enough, however, to keep Luz from going bankrupt in 1991. The Luz plants continue to operate today, under a variety of owners.

Together, the Luz plants have a peak capacity of 354 MW and provide the majority of the world's solar thermal electricity. The first plant was a 14-MW facility, with the final two achieving 80 MW each. All the plants use trough concentrators similar to the one shown in Figure 9.14c. The troughs' long axes run north-south, and their single-axis tracking systems swing them to follow the Sun as it moves from east to west across the sky. At the focus of each trough is a black metal pipe containing a hydrocarbon-based heat-transfer fluid; the pipe is surrounded by an evacuated glass pipe whose vacuum limits energy loss by convection and conduction. The maximum fluid temperature is some 400°C, which is not dramatically less than the dish concentrator systems we considered earlier. To ease output variations at night and during cloudy times, the Luz plants

Figure 9.19
A trough-concentrator solar power plant in California's Mojave Desert.

include gas-fired backup systems. But instead of being a separate power plant, the backup consists of gas boilers that produce steam in the same turbine-generator loop as does solar energy from the concentrating collectors. Thus the Luz plants are hybrids, able to run their turbine-generators from either solar or natural gas energy.

The Luz experience in California provides a rich base for further technological development of trough-concentrator solar power systems. Among the advances being considered are systems in which steam boils directly at the concentrator focus, eliminating the heat-transfer loop and heat exchanger; integration with fossil-fueled combined-cycle systems (recall Section 5.4) to enhance overall efficiency and reduce fossil fuel consumption; and a host of mechanical improvements in the concentrators, their tracking mechanisms, and the flexible connections necessary for the flow of heat-transfer fluids from the tracking collectors to stationary heat exchangers and turbines. Clearly the Luz bankruptcy has brought a pause, but not an end, to the construction of large-scale electric power plants using solar trough concentrators.

9.5 Photovoltaic Solar Energy

Imagine an energy technology that has no moving parts, is constructed from an abundant natural resource, and is capable of converting sunlight directly into electricity with no gaseous, liquid, or solid emissions whatsoever. Too good to be true? Not at all: This is a description of photovoltaic (PV) technology.

The basis of PV devices is the same semiconductor technology that fuels the ongoing revolution in microelectronics, bringing us computers, cell phones, high-definition TV, digital photography, "smart" cars, DVD players, iPods, robotic vacuum cleaners, and the host of other electronic devices that seem essential to modern life. The only significant difference is that the semiconductor "chips" in our electronic devices are small—typically a few millimeters on a side—whereas PV devices must expose large areas to sunlight to capture significant solar power.

Silicon and Semiconductors

Silicon is the essential element in semiconductor electronics, and it's widely used in solar PV cells. Silicon is the second-most abundant element on Earth (after oxygen); ordinary sand is largely silicon dioxide (SiO_2). So silicon is not something we're going to run out of, or need to obtain from despotic and unstable foreign countries, or go to war over.

Silicon is an example of a **semiconductor**—a material whose electrical properties place it between insulators, such as glass and plastics, and good conductors, such as metals. Whereas a metal conducts electricity because it contains abundant free electrons, a semiconductor conducts electricity because a few electrons are freed from an otherwise rigid crystal structure. Such a free electron leaves behind a **hole**—the absence of an electron, which acts as a positive charge and is able to move through the material and carry electric current. Thus a semiconductor, unlike a metal, has *two* types of charge carriers: negative electrons and positive holes. Electron-hole pairs can form in silicon and other semiconductors when random thermal energy manages to free an electron from the crystal structure or, more important in photovoltaics, when a photon of light energy dislodges an electron. Furthermore, we can manipulate the structure of silicon to give it abundant free electrons or holes, thus controlling its electrical properties.

The silicon atom has four outermost electrons that participate in bonding to other silicon atoms, and in pure silicon only a tiny fraction of those electrons manage to escape their bonds. But by "doping" the silicon with small amounts of impurities—typically 1 part in 10 million or less—we can radically alter its electrical properties. Adding an impurity such as phosphorus, which has five outermost electrons, leaves one electron from each phosphorus atom with no way to participate in the interatomic bonding. These electrons are therefore free to carry electric current. A semiconductor with excess free electrons is called an **N-type semiconductor**, because its dominant charge carriers are negative electrons. Dope silicon with boron, on the other hand, and you create a **P-type semiconductor**, whose dominant charge carriers are positive holes. That's because boron has only three outer electrons, so when it fits into the silicon structure, there's a missing electron—a positive hole.

The key to nearly all semiconductor electronics, including PV cells, is the **PN junction**. Joining pieces of P- and N-type silicon causes electrons from the N side to diffuse across the junction into the P-type material, while holes from the P-type side diffuse across into the N-type material. This sets up a strong electric field across the junction, pointing from the N side to the P side (Fig. 9.20). In silicon, the corresponding voltage difference is about 0.6 volt. The PN junction has some remarkable properties; for example, it conducts electricity in only one direction—from P to N, but not from N to P. These properties are exploited in transistors and the other "building blocks" of modern electronics.

Photovoltaic Cells

Figure 9.21 shows a **photovoltaic cell**, which consists of a PN junction whose upper N layer is thin enough—typically 1 μm—for light to penetrate to the junction. Photons of sufficient energy can eject individual electrons from the silicon crystal structure, creating electron-hole pairs. If that occurs near the junction, the electric field pushes the positive holes downward, into the P-type material. Electrons, being negative, are pushed in the opposite direction, into the N-type material. Metallic contacts on the bottom and

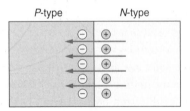

P-type   N-type

Figure 9.20

A *PN* junction. When the junction is first formed, electrons (minus signs) migrate from the *N*-type side to the *P*-type side, and holes (plus signs) migrate from *P* to *N*. This sets up a strong electric field (arrows) pointing from *N* to *P*.

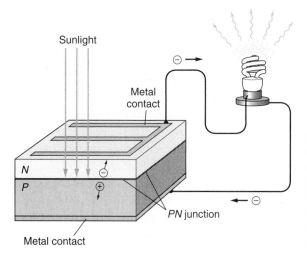

Sunlight

Metal
contact

N

P

PN junction

Metal contact

Figure 9.21
A photovoltaic cell. The energy of sunlight photons creates electron-hole pairs at the PN junction, and the electric field at the junction sends electrons into the *N*-type semiconductor and holes into the *P*-type semiconductor. Metal contacts at the top and bottom become negatively and positively charged, respectively, and as a result a current flows through an external circuit. Here the circuit includes an energy-saving compact fluorescent light.

top of the cell thus become positively and negatively charged, respectively—just like the positive and negative terminals of a battery. In fact, the PV cell is essentially a solar-powered battery, and connecting the cell's contacts through an external circuit causes a current of electrons to flow and provide energy to the circuit.

Several factors determine the efficiency of a PV cell. The most significant relates to the distribution of wavelengths in the solar spectrum that I described in Section 9.1 and Figure 9.7. Semiconductor physics shows that there's a minimum energy—called the *band-gap energy*—that's required to eject an electron from the crystal structure, creating an electron-hole pair. In silicon that energy is 1.14 electron volts (eV), or 0.18 aJ. So it takes a photon with at least this much energy to create an electron-hole pair. Equation 3.1 relates photon energy and wavelength, and it shows that an energy equal to silicon's 0.18 aJ band gap corresponds to a wavelength of 1.1 μm, which is in the infrared but not far from the boundary of the visible region. As a result, roughly one-fourth of the total solar radiation lies at wavelengths too long to create electron-hole pairs in silicon. Furthermore, photons with energies above the band gap give up their excess energy as heat. And some photon-created electron-hole pairs recombine before the junction's electric field has a chance to separate them; their energy, too, is lost as heat. The result, for silicon, is a theoretical maximum PV cell efficiency of around 33 percent.

Other losses reduce the efficiency of commercially available PV cells to around half the maximum value. Some light is reflected off the top surface, although this can be reduced but not eliminated with special antireflection coatings or textured surfaces that help "trap" light within the cell. The necessary metal contacts on the Sun-facing surface block solar radiation, but the contact area can't be made too small without reducing the amount of electric current the cell can supply. Some electrons fail to reach the metal contacts, and their energy is lost. Varying the doping near the contacts helps steer electrons into the metal and thus reduce the loss. Such improvements enable the production of silicon PV cells with efficiencies as high as 25 percent, but with the downside of higher production costs.

Another approach to increasing PV efficiency is to use materials with different band gaps. A low band gap uses more of the solar spectrum, but then most of the photons have excess energy that is lost as heat. A high band gap minimizes this loss, but then most photons can't produce electron-hole pairs. So there's a compromise band gap that optimizes efficiency. For ground-based applications, the distribution of solar wavelengths puts this optimum gap at about 1.4 eV; in space it's closer to 1.6 eV. Both of these gaps are somewhat larger than that of silicon. Semiconductors with more appropriate band gaps are indium phosphate (InP; 1.35 eV), cadmium selenide (CdSe; 1.74 eV), and gallium arsenide (GaAs; 1.43 eV). However, the use of toxic elements such as arsenic and cadmium makes these less attractive because of possible environmental contamination.

One way around the band-gap compromise is to use multilayer cells, so that light encounters several *PN* junctions with different band gaps. The topmost junction has the highest gap, allowing lower-energy photons to penetrate deeper before giving up their energy at a more appropriate gap. Obviously, multilayer PV cells are more expensive to produce. They also suffer from a subtler problem: Misalignment of the crystal structures of different materials can seriously compromise cell performance. As a result, the most effective multilayer cells produced to date have only two layers, yielding efficiencies up to 36 percent.

Although silicon is abundant, semiconductor-grade silicon is very expensive because semiconductor applications require exquisitely pure, crystalline silicon. In the semiconductor industry, large crystals are grown slowly from molten silicon; then they're sliced up into thin circular wafers that hold several hundred microcircuits such as computer processing units or memory chips. Growing crystals is expensive, and the slicing process wastes a lot of pure silicon; but given the tiny size and high profitability of the individual microcircuits, this is not an economic problem for semiconductor electronics. Solar cells require large areas of silicon, so the traditional semiconductor manufacturing process means high costs for PV power systems. As a result, PV cells are often made from **polycrystalline silicon**, whose structure includes many individual crystals with different orientations. This characteristic reduces efficiency somewhat, but it permits PV cells to be manufactured with far less expensive processes that coat thin films of polycrystalline silicon onto a substrate. Even **amorphous silicon**—which, like ordinary glass, has no crystal structure—can be used to make inexpensive but low-efficiency PV cells. Most of today's commercial PV cells are polycrystalline, and a close look at the cell surface shows the multicrystal structure (Fig. 9.22).

Photovoltaic Modules and Distributed Applications

The properties of the *PN* junction limit the voltage of a single cell to around 0.6 volt. To achieve voltages capable of driving significant currents in most circuits, individual cells are connected in series and mounted together to make **photovoltaic modules**, also called **solar panels** (Fig. 9.23). Individual panels can be joined to make large arrays that

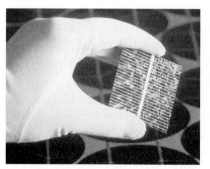

Figure 9.22
A polycrystalline solar cell. The mottled appearance results from silicon crystals with different orientations. Horizontal structures and the wider vertical bands are the metallic contacts that channel current to an external circuit. This particular cell measures 12.5 cm by 12.5 cm and produces just over 2 W of electric power, with a 14 percent conversion efficiency.

are usually fixed in place but can also have a Sun-tracking mechanism for optimal performance. A "light funnel," like the one shown in Figure 9.15, is sometimes used in conjunction with PV cells, allowing a smaller area of expensive solar cells to convert sunlight from a larger area at the funnel's entrance.

Photovoltaic panels have been used for many years in remote locations requiring reliable electric power. The most high tech of these applications is in space; nearly all

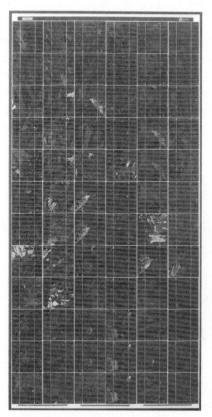

Figure 9.23
This SX170B panel from BP Solar joins together seventy-two individual polycrystalline silicon-nitride PV cells to produce 170 W of solar power. It's designed for household and commercial PV power applications. The panel measures 1.6 m by 0.79 m. (BP Solar is a unit of the company formerly known as British Petroleum, one of the world's largest oil companies. Over the past few years BP has been getting serious about developing alternatives to fossil fuels.)

spacecraft in Earth orbit or in the inner reaches of the Solar System are powered by solar PV panels. But lower-tech applications abound on Earth, wherever connections to the power grid aren't readily available. Solar-electric water pumps enhance life in the developing world, where women traditionally spend much of their time fetching water from distant sources. PV-powered lighting systems gather solar energy during the day and convert it back to light at night; they range from residential walkway lights to highway construction lighting. Field labs, remote weather stations, air-quality sensors, and a host of other scientific facilities use photovoltaics for remote power. For a home more than a few kilometers from the power grid, it's often more economical—not to mention environmentally friendly—to construct a PV system than to string wires. Environmentally conscious owners of transportation fleets use PV systems to charge electric vehicles, validating claims that they operate solar-powered cars and trucks. And PV cells find their way into novelty applications such as solar-powered calculators and watches. PV enthusiasts regularly race impractical but fully PV-powered cars (recall Fig. 9.2), and NASA's Helios solar-powered airplane has reached altitudes of nearly 100,000 feet—a world record for propeller-driven aircraft. Back on Earth, a PV-powered electric fence even protects my garden from woodchucks.

Grid-Connected Photovoltaics

Increasingly, PV panels are being used to supplement electrical energy from the power grid, and even to generate grid power. A typical home in most parts of the world can count on PV power to supply much of its electrical energy (Fig. 9.24). Some grid power may be needed on cloudy days, but at other times the PV system generates excess energy that's stored in batteries or sold back to the grid. One issue with grid-connected photovoltaics is that PV cells produce direct current (DC), whereas power grids use alternating current (AC). However, the same semiconductor technology that gives us computer

Figure 9.24
Maine hardly qualifies as a "sunshine state," but the PV panels on this Maine home generate more electric power than the home uses; the excess is sold to the local power company.

chips and PV cells now enables highly efficient **inverters** to convert DC to AC. This technology allows the house in Figure 9.24 to stay connected to the grid and sell excess energy to the power company. And it makes possible the use of photovoltaics for large-scale power generation.

The first grid-connected PV power plant was a 1-MW facility built in 1980 in Saijo, Japan. By 1984 ARCO Solar—another subsidiary of a major oil company exploring alternative energy (see Fig. 9.23)—had built a 6.5-MW PV plant in California. Although both of these pioneering plants have since been dismantled, there are today more than half a dozen PV plants of similar capacity operating around the globe. Most are in Germany, but Tucson Electric Power's Springerville plant in Arizona (Fig. 9.25) is probably the most productive, and when completed it will produce nearly 9 MW of PV power. These numbers are still small compared to the 1 GW (1,000 MW) of a typical large fossil or nuclear plant, but as Figure 9.25 and Example 9.4 show, there's plenty of room to expand PV power generation. As Figure 9.26 shows, the global production of PV energy is growing exponentially.

Example 9.4   A Photovoltaic-Powered Country?

The average rate of U.S. electrical energy consumption is about 420 GW. Assuming commercially available PV cells with 15 percent efficiency, an average insolation of 254 W/m^2 (year-round, round-the-clock average for Arizona with collectors tilted at the local latitude), and a land-area requirement twice that of the actual PV modules, find the total land area needed to supply all U.S. electrical energy needs with Arizona-based photovoltaics. Compare with Arizona's total land area of 294,000 km^2.

Figure 9.25
Tucson Electric Power's Springerville Generating Station in Arizona is expanding to an ultimate capacity of nearly 9 MW. At the time this photo was taken, the plant had some 35,000 PV modules covering 44 acres, with a peak output of 5 MW.

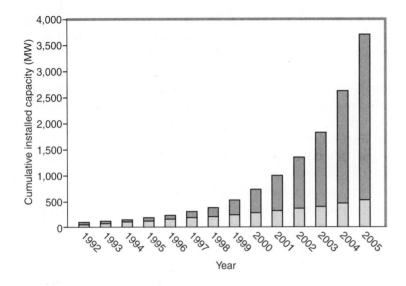

Figure 9.26
Global PV capacity has been growing rapidly in the late twentieth and early twenty-first centuries. Light shading represents off-grid applications, darker shading indicates grid-connected PV systems.

Solution

With 15 percent efficiency we average only $(254 \text{ W/m}^2)(0.15) = 38.1 \text{ W/m}^2$ from the PV modules. Accounting for the factor of 2 required for the extra land gives 19 W per square meter of land used for PV power plants. With U.S. consumption at 420 GW, we then need a total area given by $(420 \times 10^9 \text{ W})/(19 \text{ W/m}^2) = 2.2 \times 10^{10} \text{ m}^2$. This sounds like a huge area, but with 1,000 m per kilometer, and therefore 10^6 m^2 per square kilometer, it's only 22,000 km^2. This is 22/294 or 7.5 percent of Arizona's land area. Of course I'm not advocating converting this much of Arizona's land to supply the United States with electricity. However, we've already flooded nearly 1,000 km^2 of Arizona to generate a paltry 2.7 GW from hydroelectric dams. In any event, the result of this entirely realistic example suggests it would take only a small fraction of the desert Southwest to power the entire United States with nonpolluting, climate-neutral solar PV energy.

Energy Storage, Global Photovoltaic Plants, and Solar Power Satellites

Can you spot the one flaw in Example 9.4? The average insolation in the example correctly accounts for nighttime and clouds, so the total energy produced from that 7.5 percent of Arizona is indeed enough for the entire country. The problem is that it's produced only during sunny daylight hours. Without an effective way to store electrical energy, we'll need a backup energy source at night. We'll meet one such storage method in the next chapter (see Section 10.7), but it isn't viable in Arizona and is probably unworkable on a national scale.

One solution to the energy storage problem is not to rely on any single energy source, even one as environmentally attractive as solar energy. One possibility is to operate the

PV systems during the day, and open hydroelectric dams or fire up combined-cycle fossil plants at night. Another is to build hybrid plants that use solar energy when the Sun shines and fossil sources at other times. But a global all-solar solution would be to site PV plants around the globe with enough capacity to meet the world's electrical needs from the daytime side of the planet alone. This strategy requires an international grid capable of shipping power under the oceans—a major technical challenge. Furthermore, it demands a more stable and cooperative international political regime than the world knows today. But it may someday be a way to a PV-powered planet.

Yet another approach is to put PV arrays in orbit, where they would see nearly constant sunshine at levels undiminished by Earth's atmosphere or the night/day cycle that reduces average insolation on Earth. The electrical energy would be converted to microwaves and beamed to Earth for conversion back to electricity. The microwave intensity at Earth would actually be lower than the intensity of sunlight, but it could be converted to electricity with much higher efficiency than PV cells using large, open-wire antennas. Unlike the other solar schemes I've described in this chapter, solar power satellites are a long way off and face stiff technological and economic challenges. Nevertheless, their advocates see them as a near-perfect long-term energy solution for humankind, and the idea is taken seriously enough that NASA maintains a Space Solar Power program with the goal of developing gigawatt-scale orbiting solar power stations.

9.6 Other Solar Applications

Solar heating and electric power generation are the most important direct uses for solar energy, but there are plenty of other direct solar applications. Ironically, solar energy can be used for cooling. So-called absorption refrigeration systems use a heat source—in this case solar energy—instead of mechanical energy to drive a refrigeration cycle. Portable solar cooling systems of this sort have been developed to preserve medicines in developing countries that lack reliable electric power. However, large-scale development of such solar cooling for buildings is not yet commercialized.

On smaller scales, **solar ovens** have long been used in sunny climates to concentrate sunlight to temperatures appropriate for cooking. Such devices could stem the tide of deforestation in developing countries where wood for cooking is increasingly scarce. **Solar stills** use sunlight to evaporate seawater, leaving the salt behind and providing fresh drinking water. Small, inflatable versions are commercially available as emergency water supplies for boaters, and larger stills are supplying fresh water to communities on the Texas-Mexico border. **Solar ponds** are basically thermal collectors that use layers of increasingly salty water to inhibit convective heat loss; solar-heated water from such ponds can be used for conventional heating, industrial process heat, agriculture, or to run solar cooling units. A more high-tech application is **photolysis**, the use of solar radiation to break water into hydrogen and oxygen, producing clean-burning hydrogen fuel (more on this in Chapter 11). A final use is so obvious it hardly bears mentioning: **solar lighting**. As with heating, passive solar design can eliminate much of a building's need for electric lighting during daytime hours. More sophisticated systems include optical fibers that "pipe" sunlight throughout a building. So-called hybrid lighting fixtures now being developed will merge electric light sources with fiber-optic sunlight, using only as much electricity as needed to maintain a chosen lighting level.

9.7 Environmental Impacts of Solar Energy

No energy source is completely benign, but solar energy comes close. Its most obvious impact follows from the roughly 200 W/m² of average insolation at Earth. That means large-scale solar energy installations necessarily occupy large land areas—although, as Exercises 14 and 15 show, the areas of open-pit coal mines or lakes backed up by hydro-electric dams can be considerably greater for the same power output. And large-scale solar installations aren't the only—or even the most desirable—approach to solar power. Instead, we can use the tens of thousands of square kilometers of building roofs without requiring one additional acre of open land for solar energy.

While operating, solar energy systems produce no pollutants or greenhouse gases, although solar thermal electric power systems, like other thermal power plants, need to dispose of waste heat. Probably the greatest environmental impact from solar energy systems comes in their manufacture and eventual disposal. Passive and simple active heating systems use common building materials, and their environmental impacts are not much different than with other construction projects. Photovoltaics, on the other hand, may require a host of harmful materials for their construction. Even benign silicon is doped with arsenic and other toxins, albeit at very low concentrations. During manufacture, however, these and other toxic substances are present in greater concentrations. The same, of course, is true throughout the semiconductor industry. Despite its high-tech reputation and its "clean rooms," this industry contends with serious pollution issues. Some alternative semiconductors, such as gallium arsenide or cadmium selenide, are downright toxic themselves. A fire in a house roofed with cadmium selenide PV cells would spread toxic cadmium into the environment, possibly resulting in significant contamination.

A subtler question—and one that's germane to every manufactured energy technology—involves the energy needed to produce a solar water heater or a PV electric system, or to build and launch a solar power satellite. We'll compare this production energy and its impacts for different energy sources in Chapter 16. If the production energy exceeds what a solar system will produce in its lifetime, then there's no point in going solar. Solar skeptics sometimes make this claim as a way of discounting solar energy, but they're wrong: Studies of the energy involved in the production of solar cells and other solar components are actually quite favorable to solar. The **energy payback time** quantifies the energy needed to produce a solar system and the time required for the system to produce as much energy as it took to manufacture. As long as the energy payback time is shorter than the system's anticipated lifetime, a solar energy system is energetically—although not necessarily economically—viable. Several studies in the United States and Europe point

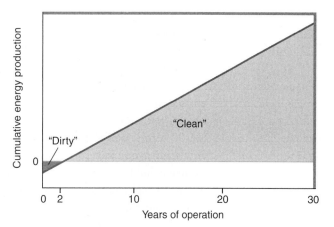

Figure 9.27
Energy payback time for a typical PV system is about two years. If the system is manufactured using fossil energy, then a system with a 30-year lifetime produces two years' worth of "dirty" energy followed by 28 years of "clean" energy. Note that the cumulative energy production after two years is zero, since at that point the system has produced as much energy as it took to manufacture.

to an energy payback time for PV systems of typically two years, comparable to that of fossil energy facilities. This means the total energy budget for a PV system with a 30-year lifetime will be 28/30, or 93 percent pollution-free (Fig. 9.27)—unless, of course, the factory that manufactures the PV system is itself solar powered.

Finally, there's a question of possible climate effects in a world that makes wholesale use of solar energy. That's because solar collectors are necessarily black, and therefore they absorb more solar energy than most surfaces they cover. This alters Earth's net energy balance and could, therefore, affect climate. But with ten thousand times as much solar energy coming in as we humans use, it would take an enormous growth in energy consumption before this effect could be globally significant or even comparable to the climatic effects of fossil CO_2 emissions. There could be local warming effects from solar installations, but we're used to those already as we've covered vast land areas with black pavement and urban roofs. This dense urban development, along with the heat that all our energy ultimately degrades into, produces the well-known "heat island" effect, whereby cities are warmer than the surrounding countryside. By the way, this same concern applies to *any* energy source—particularly fossil and nuclear—that brings to Earth's surface additional energy beyond what's carried in the natural solar flow.

9.8 Policy Issue: Encouraging a Solar Industry

Solar energy is abundant, widely available, and environmentally friendly. So why don't we live in a solar-powered world? The answer, in a nutshell, is economics. Most solar applications simply can't compete with traditional energy sources, in particular fossil fuels—at least under today's economic conditions and with the subsidies given to the fossil fuel industry. Where solar can compete, such as in space and in remote power installations, it's been used for decades. One way to quantify the cost of energy systems is in terms of capital cost—dollars per watt, or the total cost needed to construct an energy plant divided by its power output in watts. Another is operating cost per unit of energy produced, typically in dollars per joule or kilowatt-hour. Solar costs continue to drop on both scores, but they still have a way to go before they're competitive in our present-day economic situation (Fig. 9.28).

But economics is a human enterprise, not immutable like the laws of physics, and the costs of energy sources ultimately involve value judgments. Today's fossil energy technologies enjoy the benefit of more than a century of development, with few constraints on their environmental impacts for much of that time. They also enjoy—even today—massive government subsidies for research and development and for tax relief (Fig. 9.29). And the cost of fossil energy doesn't include realistic measures of its environmental and health impacts or the cost of adapting to climate change resulting largely from fossil fuel combustion. Solar technologies, in contrast, have seen less government support, and they haven't had the time or the sales volumes to mature either technologically or economically. This situation could change if societies—governments, industries, and citizens—had the will to make it happen.

Why is the successor to the Solar One and Solar Two power towers being built in Spain? Because the Spanish king in 1999 decreed that solar-generated electricity must be purchased at a price equivalent to 24¢ per kilowatt-hour. That's far above the average cost of conventional power and is therefore "uneconomical," but the goal is to encourage a viable solar industry whose costs fall even as fossil fuels become scarcer and more costly. Why are most of the world's largest PV systems in Germany? Because in 2000 the German Bundestag passed an innovative Renewable Energy Act intended to double the contribution of renewables to Germany's energy supply within a single decade. Why is the Springerville Generating Station PV plant located in Arizona? Because in 2000 Arizona became one of the first states to require a small percentage of its electrical energy be generated from renewable sources—a goal that other states have since surpassed. Why do I have a solar water-heating system at my home? In part because, when I built the system, the U.S. government

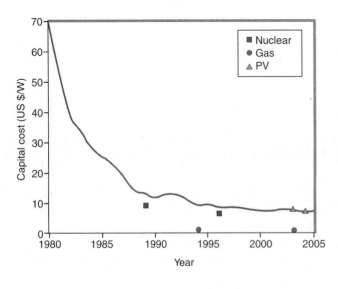

Figure 9.28

The black curve shows the approximate capital cost of solar PV power stations, in dollars per watt, for 1980 to 2005. Photovoltaic energy will be competitive with fossil fuels at somewhere around $1.50 per watt. Values for PV module costs were multiplied by 2 as an estimate of the complete costs of a PV power plant. Also shown are costs per watt of six individual power plants, including two of the last nuclear plants to come online in the United States, two PV plants, and two combined-cycle gas plants with capital costs of only 50¢ per watt. Note that PV electricity may already be competitive with nuclear-generated electricity, at least in the United States.

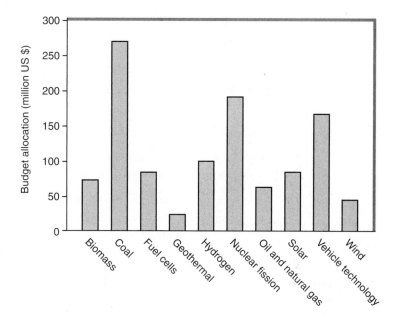

Figure 9.29
Priorities in the U.S. Department of Energy's fiscal 2006 budget for research and development favor coal and vehicle-related research (vehicle technologies, hydrogen, fuel cells). In fact, the total for fossil fuel research and development is $523 million, whereas the total for solar, geothermal, and wind combined is only $152 million.

and the State of Vermont thought it was a good idea to give tax credits for solar installations—not as a government handout but as a way of encouraging a viable solar industry that could eventually compete economically with traditional energy sources. Unfortunately, a political change brought a premature end to those credits, with the result that the solar industry virtually collapsed. Why do all Israeli homes have solar domestic hot-water systems? Because a sunny desert climate and a government serious about renewable energy combine to make solar hot water an obvious choice. Why is Australia at the forefront of solar research with its audacious plan for a 1-km solar chimney? Because Australia's 2000 Renewable Energy (Electricity) Act mandates the generation of an additional gigawatt of renewable electricity each year from 2000 through 2010.

All these "because" answers suggest that abundant, nonpolluting, renewable solar energy is getting serious attention from governments, industry, and citizens. But it's going to take a great deal more individual and collective political will before solar energy makes a significant contribution to the world's energy mix.

Chapter 9   Chapter Review

BIG IDEAS

9.1   Solar energy reaches Earth at some ten thousand times the rate at which humankind uses energy. The available energy varies with location and season. We can capture solar energy directly as heat or to generate electricity.

9.2 **Passive solar heating** avoids active devices such as pumps or fans.

9.3 **Active solar heating** captures solar energy with **solar collectors**, and uses active devices to move the energy. Active systems can provide both space heating and hot water.

9.4 **Solar thermal power systems** use reflectors to concentrate sunlight, thereby heating a fluid that operates a heat engine to generate electricity.

9.5 Photovoltaic devices use semiconductors to convert solar energy directly into electricity.

9.6 Other solar applications include cooking, lighting, and **photolysis**—the use of solar energy to break water into hydrogen and oxygen.

9.7 No energy source is environmentally benign. Solar energy's impact includes land use, the energy used in manufacturing solar energy systems, and the use of toxic materials—particularly in some advanced PV devices.

9.8 Most solar applications aren't yet economically competitive with fossil fuels. Policy decisions to encourage solar energy can help establish an economically viable solar industry.

TERMS TO KNOW

absorber plate (p. 263)
active solar heating (p. 263)
amorphous silicon (p. 278)
beadwall (p. 261)
compound parabolic concentrator (p. 270)
concentration ratio (p. 269)
diffuse insolation (p. 255)
direct insolation (p. 255)
energy payback time (p. 284)
flat-plate collector (p. 263)
focus (p. 270)
heliostat (p. 271)
hole (p. 275)
indirect gain (p. 262)
insolation (p. 252)
inverter (p. 281)
N-type semiconductor (p. 276)
parabolic dish (p. 270)
parabolic reflector (p. 269)
parabolic trough concentrator (p. 270)
passive solar heating (p. 259)
photolysis (p. 283)
photovoltaic cell (p. 276)

photovoltaic module (p. 278)
PN junction (p. 276)
polycrystalline silicon (p. 278)
power tower (p. 271)
P-type semiconductor (p. 276)
selective surface (p. 264)
semiconductor (p. 275)
single-axis concentrator (p. 270)
solar chimney (p. 273)
solar collector (p. 263)
solar constant (p. 252)
solar lighting (p. 283)
solar oven (p. 283)
solar panel (p. 278)
solar pond (p. 283)
solar still (p. 283)
solar thermal power system (p. 269)
stagnation temperature (p. 264)
stratification (p. 266)
thermal mass (p. 259)
thermosiphon (p. 267)
Trombe wall (p. 262)
two-axis concentrator (p. 270)

GETTING QUANTITATIVE

Solar constant: ~1,368 W/m^2 above Earth's atmosphere

Solar power incident on Earth: 1.74×10^{17} W (174 PW), more than ten thousand times humankind's energy-consumption rate

Midday sunlight at Earth's surface: ~1 kW on each square meter oriented perpendicular to incident sunlight

Average insolation at Earth's surface: 235 W/m^2

Photovoltaic cell efficiency: typical ~15 percent; best ~25 percent; theoretical maximum ~33 percent

QUESTIONS

1. Explain why the average rate per square meter at which solar energy reaches Earth is one-fourth of the solar constant.

2. Explain why we can't concentrate sunlight to temperatures hotter than the 5,800-K solar surface. *Hint*: Think about the second law of thermodynamics.

3. Why does stratification in a solar hot-water tank help increase the efficiency of solar energy collection?

4. At what time of year would you expect maximum midday insolation on (a) a solar collector tilted at latitude and (b) a horizontal collector?

5. How is a parabolic trough concentrator simpler than a parabolic dish?

6. In what way is the electrical output of a PV module not compatible with the electric power grid?

7. Some people claim that stone houses, because of their large thermal mass, don't need insulation. Discuss this claim.

8. What's the advantage of adding more layers of glass to a solar collector? What's the disadvantage?

9. Light with too long a wavelength can't excite electron-hole pairs in a PV cell. Why not? And why isn't light with too short a wavelength very efficient in producing PV electricity? Explain both answers in terms of the band gap.

10. Why isn't it possible to give a single number to describe the efficiency of a flat-plate solar collector?

EXERCISES

1. A more precise value for Earth's average orbital distance is 149.6 Gm, and the best value for the Sun's power output is 3.842×10^{26} W, ± 0.4 percent. Calculate the range of values for the solar constant that these figures imply.

2. Use the values of Earth's radius and the solar constant given in the text to verify the figure 174 PW for the total solar power incident on Earth's sunward side.

3. How long does it take the Sun to deliver to Earth the total amount of energy humankind uses in a year?

4. A passive solar house requires an average heating power of 8.4 kW in the winter. It stores energy in 10 tonnes (10,000 kg) of Glauber salt, whose heat of fusion is 241 kJ/kg. At the end of a sunny stretch, the Glauber salt is all liquid at its 32°C melting point. If the weather then turns cloudy for a prolonged period, how long will the Glauber salt stay at 32°C and thus keep the house comfortable?

5. A solar hot-water system stagnates at 85°C under insolation of 740 W/m² and an ambient temperature of 17°C. Find its loss rate in W/m²/°C, assuming its glazing transmits 81 percent of the incident sunlight.

6. A household uses 100 gallons of hot water per day (1 gallon of water has a mass of 3.78 kg), and the water temperature must be raised from 10°C to 50°C. (a) Find the average power needed to heat the water. Then assume you have solar collectors with 55 percent average efficiency at converting sunlight to thermal energy when tilted at the latitude. Use data from Table 9.1 to find the collector area needed to supply this power in Albuquerque in June.

7. Repeat part (b) of the preceding problem for Minneapolis in January, assuming a lower efficiency of 32 percent (because of higher losses at the lower temperature).

8. Find the average temperature in the passive solar house of Example 9.2 if it's located in (a) Fairbanks and (b) San Francisco.

9. A passive solar house has a thermal mass consisting of 20 tonnes of concrete. The house's energy loss rate is 120 W/°C. At the end of a sunny winter day, the entire house, including the thermal mass, is at 22°C. If the outdoor temperature is −10°C, how long after the Sun sets will it take the house's temperature to drop by 1°C?

10. My college's average electric power consumption is about 2 MW. What area of 15 percent efficient PV modules would be needed to supply this power during a summer month when insolation on the modules averages 225 W/m²? Compare with the area of the college's football field (55 m by 110 m overall).

11. The Solar Two power tower used 42 MW of solar power to bring its molten salt heat-transfer fluid to 566°C, and it produced 10 MW of electric power. Find (a) its actual effi-

ciency, and (b) compare to its theoretical maximum thermodynamic efficiency, assuming T_c for its heat engine is 50°C.

12. The maximum temperature achievable with concentrated sunlight is the 5,800-K temperature of the solar surface. Find the maximum thermodynamic efficiency of a heat engine operating between this temperature and the 300-K ambient temperature typical of Earth.

13. Estimate the efficiency of the proposed Australian solar chimney, given its 5-km-diameter roof and its 200-MW electric power output. Assume midday sunshine, with insolation of 1 kW/m^2.

14. The Hoover Dam on the Colorado River generates hydroelectric power at the rate of 2,080 MW. Behind the dam, the river backs up to form Lake Mead, whose surface area is 63,900 hectares (1 hectare = 10^4 m^2). Using the assumptions of Example 9.4 (average insolation of 254 W/m^2, 15 percent PV efficiency, and an area twice that of the actual PV cells), find the area that would be occupied by a PV plant with the same output as Hoover Dam, and compare that number with the area of Lake Mead.

15. The controversial Cheviot open-pit coal mine in Alberta, Canada, occupies 7,455 hectares and produces 1.4 million tonnes (1 tonne = 1,000 kg) of coal per year. (a) Use the energy content of coal from Table 3.3 to find the power in watts corresponding to this rate of coal production, assuming it's burned in power plants with 35 percent efficiency. (b) Again using the assumptions of Example 9.4, find the area of a PV plant with this power output, and compare that number with the area of the Cheviot mine.

RESEARCH PROBLEMS

1. Download the latest version of the U.S. Energy Information Administration's "Existing Electric Generating Units in the United States" spreadsheet. Find out if your state has any PV or solar thermal power sources and, if so, determine the total capacity of each. If not, where are the closest solar thermal electric power plants?

2. Explore the National Renewable Energy Laboratory's PVWATTS solar calculator at http://rredc.nrel.gov/solar/codes_algs/PVWATTS/version1/, and use it to find the performance of a PV system at your location. You can use the calculator's default values for the different parameters, or enter them yourself. Either way, prepare a graph showing monthly values of the AC energy produced, in kilowatt-hours. What is the value of the energy produced over the year?

3. If you're not in the United States, download the International Energy Agency's report *Trends in Photovoltaic Applications* (available at www.oja-services.nl/iea-pvps/products/rep1_11.htm). If your country is listed, use the data (a) to find the total installed PV power in your country and (b) to prepare a plot of installed PV power over time in your country.

Chapter 10

INDIRECT FROM THE SUN: Water, Wind, Biomass

Solar energy powers many of Earth's natural processes. As I showed in Figure 1.8, some 23 percent of the incident solar energy goes into evaporating water, which condenses to fall as rain, some on the continents where it returns via rivers to the oceans. Another 1 percent goes into the kinetic energy of atmosphere and oceans—the winds and ocean currents. A mere 0.08 percent of the incident solar energy is captured by photosynthetic plants, which store the energy chemically and provide the energy supply for nearly all life on Earth. These three—the energy of flowing water, the energy of moving air, and the **biomass** energy stored in living organisms—are thus indirect forms of solar energy. So is the thermal energy stored in soils and ocean water. Today, these indirect solar forms provide some 8 percent of humankind's energy, predominantly in the form of hydro-electricity (Fig. 10.1). In the future that figure could go much higher.

10.1 Hydropower

Humankind's use of **hydropower**—the energy associated with flowing water—has a long history. More than two thousand years ago the Greeks developed simple waterwheels to extract mechanical energy from moving water to grind grain. Hydropower remained the unrivaled source of mechanical energy for industry until the development of steam engines in the eighteenth century. Waterwheels drove mechanisms for grinding grain, hammering iron, sawing wood, and weaving cloth. The factories of the industrial revolution often had elaborate belt-and-pulley systems to transfer mechanical energy from a waterwheel to machinery located throughout the factory. It's no coincidence that industrial towns sprang up along rivers with ample and reliable flows of water, especially where natural waterfalls concentrated the water's mechanical energy.

The invention of the electric generator in the late nineteenth century provided a natural application for hydropower. Recall from Section 3.2 that a generator uses electromagnetic induction to convert mechanical energy into electrical energy; thus, connecting

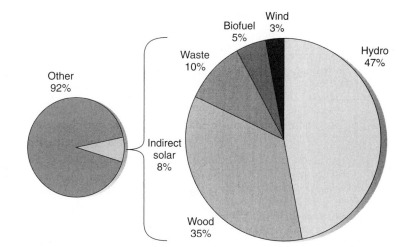

Figure 10.1
Small pie graph at left shows that indirect solar energy supplies 8 percent of U.S. net energy production. Larger pie shows the contributions from different forms of indirect solar energy.

a waterwheel to a generator transforms the energy of moving water into electrical energy. Hydropower has competed with fossil-fueled generators since the inception of the electric power industry, and in the early twentieth century hydropower supplied 40 percent of U.S. electricity. By mid-century that share had slipped to about one-third, and in the early twenty-first century only 7 percent of U.S. electricity came from hydropower. Globally, however, hydroelectric power accounts for some 17 percent of total electric power generation, and in some countries it's the main source of electrical energy (Fig. 10.2). Despite its low percentage of hydroelectric power, the United States' huge energy consumption still makes it the world's fourth-largest producer of hydropower after Canada, Brazil, and China.

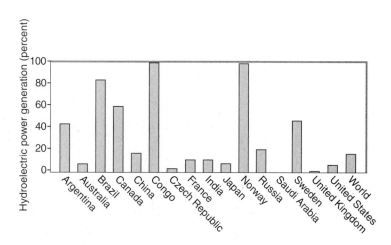

Figure 10.2
Percentage of electrical energy generated from hydropower for selected countries.

The Hydropower Resource

Hydropower's source is the 23 percent of solar energy incident at Earth that evaporates water. Averaged over Earth's surface, this amounts to some 78 W/m². As you can show in Exercise 1, the vast majority of that energy goes into the phase change from liquid water to vapor (more than 2 MJ/kg, as outlined in Section 4.5). Most of this energy heats the atmosphere when water vapor condenses to form clouds and precipitation, but a small fraction becomes gravitational potential energy of the airborne water (again, see Exercise 1). Most of this potential energy is dissipated as precipitation falls, but precipitation reaching higher ground still has considerable potential energy that becomes the kinetic energy of flowing waters.

The **hydrologic cycle** describes the ongoing process of evaporation, precipitation, and return flow. Because most of Earth's surface is covered with water, this is where most of the evaporation occurs. Much of the evaporated ocean water falls back into the oceans, but some wafts over the continents to fall as precipitation. About half of that moisture returns to the atmosphere via evaporation from rivers and lakes and via **transpiration**—the process whereby plants extract moisture from the soil and release it to the atmosphere. The remainder—some 45,000 cubic kilometers (km³) per year, or about 8 percent of the total water cycle—returns to the oceans. Rivers carry most of that water, helped by underground flows. Figure 10.3 shows these quantitative details of the hydrologic cycle.

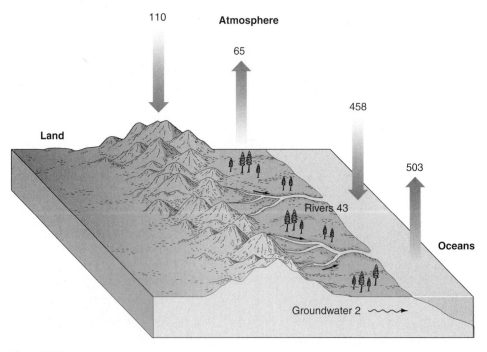

Figure 10.3

The hydrologic cycle, showing evaporation, precipitation, and return flows in thousands of cubic kilometers per year.

Example 10.1   Hydroelectric Power

Figure 10.3 shows river flows of some 43,000 km^3 per year. Suppose we could harness 10 percent of that flow with hydroelectric dams that have water dropping through a distance of 300 m. Assuming 90 percent efficiency in converting the water's energy to electricity, what would be the total electric power output in watts?

Solution

Water's density is 1,000 kg/m^3; there are (365 days/year)(24 hours/day)(3,600 seconds/hour), or about 3.15×10^7 s in a year; and there are 10^9 m^3 in 1 km^3 (this follows because the number of cubic meters in a cube 1 km on a side is 1,000^3, or 10^9). So the 43,000 km^3 per year is (43,000 km^3/y)(10^9 m^3/km^3)(1,000 kg/m^3)/(3.15×10^7 s/y) = 1.37×10^9 kg/s. Now, the potential energy change of a mass m falling through a distance h is mgh, where $g = 9.8$ m/s^2 (Equation 3.4). Here we have 1.37×10^9 kg falling every second through 300 m; multiplying those two figures together with g then gives the *rate* of potential energy change. We're told we can extract 10 percent of that as electric power, with 90 percent efficiency, giving

$$P = (1.37 \times 10^9 \text{ kg/s})(9.8 \text{ m/s}^2)(300 \text{ m})(0.1)(0.9) = 4.0 \times 10^{12} \text{ kg·m}^2/\text{s}^3 = 0.36 \text{ TW}$$

This number is about 20 percent of the average global electrical energy-consumption rate of 1.75 TW. Since 17 percent of the world's electricity actually does come from hydropower, this example suggests that we're probably using somewhere on the order of 10 percent of all the world's river flows to generate electricity—possibly more because the dam height I specified applies only to the largest dams, and possibly less because many rivers go through more than one hydroelectric power plant. Those rivers are also cooling our fossil and nuclear power plants, so you can see that the energy industry really works the hydrologic cycle!

Example 10.1 confirms that global hydropower resources are certainly adequate for the 17 percent of the world's electricity that now comes from flowing water. The numbers suggest further that there's some room to increase hydroelectric power production, although not by orders of magnitude. In fact, the United Nations estimates that the world's technically exploitable hydroelectric resources are roughly equal to today's global electrical energy production—that is, a little over five times current hydroelectric power generation. This suggests that hydropower alone probably won't supply all the world's electricity, let alone meet the planet's entire energy demand. But it could come closer in some regions. As Figure 10.2 suggests, hydroelectric resources are far from evenly distributed. Regions with large land areas, mountainous terrain, and ample rainfall are blessed with substantial hydroelectric resources—so much so that even some energy-intensive industrialized countries such as Canada and Norway can count on hydropower for most of their electrical energy (60 percent in Canada, 99 percent in Norway). But the hydroelectric resources of most industrialized countries are already substantially developed, and further increases are unlikely because of environmental objections that I'll describe at the end of Section 10.1. Figure 10.4 shows estimates of hydroelectric potential for five continents, compared with what's already been exploited in each. It's clear that hydropower has great potential in the developing world, but much less in developed countries.

Energy Quality

By now you're well aware that *amount* isn't the only criterion in evaluating energy sources; so is energy *quality*. Over and over I've stressed that the second law of thermodynamics limits our ability to convert lower-quality thermal energy into mechanical energy and electricity—whether that conversion takes place in an old, dirty coal-fired power plant or a clean, renewable solar thermal plant. What about hydropower? Does the second law's efficiency limit apply to it, too? The answer is "yes and no," with "no" being the more relevant answer when we're talking about meeting humankind's energy needs.

The energy of moving water powers our hydroelectric generating stations with pure mechanical energy, the highest-quality form, which is on a par with electricity. So in principle we can convert water's mechanical energy to electrical energy with 100 percent efficiency. We don't have to worry about thermodynamics because there's no thermal energy involved. In practice there are minor losses associated with mechanical friction, fluid viscosity, and electrical resistance, but these don't pose a fundamental limitation and clever engineers can reduce them toward zero. The result is that modern hydroelectric power plants achieve conversion efficiencies in excess of 90 percent—the highest of any large-scale energy-conversion technology.

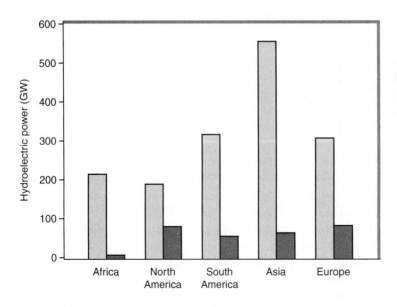

Figure 10.4

Estimates of average hydroelectric potential (tall bars) and actual average production (short bars) for five continents.

So why did I answer "yes and no" to the question about second-law limits on water-power, given that hydroelectric energy conversion isn't subject to the second law? Because it's thermal energy, in the form of solar radiation, that ultimately drives the water cycle. In that sense the water cycle is a huge, natural heat engine. Like every other heat engine, it's subject to the second law. Because it operates over a rather low temperature range and isn't specifically engineered to convert thermal to mechanical energy, the water cycle isn't a particularly efficient heat engine. But that doesn't matter when it comes to power generation, because we start with the mechanical energy of moving water that results after the natural heat engine has done its job.

Harnessing Water Power

The simplest approach to capturing the energy of flowing water is to place a paddle wheel in a river or stream. But that's not very efficient, since the wheel samples only a small portion of the flow. It's better to channel the flow to the energy-conversion wheel. And it's better still to place the wheel at a natural waterfall, where it captures some of the considerable energy gained by the falling water. Water-powered grain mills and industrial plants often used this approach (Fig. 10.5). A hydropower installation that makes only little modification to the stream flow is called a **run-of-the-river** facility. Suitable run-of-the-river locations are limited in number and in power, and many of the best run-of-the-river sites have long since been developed. Furthermore, run-of-the-river installations are subject to substantial natural variations in stream flow, and their power output varies accordingly. For these reasons, run-of-the-river hydroelectric power generation today is largely limited to smaller installations. Exceptions include a few sites where large natural waterfalls have been harnessed to produce substantial hydroelectric power. At the 4-GW Niagara installation, for example, much of the Niagara River's flow is diverted upstream of the falls into tunnels that carry water to turbine-generators located below the falls. An international agreement ensures sufficient flow over the falls to keep tourists happy.

The great majority of hydroelectric installations include dams that create upstream lakes and provide a drop through which water gains energy. Figure 10.6 shows a cross section through a typical large hydroelectric dam. Although turbine designs vary, most are vertical-axis units in which water spins a horizontal waterwheel; the inset in

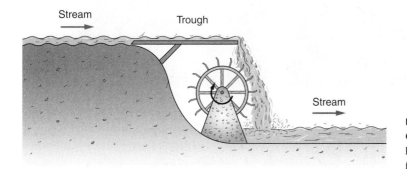

Figure 10.5
Overshot waterwheel, used historically to harness waterpower for grain mills and factories located at natural waterfalls.

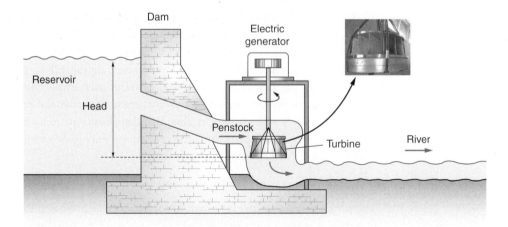

Figure 10.6

Cross section of a hydroelectric dam, showing one of the turbine-generators. Penstocks are huge pipes that deliver water to the turbines. The photo shows a turbine being installed at South Carolina's J. Strom Thurmond hydroelectric plant. (The figure is not to scale; the turbine-generator system would be considerably smaller.)

Figure 10.6 shows one such design. You saw in Example 10.1 that two factors determine the total power in a hydroelectric installation: the total flow and the vertical distance, called the **head,** through which the water drops. A given power output can come from either a large flow and low head or a high head and lower flow; Figure 10.7 and Exercises 4 and 5 explore this distinction.

(a)   **(b)**

Figure 10.7

(a) The John Day Dam on the Columbia River in Oregon is a low-head dam with a high flow rate. (b) The Hoover Dam on the Colorado River at the Arizona/Nevada border is a high-head dam with a lower flow rate. Each produces just over 2 GW of hydroelectric power. Note the lock on the left side of the John Day Dam that permits navigation.

Table 10.1 Ten Largest Hydroelectric Power Plants

Name	Country	Capacity when complete (GW)	Year of first power generation
Three Gorges	China	18.2	2003
Itaipu	Brazil/Paraguay	14.0	1983
Guri	Venezuela	10.0	1986
Grand Coulee	United States	6.5	1942
Sayano-Sheshuensk	Russia	6.4	1989
Krasnoyarsk	Russia	6.0	1968
Churchill Falls	Canada	5.4	1971
La Grande 2	Canada	5.3	1979
Bratsk	Russia	4.5	1961
Moxoto	Brazil	4.3	1977

Hydroelectric systems come in all sizes, starting with backyard "microhydro" units ranging from 100 W to a few hundred kilowatts. Small-scale hydroelectric plants up to a few tens of megawatts were built in abundance in the early and mid-twentieth century, and many are still in operation (see Research Problem 1). The largest hydroelectric installations contain multiple generators of typically several hundred megawatts each, and their total output exceeds that of the largest fossil and nuclear plants. Table 10.1 lists the world's largest hydroelectric generating facilities. Compare their power capacities with the typical 1 GW of a large fossil or nuclear plant, or even the 4 to 8 GW of some multiple-unit nuclear facilities, and it's clear that large hydroelectric installations are the world's largest electric power plants.

Peaking and Pumped Storage

It takes many hours to start up a nuclear or coal-fired steam power plant, and these plants are most efficient when run at a fixed design power. Such installations are therefore **baseload** facilities, run continuously to supply the minimum electrical load that's always present. But electrical demand varies. At the start of the workday; or in the early evening when lights, ovens, and TVs all come on; or on especially hot days that maximize the demand for air conditioning, there's need for additional power. So electric utility companies require additional sources of **peaking power** that can be turned on and off rapidly. Hydroelectric plants are ideal for supplying peaking power because all it takes to get them going is to open a valve; there's nothing to warm up.

The water behind a hydroelectric dam represents stored gravitational potential energy. It's available, on demand, for conversion to electricity in the dam's turbine-generators. In that sense, impounded water is like a fuel, but unlike a fuel burned in a thermal power plant, water's stored energy can be converted to electrical energy with nearly 100 percent efficiency.

There are times when a utility finds itself with a surplus of even baseload power. Electrical energy is ephemeral; it's not something you can bottle up for sale later. One solution is to sell excess power to a region that needs more, which is one motivation for today's large, interconnected power grids. Another approach is to convert excess electrical energy to another form for storage, but few storage methods are both technologically and economically feasible at the large scales needed in the power industry. The development of inexpensive and technologically efficient energy-storage systems would at once reduce the need to construct new power plants and would make intermittent sources like wind and solar-generated electricity more attractive.

The one electrical-energy storage method in significant use today is **pumped storage**, a technology closely related to hydroelectric power generation. Strictly speaking, pumped storage may not belong in a chapter on energy derived from the Sun, because the stored energy likely comes from a nuclear or coal-fired plant. But we'll treat it here because of its similarity to conventional hydroelectricity.

Pumped storage plants use electrical energy to pump water into an elevated reservoir, converting electrical energy into gravitational potential energy. The electric motors doing the pumping operate at over 90 percent efficiency. Pumping occurs at times of low electrical demand, when there's surplus power to run the pumps. When additional peaking power is needed, water is drawn from the reservoir to produce electricity.

Here's the beauty of the system: The same electric motors and pumps that convert electrical to gravitational energy also do the reverse conversion. That's because a pump and a turbine are basically the same thing. When the device is placed in a fluid stream, it converts the energy of fluid motion to mechanical rotation. When energy is applied to rotate the turbine, it converts mechanical rotation into fluid motion. Similarly, an electric motor and a generator are basically the same device: Put in electrical energy and you get out mechanical energy; put in mechanical energy and you get out electrical energy. I made this point back in Chapter 5's discussion of hybrid cars, whose motor-generators convert energy in both directions. The motors/generators and pumps/turbines of a pumped storage system serve a similar dual purpose. Figure 10.8 shows the workings of a typical pumped storage facility.

Today, some 140 pumped storage systems operate in the United States; their combined generating capacity is about 18 GW. A typical example is the Northfield Mountain Pumped Storage Project in Massachusetts, shown in Figure 10.9. This facility stores some 44 TJ of energy, and 75 percent of the energy used to pump the water is recovered as electrical energy. Example 10.2 explores the Northfield Project.

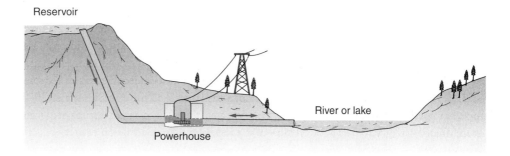

Figure 10.8
Cross section of a pumped storage facility showing the upper reservoir, piping, and the underground powerhouse containing the turbine/generators that also serve as pumps/motors.

Figure 10.9
The Northfield Mountain Pumped Storage Project in Massachusetts can generate just over 1 GW of electric power from the gravitational potential energy of water stored in its mountaintop reservoir.

Example 10.2   Pumped Storage

The Northfield Mountain Pumped Storage Project shown in Figure 10.9 stores 21 million m³ of water about 214 m above its turbine-generators. (a) Find the total gravitational potential energy stored in this water. (b) If all this energy could be converted back to electricity, how long could Northfield Mountain supply its rated power of 1.08 GW before emptying the reservoir?

Solution

(a) Equation 3.4 shows that the gravitational energy stored when a mass m is lifted through a height h is mgh, where $g = 9.8$ m/s^2 is the acceleration of gravity. Water's density is 1,000 kg/m³, so Northfield Mountain's 21 million m³ of water amounts to 21 billion kg. So the stored energy is

$$mgh = (21 \times 10^9 \text{ kg})(9.8 \text{ m/s}^2)(214 \text{ m}) = 4.4 \times 10^{13} \text{ J} = 44 \text{ TJ}$$

(b) At 1.08 GW, or 1.08 GJ/s, this energy would last for a time given by

$$t = \frac{44 \times 10^{12} \text{ J}}{1.08 \times 10^9 \text{ J/s}} = 40{,}740 \text{ s}$$

or about 11 hours. In fact, Northfield Mountain has four 270-MW turbine generators that can be switched in or out as needed, so it could supply lower power levels over longer periods.

Environmental Impacts of Hydroelectric Power

The actual generation of hydroelectric power is a clean, quiet, nonpolluting process that uses a completely renewable energy resource. Nevertheless, hydropower has substantial environmental impacts, nearly all of which are related to the dams that store water and energy for hydroelectric power generation. Some of those impacts are obvious, while others are subtle and may surprise you.

The most obvious effect of dams is to block naturally flowing rivers and thus form lakes. In countries with heavily developed hydroelectric facilities, many rivers have become nothing more than series of lakes separated by hydroelectric dams. Rivers best suited for hydroelectric power are often among the most scenic waterways, prized for their wilderness qualities and for recreational uses such as fishing and whitewater rafting. Dams transform wild rivers into placid lakes that offer other recreational opportunities—most prominently fossil-fueled powerboats (see Research Problem 2 to explore the environmental impact of this aspect of hydroelectric power). The wild river issue alone has often pitted environmentalists against the development of clean, nonpolluting hydropower.

Large power dams inundate vast land areas and often necessitate the relocation of entire towns and villages. (Recall Exercise 14 in Chapter 9, which shows that the area inundated by a hydroelectric dam can substantially exceed that needed for a photovoltaic plant of the same capacity.) As recently as the late twentieth century, the Canadian government relocated Cree Indians and disrupted their hunting and fishing grounds in order to develop the vast hydroelectric resources of northern Quebec. China's massive Three Gorges hydroelectric project required the relocation of more than a million people. Worldwide, some 40 to 80 million people have been displaced by large dam projects.

Another obvious impact is the blockage of rivers for the passage of wildlife and for navigation. These are less problematic with low-head dams, where fish ladders and locks are feasible. But fish used to the oxygen-rich waters of fast-flowing rivers may not find dam-impounded lakes all that hospitable—just one instance of the many ways in which hydroelectric development can alter river ecology. Indeed, hydroelectric development in the American Northwest and elsewhere is directly at odds with efforts to preserve or enhance the populations of salmon and other fish. Some dams in the Northwest have elaborate fish-transport systems, including ladders that enable adult salmon to travel upstream to spawn, and barges or trucks for transporting juveniles downstream around the dams. Other U.S. power dams are being removed to alleviate their impacts on fish.

Flowing rivers carry sediment that is deposited to form beaches and deltas. But when water stops behind a dam, it drops nearly all of its sediment load there and thus gradually reduces the reservoir's capacity. For this reason, large hydroelectric dams have limited lifetimes, on the order of 50 to 200 years, before their reservoirs become useless due to siltation. Meanwhile, evaporation of impounded water lowers the overall flow and increases salinity, with impacts on river ecology and the water's suitability for agricultural irrigation. Damming of the Colorado River has reduced the number of sandbars in the Grand Canyon, which park visitors use as campsites. Cessation of natural floods also has resulted in silt buildup that adversely affects fish habitat in the canyon. In 1996 and 2004 the U.S. Bureau of Reclamation released flood-like flows from its Glen Canyon Dam, attempting to restore natural conditions in the Grand Canyon. In Egypt's Nile Valley, silt buildup behind the Aswan High Dam had another consequence: Cutting off the flow of rich silt to the lower valley starved agricultural land of needed nutrients and forced farmers to turn to chemical fertilizers. Ironically, much of the dam's electric power output eventually went to plants producing the fertilizer that was needed to replace the Nile's natural nutrient-delivery system. The loss of silt also reduced the replenishment of the Nile Delta system, shrinking an area that is home to 30 million Egyptians and most of Egypt's agriculture. It reduced the nutrient flow into the eastern Mediterranean, harming local fisheries. And water losses from the dam and agricultural diversion lowered the freshwater flow into the Mediterranean, increasing salinity and affecting the marine ecosystem.

Dam building can also have an impact on human health. No longer subject to the Nile's regular flooding, Egypt's irrigation canals became breeding grounds for snails hosting parasitic worms that cause the debilitating disease schistosomiasis, whose prevalence in the Egyptian population increased dramatically following construction of the Aswan Dam. Another human impact comes from dam failures, which can cause catastrophic loss of life in downstream communities. In 1928, failure of the St. Francis Dam north of Los Angeles sent some 45 million m^3 (12 billion gallons) of water surging downstream, killing more than four hundred people. A pair of dam failures in India in 1979 and 1980 killed 3,500 people. Large dam failures, like catastrophic nuclear accidents, are rare events, but like nuclear accidents they put large populations at risk. Indeed, the U.S. Federal Emergency Management Agency estimates that one-third of the 74,000 dams in the United States pose significant hazards to life and property should they fail. Causes of dam failure might include inadequate engineering or construction, earthquakes, excessive rainfall, and even terrorism.

A final impact of hydroelectric dams comes from a surprising source: greenhouse gas emissions. Indeed, advocates have long cited the lack of climate-changing emissions among hydropower's virtues. But research now suggests otherwise. The construction of a hydroelectric dam floods terrestrial vegetation, and when submerged vegetation decays under anaerobic conditions, it produces methane, which is a much more potent climate-change agent than CO_2. Rivers bring additional carbon into the reservoir, resulting in further greenhouse emissions. These emissions continue for decades. Research in Canada suggests modest greenhouse emissions from dams in cool climates, but tropical dams are another story. One study of a Brazilian hydroelectric dam indicates that its emissions may have more than three times the climate-warming potential of a fossil power plant of similar capacity. Emissions vary dramatically from dam to dam, depending on location, altitude, local ecology, and other factors. It will take further research, and probably site-specific climate impact studies, before we understand fully the climatic effects of existing and proposed hydroelectric installations.

10.2 Wind

Like waterpower, wind is an indirect form of solar energy. Winds range from local breezes to the prevailing air currents that encircle the globe, and all result from the differential solar heating of Earth's surface. This produces both horizontal and vertical air motions that, in the large scale, couple with Earth's rotation to produce regular patterns of airflow (Fig. 10.10). Figure 1.8 showed that a mere 1 percent of the solar input goes into wind and ocean currents, but that's still enough to give wind a serious place among humankind's energy sources.

Wind, like water, has a long history as an energy source. An Egyptian vase dating from 3500 B.C.E. depicts a sailing vessel, showing that wind-powered boats were in use more than 5,500 years ago. Millennia of technological advances culminated in the tall ships of the nineteenth century, which used wind energy at rates approaching 7.5 MW—50 percent more than today's largest wind turbines. Land-based windmills for pumping water and grinding grain were operating four thousand years ago in China and eight hundred years ago in Europe; by 1750 there were ten thousand windmills in England alone. Picturesque four-blade Dutch windmills pumped water to help reclaim the Netherlands from the sea, and the familiar multivane windmills on American farms brought up water for agricultural use before rural electrification in the 1930s. Utility-scale electric power generation from wind began with a 1.25-MW unit on Grandpa's Knob in Vermont in the 1940s, and serious interest in wind-generated electricity followed the oil shortages of the 1970s. Today wind is the world's fastest-growing source of electrical energy, and it's beginning to make significant contributions to the electrical energy supply in Europe and California.

The Wind Resource

The 1 percent of solar energy that becomes the kinetic energy of the wind amounts to about 2 PW—more than one hundred times humankind's energy consumption. But its uneven distribution in both space and time limits wind's potential, and both practical and technological considerations further proscribe our use of wind energy. Nevertheless, wind is clearly capable of providing a substantial portion of the world's electrical energy, even in energy-intensive Europe and North America.

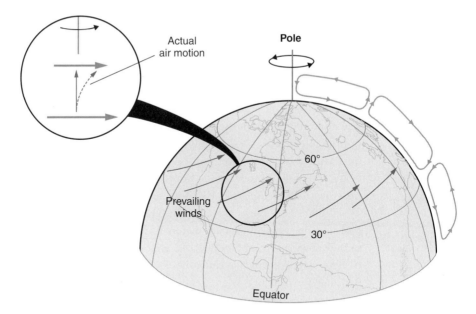

Figure 10.10
Origin of the prevailing westerly winds in the mid-latitudes of the northern hemisphere. Surface air is heated at the equator, rises, and then cools as it flows northward, sinking back to the surface at about 30° north latitude. It then flows southward to the equator, forming a cell of circulating air. A second cell forms in the temperate zone between about 30° and 60° latitude, this one with its surface flow moving northward. There's a third cell in the polar region. The inset shows the additional effect of Earth's rotation, whose actual speed (horizontal arrows) is greatest at lower latitudes and lowest at the poles so that all points on the planet rotate with the same 1-day period. Air moving northward begins with the higher speeds of the lower latitudes, so as it passes over more slowly moving regions it deflects eastward because of its greater relative speed. This produces a west-to-east flow in the temperate zone. Analogous east-to-west flows (not shown) occur in the tropics and polar region.

The energy available from wind increases rapidly with wind speed—in fact, as the cube of the speed. To see why, consider a cubical volume of air moving with speed v (Fig. 10.11). The air has mass m and so its kinetic energy is $\frac{1}{2}mv^2$. The faster it moves, the more rapidly it transports this energy. Since the energy itself scales as the *square* of the speed, this means the rate of energy flow increases as the *cube* of the speed. Suppose our cubical volume has side s and that the air's density—mass per unit volume—is

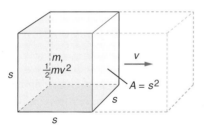

Figure 10.11
A cubical volume contains a mass m of moving air and therefore has kinetic energy $\frac{1}{2}mv^2$. The cube moves its length s in a time proportional to its speed v; dashed lines show the new position. Therefore, the rate at which the wind transports energy across the area A is proportional to the cube of the wind speed v.

ρ kg/m^3. Then the mass of air is $m = \rho s^3$, and so its energy is $\frac{1}{2}\rho s^3 v^2$. Moving with speed v, it takes time s/v for the entire cube to pass a given point. Thus the rate at which this cubical volume transports energy is $\frac{1}{2}\rho s^3 v^2/(s/v) = \frac{1}{2}\rho s^2 v^3$. More significant is the wind power per unit area. Since the cube's area is s^2, this is

$$\text{Wind power per unit area} = \frac{1}{2}\rho v^3 \qquad (10.1)$$

Equation 10.1 gives the actual wind power passing through each square meter of area; in principle, this is the maximum power we could extract for each square meter swept by a wind turbine's blades. As I'll discuss in the section "Harnessing Wind Energy" below, practical considerations limit the actual maximum power extraction to levels a little over half of Equation 10.1's value.

Figure 10.12 shows the annual average wind power available in the United States, measured in watts per square meter. Friction reduces wind speeds close to the ground, so the values shown in Figure 10.12 are taken at the 50-m height typical of a large wind turbine. Average winds in categories 3 or 4 and above (greater than 6.4 m/s or 14.3 miles per hour for category 3) are generally considered sufficient for large-scale wind power generation with current technologies. A glance at Figure 10.12 shows that large areas of the American Midwest meet this criterion, with smaller regions of high average winds in mountainous regions, on the Great Lakes, in the Northeast, and offshore on both coasts.

Example 10.3 Wind Power Classes

Wind power class 3 has a minimum average speed of 6.4 m/s at 50 m elevation, whereas the minimum for class 6 is 8.0 m/s. (a) Verify that the corresponding wind powers in the legend for Figure 10.12 are in the right ratio. (b) Assess the maximum power available to an 82-m-diameter wind turbine at the class 6 speed of 8.0 m/s.

Solution

(a) Equation 10.1 shows that wind power scales as the cube of the wind speed. Therefore the ratio of wind powers for classes 6 and 3 should be the cube of their speed ratio— that is, $(8.0/6.4)^3 = 2.0$, to two significant figures. The actual powers listed in the legend are 600 W/m^2 and 300 W/m^2, respectively, in agreement with our factor-of-2 calculation. (b) An 82-m-diameter turbine sweeps an area $\pi r^2 = (\pi)(41 \text{ m})^2 = 5{,}281 \text{ m}^2$. At 600 W/m^2 for the class 6 winds, the total power carried by the wind is $(600 \text{ W/m}^2)(5{,}281 \text{ m}^2) = 3.2$ MW. An actual turbine with an 82-m blade diameter, the Danish Vestas V82, is rated at 1.65 MW in a 13-m/s wind.

Serious analysis of wind energy's potential requires knowing not only the wind's average speed but also its variability on time scales ranging from minutes to seasons. Because wind power varies as the cube of the wind speed, the average speed isn't a very good

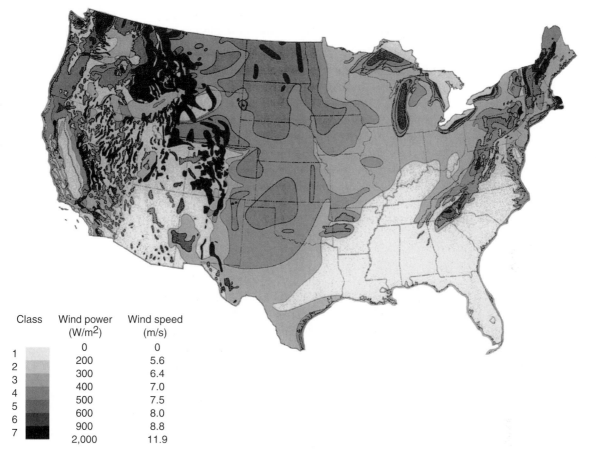

Class	Wind power (W/m^2)	Wind speed (m/s)
1	0	0
2	200	5.6
3	300	6.4
4	400	7.0
5	500	7.5
6	600	8.0
7	900	8.8
	2,000	11.9

Figure 10.12
Wind energy resources in the United States, expressed as yearly
average power per square meter and as wind speed at a height of
50 m. The wind industry recognizes seven wind energy categories.
The average winds in class 3 and above are considered sufficient for
large-scale power generation.

measure of the power one can expect from a wind turbine. A rough rule of thumb suggests that wind variability reduces the average output of a wind turbine to about 20 percent of its rated capacity in a region with category 3 wind resources, a figure that rises
to more than 40 percent at category 7. Accounting for wind variability, the required
spacing of wind turbines so they don't interfere with each other, and the wind resource
as described in Figure 10.12 suggests that U.S. wind energy potential may be several
times our total energy-consumption rate. A more practical estimate, with wind turbines
on only 6 percent of the best wind areas in the United States, could still yield 1.5 times
our electrical energy demand. However, wind's variability means that utilities today can't
rely on wind to provide more than about 20 percent of their total capacity—a figure that

could increase with the development of large-scale energy-storage technology. To supply 20 percent of U.S. electricity with wind would require wind farms on only 0.6 percent of the total U.S. land area, and most of that land would still be available for farming and ranching. Similar considerations apply to worldwide wind resources. A study by the European Wind Energy Association suggests that wind could supply some 12 percent of the world's total electric power by the year 2020.

Harnessing Wind Energy

Most modern wind turbines are propeller-like devices with three blades mounted on a horizontal shaft (Fig. 10.13a). This design puts the blade system and electric generator atop a tower, and the entire turbine-generator and blade assembly can rotate as the wind direction changes. A few of today's wind generators are vertical-axis designs (Fig. 10.13b) that respond to all wind directions; however, these tend to be less efficient.

(a)

(b)

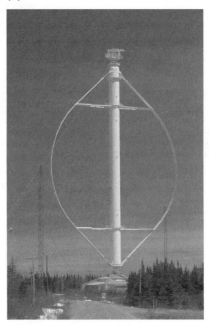

Figure 10.13
(a) This horizontal-axis wind turbine at Carleton College in Minnesota sweeps out an area 82 m in diameter and is rated at 1.65 MW. The first commercial-scale wind installation at an American college, it began operation in 2004 and supplies some 40 percent of Carleton's electrical energy. (b) Hydro-Quebec's experimental 4-MW vertical-axis Darrieus rotor is the largest of its type ever built; it operated from 1988 through 1993.

Equation 10.1 gives the actual power per unit area carried in the wind. But no wind machine can extract all that power, for a simple reason: If it did, then the air behind the turbine would have no kinetic energy left. The air would be stopped, and there would be no place for incoming air to go. Instead, all the air would deflect around the turbine, which would then produce no power. You could avoid that deflection completely if you didn't extract any energy from the flow, but this of course defeats the purpose of your wind turbine. So there's a tradeoff: As you try to increase the energy extraction, you slow the air behind the turbine and thus increase the amount of air deflected around it.

A detailed analysis, first done by the German physicist Albert Betz in 1919, gives the maximum power extractable from the flow in terms of the ratio, a, of the air speed behind the turbine to the incoming wind speed. For the extreme cases discussed in the preceding paragraph, a takes the values 0 and 1, respectively. In general, Betz showed that the absolute maximum power of Equation 10.1 is tempered by a factor of $4a(1 - a)^2$, giving

$$\text{Maximum extractable wind power} = 2a(1 - a)^2 \, \rho v^3 \qquad (10.2)$$

In both extreme cases, where $a = 0$ or $a = 1$, the extractable power is zero. Figure 10.14 is a plot of Equation 10.2, and it shows that the maximum extractable power is about 59 percent of Equation 10.1's total wind power. If you've had calculus, you can show in Exercise 11 that the exact value is 16/27 of the total power contained in the wind.

Real wind machines fall short of the 59 percent theoretical maximum for several reasons. Wind turbines have plenty of space between the blades where air can "slip through" without giving up its energy. Increasing the number of blades would help, in principle, but would make the device more expensive and more massive. Only a turbine with infinitely many blades would have the Betz **power coefficient** of 59 percent. Today's common three-blade design is a compromise that can achieve power coefficients in excess of

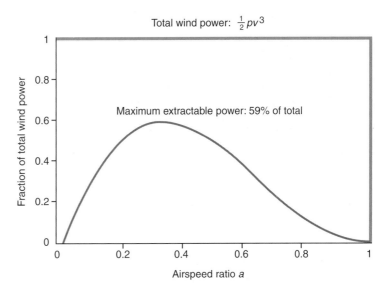

Figure 10.14

The fraction of extractable wind power as a function of the ratio a of the airspeed behind a turbine to the incident wind speed. The peak of the curve, at 59 percent, represents the maximum power any wind machine could extract.

45 percent. The Darrieus vertical rotor design shown in Figure 10.13b has a somewhat lower power coefficient than horizontal-axis machines, typically 30 percent.

The power coefficient also varies with wind speed—in particular, with the ratio of the blade tip speed to the wind speed. Practical turbine designs have an optimal wind speed at which their power coefficient is highest, and a designer will choose a turbine whose power coefficient peaks near the expected average wind speed. The longer the turbine blades are, the higher the blade tip speed is for a given rotation rate. Therefore larger turbines can turn more slowly and still achieve their optimal power coefficients. This is the reason why you'll see small, residential wind turbines spinning furiously while their utility-scale counterparts turn with a slow, stately grace.

High wind speeds could overpower a wind machine's electric generator or even damage the turbine blades, so it's necessary to shed wind—usually by changing the blade pitch—or even shut down the machine altogether in the highest winds. For that reason a practical wind turbine actually produces more power—in addition to having a higher power coefficient—at speeds lower than the maximum it's likely to experience. There's also a minimum speed at which a turbine begins generating power. Putting these factors together gives a typical **power curve** like the one shown in Figure 10.15. Combining the power curve with a site's wind characteristics—average wind speed and its variability—lets engineers assess the expected power output from proposed wind installations.

Environmental Impacts of Wind Energy

Like hydropower, wind is a clean, nonpolluting energy source. Unlike hydropower, it doesn't require damming natural flows. So with wind there's nothing like the considerable environmental impact of hydropower dams, and this makes wind a far more benign energy source. As with photovoltaic and other solar technologies, it's important to con-

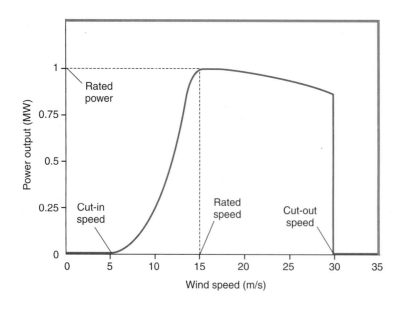

Figure 10.15

Power curve for a typical wind turbine. This unit has a rated power output of 1 MW, which it achieves at its rated wind speed of 15 m/s. The cut-in speed is 5 m/s, and power generation ceases at the cut-out speed of 30 m/s.

sider wind's energy payback—the time it takes a wind turbine to generate as much energy as it took to manufacture. Here the picture is even better than with solar: A study by the Danish Wind Energy Association suggests that the energy required to manufacture, install, maintain, and eventually decommission a wind turbine is produced in the first two or three months of turbine operation.

Wind turbines generate some noise, although large, slowly rotating units can be surprisingly quiet. Standing directly below the blades of the 1.65-MW turbine in Figure 10.13a, for example, one hears only a low, gentle "whoosh" as each blade swings through its lowest position. Turbines create a hazard for birds and bats, as studies of wildlife kills at wind farms attest. However, these losses are orders of magnitude below what we tolerate in bird kills from collisions with the glass windows of our homes and commercial buildings or with vehicles, power transmission lines, and cell-phone towers. Icing of turbine blades in the winter can result in "ice throws," whereby chunks of ice fly off the moving blades. This is more likely in mountain locations, where moisture condenses directly into ice on the turbine blades. Although ice throw is a safety issue worth considering, it's probably been exaggerated by wind opponents. Icing alters the aerodynamic properties of turbine blades, just as it does to airplane wings, and it slows the rotation of any turbine experiencing ice buildup. This, in turn, reduces the range of any ice throw. Studies further indicate that ice tends to break into smaller chunks on leaving the blades, so scary scenarios of "car-size" ice chunks probably aren't realistic. Finally, more detailed analysis suggests that the maximum ice throw distance is on the order of twice the blade diameter, making the danger zone around a turbine rather small.

Perhaps the greatest environmental impact of wind power is on wilderness values and aesthetics. In large installations, tens, hundreds, or even thousands of turbines are grouped in large wind farms like those shown in Figure 10.16a. These obviously

(a)

(b)

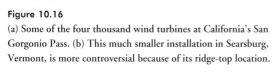

Figure 10.16
(a) Some of the four thousand wind turbines at California's San Gorgonio Pass. (b) This much smaller installation in Searsburg, Vermont, is more controversial because of its ridge-top location.

change the look and character of the landscape, although on open lands such as in the American Midwest they don't preclude continued farming or ranching operations. Offshore wind farms are more problematic, generating strong "not in my backyard" reactions from opponents who don't want their ocean views marred. In the eastern United States, the best wind sites are often on mountain ridges. Installing and maintaining turbines then means introducing roads and transmission lines into what was unspoiled wilderness, and ridge-top siting obviously maximizes the visual impact (Fig. 10.16b). The largest turbines, furthermore, require flashing lights to ward off aircraft, and this increases the visual insult. Similar considerations apply to offshore wind farms, which clearly affect scenic ocean views. In my home state of Vermont, proposals for ridge-top wind farms have split the environmental community down the middle, between those who welcome clean, nonpolluting power and those who see industrial-scale desecration of wilderness. Both sides have their points. I wonder, though, if it might be better to see where our energy comes from than to have it out of sight in a distant coal-fired plant or hidden within the concrete containment structure of a nuclear reactor.

Prospects for Wind Energy

Although wind has a long history of human use, the first serious push for large-scale electric power generation using wind came in the 1970s. California led the world with the first wave of large wind farms in the early 1980s, using turbines with outputs measured in hundreds of kilowatts. Wind growth declined in the late 1980s following changes in government policies, but revived in the 1990s and early 2000s, primarily in Europe. Denmark now holds the lead in wind-turbine manufacture, producing commercial turbines that have up to several megawatts capacity. The Danes now generate over 20 percent of their electricity from wind. Germany, with more than 16,000 wind turbines, is the world leader in wind energy capacity. Figure 10.17 shows the top wind energy–

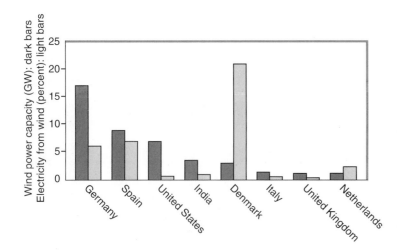

Figure 10.17
Wind power capacity in gigawatts (dark bars) and by percentage of electric power from wind (light bars) in the top eight wind-powered countries as of 2005.

producing countries, and Figure 10.18 shows wind's phenomenal growth rate—some 26 percent annually—since the mid-1990s.

What has powered the recent surge in wind energy? Government incentives, like those for solar energy described in Chapter 9, certainly help. So does serious public and institutional concern over global climate change. In addition, the economics of wind are more favorable than for solar and other nontraditional renewable energy sources. Increases in turbine size and height, advances in control systems and blade aerodynamics, and the emergence of a healthy wind industry have all contributed to a remarkable economic picture that has seen costs of wind power drop by 90 percent between 1980 and 2000; today the capital costs and energy-production expenses for wind are close to those of fossil-fueled electric power. It's hard to pin down exact figures in a world of rapidly fluctuating energy prices, but capital construction costs for wind farms in areas of class 4 winds are around $1.30 per watt. This is higher than the subdollar capital cost for the gas-fired facilities shown in Figure 9.28, but it's still far below the solar and nuclear costs in the same figure. The cost of power production once a wind facility is constructed can be well under 4¢ per kilowatt-hour, which is comparable to and even sometimes below that of natural gas. These figures suggest that wind has a promising future with the potential to contribute significantly to the world's electrical energy supply. However, until we develop large-scale energy storage technologies, wind's variability will prevent it from replacing conventional baseload electric generation.

10.3 Biomass

Waterpower is old, wind is older still, but biomass—fuel derived from recently living matter—is humankind's oldest energy source beyond our own bodies. Our ancestors domesticated fire somewhere between 1.6 million and 780,000 years ago, exploiting our first "energy servant" and starting us on the road to our modern high-energy society.

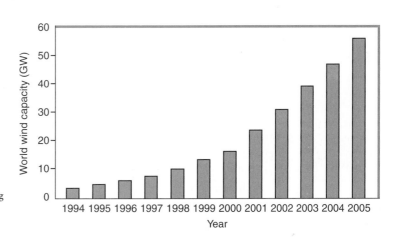

Figure 10.18
The world's installed wind-generating capacity has been growing at some 26 percent annually.

Biomass, predominantly in the form of wood, supplied nearly all of humankind's energy for most of our history. Other sources included dung, plant and animal oils, dried grasses, and agricultural waste. Wood remained the main energy source in the United States until coal took over in 1880. Even today, as Figure 10.1 shows, biomass in the form of wood, waste, and biomass-derived fuels accounts for half the energy we get from indirect solar sources.

The Biomass Resource

Much of Planet Earth is covered with natural solar energy collectors operating at a few percent efficiency—namely, the photosynthetic plants and bacteria. Globally, photosynthesis stores solar energy at the rate of about 133 TW. That's almost ten times humankind's average power consumption, and it amounts to some 0.26 W/m^2 over Earth's surface. More than half the energy—76 TW—comes from land plants, which store energy at about 0.51 W/m^2. At any time, the energy stored in Earth's biomass is some 1.5 ZJ—1.5×10^{22} J, or about thirty-five times the world's annual energy consumption.

Photosynthesis is a complex process that begins with a red chlorophyll pigment absorbing solar energy. The energy causes electron transfers that initiate a sequence of chemical reactions involving a host of intermediate chemicals. This sequence includes a second energy absorption, this time involving a blue pigment. Details vary with different plant and bacterial species and their environments, but the net result is simple: Photosynthesis uses 4.66 aJ (4.66×10^{-18} J) of solar energy to combine six molecules of water and six of carbon dioxide, producing the sugar called glucose ($C_6H_{12}O_6$) and, as a byproduct, oxygen:

$$6H_2O + 6CO_2 + 4.66 \text{ aJ} \rightarrow C_6H_{12}O_6 + 6O_2 \tag{10.3}$$

That oxygen byproduct is, of course, the atmospheric oxygen that, as described in Chapter 1, began to accumulate after the appearance of the first photosynthetic organisms. It forever changed Earth's atmosphere and therefore life's subsequent evolution.

Photosynthesis is not particularly efficient. Photon energy thresholds for chemical reactions limit the amount of the solar spectrum plants can use, just as band-gap energies limit the range of sunlight available for photovoltaic devices. Different plant and bacterial species respond to different parts of the spectrum, but for most green plants the energy absorption has two peaks, at blue and red wavelengths around 400 nm and 700 nm, respectively. There's less absorption in between, in the green region, so plants reflect these intermediate wavelengths and therefore appear green. In addition to spectral efficiency limits, chemical reactions themselves produce waste heat as well as the chemical energy stored in glucose. The net result is a theoretical maximum efficiency of around 14 percent. Most plants fall well short of this limit.

The actual rate at which plants store solar energy is called **primary productivity**, and it varies with plant type, insolation, temperature and other climatic factors, the availability of water and nutrients, and characteristics of the broader ecosystem. The total of all photosynthetic activity is Earth's **gross primary productivity**—the 133 TW figure just introduced. Although the planet's gross primary productivity exceeds humankind's energy demand, it's not possible to divert more than a fraction of this energy resource to power industrial society. For one thing, photosynthesis drives the entire biosphere—all of the living plants and animals except for rare communities that thrive on geochemical energy. In fact, plants themselves use about half the energy they capture, making the **net primary productivity** about half of the gross. Diverting too much of the planet's primary productivity to human uses endangers the rest of the biosphere. For another thing, we humans have to eat to fuel our own bodies' energy and, unlike electric utilities, we can't eat uranium or coal or make direct use of sunlight, wind, or flowing water. So we need some of the planet's primary productivity to feed ourselves. We also use biomass for shelter (e.g., wood), clothing (e.g., cotton, leather), and a host of other products. In fact, it's estimated that humans already appropriate nearly 40 percent of the planet's net primary productivity. That's why, a few sentences ago, I didn't say that diverting too much of the planet's primary productivity "*would* endanger the rest of the biosphere" but "*endangers* the rest of the biosphere."

Example 10.4   Photosynthetic Efficiency

Insolation on a tropical rainforest averages around 210 W/m² year-round. For a temperate forest during the growing season of spring, summer, and fall, average insolation is around 190 W/m². Use these values, along with Figure 10.19, to estimate the photosynthetic efficiency of tropical rainforests and temperate forests.

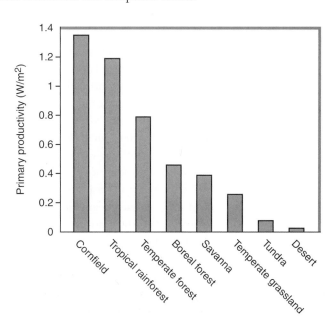

Figure 10.19
Primary productivity for seven natural environments and a typical cornfield managed with modern agricultural techniques. The productivity of the cornfield is deceptive, because considerable energy goes into fertilizers, pesticides, and farm equipment.

Solution

Figure 10.19 shows that the rainforest's productivity is about 1.2 W/m^2 and the temperate forest's is about 0.80 W/m^2. Their efficiencies are therefore 1.2 W/m^2 / 210 W/m^2 = 0.57 percent and 0.80 W/m^2 / 190 W/m^2 = 0.42 percent, respectively. These low numbers are typical of natural ecosystems, although they're underestimates for the actual photosynthetic process because they assume that all the incident sunlight manages to strike plant leaves.

The energy stored in dry biomass is nearly the same regardless of its origin. Pure glucose contains about 16 MJ/kg, roughly half the energy density of coal. Most wood contains very nearly 20 MJ/kg, while grasses and crop plants range from 16 to 19 MJ/kg. The adjective *dry* is important because the considerable energy it takes to drive out water reduces the net energy content of fresh biomass (see Exercise 16).

Harnessing Biomass Energy: Burning

The simplest and oldest way to harness biomass energy is to burn the biomass. Our ancestors used open fires for cooking and warmth, and individual fireplaces or woodstoves continued to be a major source of home heating until the widespread adoption of central heating systems in the late nineteenth and early twentieth centuries.

Burning wood in open fires is extremely inefficient. Heated wood exudes flammable gases that contain much of the wood's energy, and in an open fire these don't burn. Incomplete combustion of bulk wood results in a host of particulate and gaseous emissions, many of which are toxic or carcinogenic. The resulting indoor air pollution is a serious health threat in developing countries, where open fires are still used for cooking. The invention of the chimney around the twelfth century reduced indoor pollution, and the updraft created in the chimney helps to ensure a flow of fresh oxygen to the fire. The earliest stoves appeared around 1500, and in 1740 Benjamin Franklin invented his famous Franklin stove. Its controllable air intakes marked a significant advance in biomass burning efficiency, because a limited airflow keeps the stove interior hotter, allowing exuded gases to burn. Today's airtight woodstoves incorporate secondary combustion systems, catalytic converters, and other improvements engineered to ensure more complete combustion, thereby increasing efficiency and reducing pollution. Such technological improvements, along with regulations that specify the proper seasoning of firewood and the maintenance of steady combustion, can make wood burning a relatively clean, efficient process. Such regulations are in place today in Europe, but in the United States they're not nearly as strong.

Wood burning is more efficient when the wood is in the form of uniform chips or pellets, especially in systems designed to gasify as much of the wood as possible prior to combustion. Wood chips and pellets can be burned much like coal to power woodstoves, industrial boilers, or turbine-generators. Many industries that use wood as a raw material obtain their heat, process steam, and even some electricity from wood. Today there are nearly two hundred wood-fired electric power plants in the United States, ranging from 500-kW to 72-MW units, and the total capacity is 2.8 GW. Most have been built since 1980, and many are owned by companies that manufacture wood products or paper, which burn wood waste and by-products to produce power for their manufacturing operations. Others supply the power grid; the largest wood-fueled public utility plant is the Burlington (Vermont) Electric Company's 54-MWe McNeil Generating Station, which supplies the city of Burlington with one-third of its electric power (Fig. 10.20). In Europe, wood-fired electric power plants are often combined heat and power units that also supply hot water for community heating.

Today, wood and wood by-products account for about 2 percent of U.S. primary energy consumption, including heating, industrial process heat, and electricity generation. Three-quarters of that wood is used in industry. About 1 percent of U.S. electricity comes from wood, most of which is generated by industries for their own use.

Energy from Waste

Many materials we commonly discard as waste contain substantial energy, much of it derived from biomass. Wood-fired power plants often burn chips made from smaller branches that remain after logging. Animals don't extract all the energy from the food they eat, so manures and human sewage are sources of energy. Indeed, animal dung is an important cooking fuel in developing countries. And the solid waste that the industrialized world generates in vast quantities contains, when dry, about as much energy as an equivalent mass of wood, and nearly half as much as oil on a per-kilogram basis. Most of that waste is paper, food scraps, and other materials derived from biomass; a smaller fraction (typically 20 percent) is plastic made from fossil fuels. The United States generates some 230 million tonnes of solid waste a year; as you can show in Exercise 12,

Figure 10.20
Burlington, Vermont's 54-MW McNeil Generating Station is the largest wood-fired power plant in the United States operated by a public utility. Here the plant is seen beyond a pile of its wood-chip fuel, much of which comes from logging waste.

this is equivalent to 750 million barrels of oil—enough to displace two months' oil imports. Burned in power plants at 20 percent efficiency, this waste could produce nearly 7 percent of the United States' electrical energy (see Exercise 13).

One change in the U.S. energy mix over the past few decades has been the emergence of municipal waste as an energy source; by the early twenty-first century, so-called garbage-to-energy plants contributed about 0.5 percent of the United States' total energy and 0.6 percent of the country's electricity.

Even when solid waste is buried in landfills, it can still provide energy. Anaerobic decomposition of the buried waste produces methane (natural gas), which can be captured and burned. Some landfills use their methane for heating and electric power, either for their own facilities or to sell to electric utilities. Others simply let methane vent to the atmosphere—an unfortunate approach, both because of the wasted energy and because methane is a potent greenhouse gas.

Municipal sewage is another source of methane, and many sewage treatment plants, like landfills, use sewage-produced methane to supplement their own energy needs. Cow manure in agricultural areas represents yet another methane source; my own home, in rural Vermont, gets half of its electricity from "cow power"—methane-generated electricity produced at local dairy farms. The fifteen hundred cows on one nearby farm yield electrical energy at the average rate of 200 kW. My electric company encourages this renewable energy production by paying farmers a premium for cow power and giving environmentally conscious consumers the option to purchase the power for a few cents more per kilowatt-hour than the standard rates.

Biofuels

With the exception of the fossil fuels, the energy sources I've described in Chapters 5 to 9 are used primarily to power stationary systems such as homes, industries, and electric power plants: Nuclear and tidal plants produce electricity, geothermal energy generates electricity and supplies heat, direct solar energy heats homes and buildings or is converted to electricity, and hydropower and wind are largely sources of electrical energy. With the exception of wind-powered sailing ships, none of these energy sources is particularly useful for transportation. In fact, Example 9.1 shows the absurdity of using solar energy directly to power cars. Yet transportation in the United States consumes well over one-fourth of our total energy.

Biomass is different. It's the one source today that can directly replace fossil fuels for transportation, because biomass readily converts to liquid **biofuels**. Gaseous biofuels such as methane are also possible transportation fuels, but their low density makes them less desirable than liquids. Today, 2 percent of U.S. gasoline derives from biofuels; in Brazil the figure is 30 percent.

A range of fuels, as well as other useful chemicals, derive from biomass through chemical or biochemical processes. This is hardly surprising; biomass is, after all, the precursor to the fossil fuels. The most important biofuels today are **ethanol**—ethyl alcohol (C_2H_5OH), the same chemical that is found in alcoholic beverages—and **biodiesel**.

Ethanol results from the biological fermentation of plant sugars, or of other plant substances that can be converted to sugars. Yeasts and other microbes do the actual work of fermentation, and in addition to ethanol the process may yield coproducts such as sweeteners, animal feed, and bioplastics.

Gasoline mixed with as much as 10 percent ethanol by volume burns in today's gasoline engines without modification; specially designed engines can run on stronger mixtures or even 100 percent ethanol. The 10 percent mixture, called *E10* or *gasohol*, has slightly less energy content than gasoline because ethanol's energy content, on a per-volume basis, is only two-thirds that of gasoline. Powering cars with ethanol is not a new idea; the original Ford Model T could run on any ethanol/gasoline mix, and gasoline with up to 12 percent ethanol was available in the 1930s. Today, ethanol in the United States comes from corn, and tax breaks encourage its production. Absent this government subsidy, gasohol would probably not be an economically viable product in the United States. A controversial energy bill passed in 2005 increases ethanol subsidies, with the goal of doubling U.S. ethanol production by 2012.

More important than ethanol's economic balance sheet is its energy balance. Producing ethanol in the United States today means growing corn using fossil-powered mechanized agriculture, along with fossil-produced fertilizers and pesticides. Grinding corn and cooking it to convert the starch to sugar requires still more fossil fuel. So do fermentation, distillation, and removal of water. Quantitative studies suggest that the ratio of ethanol energy output to fossil fuel input in corn-to-ethanol conversion ranges from a low of about 0.75 to a high of around 1.7. Ethanol production with an energy ratio below 1 doesn't make much sense, since you have to put in more energy than you get out. A comprehensive study of the entire corn-to-ethanol production cycle puts the ratio at about 1.1, meaning that ethanol yields on average only 10 percent more energy than is used to make it. However, much of that energy comes from coal and gas, so ethanol ends up displacing a greater quantity of oil. Given these figures, you can see why ethanol policy in the United States today is controversial, with some decrying it as a "pork barrel" subsidy to agribusiness while others hail it as a step toward energy independence.

Ethanol, however, need not be produced from corn. A far larger bioresource is the cellulose that comprises much of the green parts of plants. A study led by the Natural Resources Defense Council argues that a vigorous program to produce ethanol from cellulosic plants—in particular, common native grasses—might by the year 2050 displace half the oil used in transportation. The energy balance of cellulose-derived ethanol could be substantially better than for the corn-derived ethanol produced today.

In Brazil the ethanol situation is quite different. Lacking in fossil fuels but with a warm climate and abundant sunshine, Brazil in 1975 committed to ethanol-based transportation fuels. The Brazilian government now mandates at least a 25 percent ethanol content in gasoline, and many Brazilian "flex-fuel" vehicles can burn any ethanol-gasoline blend, including 100 percent ethanol. Brazil's substantial investment in home-grown ethanol production has paid off: When international oil prices climbed to over

$60 per barrel in 2005, Brazil's ethanol exports soared. India and the United States are the leading importers of Brazilian ethanol.

Brazil's ethanol comes largely from sugarcane, which eliminates several fossil fuel–intensive steps needed to process starch or cellulose into sugar for fermentation. As a result, Brazilian ethanol has a much better energy balance, with the ratio of ethanol energy to fossil energy input in the range of 4 to as high as 10. Many Brazilian ethanol-conversion facilities use electrical energy generated on-site from burning sugarcane waste, thus wringing out still more energy and incidentally sending surplus electricity into the power grid. In Brazil, the cost of ethanol today is essentially competitive with fossil gasoline. Brazil's experience bodes well for other sunny, tropical countries. India, in particular, has significant potential to transport its billion-plus population with ethanol.

Ethanol use is significant in Brazil and North America, and is low in Europe and elsewhere, but ethanol production has begun growing rapidly worldwide since the turn of the twenty-first century (Fig. 10.21).

Biodiesel is the second biofuel in significant use today. In contrast to ethanol, it's most popular in Europe. That's not surprising, because half the cars sold in Europe are diesels, and biodiesel can replace conventional fossil-based diesel fuel without engine modifications. Figure 10.22 shows the growth in biodiesel production since the early 1990s. Comparing Figures 10.21 and 10.22 shows that biodiesel production is less than 10 percent that of ethanol—a situation that isn't likely to change, and for good reasons.

Biodiesel, like the oils used in cooking, derives from natural plant oils and fats. All plants contain these substances, which are concentrated in seeds and nuts. In most plant species this means the majority of the plant material is not available for biodiesel production, although nearly the entire plant can be made into ethanol. Therefore, any large-scale effort to capture solar energy for biofuels is likely to concentrate on ethanol, not biodiesel. Still, biodiesel remains a valuable and viable alternative fuel, originating not only in fresh plant

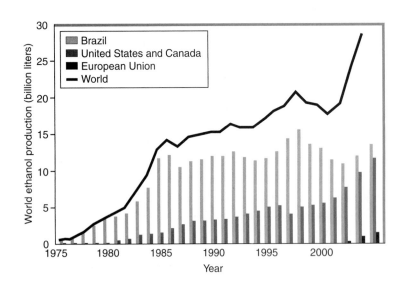

Figure 10.21
World ethanol production has increased dramatically since 1975.

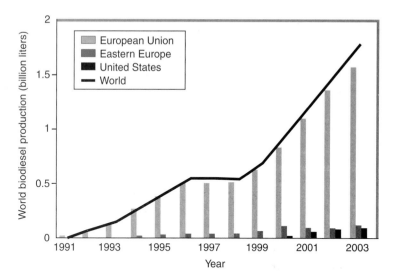

Figure 10.22
World biodiesel production is rising, especially in Europe. Compare the vertical scale with the one in Figure 10.21.

material but also from waste oils and fats. In the United States, for example, the main sources of biodiesel are soybeans and so-called yellow grease, comprised of recycled cooking oils. Many U.S. biodiesel plants can switch among these two feedstocks with no equipment changes. Biodiesel's energy ratio is somewhat better than that of corn-based ethanol, with biodiesel returning about twice the fossil energy it takes to make it.

It's possible to burn conventional cooking oils in modified diesel engines, and alternative-energy advocates are fond of vehicles powered by used oils from the deep-fat fryers of fast-food restaurants. However, cooking oils are more viscous than diesel fuel, due to the presence of glycerin, which gums up engines and reduces performance. Commercial biodiesel production therefore uses a process called *transesterification*, which reacts plant oils and fats with methanol or ethanol to remove the glycerin (itself a useful byproduct). Once broken down into smaller individual molecules, the result is a less viscous fuel with properties very similar to petroleum-derived diesel fuel. Biodiesel is often blended with conventional petrodiesel to produce the fuels B5 (5 percent biodiesel), B20 (20 percent), and B100 (pure or "neat" biodiesel). Some engine manufacturers currently approve blends up to B20, although many individuals and vehicle fleet owners use B100 with no harm to their engines.

Environmental Impacts of Biomass Energy

Burning biomass and its derivatives is like burning fossil fuels—converting hydrocarbons to carbon dioxide, water, and pollutants that include products of incomplete combustion and oxides of sulfur, nitrogen, and other substances. There are also substantial differences, however, nearly all of which favor biomass.

The most important distinction is in carbon emissions. Biomass burning necessarily produces CO_2, just like fossil fuel. But the carbon from biomass was only recently removed from the atmosphere by growing plants, and natural decay would have soon returned it to the atmosphere even if the biomass hadn't been burned. For biomass harvested on a sustainable basis and without fossil fuel required in harvesting and processing, therefore, the rate at which new plants remove carbon from the atmosphere would be equal, on average, to the rate of carbon emission from such biomass burning, which would make the biomass **carbon neutral**. *Sustainable* is the key word here; wholesale burning of biomass without replanting does result in net carbon emissions to the atmosphere—a continuing problem with "slash and burn" agriculture still practiced in tropical rain forests. Even sustainable practices initiated in forest ecosystems can take decades to achieve true carbon neutrality. Finally, as we'll see shortly, the caveat of no fossil fuel use during biofuel production considerably dampens the carbon emissions picture, especially for the more processed biofuels.

The combustion of wood in bulk can be a messy process unless burning is carefully controlled. Bulk wood is used mostly for residential heating, where high levels of particulate matter and organic chemicals can result from incomplete combustion, exceeding even the levels from coal. In the United States, newer stoves and fireplace inserts meeting EPA standards cut particulate emissions in half, to levels somewhat below coal but still far exceeding other fossil fuels. European standards go much further, mandating proper seasoning of firewood, thermal storage systems to minimize variations in combustion rate, separate heat-exchange and combustion chambers, and other design and operating techniques that make European wood burning a much cleaner process. Because stoves and fireplaces last a long time, it will be decades before all domestic wood burning is done in devices meeting modern standards.

Harvesting and transporting firewood consumes some fossil fuel, but on average wood used for heating returns some twenty-five times as much energy as its production requires in fossil fuels; thus fossil emissions associated with wood burning are minimal. However, cutting and burning of older trees can result in a rapid release of stored carbon that even vigorous new growth can't compensate for on time scales shorter than decades; as a result, firewood harvesting as it's now done in temperate forests isn't necessarily carbon neutral. One study suggests that overall net carbon emissions associated with harvesting firewood are about one-fourth those from coal, and half those from natural gas, for equal quantities of heat produced from each fuel.

Wood burned as chips or pellets in industrial applications is another matter; more regulated burning conditions, wood gasification technologies, and the use of pollution-control devices reduces particulates and ensures more complete combustion. Wood burning—whether in domestic stoves or industrial boilers—has lower sulfur emissions than any of the fossil fuels, some 40 percent less than natural gas and over 90 percent less than coal. On the other hand, although chipping and pelletizing branches or other logging waste may sound like a good idea, taking entire trees robs the forest of nutrients that would be returned to the soil through natural decay, so the use of logging waste for energy may not be sustainable in the long term.

Liquid biofuels have clear environmental benefits. Unlike firewood, they're made from short-lived crops that quickly reach carbon neutrality through sustainable agriculture.

Even modest blends such as gasohol (10 percent ethanol in gasoline) decrease many automotive emissions. The extra oxygen in ethanol (C_2H_5OH), compared with gasoline's octane (C_8H_{18}), helps to promote more complete combustion, thereby reducing carbon monoxide emissions by some 25 percent or more. In fact, the primary motivation for ethanol blends in the United States is not so much replacement of fossil fuels as controlling emissions. Ethanol also reduces particulate emissions, tailpipe emissions of volatile organic compounds (VOCs), and several carcinogenic compounds such as benzene. However, ethanol itself is more volatile (easily evaporated) than gasoline, so the use of ethanol increases somewhat the total emission of VOCs. That's because decreased VOC tailpipe emissions are more than offset by evaporation in fuel handling, dispensing, and storage.

Automotive emissions of NO_x don't change much with ethanol, but nitrogen-based fertilizer used to grow ethanol feedstock crops results in a net increase in NO_x emissions over the entire ethanol fuel cycle. Considering the entire fuel cycle also makes the greenhouse gas picture for corn-produced ethanol less rosy. The considerable fossil fuel required to produce ethanol from corn results in substantial greenhouse emissions; one study suggests that automotive consumption of the E85 blend reduces carbon emissions by only about 30 percent compared with fossil gasoline. Sugarcane-based Brazilian ethanol does a lot better, reducing greenhouse emissions by some 70 percent.

The emissions benefits of biodiesel are more pronounced. Biodiesel contains virtually no sulfur, whereas even low-sulfur petrodiesel contains up to 350 ppm of sulfur. Particulates drop substantially, along with carbon monoxide and other products of incomplete combustion. Biodiesel is a better lubricant than petrodiesel, reducing engine wear. And biodiesel has a higher cetane number (the equivalent of the octane number in gasoline), meaning it provides more reliable ignition in diesel engines. Producing biodiesel from waste grease disposes of a significant waste product; indeed, New York City alone could produce enough biofuel from its waste grease to power its municipal bus fleet five times over. One downside is slightly increased NO_x emissions with biodiesel, a level already high in diesels because of their higher operating temperature. And both biodiesel and ethanol can corrode the rubber seals used in automotive fuel systems. Finally, today's biofuel production process requires fossil energy and therefore entails carbon emissions. For this reason the net effect of biodiesel is to reduce carbon emissions by about 78 percent compared with conventional petrodiesel. This is a substantial reduction, but again this biofuel isn't carbon neutral.

Use of biomass energy today is sufficiently low that plantings managed specifically for energy production represent only a small fraction of agricultural land. However, any expansion of land use for energy crops runs up against the food needs of the world's still growing population, and developing new croplands brings all the attendant environmental problems of modern agriculture—increased use of fertilizers and pesticides, soil erosion, appropriation of limited water supplies, water pollution from runoff, and loss of biodiversity. If biomass were to make a serious contribution to the world's energy supply, then land-use issues would quickly become the primary environmental concern. You can show in Research Problem 4 that replacement of all U.S. gasoline with corn-based ethanol would require more than the country's entire cropland; Exercise 14 shows the same for European biodiesel. More efficient conversion processes, the use of grasses in place of corn, and greater use of plant waste from food-producing agriculture could all reduce the land area needed. Nevertheless, a realistic look at land requirements for biofuel production suggests that a plan to replace 10 percent of U.S. gasoline and diesel consumption by 2020 with ethanol and biodiesel might require more than 40 percent of U.S. cropland. Even the more optimistic projections run into land-use limitations unless they're coupled with substantial gains in vehicle energy efficiency. Table 10.2 summarizes some environmental impacts of biofuels as compared with fossil fuels.

Prospects for Biomass

Biomass enthusiasts see biomass energy, especially biofuels for transportation, as a major player in a renewable energy future. Maybe, but at least with current technologies the fossil energy needed to produce most liquid biofuels means that these fuels aren't truly renewable. And land-use conflicts cloud the future for biomass energy from plants grown exclusively as energy crops. It seems more likely that biomass will

Table 10.2 Environmental Impacts of Biofuels Compared with Gasoline and Petrodiesel

	Corn ethanol	Sugarcane ethanol	Biodiesel from oilseed rape
Particulate emissions	Lower	Lower	45 percent lower
Carbon monoxide emissions	25 percent lower with E10 blend	Lower	45 percent lower with B100
Volatile organics	Tailpipe: lower Fuel handling: higher Net: higher	Net higher	Lower
Sulfur emissions	~0	~0	~0 (much lower than the 350 ppm for low-sulfur petrodiesel)
Nitrogen oxide emissions	Higher	Higher	Higher
Toxicity	Lower	Lower	Lower
Energy gain $\left(\dfrac{\text{Energy from biofuel}}{\text{Fossil energy to produce biofuel}}\right)$	1.1	4	2
Greenhouse emissions (equivalent CO_2)	30 percent lower	70 percent lower	55 percent lower

Note: All numbers are approximate.

play a modestly significant role, more in some countries than in others. In the vital area of transportation fuels, we'll be well served by continuing development of more efficient technologies for production of liquid biofuels. But in the end, the low efficiency of photosynthesis on an already-crowded planet means that biomass can't be our sole path away from fossil fuels.

10.4 Other Indirect Solar Energy

The same solar-induced temperature differences that power the wind also, in conjunction with differences in salt concentration and Earth's rotational energy, drive the great ocean currents. And wind itself produces ocean waves, providing a doubly indirect form of solar energy. I briefly described schemes for harnessing the kinetic energy of waves and currents in Chapter 8's section on tidal energy, because these sources of mechanical ocean energy have much in common with tidal energy, except that their ultimate origin is primarily in sunlight rather than the mechanical energy of the Earth-Moon system. Similarly, solar heating of the Earth's outermost layers is the energy source for geothermal heat pumps that I also included in Chapter 8. Properly speaking, wind, currents, and geothermal heat pumps all belong here, but it made more sense to discuss them in Chapter 8.

Ocean Thermal Energy Conversion

As the Sun warms the tropical ocean, it creates a significant temperature difference between the surface waters and the deeper water to which sunlight doesn't penetrate. As shown in Chapter 4, any time there's a temperature difference, there's the potential to run a heat engine and extract mechanical energy. **Ocean thermal energy conversion** (OTEC) schemes propose to harness this energy, in most cases to generate electricity. The thermodynamic efficiency limit expressed in Equation 4.5, $e = 1 - T_c/T_h$, shows that we get the highest efficiency with the largest possible ratio of T_h to T_c. Practically speaking, this limits the OTEC energy resource to the tropics, where the temperature difference from surface to depth is greatest. Tropical surface temperatures can exceed 25°C, while a few hundred meters down the temperature is around 5°C to 6°C. You can show in Exercise 15 that the maximum efficiency of a heat engine operating between these temperatures is only about 7 percent, but with no fuel to pay for, this number isn't the liability it would be in a fossil-fueled or nuclear power plant. And the OTEC energy resource is vast; after all, the oceans absorb much of the 174 PW of energy that the Sun delivers to Earth. Most of that energy is out of our reach, but one serious estimate suggests that OTEC has the potential to produce as much as 10 TW, which is close to humankind's total energy-consumption rate. However, practical considerations suggest a far smaller OTEC contribution. The one significant advantage that OTEC has over solar-based schemes such as photovoltaic conversion and wind is its nearly constant availability. The large heat capacity of water smoothes out day/night and seasonal temperature variations, which would allow OTEC plants to operate continuously to supply baseload power.

Ocean Thermal Energy Conversion Technology

An OTEC plant, in principle, operates like any other thermal power plant: An energy source heats a working fluid, which boils to form gas that drives a turbine. Contact with a cool reservoir recondenses the working fluid, and the process repeats. With OTEC the heat source is warm surface water, and the cool reservoir is the colder water pumped from depths of up to 1 km. Because water boils at 100°C at normal pressures, it's necessary either to operate the heat engine at lower pressure or to use a working fluid with a lower boiling point.

Future OTEC plants have been envisioned as offshore platforms similar to oil-drilling platforms, or as free-floating structures (Fig. 10.23). Another possibility is OTEC-powered factory ships that would "graze" the tropical oceans, seeking out regions with the greatest temperature differences. The energy generated would be used for energy-intensive manufacturing done onboard the ship. Or OTEC-generated electricity could split water into hydrogen and oxygen, producing fuel for a hydrogen-powered economy. If used to desalinate seawater, OTEC technology could help the world satisfy its growing thirst for freshwater. On tropical coasts adjacent to deep ocean water, OTEC desalinization or electricity-generating plants might be constructed onshore, with warm and cold water piped in from the ocean.

OTEC faces numerous challenges that affect its practicality and economic viability. Designing structures to withstand the corrosive marine environment is difficult. Mundane problems include marine organisms clogging pipes. The energy required to pump cold water from great depths reduces practical OTEC efficiencies to a mere 2 percent or so. Although fuel costs aren't an issue, such a low efficiency means a huge investment in

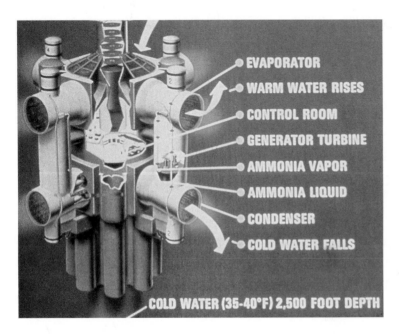

EVAPORATOR
WARM WATER RISES
CONTROL ROOM
GENERATOR TURBINE
AMMONIA VAPOR
AMMONIA LIQUID
CONDENSER
COLD WATER FALLS

COLD WATER (35-40°F) 2,500 FOOT DEPTH

Figure 10.23
Artist's conception of a floating OTEC power station. Not shown is the long pipe extending far downward to bring up cool water.

physical structure returns a relatively paltry flow of energy. For all these reasons, OTEC development to date has been limited to small (tens of kilowatts) pilot projects constructed, and mostly abandoned, in the late 1970s to early 1990s. Still, OTEC may someday provide a viable energy source, especially for tropical islands.

Environmental Impacts of Ocean Thermal Energy Conversion

Isolated OTEC facilities would have little environmental impact, although the mixing of warm and cool ocean waters changes the local temperature and could affect marine life. Nutrients brought up with the cooler deep water might enhance fish production, or lead to algae blooms. But large-scale OTEC power production could have serious effects by lowering the average ocean surface temperature and thus affecting weather, climate, and ocean circulation. A total of 60 GW of OTEC power worldwide—three times the power output of China's Three Gorges dam—would entail cold-water flows equal to all the world's rivers. We're unlikely to see anything like that level of OTEC development, which means we probably don't need to worry about global environmental impacts. But at the same time, it means that OTEC won't contribute substantially to the global energy supply anytime soon.

Chapter 10   Chapter Review

BIG IDEAS

Indirect solar energy includes flowing water, wind, and biomass.

10.1   Nearly one-fourth of the solar energy incident on Earth goes into evaporating water, and this is the source of the flowing water whose energy we capture with hydropower technologies. Hydropower has a long history, and today it supplies some 17 percent of the world's electrical energy, mostly from dams built on rivers. Hydropower represents energy of the highest quality, and hydroelectric power generation produces essentially no pollution. But hydropower dams alter river ecosystems, and tropical installations may emit methane that contributes to climate change. The hydroelectric resource is almost fully utilized, especially in the developed world.

10.2   Roughly 1 percent of the incident solar energy goes into wind. The power available from wind increases as the cube of the wind speed, so high-wind areas are particularly promising sites for wind installations. Modern wind turbines have electric power outputs of up to several megawatts, and they're often grouped into wind farms with tens, hundreds, or thousands of turbines. Wind's predominant environmental impact is aesthetic, although wind turbines may also kill migrating birds and bats. Today, wind is the fastest-growing energy source, and it provides a significant portion of the electrical energy in several European countries. Unlike direct solar alternatives, wind is often economically competitive with conventional energy sources.

10.3 **Biomass** contains solar energy captured by plants in the process of photosynthesis. Earth's plants capture solar energy at about ten times the rate of human energy consumption. Biomass may be burned directly for heat or electric power generation. Decomposition of organic waste biomass releases combustible methane. Fermentation and chemical processing yield **biofuels,** of which ethanol and biodiesel are most important. Biomass and biofuel combustion produce air pollutants, although for some pollutants the amounts are lower than with fossil fuels. Depending on how it's grown and harvested, biomass may be **carbon neutral,** contributing no net carbon to the atmosphere. But unsustainable or energy-intensive harvesting and processing mean that some biomass operations entail significant carbon emissions.

10.4 Other indirect solar energy sources include waves and ocean currents, as described in Chapter 8. **Ocean thermal energy conversion** schemes operate heat engines between warm tropical surface waters and cooler deep water. None of these approaches has yet demonstrated large-scale practicality.

TERMS TO KNOW

baseload (p. 299)
biodiesel (p. 318)
biofuel (p. 318)
biomass (p. 292)
carbon neutral (p. 322)
ethanol (p. 318)
gross primary productivity (p. 315)
head (p. 298)
hydrologic cycle (p. 294)
hydropower (p. 292)

net primary productivity (p. 315)
ocean thermal energy conversion (p. 325)
peaking power (p. 299)
power coefficient (p. 309)
power curve (p. 310)
primary productivity (p. 314)
pumped storage (p. 300)
run-of-the-river (p. 297)
transpiration (p. 294)

GETTING QUANTITATIVE

Solar energy used to evaporate water: 23 percent

Solar energy driving wind and currents: $\sim$1 percent

Solar energy captured by plants: $\sim$0.08 percent

Hydroelectricity: $\sim$8 percent of world energy, 17 percent of world electricity

Power output of largest hydroelectric dams: 10–20 GW

Wind power per unit area: $\frac{1}{2}\rho v^3$ (Equation 10.1; p. 306)

Maximum power extractable with wind turbine: $2a(1-a)^2\rho v^3$ (Equation 10.2; p. 309)

Efficiency of wind turbine, theoretical maximum: 59 percent

Power output of single large wind turbine: ~5 MW

Photosynthesis, net result: $6H_2O + 6CO_2 + 4.66$ aJ $\rightarrow C_6H_{12}O_6 + 6O_2$ (Equation 10.3; p. 314)

Gross primary productivity: 133 TW

Net primary productivity: about half of gross

Energy content of dry biomass: 15–20 MJ/kg

QUESTIONS

1. As used today, are the indirect solar energy sources of water, wind, and biomass truly renewable? Answer separately for each.

2. Hydroelectric dams can, surprisingly, have significant greenhouse emissions. What's the source of these emissions?

3. Why can't a wind turbine extract all of the wind's kinetic energy?

4. Why is the curve in Figure 10.15 so steep in the region below the rated speed?

5. Tailpipe emissions of VOCs are lower for ethanol than for gasoline, but overall VOC emissions are higher with ethanol fuels. Why?

6. Why aren't biofuels 100 percent carbon neutral?

7. Should wind turbines be allowed on undeveloped mountain ridges?

EXERCISES

1. It takes approximately 2.5 MJ to convert 1 kg of liquid water to vapor at typical atmospheric temperatures (this is the latent heat introduced in Section 4.5). Compare this quantity with the gravitational potential energy of 1 kg of water lifted to a typical cloud height of 3 km.

2. Given that 78 W/m^2 of solar energy goes into evaporating water, use the latent heat given in the preceding problem to estimate Earth's total average rate of evaporation in kilograms of water per second.

3. The average flow rate in the Niagara River is 6.0×10^6 kg/s, and the water drops 50 m over Niagara Falls. If all this energy could be harnessed to generate hydroelectric power at 90 percent efficiency, what would be the electric power output?

4. Estimate the water head at Hoover Dam shown in Figure 10.7b, given that it produces 2.1 GW of electric power from a flow of 1,800 m^3/s. Assume an energy conversion efficiency of 75 percent.

5. Repeat the previous problem for the John Day Dam (Fig. 10.7a), assuming comparable electric power, a flow rate of 9,100 m^3/s, and 82 percent conversion efficiency.

6. By what factor must the wind speed increase in order for the power carried in the wind to double?

7. What power coefficient would a wind turbine need in order to extract 350 W from every square meter of wind moving past its blades when the wind is blowing at the minimum speed for class 7?

8. The density of air under normal conditions is about 1.2 kg/m^3. For a wind speed of 10 m/s, find (a) the actual power carried in the wind; (b) the maximum possible power (Betz limit) extractable by a wind turbine with a blade area of 10 m^2; and (c) the actual power extracted by a wind turbine with a blade area of 10 m^2 and a power coefficient of 0.46.

9. What's the power coefficient of a 42-m-diameter wind turbine that produces 950 kW in a 14-m/s wind? The density of air under normal conditions is about 1.2 kg/m^3.

10. (a) Estimate the total energy produced by a wind turbine with the power curve shown in Figure 10.15 during a day when the wind blows at 2 m/s for six hours, at 10 m/s for six hours, at 15 m/s for six hours, and at 25 m/s for six hours. (b) What's the turbine's average power output over this day?

11. If you've had calculus, you know that you can find the maximum or minimum of a function by differentiating and setting the derivative to zero. Do this for Equation 10.2, and show that the resulting maximum occurs when $a = \frac{1}{3}$ and that the maximum is equal to $\frac{16}{27}$ of the total wind power $\frac{1}{2}\rho v^3$.

12. The United States generates some 230 million tonnes of solid waste each year, with an energy density (when dry) essentially the same as wood, around 20 MJ/kg. Find the equivalent in barrels of oil, and compare it with U.S. oil imports of around 4.5 billion barrels per year.

13. If all the solid waste of Exercise 12 were burned in waste-to-energy power plants that are 20 percent efficient, (a) how many kilowatt-hours of electrical energy could be produced in a year? Compare this quantity with the total U.S. electrical energy production of approximately 3.8 trillion kWh annually. (b) What would be the equivalent average electric power output from all those waste-burning power plants? How many 1-GW coal-fired plants could they displace?

14. Figure 10.22 suggests biodiesel production of around 2 billion liters (L) in Europe by 2005, which compares with a total diesel fuel consumption of about 150 billion L annu-

ally. European biodiesel production yields about 1.23 kL per hectare (ha) of cropland. (a) What fraction of Europe's 49 Mha of total cropland is now used for biodiesel production? (b) How much land would it take to replace all of Europe's petrodiesel with biodiesel, considering that the energy yield per liter of biodiesel is 10 percent lower than for petrodiesel?

15. Find the maximum thermodynamic efficiency of an OTEC heat engine operating between surface waters at a temperature of 25°C and deep water at 5°C.

16. Typical freshly cut "green" firewood contains about 70 percent water by weight; for seasoned wood the comparable figure is around 20 percent. Firewood is usually sold by volume, so compare the energy available from a given volume of green wood with that available from seasoned wood. Assume an energy content of 20 MJ/kg for perfectly dry wood, and 2.3 MJ/kg of energy needed to vaporize water. Also assume that the wood has essentially the same volume whether green or seasoned.

RESEARCH PROBLEMS

1. Download the latest version of the U.S. Energy Information Administration's "Existing Electric Generating Units in the United States" spreadsheet. Find (a) the largest hydroelectric installation in your state (you may need to add together multiple generating units at the same site); (b) the smallest hydroelectric installation in your state; and (c) the total hydroelectric generation capacity in your state.

2. Estimate the number of powerboats operating on Lake Mead, above the Hoover Dam, and assuming an average of 100 horsepower per boat, estimate the total power output of all these boats. Is it at all comparable to Hoover Dam's approximately 2 GW of electric power output? If it is, then the indirect environmental impact of this dam may be comparable to that of a fossil plant with the same output! You might start with the National Park Services Environmental Impact Statement for its Lake Management Plan for Lake Mead.

3. Consult the U.S. Energy Information Administration's list of power plants (see Research Problem 1), and use it to find the total capacity of all the wind and photovoltaic power plants in the United States. *Hint*: You can sort the spreadsheet as needed.

4. Find the total annual gasoline consumption in the United States, along with the country's total cropland. One study suggests that corn-to-ethanol production in the United States gives an annual yield of 71 GJ of ethanol energy per hectare (metric acre; 10^4 m^2) of cropland. Use your findings, along with Table 3.3, to estimate the percentage of U.S. cropland that would be required to replace all U.S. gasoline with corn-based ethanol.

Chapter 11

HYDROGEN FUTURES?

This chapter is about possible future energy sources that are more speculative and, if they prove workable, further off than most of the energy alternatives introduced in Chapters 7 through 10. There are only two such "energy futures" under serious consideration, and both just happen to involve the element hydrogen. The similarity ends there, however. Technological development, economics, infrastructure, and most of all resource availability make these two hydrogen-based futures as different as night and day. The fundamental difference is that one involves hydrogen as a chemical fuel, and the other as a nuclear fuel. Both remain speculative and are probably decades from implementation, but for very different reasons.

11.1 Molecular Hydrogen and the Hydrogen Economy

Molecular hydrogen (H_2) is a flammable gas that combines with oxygen to make water:

$$2H_2 + O_2 \rightarrow 2H_2O \qquad (11.1)$$

Besides energy, water is the only product, which gives hydrogen its environmental reputation as a clean, nonpolluting, carbon-free energy source. The reaction in Equation 11.1 releases 0.475 aJ (0.475×10^{-18} J) for each water molecule formed. This gives hydrogen an energy content of 142 MJ/kg—roughly three times that of petroleum-based fuels on a per-mass basis. But hydrogen, the lightest element, forms a gas with very low density, so under normal atmospheric pressure this energy content represents only 12.8 MJ/m^3—about one-third that of natural gas on a per-volume basis.

Hydrogen is used today in industry, with roughly half going into the manufacture of ammonia fertilizer and much of the rest into petroleum refining. Hydrogen energy has a long history. The nineteenth-century writer Jules Verne, in his lesser-known novel *The Mysterious Island*, envisioned a society powered by hydrogen rather than coal. Early-twentieth-century tinkerers and engineers, some inspired by Verne's vision, proposed hydrogen-powered trains, cars, and aircraft. Hydrogen provided buoyancy for the dirigibles that preceded transatlantic aircraft. Dirigible engineers also experimented with mixing hydrogen in the engine fuel—a process that improved performance but never saw commercial use. In the 1930s, both Germany and England tested hydrogen-powered buses, trucks, and trains; the hydrogen in these vehicles was burned in modified gasoline or diesel engines. Hydrogen transport never really caught on, but by the late twentieth century automobile companies and inventive individuals had built dozens of experimental hydrogen-powered vehicles. The Russians even converted one engine of a three-engine passenger jet to burn hydrogen. NASA's Space Shuttle runs on hydrogen; its large external fuel tank holds 100,000 kg of liquid H_2 (Fig. 11.1).

The Hydrogen Resource

I've started my presentation of every other energy category with a look at the resource: How much is there? How long will it last? How hard is it to extract? How is it distributed around the planet? For hydrogen, this section is blissfully short, because *there is no hydrogen energy resource.*

Figure 11.1

A hydrogen-powered vehicle: NASA's Space Shuttle rockets into orbit on the chemical energy of 100,000 kg of liquid hydrogen stored in its huge external fuel tank.

On Earth, that is. Hydrogen is the most abundant element in the universe, and there's plenty of H_2 floating around in interstellar clouds. Closer to home, the planet Jupiter is predominantly hydrogen, much of it H_2. Even the early Earth may have had substantial hydrogen in its atmosphere. But smaller planets like Earth don't have strong enough gravity to hold light hydrogen molecules, which, at a given temperature, are moving a lot faster than heavier nitrogen and oxygen. So Earth gradually lost its atmospheric hydrogen. Today, hydrogen constitutes less than 1 part in 1 million of Earth's atmosphere. Exercise 2 shows that the associated energy resource is truly negligible.

Thus there is no hydrogen energy resource. I want to make this very clear, because the public frequently hears hydrogen energy discussed in the media and imagines that there's a vast new energy source out there, just waiting to be tapped. So I'll say it again: *There is no hydrogen energy resource.*

End of story? No—but the lack of a hydrogen resource means we can't think of hydrogen as a fuel we can mine from Earth's bounty, like coal. Rather, hydrogen is an **energy carrier** that we can manufacture with existing energy sources and then use in place of other fuels. If we want hydrogen to power cars or generate electricity, we need to make it, and that takes energy—in principle at least as much as we're going to get back out of the hydrogen, but in practice considerably more. That's why hydrogen isn't an energy resource.

Making Hydrogen

Hydrogen can be produced by chemically separating it from hydrogen-containing compounds, commonly fossil fuels or water. Most industrial hydrogen today comes from **steam reforming**, using a reaction of natural gas with high-temperature steam to produce a mix of hydrogen and carbon monoxide (CO):

$$CH_4 + H_2O \rightarrow 3H_2 + CO \qquad (11.2)$$

The mixture can be burned directly as a gas or processed further to convert the CO to CO_2, which is then separated from the hydrogen. Oil and coal, which contain less hydrogen than does natural gas, also serve as feedstocks for hydrogen production. Some combined-cycle power plants include integrated coal gasifiers, which react coal with oxygen to produce a mix of hydrogen and CO. In that sense, hydrogen already plays some role in energy production.

A second approach to hydrogen production in use today is **electrolysis**, in which electric current passing through water splits H_2O molecules and results in hydrogen and oxygen gases collecting at two separate electrodes (Fig. 11.2). Commercial electrolyzers operate at efficiencies of over 75 percent, meaning that more than three-quarters of the electrical energy input ends up as chemical energy stored in hydrogen. But the normally high cost of electricity relative to fossil fuels makes electrolysis economical only where cheap, abundant hydroelectricity is available. In the United States, only 4 percent of today's industrial hydrogen is produced through electrolysis; the rest comes from reforming fossil fuels.

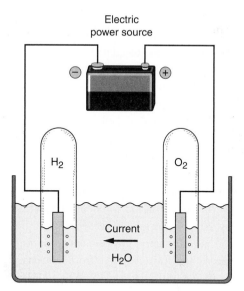

Electric
power source

H₂

O₂

Current

H₂O

Figure 11.2
A simple electrolysis cell. Electric current passes through water, and bubbles of oxygen and hydrogen form at the positive and negative electrodes, respectively. The gas bubbles rise to fill collection chambers above the electrodes.

At sufficiently high temperatures water breaks apart into hydrogen and oxygen; high-temperature nuclear reactors or solar parabolic concentrators might someday produce hydrogen directly through such **thermal splitting**. Heating to less extreme temperatures can assist the process of electrolysis, leading to a hybrid production technique that reduces the electrical energy required while increasing the thermal energy needed.

Scientists are exploring several innovative hydrogen-production schemes, although none are anywhere near commercial viability. In **photolysis**, solar energy creates electron-hole pairs in a semiconductor immersed in a water solution. Water molecules in contact with the semiconductor lose electrons to the holes and split into hydrogen and oxygen. The process is akin to electrolysis, except that solar energy creates the hydrogen directly without the intermediary of electric current. Experimental photolysis cells have achieved energy conversion efficiencies of up to 12 percent.

Biological processes also produce hydrogen; indeed, the process of photosynthesis involves splitting water molecules and rearranging the resulting hydrogen and oxygen to make carbohydrates. It might be possible to engineer organisms designed specifically to produce hydrogen gas, or to mimic hydrogen-producing biochemistry with nonbiological systems. Either way, biochemical hydrogen technology is probably a long way off.

Fuel Cells and Hydrogen Vehicles

Although hydrogen could be burned in stationary facilities such as power plants and industrial boilers, its real value is in transportation. Internal-combustion engines used in today's vehicles can be adapted to burn hydrogen, as can the gas turbines that power jet aircraft. There's good reason to burn hydrogen in such engines; not only are there fewer harmful emissions, but hydrogen-powered engines also tend to be more fuel efficient than their fossil-powered counterparts.

However, combustion engines probably aren't the future of hydrogen-powered transport. It's more efficient to convert hydrogen energy directly to electricity using **fuel cells**—devices that combine hydrogen and oxygen chemically to produce water and electrical energy. An electric motor would then propel the vehicle. A "well to wheels" comparison with conventional gasoline-powered vehicles—including the energy required to produce both the gasoline and hydrogen—suggests that fuel-cell vehicles could be as much as 2.5 times more energy efficient than conventional vehicles.

The first fuel-cell device was constructed in 1839, predating the internal-combustion engine. Fuel-cell development continued slowly through the nineteenth and twentieth centuries, and there were occasional practical demonstrations, including a fuel-cell tractor and golf cart unveiled in the early 1960s. Fuel cells got a boost when NASA adopted the technology for the Apollo Moon missions. Today there's a modest market in commercial fuel cells; by 2003 world sales exceeded $300 million and were growing at a rapid 41 percent annual rate. Industry, governments, and universities all engage in fuel-cell research, and proposed applications range from stationary power generation to transportation to mini-cells that might power cell phones and laptop computers.

You can think of a fuel cell as a battery that never runs down; whereas a conventional battery eventually depletes its finite store of chemical energy, a fuel cell is supplied continuously with the chemicals—often, but not always, hydrogen and oxygen—used to make electrical energy. You can also think of a fuel cell as the opposite of an electrolysis cell (see Fig. 11.2). In electrolysis, you put in electrical energy and get out hydrogen and oxygen; with a fuel cell, you put in hydrogen and oxygen and get out electrical energy. There's a nice symmetry here, and it shows the significance of hydrogen as an energy carrier that can store electrical energy generated by other means, then turn it back into electricity for applications such as transportation, for which the direct use of grid power isn't practical.

Figure 11.3 shows a **proton-exchange membrane fuel cell** (PEMFC), a relatively recent variety with the greatest promise for powering fuel-cell vehicles. Other types of fuel cells are conceptually similar. Two electrodes, the anode and cathode, sandwich a special membrane that permits only protons to pass through. The anode and cathode are made of catalysts that promote chemical reactions. At the anode, these reactions dissociate the incoming hydrogen fuel (H_2) into protons and electrons. Protons move through the membrane, while electrons flow out of the anode and through an external electric circuit. At the cathode, protons combine with incoming oxygen (O_2) and with electrons returning from the circuit to form water (H_2O), the cell's only material by-product.

Most of the energy released in water formation appears as electrical energy; a lesser amount is in the form of heat. Current PEMFCs are more than 60 percent efficient at converting the chemical energy of hydrogen into electricity; when coupled with electric motors that are more than 90 percent efficient, this gives a combined efficiency of well over 50 percent—more than twice the efficiency of gasoline engines and about 20 percent better than diesels. The theoretical maximum efficiency for hydrogen fuel cells alone—set ultimately by the second law of thermodynamics—is 83 percent.

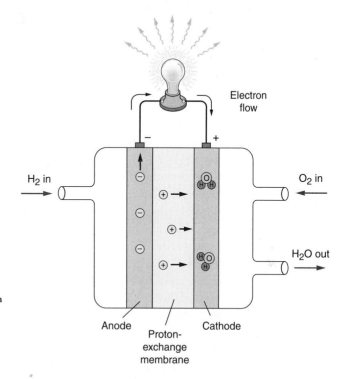

Figure 11.3

A proton-exchange membrane fuel cell. The anode splits hydrogen into protons and electrons. Protons pass through the membrane, while electrons travel through an external electric circuit. At the cathode, protons and electrons combine with oxygen to form water. Most of the energy released is available as electricity.

A complete fuel-cell system needs fuel storage, fuel-supply mechanisms, and heat removal. Catalysts and proton-exchange membranes require expensive materials. All these factors make present-day fuel cells expensive—about $3,000 per kilowatt of electric power capability. In contrast, modern gasoline engines cost about $30 per kilowatt. Given the greater efficiency of fuel-cell vehicles, fuel-cell costs would have to drop to around $100 per kilowatt to compete with gasoline.

Fuel-cell vehicles face other challenges, however. In its normal low-density gaseous state, it would be impossible to store enough hydrogen on board to propel a vehicle even as far as 1 mile (see Exercise 4). To achieve the several-hundred-mile range typical of gasoline vehicles requires compressing hydrogen to very high pressures—many hundreds of times atmospheric pressure—and that means heavy, expensive, and bulky storage tanks. Filling high-pressure hydrogen tanks presents technological, logistical, and safety challenges. An alternative is to liquefy the hydrogen, which increases its density but requires heavily insulated tanks because hydrogen's boiling point is a low 20 K. Boil-off from even the best tanks results in a steady loss of hydrogen. There's also an energy penalty; the process of liquefying consumes some 40 percent of the hydrogen's energy.

A more promising approach may be to bind hydrogen chemically with other substances. At the low-tech extreme are fossil fuels themselves; for example, pressurized natural gas could be reformed on board the vehicle, providing hydrogen to a fuel cell. Of course this strategy doesn't get us away from fossil fuels, and it leaves carbon to be disposed of, which in a vehicle means releasing CO_2 to the atmosphere. A related approach involves using methanol (wood alcohol, CH_3OH) as the primary fuel. Methanol can be used directly in fuel cells or converted on board to hydrogen. Again, CO_2 must be disposed of. Methanol could be a renewable fuel, since it can be produced from biomass as well as from fossil fuels.

Hydrogen bonds with various lightweight elements and their solid compounds, and the resulting **hydrides** represent another approach to hydrogen storage. Fueling involves flowing hydrogen through a porous structure with plenty of surface area; some heat is released as the hydrogen bonds to the solid. The hydrogen remains bound until it's needed; heating the hydride then releases hydrogen. A similar approach that shows considerable promise is to react chemical hydrides with water, producing what its developer, Millennium Cell, calls Hydrogen on Demand. Finally, nanostructures made of carbon—first discovered in the 1980s and now a hot topic in materials research—might serve as "cages" or "miniature fuel tanks" capable of absorbing and releasing significant quantities of hydrogen. Practical systems are a long way off, but theoretical calculations suggest that cylindrical carbon nanotubes might hold as much as 14 percent hydrogen by weight.

The greater efficiency of fuel cells means we won't need to duplicate gasoline's energy density to give fuel-cell vehicles the same range as today's cars. But we must come within a factor of 2 or 3. Figure 11.4 compares the energy densities, by volume, of various hydrogen storage schemes with that of gasoline.

It's technically possible to build a fuel-cell vehicle today, and many car manufacturers have done so (Fig. 11.5). These vehicles are far from being economically competitive, however, and it's going to take much more technological development to bring down costs, particularly of fuel cells and onboard hydrogen storage. Even if we can do all that, hydrogen faces some broader challenges.

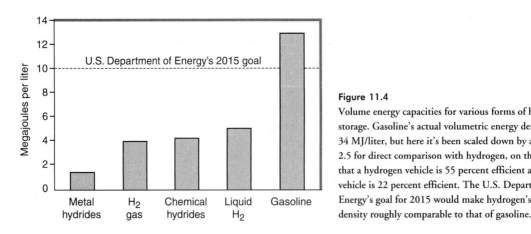

Figure 11.4

Volume energy capacities for various forms of hydrogen storage. Gasoline's actual volumetric energy density is 34 MJ/liter, but here it's been scaled down by a factor of 2.5 for direct comparison with hydrogen, on the assumption that a hydrogen vehicle is 55 percent efficient and a gasoline vehicle is 22 percent efficient. The U.S. Department of Energy's goal for 2015 would make hydrogen's storage density roughly comparable to that of gasoline.

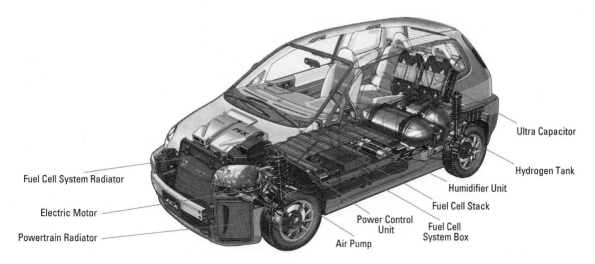

Fuel Cell System Radiator

Electric Motor

Powertrain Radiator

Ultra Capacitor

Hydrogen Tank

Humidifier Unit

Fuel Cell Stack

Power Control Unit

Fuel Cell System Box

Air Pump

Figure 11.5
Cutaway diagram of Honda's FCX fuel-cell vehicle.

Infrastructure for the Hydrogen Economy

Moving to hydrogen-based transportation presents a daunting challenge not shared by any of the other energy alternatives we've considered. Develop effective photovoltaic power stations, and you have a new source of electrical energy to plug into the existing power grid. Make solar water heating competitive, and you change a minor aspect of home construction. Design efficient, economical wind turbines, and the power grid awaits your electricity. The same holds true for advanced nuclear plants, whether powered by fission or fusion. In all of these cases only the energy source is changing. The distribution of electricity, or hot water, or whatever remains the same, as do the lightbulbs, computers, dishwashers, and showers that use the electricity or heat. But with hydrogen we're talking about changing the entire energy picture—the fuel, the distribution infrastructure, the storage mechanisms, and the vehicles themselves. And the system we're moving from—oil wells, tankers, refineries, pipelines, gas stations, and cars powered by internal-combustion engines—is highly developed, historically entrenched, and represents a huge capital investment. Even if it's no longer the best system, enormous inertia must be overcome to change it.

Hydrogen faces a kind of "chicken and egg" problem in trying to replace the fossil energy transportation system. There's no infrastructure for distributing hydrogen, and there are no hydrogen gas stations. So even if you could buy a fuel-cell car today, it wouldn't be of much use. But there aren't any fuel-cell vehicles, so if you're an entrepreneur it doesn't make much sense to pour money into hydrogen infrastructure. Somehow both the vehicles and the hydrogen infrastructure have to develop at the same time. But how?

The transition to a hydrogen economy might begin with modest growth of the existing hydrogen production and distribution system. The United States, for example, already produces some 9 million tonnes of hydrogen annually, much of it consumed on site but some distributed in liquid form by cryogenic tanker trucks. As demand grows, expensive truck-based distribution might be replaced by distributed hydrogen generation, with fossil fuel reformers located at individual filling stations. Eventually those could be replaced or augmented by electrolysis using locally generated renewable electricity from photovoltaics or wind. Another approach would have surplus power from baseload coal and nuclear plants being diverted to the production of hydrogen, which then could be delivered locally at reasonable cost. A maturing hydrogen economy might see a mix of centralized and distributed hydrogen production, with energy sources switching from fossil to renewables over several decades. Expanded use of fuel-cell vehicles would accompany the growing hydrogen infrastructure. A report by the U.S. National Research Council and National Academy of Engineering optimistically projects that 100 percent of new U.S. vehicles could be hydrogen powered by the late 2030s, and that by 2050 virtually all vehicles on the road would use hydrogen (Fig. 11.6).

Realizing this vision of a hydrogen economy won't be easy. It will take cooperation among government, industry, the public, and the research community. And it will take political courage, because it will certainly require subsidizing an energy system that, at first, won't be economically competitive with fossil fuels. At some point, rising fossil fuel prices and decreasing hydrogen costs will make hydrogen competitive—but if we wait until then to develop the technology, it may be far too late for a smooth transition.

Hydrogen Safety

For hydrogen to catch on, the public must perceive it as safe. Some people still associate hydrogen with the 1937 *Hindenburg* disaster, when the world's largest airship burst into flames at the end of a transatlantic flight. More recently, the 1986 explosion of the Space Shuttle *Challenger*, although triggered by a solid-fuel rocket leak, involved the shut-

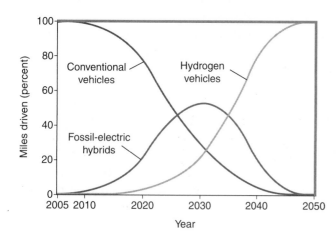

Figure 11.6
The U.S. National Research Council's optimistic projection for the transition to a hydrogen-based transportation system. The figure shows the percentage of total miles driven by conventional vehicles, gasoline and diesel hybrids, and hydrogen vehicles. In this scenario, virtually all new vehicles would be hydrogen powered by the late 2030s, but some conventional vehicles and hybrids would remain on the road until about 2050.

tle's huge tank of liquid hydrogen. Even the 2001 *Columbia* shuttle disaster was hydrogen-related, caused by a chunk of insulation needed to keep that same hydrogen tank cool. It's easy to understand why some people consider hydrogen-powered cars as possible miniature *Hindenberg*s and *Challenger*s.

Hydrogen does present unique dangers, but in other ways it's safer than gasoline, which is a familiar but dangerously flammable product. As the smallest molecule, hydrogen is much more prone to leakage. But when it leaks into the open air, the buoyant gas quickly disperses. Not so with gasoline, which forms puddles of flammable liquid. On the other hand, hydrogen burns readily in a wider range of fuel-air mixtures than most other fuels, and its flame can spread with explosive speed. The high pressures needed for practical hydrogen energy densities exacerbate the leakage problem. Clearly, widespread hydrogen use will require new safety procedures and technologies. The dangers those procedures will alleviate are different from, but not necessarily worse than, those we face with the fuels in widespread use today.

Environmental Impacts of Hydrogen

Hydrogen's main appeal is that it's a clean, nonpolluting, carbon-free fuel. So what are the environmental impacts of hydrogen? When hydrogen combines with oxygen, whether in conventional combustion or in a fuel cell, the only product is harmless water. The drinking water on the Apollo space missions actually came from the onboard fuel cells. Water from hydrogen combustion is in the form of vapor, and water vapor is a greenhouse gas, but the water cycle removes it from the atmosphere in about a week, so it's not a significant problem. In hydrogen combustion, temperatures get high enough—as they do with fossil or biomass combustion—for some atmospheric nitrogen to combine with oxygen, forming nitrogen oxides. With fuel cells, even that small impact vanishes. Hydrogen can be truly clean, nonpolluting, and carbon-free.

But remember, there is no hydrogen energy resource. The environmental impacts of hydrogen therefore include the impacts of whatever energy source was used to make it. If you insisted on making hydrogen from electrolysis with electricity generated in a 35 percent efficient coal plant, then your 75 percent efficient electrolysis coupled with a 55 percent efficient fuel-cell vehicle would give an overall efficiency of only 14 percent, and you'd be burning more fossil fuel than with a conventional gasoline-powered car. Your carbon emissions would be even higher, thanks to coal's high carbon content. If you make hydrogen by reforming natural gas, oil, or coal, then you're left with CO_2 to dispose of, and if you dump it into the atmosphere then you haven't gained much. You might gain just a little with noncarbon emissions, because it's generally easier to control pollution from stationary sources such as large-scale reforming plants. In a more technologically advanced future, you might gain on the carbon front, too, because it's easier to capture and sequester carbon from reforming fossil fuels than from combustion in a power plant or internal-combustion engine.

Example 11.1   Clean Hydrogen?

Hydrogen made by reforming natural gas (the reaction in Equation 11.2) is burned in a modified internal-combustion engine operating at 38 percent efficiency. The CO from that reaction is converted into CO_2 and dumped to the atmosphere. How do the carbon emissions in this scenario compare with those of a conventional gasoline engine operating at 22 percent efficiency? Gasoline produces about 70 grams of CO_2 per megajoule of fuel energy.

Solution

For comparison with the figure given for gasoline, we need to quantify hydrogen in terms of energy and CO_2 in terms of mass. The reaction in Equation 11.2 produces three molecules of H_2 and one of CO, which is converted to one CO_2. Burning H_2 according to Equation 11.1 produces 0.475 aJ per molecule of H_2O formed or, equivalently, of H_2 burned (see the discussion following Equation 11.1). So our three H_2 molecules contain 3×0.475 aJ = 1.43 aJ of energy, and with that we get one CO_2 molecule. Given the atomic weights of 12 and 16 for C and O, respectively, the CO_2 molecule has a molecular weight of $12 + (2 \times 16) = 44$ u, so its mass is $(44)(1.67 \times 10^{-27}$ kg/u$) = 7.34 \times 10^{-26}$ kg. So the ratio of CO_2 mass to hydrogen energy is $(7.34 \times 10^{-26}$ kg$)/(1.43 \times 10^{-18}$ J$) = 5.1 \times 10^{-8}$ kg/J, or 51 g/MJ. We're told that gasoline yields 70 g/MJ, so already we're better off with the hydrogen. Quantitatively, the lower CO_2 per unit energy coupled with the hydrogen engine's higher efficiency means that the hydrogen-associated emissions are lower by a factor of $(51/70)(22/38) = 0.42$, a reduction of 58 percent. The actual reduction would be somewhat less because we haven't accounted for the energy needed to make steam for the reaction in Equation 11.2. And the balance could well shift the other way if we used a more carbon-rich fossil fuel (see Exercise 5).

If we want carbon-free hydrogen energy, we're left with electrolysis using electricity generated from nuclear or renewable energy, or direct-conversion techniques such as photolysis. Hydrogen energy then carries whatever environmental cost goes with that electricity generation. Perhaps the most attractive hydrogen future, from an environmental if not an economic standpoint, is to generate electricity with photovoltaic cells and then produce hydrogen by electrolysis. Another but less "gentle" path to reasonably clean electricity and hydrogen is nuclear fusion.

11.2 Fusion

Chapter 7 briefly introduced nuclear fusion as the joining of lighter nuclei to make heavier ones. The curve of binding energy (recall Fig. 7.5) shows that the process releases energy. Fusion is well understood; it's what powers the Sun and other stars, and it's the main energy source for thermonuclear weapons ("hydrogen bombs").

In principle, any nuclei lighter than iron can fuse and release energy. Fusion reactions in the more massive stars join successively heavier nuclei, building all the elements up to the mass of iron. When massive stars explode, they spew these newly formed elements into the interstellar medium, where they're eventually incorporated into new stars and planets. Stellar fusion reactions are the source of all the oxygen, carbon, nitrogen, sulfur, silicon, and other elements up to iron that make up our bodies and our planet.

Modest stars, such as our Sun, "burn" only hydrogen, the most abundant element in the universe and the easiest to fuse. In the Sun, fusion joins four ordinary hydrogen nuclei ($_1^1$H, a single proton) to produce helium ($_2^4$He). Intermediate reactions convert two protons into neutrons, giving the helium its two-proton, two-neutron structure. Unlike more massive stars, the Sun simply isn't hot enough to fuse helium into heavier elements.

Easier than fusing ordinary hydrogen are fusion reactions involving the hydrogen isotopes deuterium ($_1^2$H, one proton and one neutron) and tritium ($_1^3$H). Three specific reactions are of particular interest, two of which fuse deuterium with deuterium (D-D) and the third involving deuterium-tritium (D-T) fusion:

$$\text{D-D fusion} \qquad {}_1^2\text{H} + {}_1^2\text{H} \rightarrow {}_2^3\text{He} + {}_0^1 n + 0.523 \text{ pJ} \qquad (11.3)$$

$$\text{D-D fusion} \qquad {}_1^2\text{H} + {}_1^2\text{H} \rightarrow {}_1^3\text{H} + {}_1^1\text{H} + 0.649 \text{ pJ} \qquad (11.4)$$

$$\text{D-T fusion} \qquad {}_1^2\text{H} + {}_1^3\text{H} \rightarrow {}_2^4\text{He} + {}_0^1 n + 2.8 \text{ pJ} \qquad (11.5)$$

Figure 11.7 depicts the reaction in Equation 11.3. Seeing these reactions written out will help you later in understanding the by-products of nuclear fusion. For now, note the energy yield, which is the number on the right-hand side of each reaction. The fact that we're measuring these energies in picojoules (10^{-12} J) rather than the attojoules (10^{-18} J) we used for chemical reactions should remind you of the huge energy difference between nuclear and chemical reactions. If we're to harness fusion, reactions 11.3 to 11.5 show that we'll need deuterium and tritium to fuel our fusion power plants.

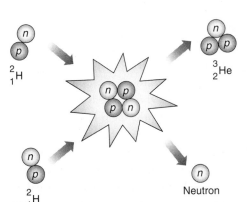

Figure 11.7
The deuterium-deuterium fusion reaction in Equation 11.3 produces helium-3, a neutron, and 0.523 pJ of energy. The energy is in the motion of the reaction products.

The Fusion Resource

So how much deuterium and tritium is there? Is fusion yet another promising energy source that we'll deplete in a matter of decades or centuries? The answer is a resounding no. There's enough deuterium on Earth to meet humankind's present energy-consumption rate for many billions of years—longer, in fact, than the Sun will continue to shine!

Hydrogen is abundant on Earth, particularly in seawater. And one out of every 6,500 hydrogen nuclei is deuterium. As Example 11.2 shows, that's enough to make each gallon of seawater the energy equivalent of over 300 gallons of gasoline!

Example 11.2 Fusion Resources: Energy from Water

Deuterium fusion releases 330 TJ of energy per kilogram of deuterium (see Exercise 6). Considering that one of every 6,500 hydrogen nuclei is deuterium, estimate the fusion energy content of 1 gallon of water, and compare it with gasoline.

Solution

A gallon of water weights a little over 8 pounds, giving it a mass of around 4 kg. Because water is H_2O with nearly all the hydrogen in the isotope ^1_1H and nearly all the oxygen as $^{16}_8O$, this gives water an average molecular weight of $(2 \times 1) + 16 = 18$, of which 2/18 or 1/9 is hydrogen. One out of every 6,500 hydrogen nuclei is deuterium; since deuterium (^2_1H) has twice the atomic weight of ordinary hydrogen, this means deuterium comprises 1/3,250 of all the hydrogen by mass. So our 4-kg gallon of water contains $(4 \text{ kg})/9/3{,}250 = 0.14$ gram of deuterium. At 330 TJ/kg, the corresponding energy content is $(330 \text{ TJ/kg})(0.00014 \text{ kg}) = 0.046$ TJ, or 46 GJ.

Table 3.3 lists the energy content of gasoline at 36 kWh/gallon; since 1 kWh = 3.6 MJ, this means 1 gallon of gasoline contains $(36 \text{ kWh/gallon})(3.6 \text{ MJ/kWh}) = 130$ MJ, or 0.13 GJ. With our 46 GJ per gallon of water, this gives 1 gallon of water the energy equivalent of 46/0.13, or about 350 gallons of gasoline!

Is it a problem that all this deuterium is tied up chemically in water molecules? This was a big issue with chemical hydrogen, making water's hydrogen unavailable as a chemical fuel because, in effect, it's already been burned. But it's not an issue for nuclear fusion. The "nuclear difference"—that factor of 10 million between the energies of nuclear and chemical reactions—means any energy needed to extract deuterium from seawater is completely negligible compared with the yield from nuclear fusion. So there's plenty of deuterium available to us at essentially no energy cost.

Although it's possible to extract energy from deuterium alone, a comparison of Equations 11.3 to 11.5 shows that the D-T reaction yields a lot more energy. So D-T appears to be the most energetically favorable route to fusion—although, as we'll see, it's not quite as environmentally friendly as D-D fusion. The distinction isn't clear-cut, however, because even with pure deuterium fuel the D-D reaction in Equation 11.4 produces tritium, which can then undergo the D-T reaction (see Exercise 6). But it's almost certain that the first practical fusion power sources will use both deuterium and tritium as fuels.

So how's Earth's tritium supply? Nonexistent! That's because tritium is radioactive, with a half-life just over 12 years. Only trace amounts exist on Earth, created when cosmic rays interact with atmospheric nitrogen and oxygen. This is not a problem, though, because tritium can be "bred" inside a fusion reactor using the neutrons that result from reactions in Equations 11.3 and 11.5. The favored scheme is to blanket the inside of the reaction chamber with a lithium compound. When lithium-6 absorbs a neutron, it can split into nuclei of helium-4 and tritium:

$$\,^{6}_{3}\text{Li} + \,^{1}_{0}n \rightarrow \,^{4}_{2}\text{He} + \,^{3}_{1}\text{H} \tag{11.6}$$

This reaction also supplies some energy. A similar reaction occurs with the more abundant isotope lithium-7; this one requires energy but produces an extra neutron. So is there enough lithium? Yes, there's plenty of it in Earth's crust and in seawater, but as an energy resource it's probably measured in thousands of years, not billions. That's more than enough, however, because once we succeed at D-T fusion it won't be long before we have fusion reactors running on pure deuterium.

The Fusion Challenge

In fusion we have a virtually unlimited energy source, outlasting even the steady stream of solar energy. So why don't we live in a fusion-powered world?

The reason is that fusion doesn't occur under normal terrestrial conditions. Fusion requires that two nuclei come together, but atomic nuclei are positively charged, so they repel strongly. Only if they get very close can the strongly attractive nuclear force (recall Section 7.1) overcome electrical repulsion. And only then can fusion occur, with its huge energy release. Making fusion happen is a little like pushing a ball up a hill to get it into a deep valley on the other side (Fig. 11.8). The hill is the repulsive electric force,

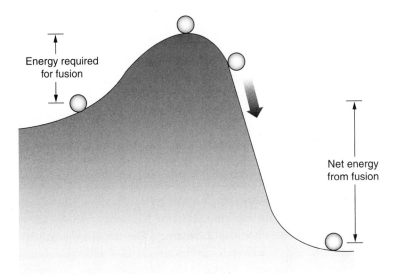

Energy required for fusion

Net energy from fusion

Figure 11.8
A hill-and-valley analogy for fusion. It takes substantial energy to get the ball up the hill, but a lot more is released when it drops into the valley.

and it takes a lot of energy to get up it. But you get even more energy back once the ball drops into the valley; that's the fusion process. Actually, quantum physics lets the approaching nuclei "tunnel" through the hill, reducing the energy required, but it's still substantial.

So how do we get the ball in Figure 11.8 up the hill? That is, how do we get two nuclei close enough to fuse? One way is to have a gas at a temperature so high that the average thermal speed is great enough for nuclei to approach closely despite their mutual repulsion. For D-D fusion the so-called **ignition temperature** is a colossal 600 million K (MK); for D-T fusion it's about 50 MK. Achieving such high temperatures is the challenge in fusion's development.

Or rather, the two challenges. How can we possibly achieve such high temperatures? And if we do, how can we possibly contain the hot gas? There's no material in the universe that can withstand a temperature of millions of kelvins. The stars solve both the heating and confinement problems with their immense gravities. Gravitational attraction compresses the stellar core, heating it to million-kelvin temperatures. Once fusion "turns on" in a star, the hot, fusing gas remains gravitationally confined. But gravitational heating and confinement aren't options here on Earth.

At fusion temperatures, molecules have long since become unstuck and atoms have lost their electrons. What we have, then, is **plasma**—a gas of separate electrons and atomic nuclei. Once we have plasma at fusion temperatures, we face the confinement problem. Here there's a choice. The denser the plasma, the more frequent are collisions among nuclei that result in fusion. So we could get the same energy by confining dense plasma for a short **confinement time**, or more diffuse plasma for a longer time. Quantitatively, what counts is the **Lawson criterion**—a condition on the product of density and confinement time. With density n in nuclei per cubic meter, and time τ in seconds, the Lawson criterion for D-T fusion is that the product $n\tau$ exceed 10^{20} s/m^3; for D-D fusion it's a factor of 100 greater, or 10^{22} s/m^3, which is another reason why we're likely to see D-T fusion developed first. The two fusion schemes take extreme opposite approaches to the Lawson criterion.

Inertial Confinement

At one extreme is **inertial confinement**, which uses high densities and short confinement times. The time is so short that the fusing particles' own inertia—their resistance to change in motion—ensures that they stick around long enough. Inertial confinement works in thermonuclear weapons, where energy from a nuclear fission explosion is focused on a deuterium-lithium mix that's then consumed in fusion reactions before it has time to blow apart. The controlled fusion that might supply us with energy takes a similar approach, focusing colossal laser systems on tiny D-T targets and inducing what are effectively thermonuclear explosions in miniature.

The most ambitious of the laser fusion experiments is the National Ignition Facility (NIF) at California's Lawrence Livermore National Laboratory. The size of a football stadium, this device will focus 192 laser beams on a millimeter-size D-T pellet, delivering infrared laser energy at the rate of 500 TW (Fig. 11.9). That's right—500 TW, some forty times humankind's total energy-consumption rate. But the lasers are only on for a few nanoseconds, discharging energy that was stored in advance at a much more gradual rate. The NIF's laser energy is designed to compress the fuel pellet until its temperature exceeds the fusion ignition temperature. The resulting fusion should release ten times as much energy as was used to power the lasers.

The NIF and similar laser-fusion experiments explore the basic physics of inertial confinement fusion, but they're a long way from a fusion power plant. The NIF, for example, fires its lasers only every 4 to 8 hours; a working power plant would need to load and blast new fuel pellets around ten times each second. And a host of practical engineering problems—from energy transfer to the overall design of a power plant—are at this point no more than ideas.

The NIF's main purpose isn't the development of a new energy source, it's nuclear weapons research. Since signing the Comprehensive Test Ban Treaty in 1996 (although never ratifying it), the United States has not conducted any nuclear weapons tests. Weapons advocates argue that laser-induced fusion can serve as a substitute for nuclear testing, verifying the reliability of existing weapons and helping to develop new ones.

Magnetic Confinement

Whereas inertial confinement exploits extreme densities and minuscule confinement times, **magnetic confinement** takes the opposite approach. Under development since the 1950s, magnetic confinement devices hold diffuse hot plasma in "magnetic bottles"— configurations of magnetic fields that keep charged particles away from the physical walls of the device. Confinement times in the latest experiments are measured in seconds, and the next generation of technology should increase that time to minutes.

Figure 11.9
Target chamber of the National Ignition Facility during installation. The chamber is 11 m in diameter and weighs 130 tonnes, yet at its center sits a millimeter-size fuel pellet that's the focus of 192 laser beams converging through the many holes in the chamber.

Early optimism for magnetic fusion was tempered by plasmas going unstable and escaping from their magnetic bottles. Instability diminishes as the size of the plasma increases, so researchers are confident that scaling up their magnetic bottles will make fusion plasmas manageable. Although a number of magnetic configurations are under investigation, the most promising is the **tokamak**, a Russian design that features a donut-shaped plasma chamber whose magnetic field spirals around without ever touching the chamber walls (Fig. 11.10).

Tokamaks solve the confinement problem with magnetic fields, but what about those 100-MK fusion temperatures? Because plasma consists of charged particles, it conducts electricity. Therefore it's possible to heat plasma by passing electric current through it, just as in the wires of a toaster or electric stove. This helps, but electrical heating alone can't reach fusion temperatures. Additional heating comes from radio waves or high-energy beams of neutral deuterium atoms. Neutral-beam heating has the advantage of delivering the fusion fuel; it's necessary to use neutral particles because charged particles can no more penetrate into the magnetic bottle than those inside can escape.

By the mid-1990s, tokamaks in the United States, Japan, and Europe had approached the so-called scientific break-even point, when the energy produced in fusion exceeds the energy required to heat the plasma. However, a lot more energy is needed for a fusion device, including energy to sustain the magnetic fields. Achieving a more practical break-even point is one goal of the International Thermonuclear Experimental Reactor (ITER), a fusion reactor being built in France by an international collaboration (see Fig. 11.10).

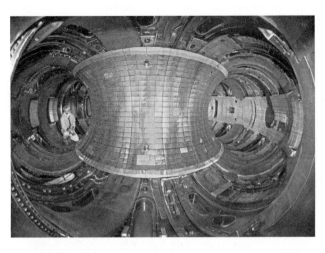
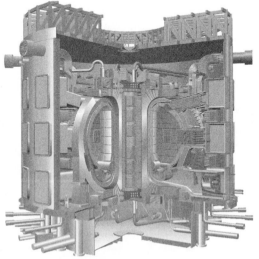

Figure 11.10
The interior of Japan's JT-60 tokamak, which measures 6.8 m across (left). The design for the ITER fusion reactor features a toroidal plasma chamber 12 m across (right); its goal is 500 MW of fusion power.

Projected to begin operation in 2016, ITER should produce fusion power of 500 MW for several minutes. It will give researchers the opportunity to explore several of magnetic confinement's other challenges, such as how to manage the high power output and the buildup of fusion-produced helium.

Early fusion power plants will probably be thermal systems, like fossil and nuclear plants. A coolant, possibly liquid lithium metal, will surround the reaction chamber and absorb fusion-produced energy. The coolant will transfer its energy to boil water and drive a conventional steam cycle (Fig. 11.11). Despite the 100-MK plasma temperature, the high temperature for the second law of thermodynamics will be that of the steam, giving efficiencies comparable to other steam power plants.

The first commercial fusion reactors will almost certainly use D-T fusion because of its lower ignition temperature, lower density-time product (Lawson criterion), and higher energy yield. But D-T fusion has disadvantages. First, it requires radioactive tritium fuel, either produced externally or more likely bred from lithium in the reactor (Equation 11.6). Second, most of the energy from D-T fusion is in high-speed neutrons that emerge from the reaction (Equation 11.5). Being neutral, those neutrons conveniently escape the magnetic confinement. But they embed themselves in the chamber walls, where they cause nuclear reactions that make the wall materials radioactive. Neutron bombardment also causes structural changes that make the walls brittle, an especially serious problem with the higher-energy neutrons from D-T fusion.

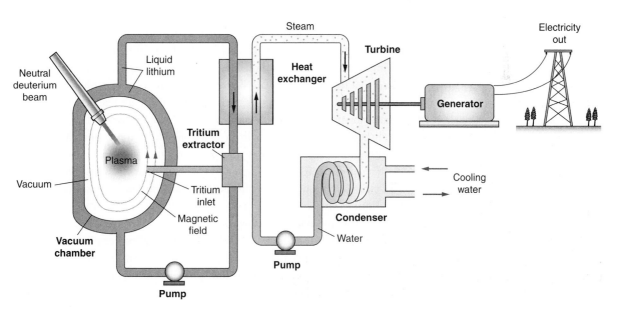

Figure 11.11
Diagram showing what a deuterium-tritium magnetic fusion power plant might look like. The D-shaped structure at left is a cross section through the toroidal plasma chamber.

If D-T fusion succeeds, we'll eventually turn to the more challenging D-D fusion. It has no radioactive fuel input, although the reaction in Equation 11.4 does produce tritium that would mostly get "burned" through the reaction in Equation 11.5. The neutron fluxes and energies are lower, reducing radioactivity and materials failure problems. And the high-energy charged particles (helium-3, tritium, and protons) produced in the reactions of Equations 11.3 and 11.4 offer a more futuristic way to generate electric power using so-called magnetohydrodynamic (MHD) generators. In these devices, charged particles flowing between electrical conductors take the place of rotating wire loops in conventional generators. The result is a direct conversion of kinetic energy to electricity with no moving parts. Higher operating temperatures could give the MHD generator an efficiency advantage over conventional steam turbines.

Environmental Impacts of Fusion

Nuclear fusion, if successful, would be cleaner than fossil fuels or nuclear fission, have zero carbon emissions, and have less visual and land-use impact than renewables such as wind, solar, and biomass. The main product of the fusion reactions (Equations 11.3 to 11.5) is helium, a benign inert gas and a useful commodity. Some tritium is also produced, and any that isn't "burned" needs to be managed as a nuclear waste; however, given tritium's 12-year half-life, waste management is nowhere near as daunting as with the long-lived wastes from nuclear fission.

The neutrons produced in fusion bombard the reaction chamber walls, inducing radioactivity and structural weakness. Therefore the fusion reactor itself becomes mildly radioactive, presenting hazards during maintenance and requiring eventual disposal as radioactive waste. Again, though, the half-lives involved are far shorter than for fission products. More important, the total amount of radioactive waste and the intensity of radiation produced in a fusion plant would be far less than in a comparable fission plant.

Fusion has the additional safety advantage that there's no chain reaction to go out of control. It's so hard to sustain the high temperature needed for fission that any failure would immediately cool the plasma and halt the reaction. Thus catastrophic nuclear accidents involving fusion are virtually impossible. And magnetic fusion is not related to the production of nuclear weapons, so there's no worry about weapons proliferation. Inertial fusion is more ambiguous in that regard, since it provides a tool for studying thermonuclear explosions in miniature.

Is there a scientific or environmental downside to fusion? Yes: It produces waste heat, which we would need to dispose of as we do with other thermal power plants. But waste heat leads to another issue we need to consider, namely the seemingly happy prospect that we may someday develop inexpensive fusion power plants using Earth's virtually unlimited deuterium resources. Human energy consumption could then grow exponentially, without the constraints of limited fuel resources, cost, pollution, or greenhouse gas emissions. Ultimately, however, all that fusion energy and its thermal waste—like all other energy we generate—ends up as heat that Earth needs to get rid of. If our fusion energy generation became even 1 percent or so of the solar input (instead of the

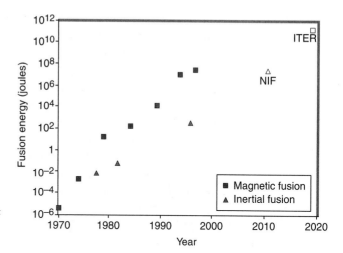

Figure 11.12

Progress toward fusion power, measured in fusion energy produced per pulse for both inertial fusion (triangles) and magnetic fusion (squares), including projections for the National Ignition Facility (NIF) and the International Thermonuclear Experimental Reactor (ITER). ITER's first fusion energy will be some eighteen orders of magnitude greater than the experiments conducted in 1970.

0.01 percent of the solar input that humans now use), then we'd have a heating problem that would dwarf today's worry about carbon-induced global warming. I don't think that scenario is likely, but it's a limitation we need to keep in mind.

Prospects for Fusion Energy

Since the 1950s, fusion enthusiasts have been announcing that fusion energy is only a few decades away. Half a century of experience with the challenges of fusion has tempered that optimism. Some think fusion will never be economically viable, even if we overcome all of the scientific and engineering problems. And national commitments to expensive fusion research have wavered; the ITER collaboration nearly fell apart in squabbles over where to site the project, and the United States withdrew from ITER for several years because of the project's high cost. Nevertheless, the fusion community has made steady, if slow, progress toward its goals (Fig. 11.12). If ITER meets its design goals, construction might start in the 2020s on DEMO, a demonstration fusion power plant that could begin experimental operation in the 2030s. But even the ITER collaboration suggests that a commercial fusion reactor is unlikely before 2050. At this time we just don't know whether fusion will be a long-term solution to our energy needs. Even if it is, it won't be in time to reduce our immediate dependence on fossil fuels.

11.3 Hydrogen Futures?

In chemical hydrogen and in hydrogen-based nuclear fusion we have two possible future energy sources whose physical basis and status today differ like night and day. There's simply no chemical hydrogen available, whereas Earth's oceans contain a virtually unlimited supply of deuterium for fusion. We know how to extract energy from chemical hydrogen using internal combustion engines, gas turbines, and fuel cells. After half a century of effort, on the other hand, we're still a long way from a fusion power plant.

Chemical hydrogen faces the daunting task of replacing an entrenched infrastructure for producing, distributing, and consuming petroleum-based liquid fuels. Fusion plants, if they're ever developed, could plug into the existing power grid with no infrastructure modifications.

At this point it's not clear whether either technology will ever see widespread use, but if both were to be developed, together they might present a solution for all our energy needs. That's because fusion would be best for large-scale stationary electric power generation, whereas chemical hydrogen's greatest value is in replacing liquid petroleum fuels for transportation. If we have inexpensive, unlimited, fusion-generated electricity, then we can use some of it to produce hydrogen by electrolysis of water. In this way, hydrogen fusion could be the source of all our energy, with chemical hydrogen the intermediate energy carrier that enables what's ultimately fusion-powered transport. Now there's an energy future! But don't count on it anytime soon.

Chapter 11 Chapter Review

BIG IDEAS

Hydrogen is a potential source of energy in two very different ways: as a chemical fuel and as a nuclear fuel.

11.1 Molecular hydrogen (H_2) is a gaseous fuel. The only by-product of hydrogen combustion is water. But *there is no hydrogen energy resource*. We have to make hydrogen, and that takes more energy than we would get out of it. Therefore chemical hydrogen isn't an energy source but an **energy carrier**. Hydrogen is extracted from fossil fuels or water. It can be burned in power plants and internal-combustion engines, whereas fuel cells combine hydrogen with oxygen to produce electrical energy. The environmental impact of hydrogen energy includes the impact of whatever energy source was used to produce the hydrogen.

11.2 Nuclear fusion is the process of combining light nuclei to make heavier ones, thereby releasing energy. Terrestrial fusion will first use the hydrogen isotopes deuterium and tritium. Deuterium in the oceans makes every gallon of seawater the energy equivalent of 350 gallons of gasoline, and there's enough to last far longer than the Sun will shine. However, fusion presents enormous technological challenges because the fusion fuel must be heated to extraordinary temperatures and then confined as it undergoes fusion. Two approaches—**intertial confinement** and **magnetic confinement**—are under study, but each is decades from commercial practicality. A successful fusion power plant will produce some radioactive waste, but far less than with nuclear fission.

11.3 Widespread use of chemical hydrogen or nuclear fusion is decades away. A hydrogen economy, with chemical hydrogen produced from nonfossil energy sources, could be an attractive energy scenario for the distant future. One such source might be nuclear fusion itself.

TERMS TO KNOW

confinement time (p. 346)

electrolysis (p. 334)

energy carrier (p. 334)

fuel cell (p. 336)

hydrides (p. 338)

ignition temperature (p. 346)

inertial confinement (p. 346)

Lawson criterion (p. 346)

magnetic confinement (p. 347)

photolysis (p. 335)

plasma (p. 346)

proton-exchange membrane fuel cell
  (p. 336)

steam reforming (p. 334)

thermal splitting (p. 335)

tokamak (p. 348)

GETTING QUANTITATIVE

Hydrogen combustion: $2H_2 + O_2 \rightarrow 2H_2O$ (Equation 11.1; p. 332)

Hydrogen energy content: 142 MJ/kg, 12.8 MJ/m^3

Hydrogen energy resource on Earth: **zero**

Steam reforming: $CH_4 + H_2O \rightarrow 3H_2 + CO$ (Equation 11.2; p. 334)

D-D fusion: $^2_1H + ^2_1H \rightarrow ^3_2He + ^1_0n + 0.523$ pJ (Equation 11.3; p. 343)

D-D fusion: $^2_1H + ^2_1H \rightarrow ^3_1H + ^1_1H + 0.649$ pJ (Equation 11.4; p. 343)

D-T fusion: $^2_1H + ^3_1H \rightarrow ^4_2He + ^1_0n + 2.8$ pJ (Equation 11.5; p. 343)

Deuterium energy content: 330 TJ/kg

Tritium from lithium: $^6_3Li + ^1_0n \rightarrow ^4_2He + ^3_1H$ (Equation 11.6; p. 345)

D-T ignition temperature: 50 MK

QUESTIONS

1. Is molecular hydrogen a fuel? A source of chemical energy? Discuss.

2. A friend argues that there's plenty of hydrogen available for fuel-cell vehicles because of all the H_2O in the ocean. Refute this argument.

3. How is a fuel cell like a battery? How is it different?

4. Why does fusion require high temperatures?

5. Describe the two main approaches to fusion now being explored.

EXERCISES

1. It takes one H_2 molecule to make one H_2O molecule. Use that fact to verify that the 0.475 aJ released in forming a water molecule corresponds to an energy content of 142 MJ/kg for H_2 when combusted with oxygen.

2. Earth's atmosphere contains 550 parts per billion of H_2 by volume. Estimate the total volume of the atmosphere, approximating it as a uniform layer 60 km thick. Then use Section 11.1's figure of 12.8 MJ/m^3 to estimate the total chemical energy content of atmospheric hydrogen. If we could do the impossible and extract all that hydrogen with minimal energy expenditure, how long would it power human civilization at our current energy-consumption rate of 15 TW?

3. What's the chemical energy content of the 100,000 kg of liquid hydrogen in the Space Shuttle's external fuel tank? How long could this energy power human civilization at our consumption rate of 15 TW?

4. A typical car's gasoline tank holds 15 gallons. (a) Calculate the energy content of this much gasoline. (b) Calculate the chemical energy content of 15 gallons of hydrogen gas under normal conditions (12.8 MJ/m^3 as described in Section 11.1), and compare with the energy in the gasoline. Your answer shows why pressurizing, liquefying, or chemically combining hydrogen is necessary in hydrogen-powered vehicles.

5. Burning gasoline releases about 10 kg of CO_2 per gallon. Compare the CO_2 emissions from a conventional car getting 20 miles per gallon and a fuel-cell car getting 0.5 mile per megajoule of hydrogen energy if the hydrogen is produced through 75 percent efficient electrolysis using electricity from a 35 percent efficient coal-fired power plant, assuming that coal is all carbon.

6. The fusion reactions in Equations 11.3 and 11.4 occur with essentially equal probability. If all the tritium produced in reaction 11.4 is consumed in the reaction in Equation 11.5, this makes a total of five deuterium nuclei involved in the three reactions. Find the corresponding deuterium energy density in joules per kilogram. (Your answer will be somewhat less than Example 11.2's 330 TJ/kg because additional reactions contribute to the energy yield.)

7. The masses of deuterium, tritium, helium-4, and the neutron are, respectively, 3.3434, 5.0073, 6.6447, and 1.6749 in units of 10^{-27} kg. Use these figures to compute the mass difference between the two sides of the reaction in Equation 11.5, and use $E = mc^2$ to find the corresponding energy. Compare with the energy yield of reaction 11.5.

8. How much pure deuterium would be required to have the same energy content as 1 tonne of coal?

9. Japan's JT-60 tokamak has a confinement time of 1 second. What density is required for the JT-60 to meet the Lawson criterion for D-T fusion?

10. At the end of this chapter I envisioned a society powered completely by fusion—including the transportation sector—using fusion-generated electricity to produce hydrogen. If that vision became reality, how much deuterium would we have to extract from the oceans each year to fuel a society powered entirely by D-D fusion, assuming today's global energy-consumption rate of 15 TW?

RESEARCH PROBLEMS

1. Compare the current U.S. Department of Energy budgets for research and development of chemical hydrogen and of fusion.

2. In my discussion of fuel cells, I noted that 2003 world fuel-cell sales exceeded $300 million and were growing at a 41 percent annual rate. Has that growth continued? Research world fuel-cell sales and make a graph covering roughly the past decade. Use your graph to estimate the average percentage annual growth rate for the period.

Chapter 12

KEEPING WARM: The Science of Climate

In previous chapters we explored humankind's prodigious energy use, amounting to many times our own bodies' energy output. Because the vast majority of our energy comes from fossil fuels, I outlined many of the environmental impacts of fossil fuel extraction, transportation, and especially combustion. But no impact from fossil fuel consumption is as potentially worrisome and global in scope as the one I have not yet discussed in detail—namely, global climate change. In this chapter and in Chapters 13 to 15, I'll describe the workings of Earth's climate and especially the role of the atmosphere; the changes resulting from fossil fuel combustion and other human activities; the evidence that Earth's climate is already changing; and the computer models and other techniques for projecting future climates. Then in the book's final chapter, I'll come full circle to explore the link between human energy use and its environmental impacts, especially climate change.

12.1 Keeping a House Warm

What keeps your house warm in the winter? A heat source of some kind—a gas or oil furnace, a wood stove, a solar heating system, a heat pump, or electric heat. You need that heat to replace the energy lost to the outdoors through the walls, windows, and roof. Your thermostat turns the heat source on and off as needed to keep energy gain and loss in balance, thus maintaining a constant temperature.

Suppose your thermostat failed, leaving the heat source running continually. Would the house keep heating up, until perhaps it ignited? No, because heat loss depends on the temperature difference between inside and outside. The hotter it gets indoors, the greater the rate at which the house loses energy. Eventually it gets hot enough that the loss rate equals the rate at which the heat source supplies energy, even if the source is on continually. At this point the house is in **energy balance**, and its temperature no longer changes. You could find that temperature by equating the known input from the heating system to the temperature-dependent loss rate, as we did in Examples 4.2 and 9.2.

12.2   Keeping a Planet Warm

A planet stays warm exactly the same way a house does: It receives energy from a source and loses energy to its environment. If the rate of energy input exceeds the loss rate, then the temperature increases and so does the loss rate, until a balance is reached. If the loss rate exceeds the input, then the planet cools. The energy-loss rate also goes down, until again a balance is reached. Once the planet reaches energy balance, its temperature remains constant unless the rate of energy input varies or the planet's "insulation"—that is, any factor that determines its energy loss—changes.

A house loses energy to the outdoors primarily through conduction, although convection and radiation are also important. But a planet's environment is the vacuum of space, and the only energy-loss mechanism that works in vacuum is radiation. Thus a planet's energy loss occurs by radiation alone. When I say "planet" here, I mean the entire planet, including its atmosphere. Other mechanisms move energy between the planetary surface and the atmosphere, as we'll see later, but when we take the big picture of a planet as an isolated object surrounded by the vacuum of space, then the only way the planet can lose or gain energy is through radiation.

Chapter 4 introduced the Stefan-Boltzmann radiation law, which gives the rate at which an object of temperature T, area A, and emissivity e radiates energy:

$$P = e\sigma A T^4 \qquad (12.1)$$

P here is for *power*, since the radiation rate is in joules per second, or watts. For a planet losing energy by radiation alone, Equation 12.1 gives the total energy loss rate. To reach energy balance, the loss rate must equal the rate of energy input. That's an easy equation to set up, but first we need to deal with the fact that planets are spheres.

For planets in our Solar System, the dominant energy source is the Sun. (Recall from Figure 1.8 that the Sun provides 99.98 percent of Earth's energy.) Figure 12.1 shows schematically the solar energy input and radiation loss for a planet, in this case Earth. As Figure 9.1 showed, sunlight reaches Earth's curved, Sun-facing hemisphere at the

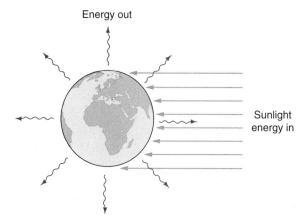

Energy out

Sunlight
energy in

Figure 12.1
Earth's energy balance, showing the energy of incoming sunlight and the energy Earth radiates back to space.

same rate it would reach a flat surface whose area is Earth's cross-sectional area, πR_E^2, where R_E is Earth's radius. The corresponding energy flow is the solar constant (S), which is 1,368 W/m^2.

On the other hand, as Figure 12.1 shows, Earth loses energy from its entire surface area. Since the surface area of a sphere is $4\pi r^2$, or four times its cross-sectional area, the energy-balance equation for Earth looks like this:

$$\pi R_E^2 S = e\sigma(4\pi R_E^2)T^4$$

Here the left-hand side of the equation is just the total energy reaching Earth—the product of the planet's cross-sectional area and the solar constant. The right-hand side is the radiation loss given by Equation 12.1, using $4\pi R_E^2$ for the area A. I've left this equation in symbols for now because we don't really need to know Earth's radius. If we divide both sides by $4\pi R_E^2$, the term πR_E^2 cancels and we're left with

$$\frac{S}{4} = e\sigma T^4 \tag{12.2}$$

This, too, is an energy-balance equation, but now it's on a per-square-meter basis. The left-hand side of Equation 12.2 gives the *average* rate at which solar energy reaches Earth. We found the same result in Chapter 9 when we recognized that the average insolation was one-fourth of the solar constant. That factor of four comes from the fact that Earth presents an effective area πR_E^2—its cross-sectional area—to the incident sunlight, but loses energy from its entire surface area, namely $4\pi R_E^2$. Physically, the factor of 4 accounts for Earth's spherical shape and for night and day.

Equation 12.2 thus expresses the average energy balance for a planet. Given the solar constant for a particular planet—that is, the intensity of sunlight at that planet's location—we can use Equation 12.2 to find the planet's temperature. There are a couple of subtleties, though. First, we need to know the planet's emissivity. For planets similar to Earth, it's reasonable to use the approximation $e = 1$, because planetary surfaces are not very reflective in the wavelength range where they emit most of their radiation. Second, not all of the input power $S/4$ is absorbed by the planet. Some is reflected directly back into space, either from the surface or from clouds, and thus plays no role in the planet's energy balance. For Earth, that reflection amounts to about 31 percent, most of it from clouds, and as a result the correct value to use for $S/4$ in Equation 12.2 is not $1,368/4 = 342$ W/m^2, but about 235 W/m^2. This latter figure represents the average rate at which Earth (including its atmosphere) actually absorbs energy, so this is the amount that Earth has to radiate back to space for the planet to be in energy balance.

So now we're ready to compute Earth's temperature. Setting the left-hand side of Equation 12.2 to 235 W/m^2, using $e = 1$, and recalling that the Stefan-Boltzmann constant σ has the value 5.67×10^{-8} W/m$^2 \cdot$K^4 we have

$$235 \text{ W/m}^2 = (1)(5.67 \times 10^{-8} \text{ W/m}^2 \cdot \text{K}^4)T^4$$

This gives

$$T^4 = \frac{235 \text{ W/m}^2}{5.67 \times 10^{-8} \text{ W/m}^2 \cdot \text{K}^4} = 4.14 \times 10^9 \text{ K}^4$$

so

$$T = (4.14 \times 10^9 \text{ K}^4)^{1/4} = 254 \text{ K}$$

Does this make sense for an average temperature of the entire planet, lumping night and day, poles and equator, all into one single number? That 254 K amounts to $-19°C$, or $-2°F$. This doesn't sound unreasonable—after all, it doesn't indicate an Earth whose oceans boil, or whose atmosphere freezes solid. But it's well below the freezing point of water, which sounds a bit low for a global average. In fact, Earth's actual average temperature is about 287 K (14°C, or 57°F)—some 33°C or 59°F warmer than our estimate. Why the difference? The answer lies in a profoundly important climate effect that I describe in Section 12.3.

The energy-balance equation we've set up here is, in fact, a very simple climate model. We solved that model to get a single quantity, Earth's average temperature. Because it treats Earth as a single point with no structure—no variation in latitudes, no oceans, no atmosphere—our model is a **zero-dimensional energy-balance model**. It's certainly not a perfect model, not only because it ignores the Earth's three-dimensional structure, but also because it doesn't even give the right value for Earth's average temperature.

Example 12.1   Martian Climate

Mars is about 50 percent farther from the Sun than is Earth. Since the intensity of sunlight drops off as the inverse square of the distance from the Sun, the solar constant at Mars is less than half that at Earth, about 592 W/m². Ignoring any reflection of solar energy from Mars, estimate the average Martian temperature.

Solution

We need to solve Equation 12.2 using 592 W/m² for the solar constant. Rather than manipulating lots of numbers, let's first solve symbolically for the temperature T:

$$T^4 = \frac{S}{4e\sigma}, \text{ or } T = \left(\frac{S}{4e\sigma}\right)^{1/4}$$

Now it's straightforward to put in the numbers:

$$T = \left(\frac{S}{4e\sigma}\right)^{1/4} = \left[\frac{592 \text{ W/m}^2}{4(1)(5.67 \times 10^{-8} \text{ W/m}^2 \cdot \text{K}^4)}\right]^{1/4} = 226 \text{ K}$$

This is cooler than our estimate for Earth, but not as cool as you might have expected given that the sunlight intensity at Mars is less than half that at Earth. That's because the fourth-power dependence of radiation loss on temperature means a relatively small change in temperature can balance a substantial change in sunlight intensity, making planetary temperatures only weakly dependent on the planet's distance from the Sun. Incidentally, this fact encourages astrobiologists looking for extraterrestrial life, because it means there's a rather broad habitable zone around Sun-like stars.

12.3   In the Greenhouse

Why does Equation 12.2 result in an estimate for Earth's average temperature that's too low? In my example of a home in energy balance, I treated the house as a single entity and ignored its detailed structure, just as I did when calculating Earth's temperature. But even a house has different parts. It may be warmer in the living room than in an upstairs bedroom, for example, and air may flow from the warm living room to the cooler upstairs—an example of convective heat transfer. These and other processes act to maintain different temperatures in the different parts of the house. The house is, in fact, a complex system, whose overall "climate" involves not only the loss of energy to the outside but also exchanges of energy among its different parts.

It's the same for Earth. Our planet has different "rooms," such as the tropical regions and the arctic—or even finer gradations: a forest, a desert, a lake, an island. It has different "floors," including the ocean depths, the land surface, and the atmosphere. The oceans and atmosphere themselves have multiple levels with different properties. All in all, Earth is a pretty complex "house." The fullest understanding of climate requires that we account for all climatologically significant variations. That's the job of large-scale computer climate models, as I'll describe in Chapter 15. But we can make a lot of progress by considering Earth to be like a simple house with just two rooms, one downstairs and one upstairs—the surface and the atmosphere. In lumping our entire planet into just surface and atmosphere, we ignore a lot of structure—for example, latitudinal variations, oceans versus land, ice versus liquid water, desert versus forest, stratosphere versus troposphere, and the gradual decline in the temperature of the lower atmosphere with increasing altitude. What we gain is a simple, easily understandable model that does a surprisingly good job of explaining the global properties of Earth's climate, including changes resulting from human activities.

Seeing the Light: Visible versus Infrared

When I introduced the Stefan-Boltzmann radiation law in Chapter 4, I pointed out that the equation for the radiated power isn't the whole story. There's also the question of wavelength. For the hot, 6,000-K Sun, most radiation is at the relatively short wavelengths of visible and near-visible light. Earth's atmosphere is, obviously, mostly transparent to visible light. I say "obviously" because we can see the Sun, the other planets, and the distant stars. This simple observation tells us that most visible light passes all the way through Earth's atmosphere and reaches the surface. Even that's a bit of an approximation; airborne particulate matter of natural or anthropogenic origin decreases atmospheric transparency, and clouds obviously block our view of the external universe. But on the whole, our atmosphere is largely transparent to visible light, which means sunlight delivers much of its energy to Earth's surface.

Because Earth is a lot cooler than the Sun, with an average temperature of about 300 K, Earth doesn't glow in visible light, but in longer-wavelength, invisible infrared. Although visible light and long-wave infrared are both forms of electromagnetic radiation, their very different wavelengths mean they interact differently with matter. In particular, some atmospheric gases absorb infrared radiation but not visible light. These aren't the dominant nitrogen and oxygen (N_2 and O_2), but gases with more complicated molecules such as triatomic water vapor (H_2O) and carbon dioxide (CO_2). The presence of these and other infrared-absorbing molecules makes Earth's atmosphere largely opaque to outgoing infrared. Clouds also contribute to infrared absorption.

The Greenhouse Effect

Those infrared-absorbing gases in Earth's atmosphere are called **greenhouse gases**. The greenhouse gases inhibit outgoing infrared radiation, making it difficult for Earth to shed the energy it gains from the Sun. This is a simplistic description of the **greenhouse effect**. The same thing happens in a greenhouse, where glass plays the role of the atmosphere in admitting visible light but inhibiting outward heat loss. However, the term *greenhouse effect* is a misnomer, because in a greenhouse the dominant effect of glass is to block convective heat loss rather than radiation. It's true that glass is much less transparent to infrared than to visible light, but in a real greenhouse this is of secondary importance. However, the terms *greenhouse gas* and *greenhouse effect* are universally accepted, so I'll stick with them.

I'm now going to present three ways to understand the greenhouse effect, in increasing order of scientific sophistication. No one explanation does full justice to the complexity of the climate system and the role of greenhouse gases, but each provides insights into the process that keeps our planet habitably warm and that will likely warm it further in the near future.

An Insulating Blanket   The most simplified description of the greenhouse effect might better be called the *blanket effect*. Atmospheric greenhouse gases have somewhat the same effect as an insulating blanket that you pull over yourself on a cold night, or extra insulation you might add to your house to cut your heating bills (and perhaps your fossil pollution and greenhouse gas emissions). The extra insulation eventually brings the system to a new state of energy balance, with the rate of energy loss again equal to the fixed energy input rate. But because a more effective insulation requires a higher temperature to achieve the same energy loss rate, the temperature rises with the increased insulation. Translation for our planet: The presence of atmospheric greenhouse gases makes Earth's surface temperature higher than it would be in the absence of those gases.

The difference between our 254-K estimate of Earth's average temperature and its actual 287-K value (a difference of 33 K, 33°C, or 59°F) is due to naturally occurring greenhouse gases, predominantly water vapor and, to a lesser extent, CO_2. (That's right: The dominant natural greenhouse gas isn't CO_2, but H_2O. Methane, ozone, and nitrous oxide also contribute.) This is the **natural greenhouse effect**, and without it our planet would be a chilly, inhospitable place. Life might still be here, but would civilization? During the last ice age, when there was a glacier 2 miles thick atop much of Europe and North America, the global average temperature was only about 6°C colder than today. Imagine a greenhouse-free planet averaging 33°C colder!

Although the "insulating blanket" description helps us understand the greenhouse effect in familiar terms, the physical mechanism of the greenhouse effect is very different from that of blankets and home insulation, which primarily reduce conductive and convective heat flow. A fuller understanding requires a more sophisticated look at the greenhouse effect.

Infrared, Up and Down   Imagine an Earth without greenhouse gases and in energy balance at that cool 254 K. Sunlight delivers energy to the planet at the average rate of 235 W/m^2. This scenario assumes no change in the amount of reflected sunlight, which is unlikely given the probable ice cover under such cold conditions, but we'll make that assumption for the sake of illustration. Since it's in energy balance, the planet is radiating infrared energy at the same rate, 235 W/m^2. Surface-emitted infrared escapes directly to space, since the greenhouse-free atmosphere is transparent to infrared as well as to visible light.

Now introduce greenhouse gases into the atmosphere. Water vapor and CO_2 begin absorbing infrared radiation coming up from the surface. The incoming solar energy hasn't changed, but now there's less infrared going out, so the planet is no longer in energy balance. As a result, the surface starts to warm up, and up goes its $e\sigma T^4$ rate of infrared emission, driving the planet back toward balance at a higher surface temperature. This is basically a more accurate description of my "blanket" explanation of the greenhouse effect, but the interactions are actually a bit more complex. As the greenhouse gases absorb infrared, they warm up and share their energy with the surrounding atmosphere. So the atmosphere warms, and then it, too, radiates infrared—both upward to space and downward to the surface.

As a result of the greenhouse-warmed atmosphere radiating infrared downward, the total rate at which energy reaches Earth's surface is considerably greater than it would be from sunlight alone. To be in energy balance, the surface has to get rid of all this energy. Since the surface energy emission rate is $e\sigma T^4$, where T is the surface temperature, this higher rate of infrared radiation from the surface requires a higher surface temperature. So we need to modify our simple planetary energy-balance diagram of Figure 12.1 to include the downward-flowing energy from the greenhouse-warmed atmosphere, as shown in Figure 12.2. Note that the upward arrows from the surface are thicker to indicate the greater rate of infrared radiation loss associated with the higher surface temperature.

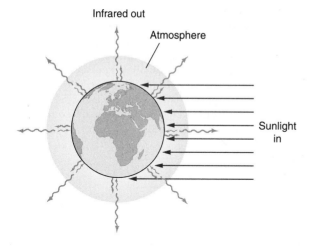

Infrared out

Atmosphere

Sunlight in

Figure 12.2
The greenhouse effect. The greenhouse gases absorb outgoing infrared, and the atmosphere radiates energy both upward and downward. The upward energy flux from the surface therefore increases, requiring a higher surface temperature. This drawing is highly simplified: The greenhouse process occurs throughout the atmosphere, which is much thinner than shown; also, not all of the incident sunlight reaches the surface.

Figure 12.2 and the discussion in the preceding paragraph might seem to suggest that we're getting something for nothing. After all, how can the total energy coming to Earth's surface exceed that of the incident sunlight, given that sunlight is, ultimately, essentially the only significant source of energy? The answer lies in the *net* energy flow. Sure, there's more energy coming to Earth's surface than there would be in the absence of greenhouse gases. But there's also more energy leaving the surface, and it all balances. We'll take a detailed, quantitative look at Earth's overall energy balance in the next section, and you'll see how this works out.

Warm Below, Cool Above Details of Earth's energy balance are complex, but concentrating for now on the loss of energy to space is sufficient to provide a still more sophisticated picture of the greenhouse effect. In this picture, I'll represent Earth's surface and atmosphere as two distinct boxes that can exchange energy with each other. In addition, energy from the Sun arrives at this combined system, and the system loses energy by infrared radiation to space. Figure 12.3 shows a simplified view of this system, emphasizing the energy loss to space. This isn't a complete energy-flow diagram, because I haven't shown reflection and absorption of the incident sunlight or all the energy exchanges between surface and atmosphere.

On the input side, Figure 12.3 shows a single arrow representing sunlight energy absorbed by the surface-atmosphere system. On the output side are two arrows, one representing infrared radiation from the atmosphere to space, the other infrared radiation from the surface. Each arrow's width reflects the size of the corresponding energy flow. At the top of the atmosphere, the widths of the two outgoing arrows combined give the same width as the incoming absorbed sunlight. This shows that the planet is in energy balance, with all the energy it absorbs from the Sun being radiated to space.

Why does the arrow representing the surface radiation taper as it passes through the atmosphere? Because the greenhouse gases absorb outgoing infrared. In fact, as the diagram

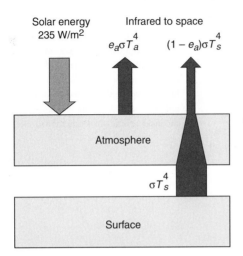

Solar energy
235 W/m²

Infrared to space

$e_a \sigma T_a^4$ $(1 - e_a)\sigma T_s^4$

Atmosphere

σT_s^4

Surface

Figure 12.3
Simple two-box climate model showing only the incoming solar energy and the outgoing infrared radiation from the surface-atmosphere system. Greenhouse gases absorb most of the surface radiation, so very little gets through the atmosphere. Surface emissivity in the infrared is very nearly 1, so e_s isn't shown in the expression for surface radiation.

suggests, very little of the surface radiation escapes directly; most is absorbed in the atmosphere. How much? That depends on the concentration of greenhouse gases. We can summarize the atmospheric absorption quantitatively with a number that indicates how effectively the atmosphere absorbs infrared radiation. Here's where fundamental physics comes in: As indicated in Chapter 4 when I introduced the Stefan-Boltzmann radiation law ($P = e\sigma A T^4$), an object's effectiveness at radiating energy—described by its emissivity, e—is the same as its effectiveness at absorbing energy. In general, emissivity depends on wavelength, and the statement about radiation and absorption effectiveness being equal holds at each wavelength. This means the emissivity may be quite different at visible and infrared wavelengths, but whatever its value is for, say, infrared, that value describes the object's effectiveness at both radiating and absorbing infrared. For that reason we can call e either *emissivity* or *absorptivity*, depending on which process we're discussing.

This is all reflected in the formulas I've put next to the arrows in Figure 12.3. The surface radiation is labeled σT_s^4, indicating that radiation from the surface depends on the *surface* temperature T_s. I didn't bother with e in this term because Earth's surface emissivity for infrared wavelengths is very close to 1. The radiation leaving the top of the atmosphere is determined by the atmospheric temperature T_a and emissivity/absorptivity e_a, whose value depends on greenhouse gases because of their absorptive effect. Consequently I've labeled that arrow $e_a\sigma T_a^4$. Finally, there's the arrow for surface radiation escaping to space. It's smaller than the radiation leaving the surface, because much of the surface-emitted infrared has been absorbed in the atmosphere. How much? If the atmospheric absorptivity were 1, then the atmosphere would be a perfect absorber and no radiation would escape to space. If the atmospheric absorptivity were 0.75, then three-fourths of the surface radiation would be absorbed and the remaining one-fourth would escape. In general, if the atmospheric absorptivity/emissivity is e_a, then the fraction of surface radiation absorbed in the atmosphere is e_a, and the remainder, namely $1 - e_a$, escapes to space (remember that e is always a number between 0 and 1). So I've labeled the surface radiation escaping to space with the term $(1 - e_a)\sigma T_s^4$. This may seem odd, because it mixes an atmospheric property (e_a) with a surface property (T_s). But it's correct: The surface radiation depends on the *surface* temperature, but the amount of that surface radiation that escapes to space depends on how much is absorbed in the *atmosphere*.

I left the area A out of all of these expressions, which means I'm keeping everything on a per-square-meter basis; the terms and arrows shown represent the average radiation from a square meter of Earth's surface or atmosphere. Not that any given square meter radiates these amounts, but if you were to divide the total radiation from the planet by its surface area, these are the numbers you would get. In nearly all discussions of climate, it's more convenient to work with per-square-meter energy flows rather than total flows. And I remind you once again about that all-important difference between energy and power; strictly speaking, the arrows and terms in Figure 12.3 describe power—the rate of energy flow, in this case in watts per square meter—rather than energy itself.

Now let's add up the two terms representing infrared radiation escaping to space:

$$\text{Power radiated to space} = e_a\sigma T_a^4 + (1 - e_a)\sigma T_s^4$$

Here the first term on the right-hand side is the infrared radiated from the atmosphere to space, and the second term is the surface radiation that made it through the atmosphere. Again, both are in watts per square meter. Now apply the distributive law to the second term:

$$\text{Power radiated to space} = e_a\sigma T_a^4 + \sigma T_s^4 - e_a\sigma T_s^4$$

Next, group the terms that involve the atmospheric emissivity, e_a, and pull out the common factor $e_a\sigma$:

$$\text{Power radiated to space} = \sigma T_s^4 - e_a\sigma(T_s^4 - T_a^4) \tag{12.3}$$

The signs are correct; check them out! I could have used a plus sign between the two main terms and reversed the terms in parentheses, but the point I want to make is clearer the way I chose to express the math here.

Equation 12.3 gives the total power radiated to space from the Earth system—that is, from the combination of surface and atmosphere. (Well, not quite—it's the *average* power radiated *per square meter*. You could multiply it by Earth's surface area if you really wanted the total power.) For energy balance, that quantity must equal the rate at which the planet absorbs energy from the Sun. Absent any changes in absorbed solar energy—which could occur if the Sun's power output varied, or the reflectivity of the Earth changed—the quantity in Equation 12.3 remains constant if Earth is in energy balance.

To get comfortable with what Equation 12.3 says, consider first the special case when $e_a = 0$. This scenario would mean that no infrared absorption is taking place in the atmosphere—in other words, there are no greenhouse gases. The energy radiated to space would all come from the surface. The rate of surface emission would be just σT_s^4, and if we equated that to the incoming solar energy, we would get back the cold 254-K average surface temperature from our earlier calculation that did not take the greenhouse effect into account. So Equation 12.3 makes sense in the absence of greenhouse gases.

Now take a good look at the second term in Equation 12.3, the term that went away when we considered the case where $e_a = 0$. Normally, Earth's surface is warmer than the atmosphere; averaged over the globe and the thickness of the atmosphere, this is always true. So $T_s > T_a$, and the second term in Equation 12.3 is *positive*. But this term is *subtracted* from the first term. So if nothing else changed, an increase in atmospheric infrared absorption—that is, an increase in e_a—would *decrease* the total outgoing infrared. This is hardly surprising, since greenhouse gases absorb infrared radiation. But the outgoing infrared has to balance the incoming solar energy, so it can't stay decreased once Earth reaches energy balance. The only way to keep the total outgoing infrared constant in the face of increasing e_a is to increase the first term on the right-hand side of Equation 12.3, namely the surface-radiation term σT_s^4. And that means increasing the surface temperature T_s. Translation from math to English: As the concentration of atmospheric greenhouse gases rises, so must the surface temperature.

The second term in Equation 12.3 is called the **greenhouse term**, because it provides a mathematical description of the greenhouse effect as a reduction in outgoing infrared radiation, a reduction that requires a compensating increase in surface temperature in order to maintain energy balance. Physically, the greenhouse gases cause that reduction, and their effect is summarized mathematically in the value of the atmospheric emissivity/absorptivity e_a.

It isn't enough to have $e_a > 0$ for the greenhouse effect to occur. It's also essential that the atmosphere be, on average, cooler than the surface. As Equation 12.3 shows, the greenhouse term can be zero if *either* $e_a = 0$ *or* $T_a = T_s$. A greenhouse atmosphere at the same temperature as the surface would absorb infrared from the surface but would radiate to space at exactly the same rate, and there would be no need for a warmer surface. It's the cooler upper atmosphere—a less effective radiator, given the σT^4 radiation-loss rate—that drives the greenhouse effect. In fact, for a given concentration of greenhouse gases (a fixed e_a), a *cooler* atmosphere actually *increases* the greenhouse effect. I emphasize this point to dispel a possible misconception about human-induced global warming, namely the notion that we should expect warming at all levels of the atmosphere. In fact, an anthropogenically strengthened greenhouse effect results simultaneously in *cooling* of the upper atmosphere and *warming* of the surface and lower atmosphere (more on this in Section 12.4).

If you looked at Earth from outer space with infrared eyes (we have satellites that do just that), you would see predominantly infrared radiation emitted from well up in the atmosphere (about 6 km altitude, on average). Very little of the radiation would come directly from the surface, because greenhouse gases absorb most surface radiation. The radiation you would see averages 235 W/m² , balancing the solar energy that's absorbed by the surface-atmosphere system at this same rate. You could infer an average planetary temperature, which would be close to our earlier 254-K estimate in the absence of greenhouse gases. But you'd be missing interesting details that depend, ultimately, on the interaction between surface and atmosphere. The greenhouse effect plays the dominant role in that interaction, and it gives us a warmer surface.

Of course, my discussion of the greenhouse effect has been based on an overly simple model—that of Earth's surface and atmosphere as being just two distinct regions, each characterized by a single temperature. Even if we continue to ignore surface variations with latitude, longitude, water, land, or ice—that is, treat the surface as a single "box"—there are still obvious variations in the atmosphere related to altitude. So my analysis of the greenhouse effect is a bit naïve, and we must regard the symbol T_a as standing for some sort of effective average temperature of the upper atmosphere throughout the region where infrared emission has a good chance of escaping directly to space. We get the big picture right, but we've glossed over many details.

Before moving on, here's just one example of the many subtleties that climate scientists need to account for in building a quantitative understanding of the greenhouse effect. As I indicated earlier, water vapor is the dominant natural greenhouse gas. In contrast to CO_2, whose concentration is more uniform, water vapor's concentration varies substantially with geographical location, altitude, and time. Water vapor in the lower atmosphere, where the temperature isn't hugely different from that of the surface, makes for a higher effective T_a in Equation 12.3, which means a less substantial greenhouse effect. The same water vapor located in higher layers of the atmosphere absorbs and emits infrared radiation at a lower effective temperature, which strengthens the greenhouse effect and causes greater surface warming. Thus, subtle details such as the specific processes that carry evaporated ocean water into the atmosphere have a direct impact on the strength of the greenhouse effect. That's just one of many complex factors that affect Earth's climate.

12.4 Earth's Energy Balance

Figure 12.1 gives a very simple picture of Earth's overall energy balance, showing just the incoming solar energy and the outgoing infrared radiated to space. It ignores the greenhouse effect and other interactions between Earth's surface and atmosphere. Figures 12.2 and 12.3 are improvements that account for the effect of greenhouse gases on infrared radiation from the surface. But none of these diagrams comes anywhere close to showing the full complexity of the interactions among incident sunlight, outgoing infrared, and Earth's surface and atmosphere. We can't hope to understand the climate system, and especially the climatic effects of human activity, unless we look more closely at these interactions.

Figure 12.4 represents our best understanding today of the global average energy flows that establish Earth's climate. Once again we ignore geographical variations, lumping Earth's entire surface into one "box" and the entire atmosphere into another. However, a little more detail is shown in the atmospheric box, with the presence of discrete clouds and processes involving them. Energy flows are labeled with the average rate of energy transport in, as usual, watts per square meter. Again, the width of a given flow reflects quantitatively the size of the flow.

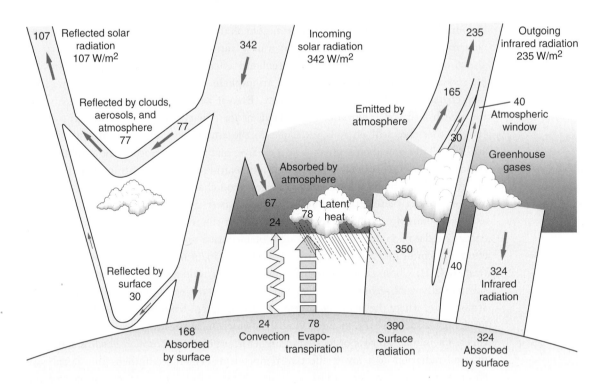

Figure 12.4
Interactions between Earth's surface and atmosphere are the determining factors in the planet's average energy balance.

At the top center of Figure 12.4 is the incoming solar radiation at 342 W/m²—a number that results, as I described earlier, from averaging the direct 1,368 W/m² above Earth's atmosphere over the full surface area of our spherical planet. I've alluded before to the 31 percent—some 107 W/m²—of the incoming solar radiation that's reflected back into space without being absorbed by the Earth or its atmosphere. In Figure 12.4 you can see that most of that reflection—77 W/m²—is from clouds and reflective particles (aerosols) in the atmosphere. Another 30 W/m² reflects from the surface, much of it from ice, snow, and desert sands. The term **albedo** describes the fraction of incident sunlight that's reflected. Figure 12.4 shows that Earth's average albedo is 107 W/m² / 343 W/m², or 0.31. This is the 31 percent figure I've mentioned before. Again, sunlight reflected back to space plays no direct role in climate, since the Earth-atmosphere system never absorbs it. However, this doesn't mean Earth's albedo isn't important climatically. Whatever sunlight isn't reflected is absorbed, meaning that any change in albedo alters the overall energy flow that drives climate. Occurrences such as the melting of glaciers or the advancing of deserts may therefore have significant effects on climate. Exercise 7 explores the effect of a changing albedo.

Take away the reflected 107 W/m² from the 342 W/m² of incident sunlight, and you're left with 235 W/m² as the rate at which the Earth-atmosphere system actually absorbs solar energy. Although the clear atmosphere is largely transparent to sunlight, the presence of clouds and aerosols makes for a significant atmospheric absorption of 67 W/m², as Figure 12.4 shows. This leaves 168 W/m²—just under half of the total incident solar energy—to be absorbed at the surface.

Absent any other effects, Earth's surface would have to rid itself of energy at that rate of 168 W/m². But there's more energy reaching the surface—namely, the downward-flowing infrared from the atmospheric greenhouse gases, which amounts to 324 W/m². So the surface has to rid itself of a total of 168 + 324, or 492 W/m². As Figure 12.4 shows, most of this energy is in the form of infrared radiation, only 40 W/m² of which escapes directly to space (from this, by the way, you can estimate the atmospheric emissivity that played such an important role in our analysis of Equation 12.3; we'll do this in Example 12.2).

However, a significant portion of the surface energy loss isn't by radiation but through two related processes that amount to bulk motion of energy into the atmosphere. Convection, which I introduced back in Chapter 4, results when heated surface air rises and takes internal energy with it. That averages some 24 W/m² over Earth's surface. The second process, **evapotranspiration**, moves energy from the surface to the atmosphere in the form of the latent heat of water vapor. Recall from Chapter 4 that it takes energy to change a substance from liquid to gas, energy that's recovered if the gas recondenses to liquid. Figure 12.4 shows that this process amounts to 78 W/m² of upward energy flow, associated with H_2O evaporated from surface waters as well as released into the air by plants in the process of transpiration, whereby soil moisture extracted by plant roots rises through trunks, stems, and leaves and evaporates into the atmosphere. When this atmospheric water vapor recondenses to form liquid droplets and clouds, the latent energy is released and heats the atmosphere.

Figure 12.4 shows that radiation, convection, and evapotranspiration from Earth's surface carry a total of 492 W/m² into the atmosphere. This energy is equal to the 492 W/m² flowing downward from both sunlight (168 W/m²) and infrared from the atmosphere (324 W/m²). So the surface is in energy balance.

What about the atmosphere? It absorbs 67 W/m² of sunlight energy, 24 W/m² coming up from the surface by convection, 78 W/m² carried upward by evapotranspiration, and 350 W/m² of infrared from the surface (the remaining 40 W/m² of the 390 W/m² total surface infrared is the small portion that escapes directly to space). That makes a total of 519 W/m² of incoming energy. On the outgoing side, the atmosphere sends infrared energy down to the surface at the rate of 324 W/m² and up to space at 165 W/m². These two numbers aren't equal because the higher levels of the atmosphere, where the radiation to space originates, are cooler than the lower levels where the downward-directed radiation comes from. Clouds radiate another 30 W/m² of infrared to space (which is distinct from the 77 W/m² of sunlight that clouds reflect as visible light). So the total rate of energy loss from the atmosphere is 324 + 165 + 30 = 519 W/m².

This number matches the atmosphere's incoming energy, so the atmosphere, too, is in energy balance.

Finally, the total upward radiation to space from the Earth-atmosphere system includes the 165 W/m² radiated upward from the atmosphere, another 30 W/m² radiated from clouds to space, and the 40 W/m² that manages to escape directly from the surface to space. Those figures add up to 235 W/m²—precisely the rate at which the Earth-atmosphere system absorbs solar energy. So the entire system is in energy balance.

Well, almost. Energy balance is never perfect—consider the tiny fraction of incoming sunlight that's stored in fossil fuels. But recent studies suggest an unusually large imbalance today, with some 0.85 W/m² more energy coming in than going out. This excess energy is warming the planet, although the surface warming is moderated because much of the excess energy goes into the oceans, where it contributes to sea-level rise (more in Chapter 14). Today's energy imbalance almost certainly results from increasing greenhouse gas concentrations, due in large part to fossil fuel consumption.

Example 12.2   Energy Balance and the Greenhouse Effect

(a) Use Figure 12.4 to determine the value of the atmospheric emissivity e_a. (b) Assuming energy balance and the actual surface temperature of 287 K, use Equation 12.3 to determine the effective atmospheric temperature.

Solution

(a) Figure 12.4 shows the surface radiating 390 W/m², of which only 40 W/m² get through the "atmospheric window" directly to space. So 350 W/m² must be absorbed in the atmosphere. This gives an absorptivity, and thus an emissivity, of 350/390, or $e_a = 0.897$. (b) In energy balance, the radiated power given in Equation 12.3 must equal the 235 W/m² of solar input, so Equation 12.3 becomes 235 W/m² $= \sigma T_s^4 = e_a \sigma (T_s^4 - T_s^4)$. We then solve for T_a to get

$$T_a = \left[\frac{1}{e_a \sigma} (235 \text{ W/m}^2 - \sigma T_s^4 + e_a \sigma T_s^4) \right]^{1/4}$$

Using our values $e_a = 0.897$ and $T_s = 287$ K gives $T_a = 249$ K. As expected, this number is significantly lower than the surface temperature. It's also close to the 254 K we got from our first simple energy-balance calculation that ignored the whole surface-atmosphere interaction. That's not surprising either: Seen from above the atmosphere, Earth is in energy balance with the 235 W/m² of incoming solar energy. Since the atmosphere radiates most of the outgoing energy, we should expect an atmospheric temperature close to that of our earlier calculation. The difference comes from the 40 W/m² that sneaks through from the surface, implying an effective emissivity of about 0.9 rather than the 1.0 we assumed in our earlier and simpler calculation.

12.5   A Tale of Three Planets

Just a Theory?

Is the greenhouse effect the only viable explanation for Earth's surface temperature? Or is it just a theory that might or might not account for Earth's climate? First, a word about scientific theories: There's a widespread misconception, fueled in part by those who don't accept Darwin's theory of evolution, that a theory is not much more than a scientific guess, hunch, or plausible but unproven hypothesis. In fact, scientific theories are solidly established and coherent bodies of knowledge, based typically on a few key ideas, and reinforced by diverse observations that provide credence and self-consistency. In physics, the theories of relativity and quantum electrodynamics provide our most firmly established understandings of physical reality. In biology, the theory of evolution gives the most solidly consistent picture of how Earth's diverse species arose. In geology, the theory of plate tectonics—a relative newcomer to the great theories of science—ties together widely diverse geological evidence to give a clear and solidly established picture of Earth's history. And in climate science, the theory of the greenhouse effect, which is grounded in long-established, basic principles of physics and chemistry and corroborated with observations, explains why our planet's surface temperature is what it is. Are these theories guaranteed to be correct? Of course not—nothing is 100 percent certain. But they're the best we have, and in the case of long-established theories, "best" means very certain indeed.

The older a theory is, the more confidence scientists generally have in it. That's because an ever-increasing body of observations and experiments proves consistent with the theory, so an older theory has more corroborating evidence. If observations or experiments arise that are inconsistent with a theory, and if those observations prove repeatable, then this spells the end of the theory and the scramble is on for a better, more encompassing, and more correct theoretical framework.

The theory of the greenhouse effect is, in fact, quite old. In the early nineteenth century, the French mathematician Jean-Baptiste Fourier was the first to recognize the effect of atmospheric greenhouse gases in warming Earth's surface. (Fourier is better known for his technique of analyzing waves into their constituent frequencies.) Later in the century, the Irish-born John Tyndall—a versatile scientist who studied everything from the germ theory of disease to glaciers to the dispersion of light in the atmosphere—made actual measurements of infrared absorption in CO_2 and water vapor and explored the possibility that a decline in the greenhouse effect might be the cause of ice ages. Late in the nineteenth century the Swedish chemist Svante Arrhenius, better known as the 1903 Nobel laureate for his work on electrolytic chemical decomposition, quantitatively explored the effect of increased CO_2 concentration on Earth's temperature. Arrhenius carried out tens of thousands of hand calculations and reached the conclusion that Earth's temperature might increase by 5°C to 6°C with a doubling of atmospheric CO_2—a result generally consistent with modern predictions. However, Arrhenius erred in his time scale, believing it would take humankind 3,000 years to double atmospheric CO_2. His error was in predicting human behavior, though, not in the science underlying the greenhouse effect. The fact is that, by the beginning of the twentieth century, the greenhouse effect was well-established science, although hardly common knowledge.

Nature's Experiment

Still, it would be nice if we could do controlled experiments to verify the greenhouse theory. That might involve dumping some greenhouse gases into the atmosphere, observing the climatic response, removing the greenhouse gases and then starting a new experiment, which is of course impossible. We've got only one Earth, and it's a complex system that's also our home, so we can't use it for controlled experiments. We are, however, doing an uncontrolled experiment with the greenhouse gas emissions from fossil fuel combustion. The problem is that we may not know the outcome until it's too late.

Nature, however, has provided us with a ready-made greenhouse climate experiment in the form of the three planets Venus, Earth, and Mars (Fig. 12.5). Being different distances from the Sun, each has a different value for the rate of incoming solar energy. But as Example 12.1 indicates, that may not make as big a difference as one might expect. Considering Earth in the absence of greenhouse gases, but accounting for the reflection of 31 percent of the incident sunlight, we found a predicted temperature of 254 K. For Mars, Example 12.1 gives 226 K in the absence of reflection. Including reflection with Mars' planetary albedo at 0.25 gives 210 K for the predicted Martian temperature (see Exercise 2). For Venus, as you can show in Exercise 3, a calculation analogous to

Figure 12.5
Venus, Earth, and Mars are different distances from the Sun and have different atmospheric compositions. Together they provide a natural "experiment" that verifies the theory of the greenhouse effect.

Example 12.1 yields 328 K for Venus' temperature without reflection and only 232 K when Venus' substantial albedo of 0.75 is taken into account.

However, the three planets differ in a more climatically significant way than their different distances from the Sun and their different albedos: They have very different atmospheres. Earth, as we know, has an atmosphere that's largely nitrogen and oxygen, but with enough greenhouse gases to raise the average temperature by 33°C. Mars, in contrast, has an atmosphere with only 1 percent the density of Earth's. Mars' atmosphere is mostly CO_2, but it's so diffuse that the Martian greenhouse effect is negligible. Venus, on the other hand, has an atmosphere one hundred times denser than Earth's, and it's 96 percent CO_2. Venus' surface temperature is a sizzling 735 K—way beyond the boiling point of water and hot enough to melt lead! Table 12.1 summarizes the very different conditions on these three neighboring planets. I've listed predicted temperatures that include the effect of reflection as calculated from the tabulated planetary albedo, since that's how we did the calculation for Earth. Note that the greenhouse effect column lists the temperature rise only in Celsius degrees; that's because a temperature *change* has the same numerical value in both kelvins and degrees Celsius.

The huge greenhouse effect on Venus demands a closer look. How did a planet that in many ways is much like Earth end up with such a huge surface temperature? The answer lies in two factors: Venus' proximity to the Sun and a positive feedback that enhanced its greenhouse effect. I've shown that Venus' smaller distance from the Sun wouldn't make it all that much hotter than Earth in the absence of an atmosphere (328 K for Venus versus 279 K for Earth, not accounting for reflection). But the difference is enough that in Venus' case all surface water evaporated early in the planet's history. Water vapor is a strong greenhouse gas, so as Venus' atmospheric water vapor increased, so did the surface temperature, which drove greater evaporation. As a result, Venus suffered a runaway greenhouse effect that led rapidly to its inhospitable temperature. (See Research Problem 1 for more on Venus' climate.) Today, Venus' CO_2 atmosphere continues the strong greenhouse effect that resulted at first from water vapor. On the cooler Earth, in contrast, surface water maintains equilibrium with atmospheric water vapor, limiting further evaporation and making a runaway greenhouse effect impossible

Table 12.1 Three Planets

Planet	Albedo	Calculated temperature	Actual temperature	Greenhouse effect
Venus	0.75	232 K −44°C	735 K 462°C	503°C
Earth	0.31	254 K −19°C	287 K 14°C	33°C
Mars	0.25	210 K −63°C	210 K −63°C	0°C

on our planet. This doesn't mean we shouldn't worry about greenhouse warming, only that we aren't going to drive the planet to the melting point of lead or boil away the oceans.

So our neighboring planets confirm what basic science already tells us: The composition of a planet's atmosphere has a major influence on climate. In particular, the presence of greenhouse gases in the atmosphere necessarily leads to higher surface temperatures. Anything that changes the concentration of greenhouse gases therefore changes climate. On present-day Earth, the greenhouse gas concentration is indeed changing, increasing largely as a result of CO_2 emissions from our fossil fuel combustion. So basic science suggests that climate should be changing as well. Just how much and how fast are more subtle questions that we'll explore in the next three chapters.

Chapter 12 Chapter Review

BIG IDEAS

12.1 Keeping a house warm involves a balance between energy input from the heating system and energy lost through the walls.

12.2 Keeping a planet warm involves a balance between incoming solar energy and infrared radiation to space. Since the rate of energy loss increases with increasing temperature, a planet naturally achieves this state of **energy balance**.

12.3 Earth is warmed further because of infrared-absorbing **greenhouse gases** in the atmosphere. These make the atmosphere partially opaque to outgoing radiation, and therefore increase the surface temperature needed to maintain energy balance. This is the **greenhouse effect**.

12.4 Earth's complete energy balance involves additional processes, including reflection of sunlight; convection, evaporation, and evapotranspiration; and the role of clouds.

12.5 The three planets Venus, Earth, and Mars corroborate the theory of the greenhouse effect. Mars has a tenuous atmosphere and very little greenhouse effect; Earth's atmosphere contains enough greenhouse gases to increase its temperature by 33°C; and Venus has a runaway greenhouse effect of some 500°C.

TERMS TO KNOW

albedo (p. 368)

energy balance (p. 356)

evapotranspiration (p. 369)

greenhouse effect (p. 361)

greenhouse gas (p. 361)

greenhouse term (p. 366)

natural greenhouse effect (p. 362)

zero-dimensional energy-balance model (p. 359)

GETTING QUANTITATIVE

Stefan-Boltzmann radiation law: $P = e\sigma AT^4$ (Equation 12.1; p. 357; introduced in Chapter 4)

Stefan-Boltzmann constant: $\sigma = 5.67 \times 10^{-8}$ W/m²·K⁴

Solar constant: $S = 1,368$ W/m² (introduced in Chapter 9)

Earth's energy balance, no atmosphere: $\dfrac{S}{4} = e\sigma T^4$ (Equation 12.2; p. 358)

Natural greenhouse effect, Earth: 33°C = 33 K = 59°F

Power radiated to space: $\sigma T_s^4 - e_a\sigma(T_s^4 - T_a^4)$ (Equation 12.3; p. 365)

Earth's albedo: ~0.31

QUESTIONS

1. Ignoring the complications of the greenhouse effect, explain why changing a planet's distance from its star has only a modest effect on temperature.

2. In what sense is radiation the only heat-transfer process affecting a planet's energy balance? In what sense do other processes play a role?

3. Explain why the greenhouse effect requires that the atmosphere be cooler than the surface.

4. Which arrow in Figure 12.4 is most directly related to the presence of greenhouse gases?

5. How does the study of Mars and Venus help establish the validity of the greenhouse effect?

EXERCISES

1. Use a simple zero-dimensional energy-balance model without atmosphere to estimate Earth's average temperature, but also assume there is no reflection of solar energy.

2. Rework Example 12.1, now assuming a Martian albedo of 0.25.

3. Use a simple zero-dimensional energy-balance model without atmosphere to estimate Venus' average temperature, (a) ignoring reflection and (b) assuming an albedo of 0.75. Venus is 108 million km from the Sun, versus Earth's 150 million km. *Hint*: Since sunlight intensity falls as the inverse square of the distance, the solar constant at Venus is a factor of $(150/108)^2$ larger than at Earth.

4. What's the solar constant at Jupiter, which is 5.2 times as far from the Sun as is Earth? See the hint in Exercise 3.

5. What should be the temperature on a spherical asteroid located between Mars and Jupiter, twice as far from the Sun as Earth? The asteroid has no atmosphere, and its albedo is 0.15. See the hint in Exercise 3.

6. What albedo would give a body at Earth's distance from the Sun a surface temperature of 220 K, assuming it had no atmosphere?

7. If Earth's albedo dropped from 0.31 to 0.29, what would happen to the surface temperature? Ignore the greenhouse effect (this makes your absolute temperature incorrect but still gives a reasonable estimate of the change).

8. A hypothetical planet receives energy from its star at the average rate of 580 W/m^2. Its surface temperature is 340 K and its surface emissivity in the infrared is 1. Its atmospheric temperature is 290 K. Find (a) the rate at which the surface radiates energy and (b) the atmospheric emissivity.

9. A planet with a surface temperature of 380 K and a surface emissivity of 1 has an atmospheric temperature of 320 K and an atmospheric emissivity of 0.82. If it's in energy balance, what's the average rate of energy input from its star?

10. A planet's atmospheric emissivity is 0.25, and its atmospheric temperature is 240 K. It receives energy from its star at the rate of 47 W/m^2. (a) Does it have a greenhouse effect? (b) What is its surface temperature?

RESEARCH PROBLEMS

1. Read the article "Global Climate Change on Venus" by Mark A. Bullock and David H. Grinspoon in the March 1999 issue of *Scientific American*. Write a brief paragraph contrasting Venus' climate history with Earth's. Include an interpretation of Bullock and Grinspoon's comment that Venus' "atmospheric processes are one way" (in the caption on page 57).

2. Earth's albedo plays a major role in climate because it determines the proportions of solar energy reflected and absorbed by the Earth-atmosphere system. Table 12.1's albedo value of 0.31 is a global average; albedo varies with latitude, surface conditions, season, and other factors. Find more detailed data on Earth's albedo, and make either (a) a graph of albedo as a function of latitude (its "zonal average," meaning an average over longitude) or (b) a table of different surface conditions (desert, open ocean, ice, snow, cropland, forest, etc.) and their respective albedos.

FORCING THE CLIMATE

In Chapter 12 I showed how Earth's energy balance results from a complex set of energy flows from Sun to Earth, from Earth to space, and between Earth's surface and atmosphere. Those flows ultimately establish surface and atmospheric temperatures. Any deviation from energy balance eventually leads to a new balance, but with different values for the energy flows and therefore different temperatures. This chapter explores the causes and implications of such changes in Earth's energy balance.

13.1 Climate Forcing

Suppose some abrupt change occurs to upset Earth's energy balance. It could be an increase in the Sun's power output, a decrease in arctic ice cover that causes a change in albedo and therefore in absorbed solar energy, an increase in greenhouse gases, a volcanic explosion that dumps particulate matter into the atmosphere, the destruction of a tropical forest that affects both albedo and evapotranspiration, or any of a host of other changes to the Earth's surface or atmosphere that affect energy flows. How are we to characterize this change and understand its impact on climate?

In principle, the correct way to account for a change in Earth's energy balance and to determine its climatic implications is to explore quantitatively the physical, chemical, geological, and biological processes that are affected, and to work out how those altered processes affect the entire Earth system. This is a job for large-scale computer models, which even today have trouble handling every conceivable process of climatic importance. (More on computer models in Chapter 15.)

But a simpler, approximate approach is to characterize changes by the extent to which they upset Earth's energy balance. Suppose, for example, that the average rate of solar energy input to the Earth's system were to increase. This change would clearly result in planetary warming. Or suppose that the concentration of greenhouse gases increases; this, too, should result in warming. Although these two processes are physically very different—one involves a change in the Sun, the other a change in Earth's atmosphere—they both have the same general effect, namely, warming the planet.

An upset in Earth's energy balance is a **climate forcing**. The word *forcing* comes from the scientific idea of applying a force—a push or a pull—to a mechanical system and seeing how it responds. The force appears as a mathematical term in the equations describing the system. Even when scientists aren't talking about mechanical systems and push–pull forces, they still use the word *forcing* for analogous terms in their equations. Thus an increase in greenhouse gases or a change in solar radiation adds a forcing term to equations describing the climate system. Numerically, the forcing term is the value of the resulting energy imbalance in watts per square meter. The forcing is considered positive if the imbalance results in a net energy inflow to the Earth system and negative if it results in a net outflow. Thus an increase of 5 W/m^2 in solar input is a forcing of $+5$ W/m^2, and an increase in greenhouse gases that blocks 5 W/m^2 of outgoing radiation is also a forcing of $+5$ W/m^2. An increase in aerosol particles that reflect to space an average of 2 W/m^2 of solar energy is a forcing of -2 W/m^2. Clearly, a positive forcing results in warming and a negative forcing in cooling.

In the crudest approximation to the climate system, namely the zero-dimensional model I introduced in Chapter 12, a 5-W/m^2 forcing would have the same effect whatever its cause. But that's a gross oversimplification. For example, an increase in solar radiation warms all levels of the atmosphere; however, as you saw with the two-box climate model also introduced in Chapter 12, an increase in greenhouse gases warms the troposphere and cools the stratosphere. So we look at the positive or negative sign of a forcing to determine whether that forcing results in a warming or a cooling, but we consider the magnitude of the forcing as giving only the roughest quantitative estimate of its climatic effect.

It's simplest to think of energy balance as occurring only between the Earth–atmosphere system and space, as I did in Chapter 12's figures. However, climate scientists understand forcing to mean specifically an upset in energy balance at the boundary between the troposphere (the lower atmosphere) and stratosphere (upper atmosphere). This location gives the best correlation between forcing and effects on climate because, when the climate system is in balance, there's as much energy flowing upward across the troposphere–stratosphere boundary as there is flowing downward. Any new forcing upsets that balance, at least temporarily. You may also see what I call *climate forcing* referred to as **radiative forcing**, a term that reflects the fact that forcing amounts to an imbalance between incoming and outgoing radiation.

Suppose an instance of forcing occurs—say, an abrupt increase in the incident solar radiation to a new, constant value. Then, if nothing else changes, there is a net incoming energy flow to the Earth system, and the planet warms up. The outgoing infrared increases accordingly, and eventually we reach a new energy balance at a higher temperature, as I described in Chapter 12. The forcing is still there, in that the solar radiation has increased over its original level. The value of the forcing still describes the upset in the original energy balance, but in the new equilibrium state there's no longer an imbalance. So when I'm talking about a forcing of, say, 5 W/m², I'm talking about a change relative to some earlier state of energy balance. Depending on how much time has elapsed since the forcing "turned on," the energy flows may still be out of balance or the climate may have reached the new state in which they're again in balance.

Example 13.1   Climate Forcing in a Simple Model

Use a simple zero-dimensional model to estimate the effect on Earth's average temperature of a 5-W/m² forcing associated with an increase in solar radiation.

Solution

In Chapter 12, we used a solar input of 235 W/m² to calculate an effective temperature for Earth of 254 K. Our solar input accounted for reflection but we ignored the greenhouse effect. A forcing of 5 W/m² adds an additional 5 W/m² to the solar input. So we can set up the same equation, now including the forcing term:

$$235 \text{ W/m}^2 + 5 \text{ W/m}^2 = e\sigma T^4 = (1)(5.67 \times 10^{-8} \text{ W/m}^2\text{·K}^4)T^4$$

which gives

$$T^4 = \frac{240 \text{ W/m}^2}{5.67 \times 10^{-8} \text{ W/m}^2\text{·K}^4} = 4.23 \times 10^9 \text{ K}^4$$

or $T = 255$ K. Thus the temperature goes up about 1 K, or 1°C. I say "about" because I didn't keep any significant figures beyond the decimal point in calculating the 254-K result in Chapter 12. Exercise 1 leads to a more accurate value for the temperature increase for this example, namely 1.34°C.

The results of Example 13.1 would be the same if the 5-W/m² forcing resulted from a greenhouse-induced decrease in outgoing infrared. This equivalence occurs in the simple zero-dimensional model because it doesn't matter whether you add 5 W/m² to the left-hand side of the energy-balance equation or subtract 5 W/m² from the right-hand side. But I emphasize that this equivalence is only a crude approximation; with an even slightly more sophisticated model, the effects of the two numerically equal but physically distinct forcings aren't quite the same.

13.2 Climate Sensitivity

Climate scientists, policymakers, environmentalists, and others concerned with climate change often ask "what if" questions such as, "What if this or that action or policy were to result in a climate forcing of such-and-such a value? What would be the resulting temperature change?" To answer these questions we could repeat a calculation like the one in Example 13.1 or, better, run a computer climate model. But it's simpler to characterize the climate system's response to a forcing in terms of **climate sensitivity**. Because increased atmospheric CO_2 is the dominant anthropogenic forcing, most scientists define climate sensitivity as the temperature change expected for a doubling of atmospheric CO_2 from its preindustrial level of about 280 parts per million.

Scientists estimate climate sensitivity from experiments with computer climate models and from data on past climates. There are some subtleties involved, especially with sensitivities derived from climate models. For example, the climate response to a sudden doubling of CO_2 is different from the response to a gradual increase to the same level. After a sudden change, it takes a while for the climate to reach equilibrium at the doubled concentration. Sometimes climate sensitivity is qualified as to whether it describes the immediate response—sometimes called the *transient response*—or the long-term equilibrium response. This distinction is important: For example, if we humans were to stop all our climate-forcing activity immediately, we would still be committed to several decades of warming at a rate of around 0.1°C per decade as Earth's climate comes to equilibrium with present-day concentrations of greenhouse gases and other forcing agents.

Today's best estimates of Earth's equilibrium climate sensitivity are in the range of 2°C to 4.5°C, with values below 1.5°C considered very unlikely. However, the statistical distribution of possible climate sensitivities has a long "tail" at the high end, so there's a significant chance that the actual sensitivity is considerably higher than 4.5°C.

You may also see climate sensitivity described as the change in temperature (K or °C) per unit change in forcing (W/m²)—a more general definition that attempts to quantify the climatic effects of all forcings, not just those of CO_2. Expressed this way, climate sensitivity is believed to lie in the approximate range of 0.25 K to 0.75 K per watt per square meter. To be quantitatively useful, however, this more general definition of climate sensitivity must be adjusted for different forcings, especially those that are geographically localized. Exercises 2, 3, 4, and 6 explore this second definition of climate sensitivity.

The concepts of forcing and climate sensitivity are useful on the assumption that the climatic response to a given forcing is proportional to the magnitude of the forcing. Double the forcing, and the temperature goes up twice as much; triple the forcing, and the temperature rise triples. This is known as a **linear response**. The assumption of linearity is usually a good one for relatively small changes, but linearity isn't guaranteed.

Linear versus Nonlinear

Under what conditions would a response (e.g., a temperature rise) not be linearly proportional to its cause (e.g., a forcing)? For a simple example from everyday life, think of a light switch. The motion of the switch lever is the forcing, and the brightness of the light is the response. If you push slowly on the lever, for a while nothing happens; there's no response whatsoever. But suddenly you reach a point where the switch flips almost instantly from off to on and the light comes on immediately and at full brightness. That's a **nonlinear response**, and the switch is a nonlinear device. Figure 13.1 shows examples of linear and nonlinear responses.

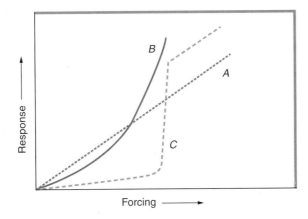

Figure 13.1
Linear (A), nonlinear (B), and highly nonlinear (C) responses to increasing forcing. At the vertical portion of curve C, a tiny change in forcing produces a huge change in the response. In climate science, the response could be temperature change.

Could the climate behave like the light switch, exhibiting a sudden, dramatic change with just a small change in forcing? Climate scientists believe it could, but it's difficult to predict extreme nonlinear effects. One example of a nonlinear climate response would be a sudden shutdown of the ocean current system that carries warm water from the tropical western Atlantic toward Europe. This event could be caused by a decrease in ocean salinity driven by the infusion of fresh water from melting arctic ice. Ironically, the now-benign climate of Britain might rapidly cool in response to warming-induced ice melt.

Many processes in nature and in technological systems are nonlinear, albeit less dramatically so than a light switch or ocean-circulation "switch." But it's almost always the case that the response to small disturbances is essentially linear. For this reason, the linear approximation is widely used throughout science and engineering, and it's used wisely as long as one keeps in mind that deviations from linearity may become significant and even dramatic with larger disturbances.

13.3 Feedback Effects

The energy balance that determines Earth's climate involves interactions among the surface, atmosphere, and incident sunlight; biological, chemical, and geological systems; human society; and a host of other components of the complex system that is Planet Earth. When one aspect of those interactions changes—for example, an increase in the surface temperature—other changes may follow. Those, in turn, may further alter the climate. These additional effects on climate are called **feedback effects**.

Negative Feedback

Feedback effects fall into two broad categories. **Negative feedback** effects tend to reduce the magnitude of climatic change. A household thermostat is an example of negative feedback: If the temperature in the house goes up, the thermostat turns off the heat and the temperature then drops. If the temperature falls, the thermostat turns on the heat and warms the house back up. The increase in Earth's radiation with increasing

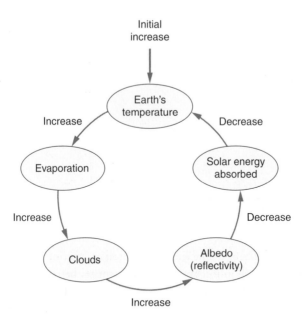

Initial
increase

Figure 13.2

Negative feedback associated with clouds. An initial increase in Earth's temperature results in increased evaporation and therefore more clouds. Because clouds are highly reflective, they increase the planet's albedo. More reflected sunlight means less solar energy is absorbed by the Earth system. The final effect, therefore, is to moderate the initial temperature increase. This is only one of many cloud feedbacks. The feedback works in both directions, so the same diagram applies if the increases and decreases are interchanged.

temperature also provides negative feedback: If the energy balance is upset by a small increase in the surface temperature, then the infrared energy loss $e\sigma T^4$ goes up, thus restoring balance, although at a higher temperature.

Many other negative feedbacks operate in the climate system. For example, increased atmospheric CO_2 can stimulate plant growth, and growing plants remove CO_2 from the atmosphere. With this feedback alone, an increase in CO_2 emissions has a lesser effect on climate than it otherwise would. Clouds can provide another example of negative feedback: When the Earth's surface warms, the result is more evaporation of surface waters and therefore more cloud formation. But clouds reflect incident sunlight, lowering the energy input to the climate system. Again, the effect of the initial warming becomes less than it would have been without the feedback effect (Fig. 13.2). (This, by the way, is only one of several cloud-related feedbacks; more are discussed in the next section.) Here's another negative feedback: An increase in surface temperature may result in decreased soil moisture in continental interiors; these dry conditions, in turn, may reduce vegetation and increase surface albedo, which means less sunlight is absorbed and therefore the initial warming is mitigated. By the way, negative feedback works for either an initial rise or a decrease. A drop in temperature, for example, means less evaporation, less cloud formation, less sunlight reflected, and therefore greater energy absorption in the climate system, which thus tends to mitigate the initial temperature drop.

Could a negative feedback effect be strong enough to reverse an initial warming or cooling? You might think so, but in fact this is not possible—at least in the small-change realm where the climate's response is linear.

Positive Feedback

Positive feedback effects exacerbate, rather than mitigate, initial changes in a system. Imagine a thermostat that, in response to a rising temperature, turned the heat on instead of off. That's an extreme example of positive feedback. The most important positive feedback in the climate system is **water vapor feedback**. Water vapor is itself a powerful greenhouse gas—the dominant factor in the natural greenhouse effect. Suppose increased atmospheric CO_2 enhances the greenhouse effect, resulting in surface warming. Again, there's more evaporation of surface waters, resulting in a still higher concentration of atmospheric greenhouse gas, in this case from the evaporated H_2O. This further enhances the greenhouse effect, and the temperature rises more than it would have without the feedback. Water vapor feedback is difficult to measure directly, but computer climate models suggest that it's very significant, increasing the response to radiative forcing changes by about 50 percent over what would occur without this feedback.

Ice and snow provide a second important feedback, the **ice-albedo feedback**. Suppose Earth's surface warms a bit, causing some Arctic sea ice to melt. The darker liquid water has a much lower albedo than the shiny white ice, so more incident solar energy is absorbed and less is reflected. The result is additional warming. As with any feedback, the opposite occurs, too: If the planet cools a little, more ice forms. Then more sunlight is reflected and the energy input to the climate system drops, exacerbating the initial cooling (Fig. 13.3). A runaway ice-albedo feedback is an essential factor in the snowball Earth episodes I described in Chapter 1.

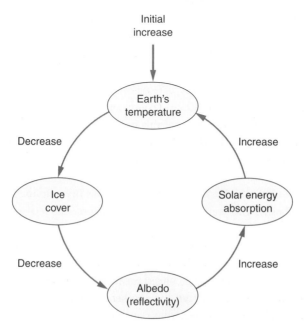

Figure 13.3
Ice-albedo feedback is a positive feedback effect. A temperature increase results in melting sea ice, replacing shiny ice with dark seawater. The Earth's albedo is reduced, resulting in more solar energy absorption. This, in turn, exacerbates the initial temperature increase. The same process would exacerbate an initial cooling.

Clouds exert positive feedback effects in addition to the negative feedback I described in the previous section. Clouds, like greenhouse gases, absorb outgoing infrared radiation. So an increase in surface temperature, leading to more evaporation and more clouds, results in a stronger greenhouse effect and still more warming. So what's the overall effect of clouds? The answer is still a toss-up. The formation of clouds and their interaction with incoming solar radiation and outgoing infrared remains an active area of climate research. The details depend on the type of cloud, its three-dimensional shape, whether it's made of water droplets or ice crystals, the size of the droplets or crystals, the cloud's altitude, and a host of other factors. Cloud effects are further complicated by the geographical and temporal distribution of clouds. For example, computer models and observations from satellites both suggest that clouds result in a positive forcing in the Northern Hemisphere during the winter and a negative forcing in the Southern Hemisphere at the same time. Therefore cloud feedback can be either positive or negative, depending on location and time of year. The overall global average effect of clouds on climate is a balance between positive and negative feedbacks that's so close we don't yet know for certain whether the net effect is positive or negative.

Here's yet another positive feedback, one that illustrates the role of human society in the climate system: When the weather gets a little warmer, people buy more air conditioners and more coal is burned to generate the electricity to run all those air conditioners, thus increasing CO_2 emissions, enhancing the greenhouse effect, and causing still more warming.

13.4 Natural and Anthropogenic Forcings

Until now, this chapter has been a rather abstract introduction of the new concepts of climate forcing, climate sensitivity, and feedback effects. What does all this have to do with Earth's real climate, with climate change that's occurring now or will occur in the future? Above all, what does it have to do with the main theme of this book—the environmental impact of human energy use, especially as it affects climate?

Let's begin by considering climatic conditions just before the start of the industrial era, around 250 years ago. We can use this preindustrial state as a baseline and ask what changes have occurred in Earth's energy balance since then—changes that we might expect would result in an altered climate. I'm not suggesting here that Earth's climate was unchanging before the industrial era began. Climate change has both natural and anthropogenic causes, and natural climate change has been an ongoing feature of Earth's history. But choosing the preindustrial state as our baseline lets us focus particularly on anthropogenic forcings.

Figure 13.4 shows some changes in forcings that climate scientists believe have occurred since about the year 1750. All but one of these forcings is anthropogenic; the lone natural forcing is a somewhat uncertain increase in the Sun's energy output since preindustrial times. Volcanic activity is another natural forcing, this one negative, but it's so highly variable that it's difficult to assign an average value for the industrial era

Figure 13.4

Changes in climate forcings since 1750. Positive forcings result in warming; negative forcings result in cooling. Error bars show uncertainties, which in some cases are relatively large. The well-mixed greenhouse gases, including the halocarbons, are stacked in a single bar. The effect of ozone is complicated, since we've added this greenhouse gas to the troposphere while depleting it from the stratosphere. Aerosols result largely from fossil fuel combustion, and their dominant effect is cooling. Land-use changes include several factors. The slight negative land-use forcing results from replacement of dark forests with lighter croplands, which gives a higher albedo; however, this forcing is quite uncertain. Only solar forcing is a natural change in energy balance. Not shown are anthropogenic forcings due to aviation-induced clouds (contrails) and the sporadic forcing associated with individual volcanic eruptions.

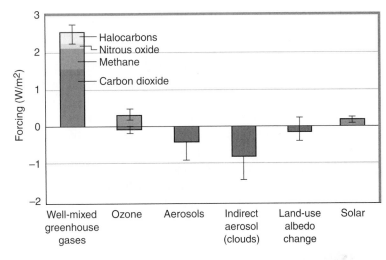

and so volcanic forcing isn't shown in Figure 13.4. The values in Figure 13.4, like much of the other climate data in this book, are from reports of the Intergovernmental Panel on Climate Change. There's considerable confusion about what this organization does and does not do, so I've described it briefly in Box 13.1.

Greenhouse Gases

The dominant forcing shown in Figure 13.4 is from the so-called **well-mixed greenhouse gases**. These gases remain in the atmosphere long enough to become thoroughly mixed and thus have nearly equal concentrations around the globe. As the figure shows, the dominant well-mixed greenhouse gases are CO_2 and methane, with CO_2 accounting for somewhat over half of the forcing in this category. Lesser contributions to well-mixed greenhouse forcing come from nitrous oxide (N_2O) and halocarbons. You've already seen how N_2O results when atmospheric nitrogen and oxygen combine during fossil fuel combustion. Industrial processes also emit N_2O, but the dominant source is the use of synthetic nitrogen-based fertilizers in agriculture. The halocarbons are synthetic chemicals; one group of these, the chlorofluorocarbons (CFCs), were widely used in spray cans and refrigeration systems until they were phased out beginning in 1987 after it was discovered that they destroy the stratospheric ozone that protects us from solar ultraviolet radiation. Halocarbons also include the hydrochlorofluorocarbons (HCFCs), which were introduced as ozone-safe replacements for CFCs. Unfortunately, the HCFCs also contribute to greenhouse warming.

Box 13.1   The Intergovernmental Panel on Climate Change

The **Intergovernmental Panel on Climate Change** (IPCC) was established in 1988 as an outgrowth of the World Meteorological Organization and the United Nations Environment Program. The IPCC's purpose is to assess the state of knowledge—in both the natural sciences and the social sciences—about changes in Earth's climate, the impacts those changes may have on human and natural systems, and the steps we humans might take to adapt to climate change or to mitigate its effects.

The IPCC comprises a broad coalition of hundreds of scientists from a wide range of disciplines, as well as policymakers from the world's governments. The IPCC compiles and evaluates the results of published scientific research, and expresses its findings in comprehensive assessment reports that are published about every six years. The IPCC also publishes a number of technical reports and other specialized documents that support its major reports.

The IPCC's reports typically include detailed volumes on aspects of climate science, climate impacts, and adaptation to or mitigation of changing climate. They also provide Technical Summaries, which are themselves comprehensive introductions to climate science, climate impacts and adaptation, and mitigation. Finally, the reports include Summaries for Policymakers, which distill the IPCC's findings succinctly in a way that's designed to inform intelligent policy decisions. The wording of each IPCC report represents a careful consensus of a broad range of participants; the Summaries for Policymakers, especially, involve delicate negotiations leading to approval by the IPCC member governments.

The IPCC itself is not a scientific research organization. It does not conduct scientific research, and it has no laboratories, climate-observing satellites, computer centers, or computer climate models. Again, the IPCC synthesizes and assesses results published in the scientific literature. The latest IPCC report, for example, compared the results of twenty-three different computer climate models run by fourteen separate research groups from around the world; the IPCC based its findings on ensemble averages of these different models.

The scientists who gather to prepare the IPCC reports are independently active in research on many aspects of climate research. They represent a broad range of expertise in areas as diverse as atmospheric chemistry, oceanography, botany, cloud physics, computer science, paleoclimatology, geology, glaciology, solar physics, fluid dynamics, remote sensing, and myriad others. But as participants in the IPCC process, their role is not to do research but to synthesize research that's already been done.

The CO_2 forcing shown in Figure 13.4 is largely the result of CO_2 emissions from fossil fuel combustion (Fig. 13.5). Changes in land use, mostly tropical deforestation, also contribute significantly, while cement production and other industrial processes add modestly to anthropogenic CO_2 forcing. We'll take a much closer look at CO_2 in Section 13.5.

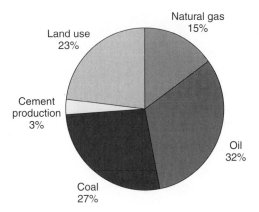

Figure 13.5

Major sources of global CO_2 emissions in the early twenty-first century. Three-fourths of the CO_2 comes from fossil fuel combustion. The CO_2 emissions from land-use changes are largely from deforestation in the tropics. All of these emissions total about 9 Gt of carbon annually.

You've met methane before; it's the principal component in natural gas, one of the three fossil fuels. Atmospheric methane emissions result from a wide range of natural and anthropogenic sources. Among the anthropogenic contributions to the roughly 0.5 W/m^2 of methane forcing shown in Figure 13.4 are natural gas releases from coal mining, oil and gas drilling, and pipeline leaks. There's considerable uncertainty in our estimates of anthropogenic methane emissions, and fossil fuel-related methane might account for more than half the total, or for a lot less. Other sources include sewage treatment plants and landfills, where methane forms when organic waste decays under anaerobic conditions. Cows and other ruminants produce methane in their digestive tracts, and the decay of animal waste contributes additional methane. Rice paddies are another significant source, again because of anaerobic conditions in the flooded paddies where rice is grown. And, as we found in Chapter 10, even hydroelectric dams result in methane emissions, so much so that a hydroelectric plant in the tropics can actually cause more greenhouse warming than a comparable fossil-fueled plant.

Global Warming Potential

Greenhouse gases differ in their ability to absorb infrared radiation. For example, a methane molecule is twenty-six times more effective at infrared absorption than is a molecule of CO_2. This means a given amount of methane released to the atmosphere has a greater forcing effect and therefore causes greater warming than the same amount of CO_2. Determining how much more is a bit subtle, for reasons having to do with **atmospheric lifetime**. Methane remains in the atmosphere for about a decade before chemical reactions destroy it; CO_2 lasts much longer (although quantifying this, too, is subtle; more about this in Section 13.5). So on short time scales, a given amount of methane emission has a much greater effect than the same amount of CO_2. But wait a hundred years and the cumulative effect of the lingering CO_2 is greater. The effectiveness of a greenhouse gas relative to CO_2 is its **global warming potential** (GWP). For the reason I've just outlined, the GWP isn't a single number but depends on one's time frame. Table 13.1 lists the GWPs of some important greenhouse gases over three different time frames. The gases include a common CFC, an HCFC, and a hydrofluorocarbon (HFC). Note that these synthetic substances have far

Table 13.1 Global Warming Potentials

Gas	Atmospheric lifetime (years)	Global warming potential relative to CO_2 Time frame		
		20 years	100 years	500 years
Carbon dioxide (CO_2)	~1,000*	1	1	1
Methane (CH_4)	11	67	23	6.9
Nitrous oxide (N_2O)	114	291	298	153
CFC-11 (CCl_3F)	45	6,700	4,750	1,620
HCFC-22 ($CHClF_2$)	12	5,200	1,800	550
HFC-23 (CHF_3)	270	12,000	14,800	12,200

*The lifetime of CO_2 is ambiguous; see Section 13.5.

higher GWPs than the more widespread but simpler CO_2, methane, and N_2O. The GWPs in Table 13.1 are per unit of mass, rather than per molecule. Since a kilogram of methane has many more molecules than a kilogram of CO_2, the GWP of methane on a per-mass basis is higher than on a per-molecule basis (see Exercise 8).

Concentration and Forcing

Because of their differing GWPs, a low atmospheric concentration of methane or a CFC can have a much greater impact on climate than a larger concentration of CO_2. On the other hand, there's much more CO_2 in the atmosphere—and in anthropogenic emissions—than there is of the other greenhouse gases. That's why CO_2, despite its lower GWP, contributes the greatest greenhouse gas forcing, as Figure 13.4 shows.

You might expect the forcing of a given gas to depend directly on its concentration, but that isn't generally the case. The reason for this discrepancy is that some specific wavelengths at which a given molecule absorbs infrared radiation are **saturated**, meaning that the atmosphere already absorbs 100 percent of the infrared at that wavelength. An increase in gas concentration can't increase absorption at that wavelength, since there's nothing left to absorb. Other wavelengths aren't saturated and so, overall, an increase in a given greenhouse gas does increase infrared absorption and thus climate forcing. But because of the saturation effect, the increase isn't linear. Exercise 5 explores the relationship between concentration and forcing for CO_2.

The forcings shown in Figure 13.4 reflect the cumulative effect of changes in greenhouse gases and other forcing agents since preindustrial times. In the case of the greenhouse gases, much of that increase has occurred very recently. Figure 13.6 shows the industrial-era increases in atmospheric concentrations of methane and HFC-23, the former from a preindustrial level of around 700 parts per billion and the latter from zero

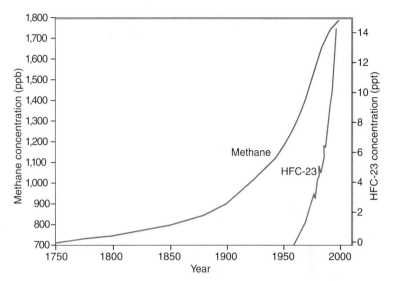

Figure 13.6

Concentrations of methane and HFC-23 since preindustrial times. The methane curve starts at its preindustrial value of around 700 parts per billion (ppb). HFC-23 starts from zero since this synthetic chemical became widespread only in the second half of the twentieth century. Note that its concentration is in parts per trillion (ppt). That's a good thing because, as Table 13.1 shows, HFC-23's global warming potential is some 12,000 times greater than that of CO_2.

because HFCs are synthetic chemicals first created only a few decades ago. Atmospheric gas concentrations are generally expressed in parts per million, billion, or trillion by volume (ppmv, ppbv, or pptv, often without the "v"). To picture methane's preindustrial concentration of 700 ppb, for example, imagine assembling a billion 1-gallon milk jugs full of air. If you separated out the preindustrial methane, it would occupy 700 of those jugs.

Figure 13.7 shows atmospheric concentrations of the most important anthropogenic greenhouse gas, CO_2, since the beginning of the industrial era. Values prior to about

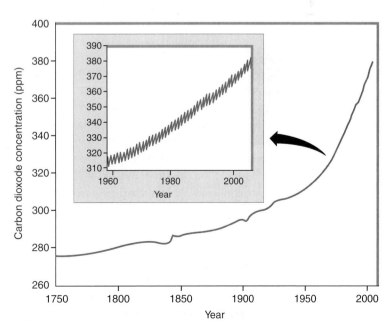

Figure 13.7

Increase in atmospheric CO_2 since preindustrial times. The inset shows the monthly averages recorded at Mauna Loa in Hawaii since 1960, with measurements sensitive enough to record the seasonal variations as plants take up CO_2 during the Northern Hemisphere's growing season and then reduce their uptake during the winter.

1950 come from gas bubbles trapped in ice cores drilled from the Antarctic ice cap. Since 1958, a monitoring station at an altitude of nearly 2 miles on Hawaii's Mauna Loa volcano has provided detailed measurements of atmospheric CO_2 in the clean, well-mixed air over the central Pacific Ocean. The Mauna Loa data are so precise that, as the inset in Figure 13.7 shows, they track the seasonal variations in CO_2 each year, reflecting the increased uptake of CO_2 by trees and plants during the Northern Hemisphere's growing season, and the corresponding reduction of this uptake during the winter. Research Problem 1 explores this effect further.

How do we know that the CO_2 rise shown in Figure 13.7 is the result of anthropogenic CO_2 emissions? Several independent pieces of evidence lead to this universally accepted conclusion. (1) Because fossil fuels are commercial commodities, we have a good quantitative handle on the rate at which we burn them. The buildup of atmospheric CO_2 corresponds to the known emissions—although, as you'll see in Section 13.5, not all of the emitted CO_2 stays in the atmosphere. (2) Carbon dioxide is a well-mixed greenhouse gas, but its concentration is marginally higher in the Northern Hemisphere, where most fossil fuel combustion takes place. (3) The ratio of cosmic ray–formed C-14 to ordinary C-12 has been decreasing, as would be expected if the atmosphere were being flooded with ancient carbon in which C-14 had long since decayed. That ancient carbon is, of course, the carbon trapped underground in fossil fuels. Additional evidence comes from the ratio of stable C-13 to the more abundant C-12; this ratio is lower in plants than in the atmosphere, but it's dropping as plant-derived fossil carbon enters the atmosphere through fossil fuel combustion. Finally, Figure 13.8 shows that the rise in atmospheric CO_2 during the industrial era has been dramatic and unprecedented on time scales of thousands to hundreds of thousands of years. Today's CO_2 concentration is nearly 40 percent higher than the preindustrial value and is thought to be significantly higher than it's been for millions of years.

Ozone

Ozone (O_3) is another greenhouse gas, but it's included separately in Figure 13.4 because it isn't mixed evenly throughout the atmosphere. Furthermore, it's listed twice because the effects of low-level (tropospheric) ozone and high-level (stratospheric) ozone are very different. Tropospheric ozone results, as I described in Chapter 6, from photochemical reactions involving air pollution. Tropospheric ozone is both a noxious, toxic pollutant and a greenhouse gas. As Figure 13.4 shows, anthropogenic ozone in the troposphere is responsible for a positive forcing of about 0.35 W/m². Stratospheric ozone is a different story altogether. It forms naturally from the action of solar ultraviolet radiation on atmospheric oxygen and, by absorbing ultraviolet, ozone protects us surface dwellers from this harmful radiation.

Chlorofluorocarbons were invented around 1930. They are powerful greenhouse gases but generally are considered chemically inert. Freon, used widely in refrigeration into the 1990s, is a CFC. Freon saved many lives by replacing the toxic ammonia, methyl chloride, and sulfur dioxide that were used in refrigerators of the early twentieth cen-

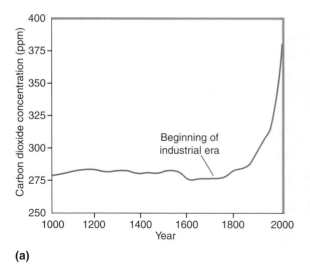

(a)

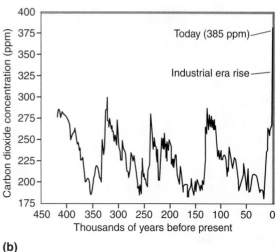

(b)

Figure 13.8
(a) Carbon dioxide concentration over the past thousand years shows a sharp rise that coincides with the industrial era. (b) The rise is even more dramatic on a longer time scale. Clearly, Earth has not seen anything close to today's CO_2 levels in at least half a million years. The regular patterns in (b) correspond to ice ages and interglacial warm periods (more on this topic in Chapter 14).

tury. The CFCs also found use as propellants in spray cans and in the manufacture of plastic foams, including energy-saving insulation. A related material (although not itself a CFC) is the nonstick coating Teflon; as with Freon and other CFCs, Teflon's value lies in its chemical inertness. In 1974, however, chemists Paul Crutzen, Mario Molina, and F. Sherwood Rowland showed that CFCs rising into the stratosphere would decompose under the influence of solar ultraviolet radiation, and that the chlorine freed in this decomposition would destroy ozone without itself being used up. Their work explained a phenomenon that was first observed in the 1970s, namely the depletion of stratospheric ozone over the Antarctic, popularly referred to as the *ozone hole* (Fig. 13.9). Crutzen, Molina, and Rowland shared the 1995 Nobel Prize in Chemistry—the first time a Nobel was awarded for environmental research. Alarm over ozone depletion led to a remarkable international agreement, the 1987 Montreal Protocol, which requires a gradual phase-out in the production and use of CFCs. As a result, stratospheric ozone should be back to its natural levels by about 2050.

Nonetheless, ozone is a greenhouse gas, and *stratospheric* ozone is especially effective because of its high altitude and hence low radiating temperature (recall the discussion "Warm Above, Cool Below" under Section 12.3). So the reduction of stratospheric ozone by anthropogenic CFCs reduces the greenhouse effect and thus contributes a negative forcing. That's why the bar for stratospheric ozone in Figure 13.4 extends downward.

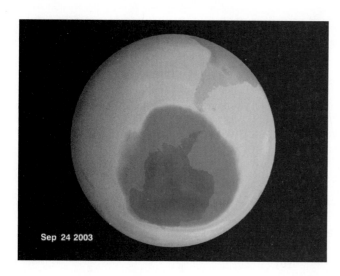

Sep 24 2003

Figure 13.9
The ozone hole—a region of depleted stratospheric ozone—shows as a gray zone around Antarctica in this image made with data from NASA's Total Ozone Mapping Spectrometer (TOMS). South America is at the upper right.

Overall, the net effect of anthropogenic processes that produce or destroy ozone is a positive forcing due to the greater influence of tropospheric O_3 as shown in Figure 13.4.

The public often equates the problems of ozone depletion and global warming. Actually, they're very different, and the remarkable international cooperation of the Montreal Protocol is resolving the ozone problem. We're far from anything remotely similar with global warming. However, ozone depletion and global warming are not unrelated, as I've just discussed. But the relationship is a subtle one, involving atmospheric chemistry and two distinct atmospheric layers.

Aerosols

As described in Chapter 6, fine particulate matter constitutes a serious form of air pollution. But particulates and liquid droplets—collectively, aerosols—also affect Earth's energy balance. The chart of industrial-era forcings (see Fig. 13.4) shows that, overall, aerosols produce a negative forcing (a cooling) of about -0.5 W/m^2. Aerosols are less well understood than greenhouse gases, so there's considerable uncertainty in this value. Aerosol forcing is further complicated by the presence of different types of aerosols that provide positive or negative contributions to the overall aerosol forcing equation.

Most significant are the sulfate aerosols, which are produced largely from coal combustion. These substances reflect sunlight back to space, decreasing the solar energy input to the climate system. Some have argued that this negative forcing might produce enough cooling to compensate for CO_2-induced warming, but this view is incorrect for two reasons. First, the overall aerosol forcing simply isn't great enough to counteract greenhouse warming. Second, sulfate aerosols have relatively short atmospheric lifetimes, so they don't get well mixed and thus their climatic effect tends to be localized. Direct forcing by sulfate aerosols may indeed cause cooling in regions downwind of heavily industrialized areas, but this can't compensate globally for greenhouse gas warming.

Other aerosols result from incomplete combustion of fossil fuels, which produces black carbon and more complicated organic carbon compounds. Black carbon absorbs sunlight and contributes to warming, whereas organic carbon exerts a cooling effect. Aerosols associated with biomass burning, nitrate fertilizers, and mineral dust from industrial activity and land development also contribute to the overall aerosol forcing, although their effects are small and not well quantified.

Sulfate and other aerosols act as nuclei on which atmospheric water vapor condenses, forming clouds. Clouds, as we know, have significant forcing effects, both positive and negative. Cloud formation is an indirect effect of atmospheric aerosols, and with it comes an indirect aerosol-caused forcing. As Figure 13.4 suggests, this indirect effect is also negative so it, too, produces cooling. Although its magnitude remains somewhat uncertain, indirect aerosol forcing is likely greater than the direct effect of aerosols.

Other Anthropogenic Forcings

Land-use changes affect climate in several ways; for example, Figure 13.4's CO_2 forcing includes CO_2 from deforestation. One direct effect of land-use changes is on Earth's albedo. Crop plants are generally more reflective than trees, so replacing forests with agriculture increases albedo and thus decreases the absorbed solar energy. This is a negative forcing. High-latitude winters exacerbate this effect, since snow cover is more reflective on cropland than when snow falls in trees. A variety of other factors related to land use also affect albedo, both positively and negatively. Overall, land-use albedo forcing is not well quantified, although, as Figure 13.4 suggests, it's most likely negative.

Aviation creates a small but potentially important climatic effect, which is not shown in Figure 13.4 because at present it's only a few hundredths of a watt per square meter. Jet aircraft fly at high altitudes where the temperature is low, and thus their emissions have a greater impact on the greenhouse effect. The jet contrails you see when planes pass overhead are clouds that wouldn't be there if the airplanes weren't flying, and we know that clouds affect climate. When commercial aviation in the United States was grounded for three days in the wake of the September 11, 2001 terrorist attacks, scientists found a clear signature of altered climate. In particular, the range of daily temperature variation throughout the United States increased during the aviation-free period. This is significant because decreasing daily temperature variation is one expected consequence of anthropogenic climate change, and in this case the removal of a particular anthropogenic forcing had a clear effect.

The Sun and Climate

Finally, there's the Sun. Our star undergoes a complex cycle of magnetic activity that results in the magnetic field of its north and south poles reversing every 11 years. The best known of the many manifestations of this solar cycle is the changing number of sunspots, those cooler, darker areas on the Sun's surface that are associated with strong magnetism. The presence of sunspots, and the areas of hotter than normal temperatures surrounding them, result in a variation of the Sun's overall energy output. This, in turn, causes changes in the solar constant—the intensity of sunlight just outside Earth's atmosphere—of about 1 W/m^2, or about 0.1 percent. The direct effect of such a small change on Earth's energy balance should produce a temperature change measured in hundredths of a degree Celsius. Solar-cycle changes in global and regional climate have been detected, but it's very difficult to isolate them from the larger fluctuations associated with natural climate variability.

The more interesting consideration for our climate is the question of long-term variations in the Sun's energy output. Astrophysicists understand very well just how stars evolve, so we know that on billion-year time scales the Sun's energy output has been increasing and will continue to do so. That long-term trend has significant climatic implications over the lifetime of our planet, but it's irrelevant on scales of decades to centuries. However, there's some evidence for solar variations on these shorter time scales. This evidence comes from a variety of sources, including records of sunspot numbers and their assumed correlation with solar activity and hence energy output; measurements of radioactive isotopes formed by cosmic rays whose intensity is affected by solar activity; and observations of Sun-like stars. Such evidence suggests a modest increase in the solar output since preindustrial times, giving the forcing of $+0.12$ W/m^2 shown in Figure 13.4. However, this quantity remains uncertain by a factor of about 2 in either direction.

Although the Sun's overall energy output changes very slightly on time scales of current climatic interest, there's much greater variability in the output of solar ultraviolet radiation. Because stratospheric ozone absorbs this ultraviolet radiation, the impact of solar variability on the stratosphere is significant. Climatologists are exploring subtle mechanisms whereby changes in the stratosphere could influence climate in the lower atmosphere and even at Earth's surface. Such effects might amplify the very modest direct influence of solar variability on climate. But given the diffuse nature of the stratosphere, with its very low mass and low energy content, any change in the stratosphere is unlikely to increase significantly the surface climate response to solar variability. As we'll see in Chapter 14, this means that solar forcing, while a significant factor in climate change even through the early twentieth century, is inadequate to explain the rapid warming observed in recent decades.

13.5 Carbon: A Closer Look

Figure 13.4 shows a variety of climate forcings since preindustrial times, but one factor stands out as the largest single anthropogenic contribution to net climate forcing: carbon dioxide. Furthermore, one can argue that CO_2 emissions are a rough measure of human energy consumption in our fossil-fueled world, and therefore of industry, economic activity, and population. We might then expect other anthropogenic forcings to scale with CO_2. For example, much of the anthropogenic CO_2 comes from burning coal, which also produces sulfate aerosol. Therefore, absent stricter air-pollution regulations, sulfate aerosol forcing should scale roughly with that of CO_2. So might other forcing agents, such as N_2O and tropospheric ozone, since these result from human activities that increase with population, industrial growth, and the spread of automobiles. So in a crude sense, CO_2 becomes a proxy for nearly all anthropogenic forcings.

However, CO_2 also has some unique properties that distinguish it from other forcing agents. In the natural world, many substances—including carbon, water, nitrogen, sulfur, phosphorus, and so forth—cycle through the system of atmosphere, land, ocean, and living organisms. Many of these natural cycles influence climate, so any human-caused disturbance can lead to anthropogenic climate change. Most important in this respect is the **carbon cycle**.

The Natural Carbon Cycle

Figure 13.10 depicts the carbon cycle in both its preindustrial state and as it is today. The figure shows several so-called **reservoirs**—parts of the Earth system where carbon

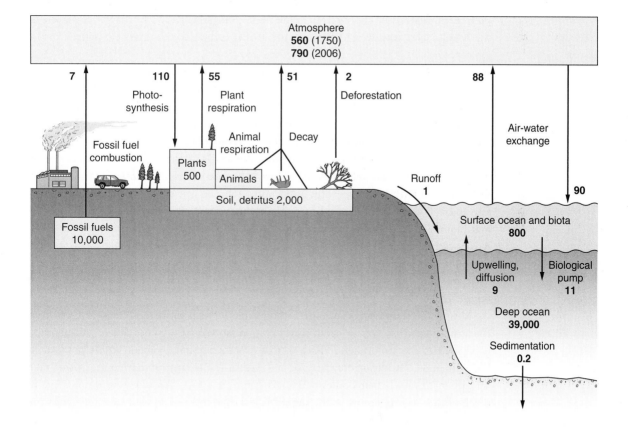

Figure 13.10

The carbon cycle. Carbon reservoirs are in gigatonnes (Gt) of carbon; flows (arrows) are in gigatonnes per year. All numbers are approximate and have been rounded to integers. Fossil fuel combustion and deforestation are the dominant anthropogenic factors upsetting the carbon cycle. Today, carbon is accumulating in the atmosphere at nearly 4 Gt per year.

is stored. The quantity of carbon in each reservoir is given in gigatonnes (Gt) of *carbon*, not *carbon dioxide*; Box 5.1 explored the difference, which amounts to a factor of 44/12, or 3.67 times as much mass of CO_2 as of carbon. The reason we measure carbon rather than CO_2 is that carbon occurs in different chemical forms as it cycles through the system. So, as Figure 13.10 shows, the preindustrial atmosphere held about 560 Gt of carbon, most of it in the form of CO_2; today there's around 800 Gt of carbon in the atmosphere. Living organisms on land hold another 500 Gt or so, nearly all of it in plants, and soils hold some 2,000 Gt. We know the atmospheric figure quite accurately, but the others are less certain. A far larger amount of carbon, some 39,000 Gt, resides in the ocean. This carbon takes various forms, including dissolved CO_2 (a small amount) and other inorganic carbon compounds (comprising most of the dissolved carbon), calcium carbonate ($CaCO_3$) in the shells of marine organisms, and organic carbon in plants and animals. However, most of that 39,000 Gt of oceanic carbon lies deep in the ocean and plays little role in the short-term carbon cycle. Only the carbon in the surface layers of the ocean cycles rapidly, and the amount of this carbon is comparable to what's in the atmosphere.

Carbon doesn't just sit in its reservoirs, but cycles around among them. Figure 13.10 also shows the associated carbon flows, quantified in gigatonnes per year. For example, terrestrial plants remove 110 Gt of carbon from the atmosphere each year; that's indicated by the downward arrow from atmosphere to plants. Through the process of respiration, both plants and animals "burn" their carbon-containing food to produce energy, water vapor, and CO_2. This process returns some 55 Gt of carbon to the atmosphere each year. The difference between these two flows is the net primary productivity I introduced in connection with biomass energy in Chapter 10. The rest of the biotic carbon eventually ends up as waste or as dead matter, although some of it temporarily becomes animal biomass. Nearly all of it eventually decays and returns to the atmosphere as CO_2. Rivers carry a small fraction (less than 1 Gt per year) to the oceans, and an almost infinitesimal amount is buried and begins the long process of turning into fossil fuels. That amount is so small, as I made clear in Chapter 5, that I'm not even showing it in Figure 13.10.

Meanwhile the oceans exchange carbon directly with the atmosphere, as CO_2 from the atmosphere dissolves in ocean water or as dissolved CO_2 comes out of solution and is released to the atmosphere. Flows in this process amount to about 90 Gt per year in both directions. Additional cycling occurs within the oceans, as marine organisms take up carbon to build their bodies and shells, and emit CO_2 through respiration. Again, much of this cycling occurs in the surface layers, but there's a modest exchange with the deeper ocean. Physical processes such as diffusion and upwelling transport carbon upward. Meanwhile a continuous "rain" of dead microorganisms and other formerly living things falls from the surface waters to the depths, and this so-called **biological pump** carries carbon downward. The upward and downward flows aren't quite balanced, giving a small net flow downward. This flow removes carbon from the ocean surface more or less permanently, contributing to the huge store of carbon in the deep ocean. Some of the deep carbon is incorporated into ocean floor sediments and is eventually returned to the atmosphere by volcanoes, but this process involves geological time scales of hundreds of millions of years. On shorter time scales, we can consider this carbon lost from the carbon cycle.

So here's the natural carbon cycle in a nutshell: Carbon is stored in the atmosphere, surface biota and soils, and ocean surface. This carbon cycles rapidly back and forth between the atmosphere and other reservoirs. How rapidly? So rapidly that a typical CO_2 molecule spends only about five years in the atmosphere before it's removed by photosynthesis or dissolved in the ocean. So why did I list CO_2's atmospheric lifetime as approximately one thousand years in Table 13.1? To resolve this question we need to look at what happens when we alter the carbon cycle.

Anthropogenic Perturbation of the Carbon Cycle

We humans have upset the balance of the natural carbon cycle, largely by burning fossil fuels, with the net effect that carbon is accumulating in the atmosphere. Figure 13.10 indicates the fossil carbon flow with an arrow pointing from the buried fossil fuel reserves to the atmosphere. Today, fossil fuel combustion (with a little help from cement production, which releases carbon stored in rocks) accounts for around 7 Gt of carbon released to the atmosphere each year—a number that continues to rise (Fig. 13.11). Another 2 Gt comes from land-use changes, mostly deforestation. Now, you might argue that 9 Gt isn't much compared with the natural flows of 110 Gt per year back and forth between the surface biota and the atmosphere, and so the human impact should be negligible. After all, isn't the anthropogenic carbon also removed on that five-year time scale? Yes, but here's the problem: The cycling of carbon among the atmosphere, land surface, and ocean takes place on roughly that same time scale. Thus, much of the carbon we add to the atmosphere continues to cycle through the system and ends up back in the atmosphere, resulting in a net increase in atmospheric CO_2. This is the reason for the rise in atmospheric CO_2 shown in Figures 13.7 and 13.8.

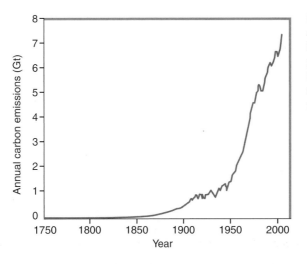

Figure 13.11
Annual global carbon emissions from fossil fuel combustion and cement production. This graph does not include land-use changes, which are responsible for another 2 Gt per year in the early twenty-first century.

Carbon Sinks

A close look at Figure 13.10 shows that the 7 Gt per year from fossil fuel combustion and 2 Gt from deforestation aren't the only imbalances in the carbon cycle. On the terrestrial side, photosynthesis removes about 110 Gt of carbon per year from the atmosphere, but the processes of plant and animal respiration and decay return only about 106 Gt per year. Given that anthropogenic deforestation sends approximately another 2 Gt per year to the atmosphere, this means there's a terrestrial **carbon sink** of somewhat over 2 Gt per year. (These numbers are approximate, and I've rounded off all decimals.) This terrestrial sink is poorly understood, but it is thought to include the incorporation of carbon into increasing forest biomass as well as directly into soils. There's a similar imbalance on the ocean side, with the ocean carbon sink removing 2 to 3 Gt more carbon each year than is returned to the atmosphere. Overall, the effect of these two sinks is to remove just over half of the carbon that we humans put into the atmosphere.

The operation of both terrestrial and ocean carbon sinks depends on climate, and there's no guarantee that the sinks will continue in a warming world. Here are two examples: A warming-caused increase in forest fires could return to the atmosphere much of the carbon that the terrestrial sink has removed through increased forest growth. And acidification of the oceans as they absorb more CO_2 will have a deleterious impact on marine organisms, many of which play important roles in the ocean carbon cycle.

Carbon's Long-Term Fate

How long would it take to return to the natural CO_2 concentration if we humans stopped dumping CO_2 into the atmosphere? We can answer this question by considering what would happen to a quantity of CO_2 injected suddenly into the atmosphere, as shown in Figure 13.12. We've seen that terrestrial and ocean sinks would remove about half of this anthropogenic CO_2 on a relatively short time scale, accounting for the steep drop seen in Figure 13.12. But the remaining carbon would continue to cycle through the atmosphere–land surface–ocean system, sustaining a higher atmospheric CO_2 concentra-

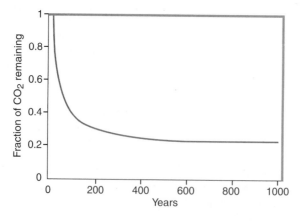

Figure 13.12
Removal of CO_2 injected into the atmosphere takes place on a variety of time scales. The concentration drops rapidly at first, but significant CO_2 remains in the atmosphere even after 1,000 years.

tion. Only on the much longer time scale of centuries to millennia does the rest of the carbon leave the system more or less permanently, mostly through processes that take it into the deep ocean. That's why the decline shown in Figure 13.12 slows, leaving a significant chunk of anthropogenic carbon in the atmosphere even after 1,000 years. So here's what's unique about carbon: Even though it's rapidly removed from the atmosphere, much of it is returned just as rapidly. It takes far longer for carbon to leave the entire system, so the relevant lifetime for anthropogenic carbon in the atmosphere is far longer than the five years that a typical carbon atom spends in the atmosphere. Because carbon-removal processes operate on different time scales, it's impossible to pin down an exact lifetime for atmospheric carbon. A rough figure of 300 to 1,000 years is often used, but this estimate is a compromise between the rapid removal by terrestrial and ocean sinks and the much longer time scales associated with processes such as the flow of carbon to the deep ocean and into ocean floor sediments.

In this way carbon is different from other forcing agents. Methane, for example, is removed from the atmosphere on a roughly 10-year time scale, largely by chemical reactions that convert it to CO_2. Once the methane is destroyed, it can't cycle back into the system. The same is true of aerosols, N_2O, and most other forcing agents. But CO_2 is different, in that we're stuck with it for a very long time. This means the climatic effects of anthropogenic CO_2 will be difficult to reverse, and in fact they may not become fully apparent until it's too late to prevent additional and very likely deleterious climate change.

In the next two chapters we'll explore the effects of anthropogenic CO_2 and other forcing agents on present and future climates.

Chapter 13   Chapter Review

BIG IDEAS

13.1   A **climate forcing** is any factor that upsets Earth's energy balance.

13.2   **Climate sensitivity** describes the climatic change that results from forcings. A common measure of climate sensitivity is the global average temperature change for a doubling of atmospheric CO_2.

13.3   Feedback effects can either exacerbate (**positive feedback**) or reduce (**negative feedback**) the response to a given forcing. Water vapor and ice albedo are important positive feedbacks; clouds may provide negative or positive feedback.

13.4   Anthropogenic climate forcings have increased throughout the industrial era, dominated by greenhouse gases. Some forcings, including greenhouse gases, lead to warming; others, such as sulfate aerosols, result in cooling.

13.5   Carbon dioxide is a particularly significant forcing agent. It alters the carbon cycle by adding carbon to the atmosphere, biosphere, and ocean; the additional carbon remains in the system for hundreds to thousands of years.

TERMS TO KNOW

atmospheric lifetime (p. 387)

biological pump (p. 396)

carbon cycle (p. 395)

carbon sink (p. 398)

climate forcing (p. 378)

climate sensitivity (p. 380)

feedback effect (p. 381)

global warming potential (p. 387)

ice-albedo feedback (p. 383)

Intergovernmental Panel on Climate
 Change (p. 386)

linear response (p. 380)

negative feedback (p. 381)

nonlinear response (p. 380)

positive feedback (p. 383)

radiative forcing (p. 378)

reservoir (p. 395)

saturation (p. 388)

water vapor feedback (p. 383)

well-mixed greenhouse gas (p. 385)

GETTING QUANTITATIVE

Preindustrial CO_2 concentration: ~280 ppm

Climate sensitivity estimate: ~2°C to 4.5°C for CO_2 doubling

Current CO_2 concentration: ~385 ppm

Water vapor feedback: ~50 percent enhancement of warming

Total anthropogenic carbon emissions: ~9 Gt per year

Anthropogenic carbon emissions from fossil fuels: ~7 Gt per year

Anthropogenic forcing due to well-mixed greenhouse gases: 2.6 ± 0.3 W/m^2

Anthropogenic forcing due to aerosols, direct and indirect: -1.2 ± 0.5 W/m^2

Net anthropogenic forcing, all sources: 1.6 W/m^2, +0.8 W/m^2, −1 W/m^2

QUESTIONS

1. Give two examples of positive feedback effects and two examples of negative feedback effects that operate in the climate system.

2. What is meant by a climate forcing? What are its units, and why?

3. Sulfate aerosols exert a cooling effect on climate, and they're produced in the same coal-burning facilities that account for a significant portion of anthropogenic CO_2 emissions. Why, then, doesn't the cooling effect of aerosols cancel CO_2-induced warming?

4. When you hold a microphone near a loudspeaker, you get feedback—a shrill screeching sound that may grow in intensity. How is this phenomenon related to the feedback mechanisms discussed in this chapter? Is this feedback positive or negative?

5. A CO_2 molecule remains in the atmosphere for only about five years. Why, then, are we stuck for centuries with anthropogenic increases in atmospheric CO_2?

6. A change of 1 W/m^2 in the solar constant results in a forcing of only 0.25 W/m^2, or less when albedo effects are included. Why is the forcing only one-fourth of the change in the solar constant?

7. Most of the greenhouse gases listed in Table 13.1 show declining GWPs as the time frame increases, but N_2O and HCFC-23 actually increase at the 100-year time frame before decreasing at 500 years. Why?

EXERCISES

1. Repeat the calculations used in Example 13.1, carrying enough significant figures that you can be sure of one figure to the right of the decimal point in your final answer. You should find a temperature rise of 1.34 K.

2. This exercise is for those of you who have had calculus. (a) The zero-dimensional model of Chapter 12 expresses energy balance through the equation $F = e\sigma T^4$, where we now take F to be the solar input plus any additional forcing. Use calculus to take the derivative dF/dT. Invert to get dT/dF, which corresponds to my second definition of climate sensitivity as the change in temperature per change in forcing. Evaluate your expression for climate sensitivity, using the value $T = 254$ K that we found in Chapter 12. (Your answer is an underestimate because it doesn't include feedback effects.) (b) The climate forcing due to anthropogenic CO_2 now in the atmosphere is about 1.65 W/m^2. Use your value for climate sensitivity to estimate the temperature rise that would have occurred since preindustrial times if CO_2 were the only forcing at work in the climate system.

3. Repeat Exercise 1, now using the climate sensitivity in kelvins per watt per square meter found in Exercise 2.

4. Using a climate sensitivity of 0.5 K per watt per square meter (in the middle of the range given in the text), estimate how much of the 0.8°C global temperature increase since 1900 could be due to solar forcing, whose value in Figure 13.4 is in the range of 0.06 to 0.24 W/m^2.

5. Because of the saturation effects described in Section 13.4, the forcing due to CO_2 doesn't increase linearly with CO_2 concentration C. Rather, the change in forcing relative to some reference concentration C_0 is given approximately by $\Delta F = \alpha \ln(C/C_0)$, where α is a constant equal to 5.35 W/m^2 and ln is the natural logarithm. (a) Use this equation to find the change in CO_2 forcing associated with the increase from the preindustrial concentration of 280 ppm to today's approximately 385 ppm, and compare it with the appropriate quantity in Figure 13.4 (your answer will be higher because the data in the figure are for the year 2004). (b) Find the forcing we can expect when CO_2 concentration reaches twice its preindustrial level.

6. Exercise 5 shows that a doubling of preindustrial CO_2 amounts to a forcing of about 3.7 W/m^2. Given a best-guess climate sensitivity of 3°C for a doubling of CO_2, find the corresponding climate sensitivity when expressed in kelvins per watt per square meter.

7. The density of gasoline is about 6 pounds per gallon, and carbon accounts for nearly all the weight of gasoline. Show that combustion of 1 gallon of gasoline produces about 20 pounds of CO_2 (the exact value is closer to 22 pounds).

8. On a per-molecule basis, methane is twenty-six times more effective as an infrared absorber than is CO_2. Calculate the corresponding ratio on a per-unit-mass basis, and compare it with the 20-year GWP for methane from Table 13.1. What's the reason for any discrepancy you find?

9. Figure 13.10 shows anthropogenic carbon emissions of about 7 Gt per year. Given that the United States accounts for about one-fourth of global emissions, estimate the U.S. annual per capita emissions of CO_2 (not carbon; see Box 5.1). Give your answer as a round (approximate) number.

10. Figure 13.10 shows that we humans have added about 230 Gt of carbon to the atmosphere during the industrial era. If the dominant removal mechanism is the transfer of carbon to the deep ocean, use the flows shown in Figure 13.10 to obtain a crude estimate of the time it would take to remove all this anthropogenic CO_2, and compare your answer with the 300- to 1,000-year CO_2 lifetime discussed in Section 13.5.

RESEARCH PROBLEMS

1. You can find atmospheric CO_2 concentration records from a variety of sources, covering many time scales, through the Carbon Dioxide Information Analysis Center web site (http://cdiac.ornl.gov). Locate the record from the Mauna Loa Observatory, and plot the monthly data for any five successive years *on the same graph*, using the same horizontal axis running from January through December. Explain the common seasonal pattern shown in each year's data, and explain why the individual years' plots don't end up right on top of each other.

2. Go to the data source for Figure 13.5 (see p. 514) and add up the total carbon emissions from fossil fuel combustion and cement production since 1750. Assuming that roughly half of this carbon stays in the atmosphere, add your result to the 560 Gt of preindustrial carbon shown in Figure 13.10 to get an estimate of total atmospheric carbon in the early twenty-first century, and compare it with Figure 13.10. (Your answer will be lower since the data in Figure 13.5 don't include carbon from land-use changes.)

3. Exercise 9 shows that annual CO_2 emissions in the United States average about 20 tonnes of CO_2 per capita. Find the corresponding figure for your state or country.

IS EARTH WARMING?

Fossil fuel consumption and other human activities have increased atmospheric CO_2 by some 40 percent since preindustrial times. We've changed a host of other climate forcing agents, too. Some changes are positive, some negative—but the net effect is a positive forcing that means more energy is coming into the Earth system than is going out. So Earth should be warming. Is it?

14.1 Taking Earth's Temperature

It's not easy to take a planet's temperature, even when we live on it. Temperature varies with geographical position, with altitude and ocean depth, and with time. Air, water, and land temperatures are generally different. So there are many measures of Earth's temperature. Most immediately relevant to us is the temperature at Earth's surface, although we've seen enough of how the climate works to know that it's also important to understand what's happening to the temperature in the atmosphere and oceans.

The Thermometric Record

Glance out the window at your thermometer and you've got a temperature measurement at a single time and place. In principle, enough such measurements, spread over the planet, could be combined to provide an average global temperature. Tracking that temperature over time would then answer our question about whether Earth is warming.

We do, in fact, have enough thermometer-based temperature measurements to calculate global average temperatures going back to about the mid-nineteenth century. Although the mercury thermometer was invented in the early eighteenth century, records before about 1850 are simply too sparse to compute a meaningful global average from direct temperature measurements. And once we do have enough data, combining those temperature measurements requires care.

The thermometric temperature record consists of observations from weather-station thermometers that measure the surface air temperature, and measurements of sea-surface temperature and marine air temperature from ships and automated buoys. Several groups around the world maintain independent analyses of global temperature records, each using different combinations of data and different techniques for calculating the global average. Each group applies corrections to the raw data to compensate for factors that could lead to artificial trends in the calculated temperature. For example, today's weather stations mount thermometers at a standard height somewhat over 1 m above the surface to avoid steep temperature gradients just above the ground. But early stations weren't standardized, which introduces uncertainties. Sea-surface temperature measurements suffer analogous inconsistencies. Early sea-surface temperature measurements were made by hauling buckets of water on board ships and then measuring the water temperature. Obviously, some time elapsed between collecting the water sample and getting it onto the ship, during which time its temperature could change. Both the distance to the ship deck and the material from which the bucket was made could affect that change. Later, ships began measuring sea temperature more directly by taking samples at the engine cooling water inlets. An accurate record of sea-surface temperature requires corrections for these changes.

Urban energy use and albedo changes (e.g., more black pavement, fewer plants) make cities warmer than the surrounding countryside. As cities grow, weather stations located on their outskirts become increasingly urbanized, which means they'll begin to record higher temperatures. Absent any real global temperature change, this **urban heat island effect** would produce a warming trend in temperature records from urban weather stations. Fortunately, we can separate urban stations from rural and marine records; when that's done, it appears that the urban heat island effect accounted for no more than about 0.05°C of warming in the twentieth century—about 10 percent of the total observed warming.

Climatologists consider these and other effects as they merge individual temperature measurements into time series of global average temperatures and associated uncertainties. It's harder to establish unambiguously the actual temperature than it is to calculate **temperature anomalies**, or deviations from an average temperature. That's partly because temperature changes are well correlated over large distances, but absolute temperatures aren't. It's also because two thermometers that might not agree perfectly on the actual temperature will nevertheless give the same response to a given temperature change. So the various time series calculated by different groups all give temperature anomalies relative to some average; for most of the data I use in this chapter, it's the average temperature for the period from 1961 to 1990. A negative anomaly indicates a global temperature below this average, and a positive anomaly indicates a higher temperature.

Figure 14.1 shows the global temperature record from the University of East Anglia's Climatic Research Unit (CRU) in the United Kingdom, one of the several groups around the world that maintains such records. You can explore the CRU's data further in Research Problems 1 through 3, in which you'll find that the Northern and Southern Hemispheres

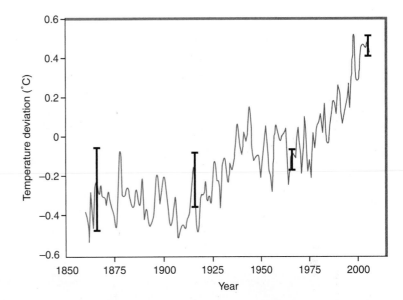

Figure 14.1
Global temperature from 1856 through 2006, using as a baseline the average temperature of the period from 1960 to 1991. The error bars show the approximate uncertainty at four different dates.

show very similar trends, and you'll see the increase in global coverage of the temperature reporting stations. In Research Problem 4 you can compare the CRU data with those of the U.S. National Climatic Data Center, and you'll find close agreement.

The temperatures in Figure 14.1 fluctuate a lot from year to year, but nevertheless the trend is obvious: There's been a significant rise over the period from 1850 to the present. A more detailed look shows distinct trends in different eras: (1) a fairly steady average temperature during the second half of the nineteenth century, (2) an obvious rise beginning around 1910 and ending around 1945, (3) a steady or slightly declining temperature from 1945 to about 1975, and (4) a steep rise beginning in the mid-1970s and continuing to the present. We'll explore the reasons for this pattern in Section 14.3; for now it's enough to note the overall picture. The Intergovernmental Panel on Climate Change (IPCC) estimates the temperature rise of the past one hundred years as being about 0.74°C, with the past 50 years experiencing a rise of 0.13°C per decade. Although such global numbers mask a lot of details, they're useful figures to keep in mind when we talk about longer-term temperature trends or try to project future climate.

How confident are we in the temperatures shown in Figure 14.1? For the first few decades, the estimated uncertainty is on the order of ±0.2°C—a little larger than typical year-to-year fluctuations. By 1950, with an increase in the number of reporting stations and their global distribution (see Research Problem 2), the uncertainty drops to roughly ±0.05°C. I've marked approximate uncertainties at four points in Figure 14.1. Although the uncertainties are significant, they're far smaller than the overall trends, which means we can have considerable confidence that those trends are real features of Earth's recent climate history.

Regional Patterns

The 0.74°C warming over the past one hundred years is a global average. The actual warming is distributed quite unevenly over the planet. In recent decades, particularly, the warming has been greater on land than over the oceans. It's been more pronounced in the Northern Hemisphere than in the Southern Hemisphere. And the Arctic has warmed at about twice the global rate, which is to be expected as the ice-albedo feedback introduced in Chapter 13 exacerbates Arctic warming. In addition, there's less evaporation in the ice-covered Arctic, so more of the incident solar energy goes directly into surface heating. That's because evaporation changes water's state rather than its temperature, so if less of the incident solar energy is being used for evaporation, then more is available for surface warming. Several other processes also contribute to enhanced Arctic warming, all of which make the Arctic an especially sensitive indicator of climate change. Figure 14.2 shows the distribution of warming as of 2005, relative to the standard 1961 to 1990 averages.

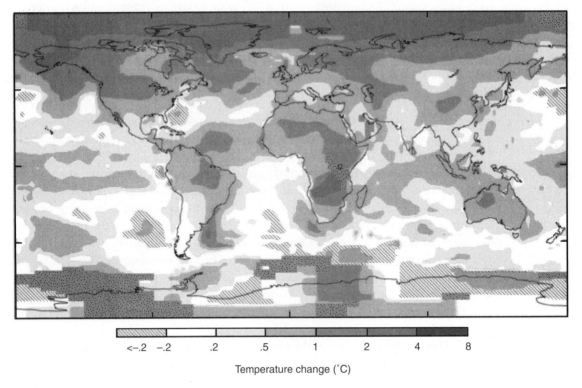

Temperature change (°C)

Figure 14.2
Average temperature change in 2005 relative to the 1961 to 1990 average. Note the greater warming over land, in the northern hemisphere, and especially at high latitudes. White areas show no significant change, hashed areas show cooling, and stippled areas with rectangular outlines have insufficient data.

Other Modern Temperature Records

The temperature record of Figure 14.1 derives from air and water temperatures measured with thermometers at Earth's surface; it's therefore a good indication of surface temperature. We can also get a record of surface temperature over time by boring into the Earth and measuring the temperature as a function of depth. That's because a year's average temperature is reflected in the near-surface soil temperature, and that "signal" of yearly temperature propagates downward as new temperature "information" flows into the Earth from above. Temperature records from such **boreholes** have the advantage of smoothing over the rapid fluctuations seen with surface temperatures, but they have the disadvantage of offering far less global coverage. And it's important to ensure that vegetation over the borehole site hasn't changed; otherwise, changing albedo can affect the rate of solar energy absorption and thus introduce artificial warming or cooling trends. Nevertheless, borehole records confirm Figure 14.1's general picture of an ongoing warming, and an analysis of nearly four hundred boreholes worldwide indicates a warming of 0.5°C during the twentieth century—in good agreement with the surface record's 0.6°C warming during the same period.

Since the 1980s, satellites using infrared detectors have been mapping sea-surface temperature around the globe. Satellite sea-surface temperature measurements generally agree with data from ships and buoys, and satellites have the advantage of covering the entire ocean at high resolution.

Like borehole temperatures, profiles of temperature with ocean depth give additional information about what's happening to global temperature. In the early 2000s researchers combined millions of individual temperature profiles to give a thermal picture of the top ocean layers. These data yield a measure of the total heat content in the upper ocean. This quantity, like temperature itself, has been increasing (Fig. 14.3), an indication that the oceans are absorbing some of the greenhouse-trapped energy. This trend provides

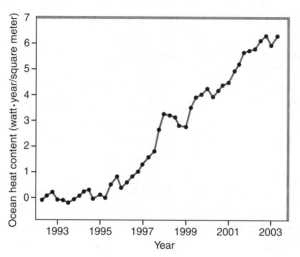

Figure 14.3
Increase in ocean heat content since the early 1990s, measured in watt-years (a unit of energy) per square meter of ocean surface. The increase results from excess energy trapped by anthropogenic greenhouse gases and other forcing agents.

supporting evidence for the 0.85 W/m² imbalance in Earth's energy flows that I described in Chapter 12.

Air temperature is harder to measure, except right near Earth's surface. For decades, balloon-borne **radiosondes** have probed the atmosphere, reporting back temperature as a function of altitude. But these devices are designed to help with immediate weather forecasting, not long-range climate studies. Variations among instrument designs casts doubt on the reliability of radiosonde-derived climate data, and improved shielding of radiosonde thermometers from sunlight has resulted in a phantom cooling trend in the radiosonde record. This is important to note, because some radiosonde analyses claim a recent cooling in the lower troposphere, especially over the tropics, which contrasts with climate model predictions that the lower troposphere should warm somewhat more than the surface. (Recall from Chapter 12 that the greenhouse effect cools the stratosphere, but here we're talking about the lower layers of the troposphere.)

A more consistent approach to air temperatures comes from satellite instruments that measure atmospheric microwave emissions. These **microwave sounding units** (MSUs) look down on vast volumes of atmosphere, and the data from different microwave wavelengths yield temperatures at different levels in the atmosphere. The satellite temperature record is short, extending back only to 1979. Extraction of temperatures from MSU data is complicated, requiring corrections for variations in satellite orbits and for the measurement time in relation to the daily cycle of atmospheric temperatures. Furthermore, the instruments that measure tropospheric temperature also sample the lower stratosphere, "contaminating" the tropospheric record with stratospheric temperatures that behave oppositely. Through the 1990s, the one available analysis of MSU data seemed to support the radiosonde data in suggesting that the troposphere was actually cooling or, if warming, was doing so at a much slower rate than climate models suggested it should be. But by 2005, additional data points from the new century along with careful reanalysis of the data and correction for satellite orbits indicated that satellite air temperature measurements are indeed consistent with climate model predictions of an enhanced warming in the troposphere relative to the surface. Figure 14.4 compares temperature records from the surface, the troposphere, and the stratosphere; the latter two are derived from satellite MSU measurements.

Going Further Back

The global temperature of Figure 14.1 shows a general rise since the record began in the mid-nineteenth century, and an especially steep rise in the past several decades. Borehole records confirm this trend, and more recent satellite-based measures of sea-surface and lower atmosphere temperatures agree that recent decades have seen substantial warming. The rise in ocean heat content is a further indication of ongoing global warming. But is this unusual? Is it related to the increasing anthropogenic climate forcings—especially the rise in greenhouse gas concentrations—that I described in Chapter 13? Or is it a

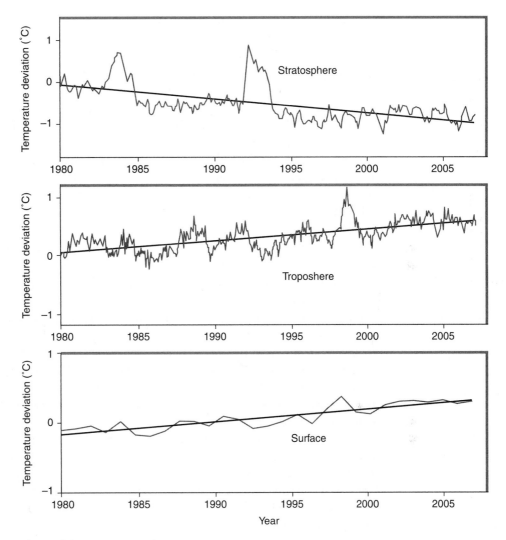

Figure 14.4
Satellite temperature records for the lower stratosphere (top) and lower troposphere (middle) relative to the 1979 to 1999 average temperature. The bottom curve is the corresponding surface temperature record, adapted from Figure 14.1. The straight lines show the general trend. Notice that the stratosphere has cooled while the lower atmosphere and surface have warmed, just what we should expect from Chapter 12's discussion of the greenhouse effect. The spike in stratospheric temperature in the early 1990s corresponds to the Mount Pinatubo volcanic eruption, which cooled the lower atmosphere but warmed the stratosphere.

natural climatic fluctuation? We can answer that question in several ways, the most obvious being to push the temperature record further back in time.

Unfortunately, we don't have enough thermometric measurements to determine an accurate global average temperature before about 1850. Instead, we use **proxies**—other quantities that "stand in" for temperature. To be useful, a proxy has to be widely available and we have to understand quantitatively how its value relates to temperature. Among the frequently used proxies are tree rings, whose width and density reflect climatic conditions on a year-to-year basis; lake sediments, whose thickness and composition reflect the thermal energy available to produce snowmelt streams that carry sediments into lakes; coral reef bleaching, which occurs when coral organisms are stressed by rising temperatures; and isotope ratios (see Box 14.1). Proxies sensitive to precipitation or other quantities may also convey information about regional temperature variations even though they don't translate directly into local temperatures.

Box 14.1   Isotopic Temperature Reconstruction

Chapter 7 introduced isotopes as versions of the same element that differ in the number of neutrons and hence in mass. Although isotopes are chemically similar, the lighter isotopes are more mobile and therefore participate more readily in processes such as evaporation and uptake into living organisms. This is why, as I noted in Chapter 13, plants take up more C-12 than C-13. Analysis of the isotopic composition of atmospheric CO_2 therefore helps us identify the increased carbon in today's atmosphere as coming from ancient plants—that is, from fossil fuels.

Oxygen, like carbon, has several stable isotopes. About 99.8 percent of natural oxygen is the common isotope O-16; nearly all of the remaining 0.2 percent is O-18, and there's a little O-17. Water made with O-16 ($H_2^{16}O$) is lighter and evaporates more readily, making atmospheric water vapor high in O-16 and low in O-18 as compared with the oceans. Conversely, the heavier O-18 water ($H_2^{18}O$) condenses more rapidly. As water that evaporated in the tropics rides the atmospheric circulation toward higher latitudes, more of the $H_2^{18}O$ condenses and precipitates out earlier, at lower latitudes, leaving precipitation in the polar regions depleted of the heavy isotope. How depleted depends on the prevailing temperature; the cooler the climate, the sooner the heavier water precipitates out, and the less O-18 ends up in polar ice. I've skipped some complications here that involve the distance the water vapor is transported before reaching the polar regions, and the changes in elevation at which precipitation reaches the ground. Nevertheless, the general conclusion holds: When the climate is cooler, high-latitude precipitation contains less O-18.

The great ice sheets of Greenland and Antarctica contain hundreds of thousands of years' accumulation of snow, which is compacted into ice by the weight of the overlying layers. In Greenland, where the deepest ice is over 125,000 years old, distinct bands mark each season's snowfall in the more recent ice and thus allow scientists to date the ice layers year by year. Antarctic ice goes back as far as 800,000 years, but it generally lacks the yearly banding, and thus the dating is not as finely resolved. In either case, though, measurement of the O-18 to O-16 ratio provides an estimate of temperature versus time. The relative amount of O-18 is specified by giving the fractional deviation of the $^{18}O/^{16}O$ ratio from a standard value, expressed in parts per thousand ($^0/_{00}$). This quantity, called $\delta^{18}O$, can be negative or positive depending on how a given $^{18}O/^{16}O$ ratio compares with the standard. The deep ocean has a $\delta^{18}O$ value of about zero, whereas warm tropical waters have a $\delta^{18}O$ value of around $1^0/_{00}$. Antarctic ice, in contrast, can have a $\delta^{18}O$ value as low as $-55^0/_{00}$, reflecting the preferential evaporation of O-16 and subsequent precipitation of O-18 in water that traveled as vapor from the low latitudes to the Antarctic before it became part of the ice sheets.

Oxygen isotope ratios provide information about another important climatic factor, namely the total amount of water locked up as ice. This is evident, again, because the water that evaporates and eventually becomes ice is enriched in O-16, leaving behind seawater with higher $\delta^{18}O$ values. Measurement of $\delta^{18}O$ in seawater therefore provides a measure of the total ice volume. Extracting $\delta^{18}O$ values from sediments containing calcium carbonate ($CaCO_3$) derived from shells of marine organisms provides a record of the ice volume over time and therefore, indirectly, of sea-level variations.

Hydrogen isotope ratios provide an alternative and sometimes complementary approach to climate reconstruction. The method is similar: Water containing the heavy hydrogen isotope deuterium (one proton and one neutron) evaporates less readily and condenses more readily than normal water molecules whose hydrogen atoms contain only a single proton. Therefore, measurement of the ratio of deuterium to ordinary hydrogen in ice cores provides a measure of temperature at the time the water fell as precipitation.

Figure 14.5 compares ten different proxy-based studies of temperature change over the past millennium. Also shown is the instrument record of Figure 14.1 for the twentieth century and beyond. It takes only a quick glance to spot the general trend in all the studies—a gradual decline for the first nine hundred years, followed by a sharp upturn from 1900 to the present. The different studies use a variety of statistical techniques to reconstruct historical temperatures from different sets of proxies. As a result, there's considerable variation among these temperature reconstructions, especially prior to the twentieth century. Nevertheless, all concur in suggesting that the warming of the last hundred years is unprecedented in both its magnitude and rapidity, at least on a millennial time scale.

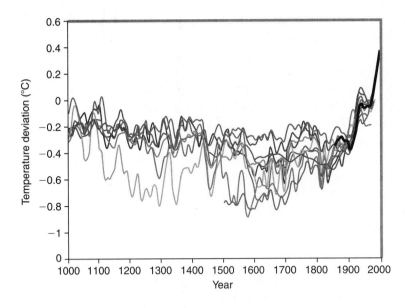

Figure 14.5
Results of ten different reconstructions of temperature for the past millennium (gray curves). The instrumental record, dating to the mid-nineteenth century, is shown in black. Deviations are relative to the 1961 to 1990 average temperature.

Going Even Further Back

The warming of the twentieth and twenty-first centuries appears unusual in the context of the millennial reconstructions shown in Figure 14.5. Does it still look unusual if we go further back? Ice-core data provide an answer. Figure 14.6 shows a temperature reconstruction of the past 160,000 years made from hydrogen isotope ratios in Antarctic ice cores. This particular data set goes back some 420,000 years; a similar core completed in 2003 extends nearly 800,000 years.

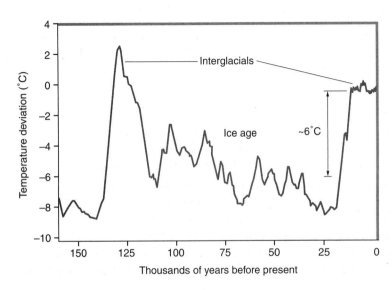

Figure 14.6
The ice-core temperature record from Vostok, Antarctica, shows a pattern of warm interglacial periods separated by ice ages lasting roughly 100,000 years. This pattern persists for nearly a million years into the past. The temperature deviation is relative to the present, but the plot shows a running average over twenty-five data points, so the final point plotted is not quite at the present. Note that only about 6°C separates the modern climate from the average ice-age temperature.

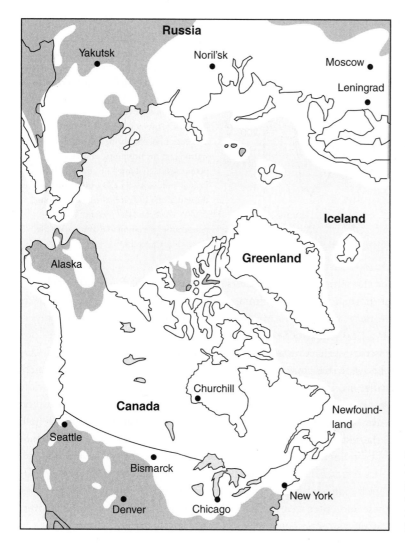

Figure 14.7

Maximum extent of northern hemisphere glaciation during the most recent ice age. Ice sheets were typically 1.5 to 3 km (about 2 miles) thick over much of the region shown in white.

There's a lot of variation in Figure 14.6, but there's also a hint of a pattern. At the modern end of the ice-core record (which is about 2,000 years ago), we're at a much higher temperature than the average. Roughly 130,000 years ago, Earth also saw a short (10,000 years or so) spell of warmer than normal temperature. Between these brief, warm **interglacials** is a much longer **ice age**. During the ice age, Earth's climate was profoundly different, with 1.5 to 3 km of ice covering a good part of North America and northern Europe (Fig. 14.7).

What causes this pattern of long, cold ice ages punctuated by briefer warm spells? The answer lies in variations in the distribution of solar radiation that result from periodic changes in Earth's orbit and the tilt of its axis. However, these effects alone are too

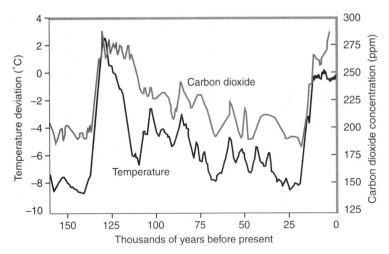

Figure 14.8
Temperature and atmospheric CO_2 show a tight correlation through the ice-age cycles, as evident in this record from the Vostok ice core in Antarctica. However, the relationship is not a simple cause and effect, but a positive feedback between temperature and CO_2 concentrations that brings about the rapid warming at the beginning of each interglacial period. Unlike Figure 13.8, this graph doesn't include industrial-era CO_2; both plots stop at the end of the CO_2 ice-core record, about 2,300 years before the present. The temperature deviation and average are the same as in Figure 14.6.

small to account for the climate variations shown in Figure 14.6. Rather, it's believed that small temperature changes resulting from orbital variations trigger feedback effects that enhance the orbitally induced heating and cooling trends. Combining the temperature record of Figure 14.6 and the CO_2 concentration from Figure 13.8b illustrates this point. Figure 14.8 is the result: a remarkable correlation between temperature and CO_2 concentration. But be aware that this isn't a simple cause-and-effect relationship in which, for example, variations in CO_2 concentration cause a similar pattern of temperature changes. Rather, you can imagine a small temperature increase associated with orbital changes. This, in turn, increases sea-surface temperatures; since warmer water can hold less dissolved CO_2, the result is a flow of CO_2 into the atmosphere. The additional atmospheric CO_2 creates a stronger greenhouse effect and exacerbates the warming. On land, warming temperatures increase the respiration rate of soil microbes, returning more CO_2 to the atmosphere and again enhancing the warming. Although the details aren't yet fully understood, these and other feedbacks are ultimately responsible for both the overall temperature and CO_2 patterns shown in Figure 14.8, and for the correlation that's so dramatically evident.

It's possible to push climate reconstructions back millions and even billions of years by examining oxygen isotope data, the shape of fossil leaves, and other proxies. But more interesting here is the question of natural climate variability in recent prehistoric times. As Figure 14.6 shows, there's been considerable variability both in the overall ice-age cycles and within the ice ages themselves. A closer look, using data from Greenland ice cores, reveals some especially rapid variations in the roughly 20,000 years since the depths of the last ice age (Fig. 14.9). A particularly dramatic example is the so-called Younger Dryas event, which swung the Arctic back into ice-age conditions for some 1,500 years starting around 12,000 years ago. This cool period ended abruptly with a 7°C warming of Greenland in less than a century. The Younger Dryas and similar rapid climate fluctuations are nonlinear effects believed to result from sudden changes in ocean circulation

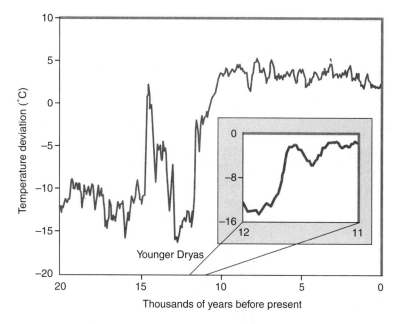

Figure 14.9
Detailed record of Greenland ice-core temperatures over the past 20,000 years, relative to the average for the period. The Younger Dryas event marked a sudden, brief return to ice-age conditions. The inset depicts the period from 12,000 to 11,000 years ago, showing the extremely rapid temperature rise at the end of the Younger Dryas period. The graph indicates that climate has been much more stable over the past 10,000 years.

that occur when an influx of fresh water from melting ice alters the salinity and hence the density of ocean water. Although the Younger Dryas fluctuation may seem even more dramatic than the late-twentieth-century warming that's so obvious in Figure 14.5, I caution you against inferring that the latter must also be a natural fluctuation. The large, rapid temperature rise at the end of the Younger Dryas is clearly evident only in Greenland—an arctic region that tends to exaggerate global trends. Although the Younger Dryas event appears in other ice cores, pollen records, and ocean sediments across the globe, its climate signature is much less obvious in these other areas, suggesting an event that was concentrated largely at higher latitudes in the Northern Hemisphere.

Figure 14.9 shows that the climate of roughly the past 10,000 years has been substantially more stable than that of the last ice age and the beginning of the current interglacial warm period. This is not to say that there haven't been significant variations, but such variations have been gradual and modest. This stable period coincides with the development of agriculture, and many would argue that a stable climate was essential for this major advance in human civilization. Although most climatologists accept predominantly natural causes for climate behavior until the past half-century or so, some argue that humans, as early as 8,000 years ago, began modifying the climate substantially in ways that enhanced stability and forestalled the slide into the next ice age. This anthropogenic modification purportedly results from the greenhouse gases (CO_2 and methane) generated by land clearing and agricultural practices. Whether natural or anthropogenic, though, the climate of recent millennia has been benignly constant compared with that of earlier times.

This picture of earlier rapid climate fluctuations followed by the more stable climate of recent times suggests two points: First, the rapid increase in global temperature during the late twentieth and early twenty-first centuries is abnormal in the present climatic context. Second, slight changes in the climate system can result in highly unstable conditions, with swings of several degrees in average temperature occurring within centuries or even decades over large regions of the globe. Such changes would not be particularly beneficial to a highly developed civilization on an overcrowded, underfed planet.

14.2 Other Climatic Changes

Temperature isn't the only measure of global climate change. A variety of other indicators suggest a general global warming trend accompanied by more subtle changes in temperature and weather patterns. All reinforce the picture of a climate that's undergoing abnormally rapid change.

Ice and Snow Cover

In 1850 the glaciers of Montana's Glacier National Park covered some 100 km^2. By 1998 that area was down to 27 km^2 and shrinking fast. At the current rate of glacier retreat, the park's name will be meaningless in several decades' time as permanent glaciers disappear altogether.

Mountain glaciers are shrinking across the globe, with many exhibiting dramatic reductions like the one shown in Figure 14.10. A few glaciers are growing, but that's because glacial behavior is governed by a balance between snowfall and melting. Snowfall depends on evaporation that occurs elsewhere, and evaporation, as we've seen, can increase with global warming. But on average, glaciers worldwide are in retreat, losing both area

Figure 14.10
The South Cascade Glacier in Washington State has shrunk
dramatically as seen in these photos taken in 1928 (left) and 2000 (right).

and thickness. As they melt, mountain glaciers leave behind an altered landscape with lower albedo and thus increased solar energy absorption. They also increase freshwater runoff that ends up in the oceans, where it contributes to rising sea levels.

Polar ice is also melting, both the floating sea ice and the large continental ice sheets of Greenland and Antarctica. On average, Arctic temperatures have increased in recent decades at about twice the global average, as shown in Figure 14.2. The picture for Antarctica is less clear, because the Antarctic climate has behaved differently in different regions and seasons. Interior Antarctica, for example, has cooled somewhat in recent decades, while the Antarctic coast has generally warmed. The Antarctic summer has seen little change, while autumn has cooled. Winter and spring exhibit complex regional patterns of warming and cooling.

In both the Arctic and Antarctic, however, the extent of ice cover is shrinking. In the Antarctic this is due primarily to excessive warming at the coastal margins; in the Arctic it occurs throughout the region. The summertime coverage of Arctic sea ice has declined some 25 percent since 1979, with more modest but still significant declines in other seasons (Fig. 14.11). It now appears that the Arctic Ocean may be ice-free in summer by the end of the twenty-first century. Sea-ice area is obvious when viewed from above, but thickness isn't. However, declassified data from U.S. Navy nuclear submarines operating under the polar ice contain information on ice thickness, and these data, along with later oceanographic studies, suggest that Arctic ice thickness has declined by as much as 40 percent in recent decades.

Loss of sea ice affects the local ecology and in particular endangers wildlife such as polar bears and seals; this in turn affects indigenous peoples of the Arctic. Melting sea ice floods the ocean with freshwater whose lower density can alter large-scale ocean circulation. This is an example of a nonlinear climate effect I introduced in Chapter 13. When the area of sea ice is reduced, its high albedo diminishes, leaving dark seawater to absorb more solar radiation. The resulting ice-albedo feedback is one of the reasons for increased warming in the Arctic. Unlike melting land ice, however, one thing melting sea ice doesn't do is raise sea level significantly. That's because floating ice displaces the same mass of ocean water whether it's in a solid or liquid state—although there is a tiny effect because of the density difference between the freshwater that comprises ice and the salt water that remains in the oceans.

Seasonal Temperature Changes

Figure 14.2 shows that global temperature increases aren't spread evenly around the globe; for example, they're greater on land and at high latitudes. They're also not evenly distributed in time. Although the average Arctic warming over the past hundred years is about twice the 0.74°C global average for that period, Arctic winter temperatures have increased as much as 3°C to 4°C. A warming climate also affects the frequency of extreme temperature excursions, such as heat waves or extreme cold. Because extremes are by nature rare, it's more difficult to obtain firm statistical evidence. However, several studies point to a decrease in the number of extremely cold days over some regions,

(a)

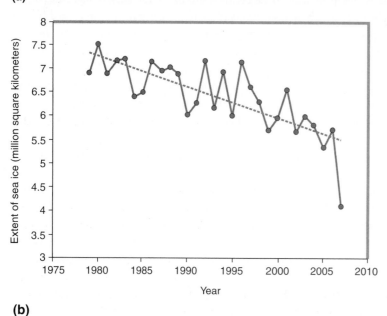

(b)

Figure 14.11
Decline in the area covered by Arctic sea ice, based on satellite measurements taken annually in September. (a) Composite satellite images of the ice cover in September 1979 and 2005. (b) A plot of the decline over time. The straight line indicates an average decline in ice area of about 9 percent per decade. Decline accelerated dramatically in 2007.

thereby lengthening the growing season (the frost-free period) by as much as several weeks. There's also evidence for increases in the number of extremely hot days. We'll take a closer look at the statistics of such events in Chapter 15.

Weather Changes

Rising temperatures should affect Earth's weather in several ways. Increasing sea-surface temperatures mean more evaporation, cloud formation, and precipitation. There's also more energy available to power tropical storms. Large-scale patterns such as El Niño—a warming in the eastern Pacific that affects global weather—could change in frequency

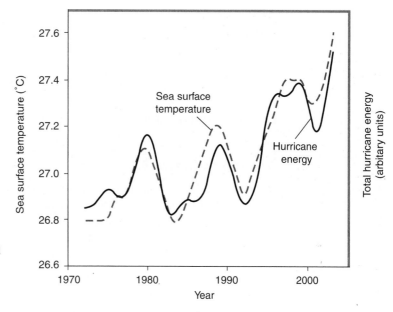

Figure 14.12

Total energy dissipated per year in all North Atlantic hurricanes (solid curve) has increased substantially in recent decades, in step with rising sea-surface temperatures (dashed curve).

and intensity. Weather's intrinsic variability in both time and space makes it difficult to confirm such changes with high levels of statistical confidence. Nevertheless, there's strong evidence that precipitation at mid-latitudes increased some 5 to 10 percent during the twentieth century, with lesser increases in the tropics and subtropics. A related change is that more precipitation falls in shorter, more intense events. Atmospheric water vapor and clouds have increased slightly, although we're less confident of those measures. Studies published during the disastrous 2005 North Atlantic hurricane season show increases in the number of category 4 and 5 hurricanes and in measures of overall hurricane energy. The latter scales as the cube of the wind speed, for reasons I discussed in Chapter 10 in connection with wind energy, and is a rough measure of the total destructive potential of a season's worth of hurricanes. Figure 14.12 shows that the rise in hurricane energy correlates closely with rising sea-surface temperatures. All of these weather changes are broadly consistent with increasing global temperature.

Species Ranges

As Earth warms, you might expect temperature-sensitive species of plants and animals to migrate so as to remain in their optimum climatic zone. However, climate is one of several factors that affect species' geographical ranges, so studies of individual species don't give conclusive evidence for global warming. But a method known as *meta-analysis*—a statistical survey of many individual studies—can reveal significant trends that indicate a biological response to global warming.

One such meta-analysis examined hundreds of different plant and animal studies, and the conclusion was that species were moving northward at an average of 6.1 km per decade, or upward in altitude at 6.1 m per decade. Significant springtime events in the lives of species—breeding, nesting, flowering, budburst, ending migratory journeys—were happening earlier in some 62 percent of species studied; in one subset of 172 species, the spring events were advancing at the rate of 2.3 days per decade. Overall, some 87 percent of species studied showed changes in timing or range consistent with global warming; only 13 percent showed contradictory trends. So it appears that plants and animals know what our thermometers are telling us: We live on a warming planet.

14.3   Are We to Blame?

It's clear that Earth's climate is changing, with a temperature rise that's been especially pronounced in recent decades. But is the observed change a result of human activities, especially our greenhouse emissions? Or is it a natural fluctuation, driven by external factors such as solar variability and volcanic eruptions? Or does recent climate change result from natural internal variations of the complex physical, chemical, and biological system that determines Earth's climate?

In the early 1990s, when concern about climate change first became widespread, the "signal" of anthropogenic effects hadn't unambiguously emerged from the "noise" of natural climate variability. But by 1995 the scientists and policymakers writing the IPCC's Second Assessment Report cautiously noted "a discernible human influence on climate." The IPCC's Third Assessment Report, published in 2001, stated more boldly, "There is newer and stronger evidence that most of the warming observed over the past 50 years is attributable to human activities." When asked for an independent assessment of the IPCC's findings, the U.S. National Academy of Sciences concluded that "greenhouse gases are accumulating in Earth's atmosphere as a result of human activities, causing surface air temperature and subsurface ocean temperature to rise." In a 2004 report to the U.S. Congress, the U.S. Global Change Research Program conceded that "North American temperature changes from 1950 to 1999 were unlikely to be due only to natural climate variations." Finally, the IPCC's Fourth Assessment Report, released in 2007, declared even more strongly, "Warming of the climate system is unequivocal" and "Most of the observed increase in globally averaged temperature since the mid-twentieth century is *very likely* due to the observed increase in anthropogenic greenhouse gas concentrations." By "very likely," IPCC scientists mean a probability between 90 and 95 percent.

So what is the evidence that today's climate change results largely from human activities? First, there's the basic physics of the greenhouse effect, confirmed by the natural "experiments" of Mars and Venus: We know that greenhouse gas concentrations have an impact on climate. In particular, increasing CO_2 in itself should result in increasing temperature. In that context, take another look at Figure 13.8b, which shows that today's atmospheric CO_2 concentration is nearly 40 percent higher than at any time during the past half-million years. And we know, from the evidence described in Chapter 13, that most of the excess carbon comes from fossil fuel combustion. So we humans have taken Earth's atmosphere into a regime that our planet probably hasn't seen for millions of years. Now look back at Figure 14.8. Although the interplay between CO_2 and temperature here is complex and not necessarily predictive, their obvious correlation suggests that we might expect a significant climatic response to the industrial-era spike in fossil-derived atmospheric CO_2.

Comparison of Figures 13.7 and 14.1 shows somewhat similar rises in temperature and CO_2 over the past 150 years. Again, this isn't enough to establish cause and effect. However, the coincidence is certainly noteworthy, especially given that these are quantities basic physics says should be related and because both increases are probably unprecedented, at least in recent millennia. To explore this relationship further, we can ask how the recent global temperature rise correlates with several known forcing agents that we expect to influence climate. This, in fact, was done for the most recent four centuries of one millennial temperature reconstruction shown in Figure 14.5. The authors of that study checked their results for correlations with variations in solar radiation, with the so-called dust veil index that measures volcanic particulates in the atmosphere, and with atmospheric CO_2 concentration taken as a proxy for all anthropogenic greenhouse gases. The results show only a modest correlation with volcanic dust over the entire period. The volcanic correlation is negative, because dust exerts a cooling effect, and it was strongest in the early 1800s—a time of exceptionally active volcanism. For the majority of the four-century interval, the global temperature correlates most strongly with variations in solar radiation. Especially significant is the so-called Maunder minimum, a time when solar activity diminished substantially and northern Europe experienced a period of unusually cool conditions sometimes called the "little ice age." The solar influence continued into the twentieth century, and it probably explains much of the rise that Figure 14.1 shows occurred in the early part of the century. But by the second half of the twentieth century, anthropogenic greenhouse gases emerge as the factor most strongly correlated with global temperature. The correlations associated with millennial temperature reconstructions are noteworthy because they involve only observational data; there are no hidden model assumptions and no need to understand complex feedback mechanisms or details of cloud physics. If anything, the correlation approach is a bit conservative because it compares temperature and forcing agents contemporaneously, without allowing for "inertia" associated with, for example, the slow warming of the oceans that should delay the temperature response relative to the forcing. So the late-twentieth-century correlation between temperature and CO_2 may be even stronger than these studies suggest.

Additional evidence for a human influence on climate comes from computer models designed to project future climate. I'll have more to say about these models in Chapter 15, but for now I'll describe one way of testing their validity, which is to start at some time in the past, run the model forward, and see if it reproduces today's climate. A variant on this test lets modelers assess the factors influencing climate. By running their models without volcanism, or without solar variations, or without greenhouse gases and other anthropogenic forcings, modelers can explore the role of each factor in determining present-day climate. The results of several distinct studies all point to the same conclusions: that solar variability and volcanism account for much of the climate change seen into the early twentieth century, but that only with anthropogenic forcings can models reproduce the climate of recent decades. Figure 14.13 shows the results of one such study.

I noted at the beginning of this section that climate change might result not from external forcings, either anthropogenic or natural, but from natural internal variability of the complex climate system. Once again, though, computer models help rule out the possibility that internal variability could explain recent climate change. By "turning off" all changes in external forcings, climate modelers can focus on internal variability alone. Thousand-year runs with different models exhibit internal variability that, although significant, never shows trends as large as the observed warming of recent decades.

A third piece of evidence is more subtle. In Chapter 13 I noted that different forcing mechanisms can have different effects on climate, even if they result in the same numerical upset in Earth's energy balance. In particular, some forcings have different geographical and temporal distributions, and their influence varies with region and season. So there are unique patterns in the climate response to different forcings, and these provide "fingerprints" that climatologists can use to identify the causes of climate change. Among the fingerprint evidence for anthropogenic climate change are regional patterns of global temperature change, including increased warming over land and regional variations in precipitation. The cooling of the stratosphere shown in Figure 14.4 is another fingerprint of anthropogenic greenhouse warming. Recall from the discussion surrounding Equation 12.3 that the greenhouse effect requires a temperature difference between the surface and upper atmosphere. Figure 14.4 shows that this difference is increasing and thus is evidence that recent warming is due to a strengthening greenhouse effect. Coupled with the anthropogenic greenhouse gas increases documented in Chapter 13, this is solid evidence for anthropogenic warming.

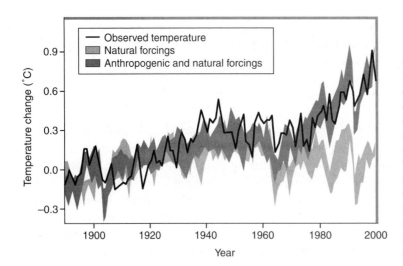

Figure 14.13
Results of climate model experiments to determine the roles of natural and anthropogenic forcings on present-day climate. Only the models that include anthropogenic effects can account for the temperature rise observed in recent decades. The two shaded bands show the results from an ensemble of four model runs conducted with and without anthropogenic forcings factored in. Natural forcings are volcanism and solar variability; anthropogenic forcings include greenhouse gases, sulfate aerosols, and ozone.

Thus climatologists have a host of evidence—statistical inferences, past climate reconstructions, computer model experiments, and pattern-based fingerprint studies—that all point to a substantial human influence on climate since the mid- to late twentieth century. That influence, as we saw as early as Chapter 2 and again in Chapter 13, results primarily from our species' prodigious consumption of fossil fuels and the resulting atmospheric greenhouse emissions. In the next chapter we'll see where the human impact on the atmosphere is likely to lead, and in the final chapter we'll address the steps we might take to moderate future climate change.

Chapter 14   Chapter Review

BIG IDEAS

14.1   Thermometers provide a global temperature record back to the mid-1800s, and their accumulated data show rapid warming in recent decades. Proxy-based climate reconstructions suggest that this warming is unprecedented over the past thousand years. Ice-core data extending back nearly 1 million years show a pattern of long, cool periods—the **ice ages**—alternating with briefer, warmer **interglacials**.

14.2   Other indicators of global warming include melting ice and decreased snow cover, changes in weather patterns and storm intensities, and poleward movement of plant and animal species.

14.3   Solar variability and volcanic activity account for much of the climate variation observed until the mid-twentieth century. But only by including anthropogenic effects, especially greenhouse emissions, can climate scientists account for the unprecedented warming of the late twentieth and early twenty-first centuries.

TERMS TO KNOW

borehole (p. 407)	proxy (p. 410)
ice age (p. 413)	radiosonde (p. 408)
interglacial (p. 413)	temperature anomalies (p. 404)
microwave sounding unit (p. 408)	urban heat island effect (p. 404)

GETTING QUANTITATIVE

Global warming over the past one hundred years: ~0.74 °C

Temperature difference between present day and ice age: ~6°C

Poleward migration of species ranges, average: ~6 km per decade

QUESTIONS

1. By roughly how much did global temperature increase during the twentieth century?

2. How do oxygen isotopes help scientists extract temperature information from ice cores?

3. What is the ultimate origin of the periodic climatic changes—ice ages and warmer inter-glacial periods—that are so obvious in Figure 14.6?

4. A friend argues that climate naturally varies, so the temperature increase of recent decades might well be a natural fluctuation. How might you counter this argument?

5. Why is climate change more rapid in the Arctic than elsewhere?

6. How does Figure 14.4 provide evidence that recent global warming results, specifically, from increasing greenhouse gas concentrations?

RESEARCH PROBLEMS

1. Go to the University of East Anglia Climatic Research Unit's web page at www.cru.uea.ac.uk/cru/data/temperature and locate the Northern Hemisphere (NH) and Southern Hemisphere (SH) temperature data from the HadCRUT3v dataset. Copy these into a spreadsheet and put each entry in its own column (in Excel use the Text to Columns function under the Data menu). Plot the annual average temperature versus time for both of these datasets, and compare.

2. Look at the CRU's global temperature dataset (GL, from the HadCRUT3v dataset; see Research Problem 1 for instructions) and note that the second line for each year gives the percentage of global coverage provided by the reporting stations. How has that increased from the early years of the dataset to the present?

3. Use the CRU's global temperature dataset (see Research Problem 2) to plot an updated version of Figure 14.1 using the latest available data. Does the warming trend continue, or does the global temperature decline? Or isn't there an obvious trend?

4. Go to the U.S. National Climatic Data Center's global temperature data site at www.ncdc.noaa.gov/oa/climate/research/anomalies/anomalies.html#anomalies and find the link entitled "The Annual Global (land and ocean combined) Anomalies (degrees C)." Copy the dataset into a spreadsheet and put the year and temperature anomaly in different columns (in Excel use the Text to Columns function under the Data menu). Add the CRU's global temperature record (see Research Problem 2) to your spreadsheet, then plot both global temperature records on the same graph and compare.

5. Figure 14.11b shows that Arctic ice cover was exceptionally low in the years 2002 to 2007, even compared with the declining average trend. Has this accelerated loss of ice cover continued? To find out, go to the National Snow and Ice Data Center's web site

(http://nsidc.org/) and find the data to update Figure 14.11b to the present day. New data are released each September, at the time of minimum ice cover.

6. Go to the Carbon Dioxide Information Analysis Center's web site (http://cdiac.esd.ornl.gov) and find the Vostok temperature data used in Figure 14.6. Plot the entire dataset. How many ice-age cycles does it show? Are the temperatures and durations of the warmer inter-glacials roughly the same, or do they vary significantly?

7. A newer ice core from the European Project for Ice Coring in Antarctica (EPICA) goes back 740,000 years. You can find the EPICA data on the Carbon Dioxide Information Analysis Center's web site (see Research Problem 6), listed under "740,000-year Deuterium Record in an Ice Core from Dome C, Antarctica." Does the earlier data show the same cyclic pattern of ice ages and interglacials as the Vostok data of Figure 14.6 and of the previous problem?

8. The U.S. Historical Climatology Network (HCN) provides monthly temperature and pre-cipitation data for more than 1,000 stations across the United States, with many station records going back over 100 years. If you're a U.S. resident or a student in the United States, go to the HCN monthly data site at http://cdiac.ornl.gov/epubs/ndp/ushcn/ monthly.html and find the station nearest you. The HCN site can prepare a plot for you of monthly data over time. Fit a straight line to the plot and estimate your station's tempera-ture rise over the twentieth century. Compare with the IPCC's estimate of a 0.6°C global average temperature rise over the same period.

Chapter 15

FUTURE CLIMATES

Chapter 13 described how humans have altered the global atmosphere, particularly through a substantial increase in atmospheric CO_2 caused largely by our fossil fuel consumption. Chapter 14 documented a century-long rise in global temperature, and recounted climatologists' arguments as to why humans are behind at least the unusual warming of the most recent half-century. So can we predict accurately the global climate response to our ongoing greenhouse gas emissions and other climate forcings?

Predict *accurately*? No, and for two main reasons. First, we don't know enough about every detail of the climate system—the myriad feedbacks, the surprising nonlinear effects, the behavior of clouds, and the complex interactions of land, sea, air, ice, and biosphere—to make precise predictions of what will happen when and where. Second, we can't predict future human behavior, yet humans are now major players in the global climate system. But despite our imperfect knowledge, we do know a great deal about the physical, chemical, and biological processes that govern climate. And we can make assumptions about human behavior that, together with our knowledge of climate science, let us project scenarios of how Earth's climate might evolve in the future.

Although the simple zero-dimensional energy-balance model I've been using in previous chapters can give rough estimates of the global average temperature change associated with a given forcing, it can't do justice to the complexity of the climate system or the geographical diversity of our planet. Nor can it account for the variations in precipitation, storm intensity, melting ice and snow, altered vegetation, and other changes resulting from shifts in global climate. To project global changes with higher confidence, or to project any regional changes at all, we need sophisticated, multidimensional models that account for the myriad interactions affecting climate. Then we need powerful computers to run those models.

15.1 Modeling Climate

Figures 12.3 and 12.4 hint at a possible next step in modeling sophistication beyond our zero-dimensional model: Treat the climate system as if it consisted of two "boxes"— surface and atmosphere—and describe the flows of energy between them. Figure 12.3 is, in fact, part of such a model, complete with equations describing some of the energy flows. Figure 15.1 is a full **two-box model** that includes all the processes shown in Figure 12.4's diagram of the surface–atmosphere energy flows. Describing each arrow with a numerical value or an equation involving surface and/or air temperature would then let us write separate equations for energy balance in both the surface and atmosphere. Solving these two equations simultaneously would give the model's values for the temperatures of surface and atmosphere. You can explore this two-box model in Research Problems 1 and 2.

We know that temperature varies with altitude, so an obvious next step is to represent the atmosphere with many boxes. This gives a **one-dimensional model**, so called because quantities vary with position in a single dimension. A one-dimensional model could, for example, explore the simultaneous warming of the lower atmosphere and cooling of the upper atmosphere in response to increasing quantities of greenhouse gases.

Climate also varies with geographical location, and **two-dimensional models** account typically for different latitudes as well as altitudes (Fig. 15.2). These models can simulate such things as seasonal changes and the atmospheric circulations that transport energy from the tropics toward the poles. For the latter, a model needs more than just energy balance at each box; modelers also have to account for flows of matter and momentum. Matter includes the air itself, along with the all-important moisture that carries latent energy and is responsible for cloud formation and precipitation; a more advanced model might also track separately the movement of greenhouse gases and other forcing agents. Models also follow momentum—the "quantity of motion" first introduced by Isaac

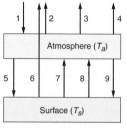

Figure 15.1

A two-box climate model featuring energy exchanges between Earth's surface, atmosphere, Sun, and space. The model is a simplified version of Figure 12.4.

1 = Incident solar energy
2 = Sunlight reflected from atmosphere
3 = Infrared transmitted through atmosphere to space
4 = Infrared emitted by atmosphere to space
5 = Sunlight absorbed by surface
6 = Sunlight reflected from surface
7 = Convection and evapotranspiration
8 = Infrared emitted upward from surface
9 = Infrared emitted downward from atmosphere

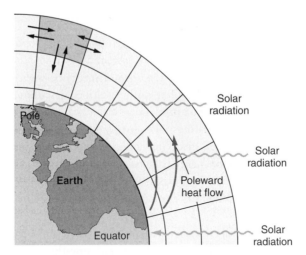

Figure 15.2
A two-dimensional climate model that accounts for the varying intensity of solar radiation from equator to poles. One of the two-dimensional grid boxes is shaded in darker gray, with arrows representing the interchanges of matter and energy between the box and its neighbors.

Newton—whose flows describe the forces one box of air exerts on its neighbors, as well as the frictional forces the ground exerts on the air.

Three-dimensional models have it all. They divide the Earth–atmosphere system into boxes whose positions vary with latitude, longitude, and altitude (Fig. 15.3). Three-dimensional models can therefore account for land and water surfaces; for the actual configurations of the continents; for snow, ice, and vegetation cover and their associated albedos; for atmospheric circulation in both latitudinal and longitudinal directions; for the transport of moisture from the oceans to the continents; and for a host of other realistic processes that affect climate. The large-scale models that include atmospheric circulation are called **GCMs**—originally standing for **general circulation model** but now taken also to mean **global climate model**.

Physical processes in the oceans are sufficiently different from those on land and in the atmosphere that they're usually handled with separate models. Ocean and atmosphere models are then coupled, with the output of each model feeding the other to describe processes at the air–water interface—processes such as evaporation; CO_2 dissolving or coming out of solution; wind and waves producing airborne salts that become cloud condensation nuclei; and ice formation and melting. A model that couples atmosphere and ocean is an **AOGCM**—an **atmosphere–ocean general circulation model**. AOGCMs are our largest, most comprehensive climate models. In some cases, other specialized models, such as carbon-cycle models, are coupled into GCMs as well.

We're now up to three-dimensional models. But there's another "dimension" to the physical universe, namely time. The simplest models, like our zero-dimensional energy-balance model of Chapters 12 and 13, or the two-box model of Figure 15.1 and Research Problem 1, are **equilibrium models**. For these, values are assigned to the various energy flows, and the models calculate one or more temperatures. These are the temperatures that would prevail only after the climate system reaches equilibrium at the conditions

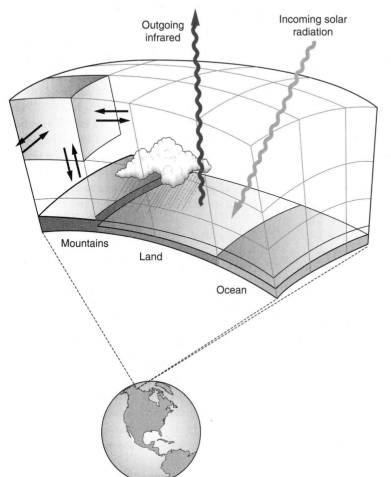

Outgoing
infrared

Incoming solar
radiation

Mountains

Land

Ocean

Figure 15.3
A three-dimensional climate model includes land and ocean surfaces. One of the three-dimensional grid boxes is shaded in darker gray, with arrows representing the interchanges of matter and energy between the box and its neighbors.

given in the model. Equilibrium models can't tell how long it would take to reach equilibrium or how the climate would respond to, say, varying concentrations of greenhouse gases over time. More sophisticated models, in contrast, are **time-dependent models**. Not only do they consider flows of energy and matter among a great many individual boxes, but they also advance the model system forward in time, solving the entire many-box model over and over as time progresses and model conditions—temperatures, greenhouse gas concentrations, ice and vegetation cover, and other quantities—change. Thus, time-dependent models not only project future climate, but they also show the path from present to future. As such, they let climate modelers experiment with different scenarios for future greenhouse emissions, solar variability, land-use changes, and other factors that might alter Earth's climate.

Model Resolution

The more boxes—also called **grid cells**—a model has, the finer its spatial **resolution**. A high-resolution model can simulate regional variations in climate and is better able to describe small-scale features of oceanic and atmospheric circulation. But high resolution comes at a price. Dozens of processes go on in each cell—absorption of solar and infrared radiation, inflows and outflows of matter and energy, condensation and evaporation, chemical reactions, photosynthesis and respiration, and many more. It takes at least one equation to describe each process, and terms in the equations link adjacent boxes. So a model consists, mathematically, of an enormous number of equations to be solved simultaneously, over and over again with each successive time step. As a result, the most elaborate climate models tax even the fastest computers. A simulation covering several centuries of climate can take months of computer time to run.

Box 15.1   The Community Climate System Model

An example of a large-scale coupled climate model is the Community Climate System Model (CCSM) of the U.S. National Center for Atmospheric Research (NCAR). The CCSM has separate model components for land, atmosphere, ocean, and ice (Fig. 15.4). In its 2004 incarnation, the highest-resolution version of this model divides the ocean and ice systems into cells spanning approximately 1° of longitude and as little as 0.3° latitude in the tropics. The model's ocean component consists of forty layers, ranging from a thickness of 10 m at the surface to 250 m at depth. All of this together adds up to some 2 million boxes for the ice–ocean systems alone. The model's surface and atmosphere have a grid resolution approximately 1.4° in latitude and longitude—giving a grid cell size on the order of 160 km, or 100 miles—for about 33,000 cells at the surface. With twenty-six atmospheric layers, this makes roughly another

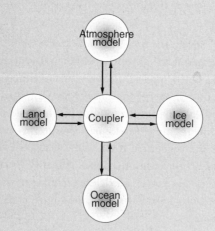

Figure 15.4
Conceptual diagram of the National Center for Atmospheric Research Community Climate System Model, which consists of four separate modules and the coupler that communicates results among the modules.

1 million cells for the surface–atmosphere systems. With its 3 million or so cells, CCSM can advance the climate about four years in one 24-hour day of computation on NCAR's supercomputers.

The word *community* in CCSM's name reflects the fact that the model is freely available to scientists in the climate-research community, and that scientists in the broader community contribute to the model's development. As a result, CCSM has become a widely used tool in climate research. The computer program that handles the coupling between the different model components is sufficiently versatile that scientists can plug in different versions of the individual models as appropriate to specific lines of research. You can learn more about CCSM at www.ccsm.ucar.edu/models/ccsm3.0/. You can even download the model, but I don't recommend trying to run it on your personal computer! (Check out Research Problem 3 if you'd like to get involved in climate modeling. This problem describes a global project that harnesses the spare computing power in tens of thousands of personal computers to run large-scale climate models.)

Even today's highest-resolution climate models can't capture every important climate process. Many clouds, such as the puffy cumulus that form in fair weather, are smaller than any grid cell. So are cities, farm fields, and many lakes and forests. Yet clouds, urbanization, local geography, and land-use variations have definite effects on climate. As a result, models use techniques known as **subgrid parameterization** to account for these and other small-scale effects. A subgrid parameterization of clouds, for example, might ascribe average cloud-related albedo and absorption to an entire grid cell, based on a calculation that suggests the formation of small-scale clouds providing only partial sky coverage.

Can't modelers handle every aspect of climate by simply shrinking their grid cells further? In principle, perhaps, but remember that today's largest climate models challenge even the fastest supercomputers. Halving each dimension of every grid cell in a three-dimensional model increases the number of cells eightfold (since $2^3 = 8$), and in a time-dependent model this requires halving the time step as well, which makes for a sixteenfold increase in computer time! Modelers do use smaller grid cells to study regional effects—the climate of, say, the United Kingdom, the northeastern United States, or the Sahel region of Africa. Embedding such a regional model in the coarser grid of a global model provides values of climate variables at the boundaries of the region under study. With its fewer but finer cells, the regional model then runs in reasonable computer time to give a detailed picture of regional climate.

However, computer models are advancing rapidly in resolution and in the number and detail of the physical, chemical, and biological effects they include. Models improve with our increasing understanding of climate processes, with advances in computational techniques, and especially with increases in raw computer power. The latter, in fact, doubles roughly every two years—a trend known as **Moore's law**. In 1965 Gordon Moore,

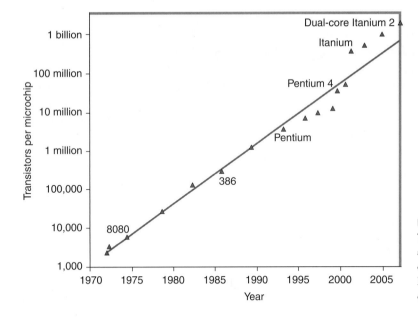

Figure 15.5
The number of transistors on a semiconductor chip has doubled roughly every two years, thus validating Moore's law to date. Computer speed has risen correspondingly.

one of the founders of Intel, the company whose semiconductor chips power most of today's personal computers, suggested that the semiconductor industry would double the number of transistors—the basic elements of computer circuitry—on a single semiconductor chip every year or two. Moore's law has held for over 40 years, and today's most advanced computer chips contain over a billion transistors (Fig. 15.5). Computer speed has risen along with the number of transistors. Moore's law is one reason we can expect continued exponential improvement in climate models, and it's also the reason your new laptop computer is obsolete almost as soon as you buy it!

A Hierarchy of Models

With supercomputers grinding out results from comprehensive coupled climate models at the world's major climate research centers, you might think there's no place for simpler models such as Figure 15.1's two-box model, the one- and two-dimensional models, or the equilibrium models that don't include time dependence. But that's not the case at all. The large-scale AOGCMs take so much computer time to run that it's simply impossible to experiment with as broad a range of climate situations as modelers might want. And since these models are nearly as complex as the real climate system itself, interpreting the results and determining the causes of modeled phenomena are sometimes difficult.

Simpler models, in contrast, run quickly and so allow researchers to carry out numerous experiments with different initial conditions, assumptions about greenhouse emissions, and other parameters. Since simpler models involve fewer details, it's often easier to interpret the results or focus on a particular physical process of interest. Sometimes simpler models are calibrated by comparing their results with those of AOGCMs and

Figure 15.6

Model runs of a simplified model (dashed curve) and a coupled AOGCM showing transient climate response to a 1 percent per year increase in atmospheric CO_2, resulting in a CO_2 doubling at about 70 years. After that point, CO_2 is held constant. The models are in good agreement until the time of CO_2 doubling, but the more realistic AOGCM takes longer to reach equilibrium temperature, at about 3.5°C above the starting temperature. That's because the AOGCM includes the heat exchange between the ocean surface layer and the deep ocean, a process that delays equilibrium. Note that the AOGCM, as with real climate, exhibits intrinsic variability.

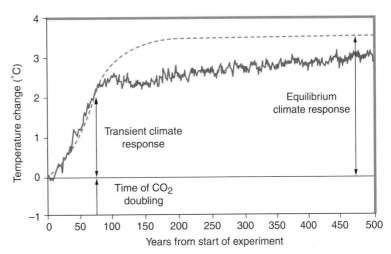

adjusting the model as needed for agreement. Then modelers can be confident that, for small changes in model conditions, the simpler model will give results similar to those of the full AOGCM (Fig. 15.6). So there's room in climate studies for a whole hierarchy of models, from the simplest energy-balance models to the full, computing-intensive coupled atmosphere–ocean models.

Validating Climate Models

Why should we trust climate models? There are several main reasons. First, climate models are based on long-established scientific principles and measured properties of basic materials—science that governs the behavior of matter and energy in realms far beyond climate. Examples include the $e\sigma T^4$ radiation law that I introduced in Chapter 4 and have used throughout the book; the infrared absorption spectrum of CO_2 that ultimately determines CO_2's greenhouse characteristics; the thermodynamic properties of water as it changes among solid, liquid, and gaseous phases; the laws of electromagnetism as they describe how electromagnetic radiation (light) scatters off small particles such as atmospheric aerosols; and even more fundamental principles such as conservation of matter and energy. Where modelers don't know all the details—for example, just how plants respond to changes in CO_2 concentration; the physics of cloud formation and cloud behavior; the strengths of some feedback effects—they incorporate the results of laboratory and field experiments as empirically based equations in their models. So climate models are based on solid science despite uncertainties in some of the details.

Second, climate models successfully reproduce present and past climates. Figure 14.13 is one example: Starting with nineteenth-century conditions and given the subsequent changes in natural and anthropogenic forcings, it reproduces quite well the observed temperature record of the twentieth century. Confidence that models work outside the present-day climate regime comes from model experiments using very different conditions from ancient times. For example, changes in Earth's orbit led to different distributions of solar radiation thousands of years ago; when these changes are incorporated into climate models, the resulting model climates show features similar to those found in paleoclimate data.

Such verification isn't limited to global temperature. Climate models reproduce the distribution of temperature changes with geographical location, and they do an excellent job of describing the observed temperature structure throughout the atmosphere (Fig. 15.7). They give reasonable agreement with observed variations in precipitation (Fig. 15.8), cloudiness, and ice cover, and the ocean components of coupled models correctly follow the spread of anthropogenic trace chemicals throughout the oceans. The models also reproduce changes in patterns of natural climate variability.

A helpful test would be to do controlled experiments with Earth's climate, making changes to the actual climate and seeing if models reflect those changes. We can't do that, of course, although our greenhouse gas and other emissions amount to an uncontrolled climate experiment. However, nature provides us with real-world experiments in the form of volcanic eruptions. The 1991 eruption of Mount Pinatubo in the Philippines provided one such experiment, putting enough dust into the atmosphere to cause a brief global cooling. Average global temperature quickly dropped about 0.5°C following this eruption, then recovered over several years. Figure 15.9 shows that climate models successfully reproduced the pattern of Pinatubo-related global temperature change.

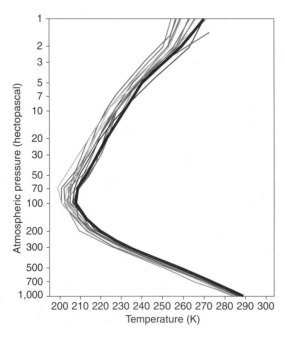

Figure 15.7
Temperature structure of the atmosphere, from actual observations (thick black curve) and thirteen different climate models (gray curves). The vertical axis is atmospheric pressure, measured in hectopascals, which climatologists use as a proxy for altitude. Decreasing pressure corresponds to greater altitude.

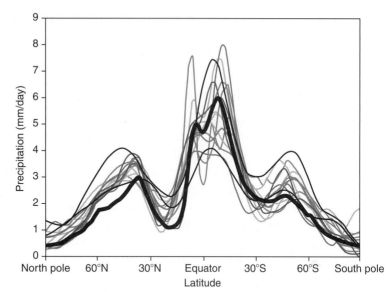

Figure 15.8
Three months of precipitation (December to February), averaged over longitude, as a function of latitude from pole to pole. The thick black curve represents actual observations; the gray curves are the results for present-day conditions from fifteen different climate models.

Finally, there are a great many climate models in use today. Climate groups around the world develop their own independent models, using different computational techniques, coupling schemes, submodels, and, where details are less certain, different assumptions about such things as feedback strengths or subtle climate processes. Yet, as the multiple plots in Figures 15.7 and 15.8 show, the diverse climate models exhibit good agreement, even on such details as precipitation and atmospheric structure. Our most robust

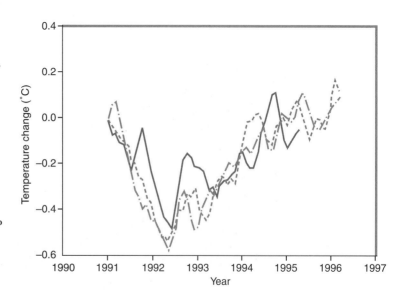

Figure 15.9
Global temperature change following the 1991 eruption of Mount Pinatubo in the Philippines. The solid curve is the observed temperature, and the dashed curves show two model results. The temperature changes are averages over April, May, and June of the years shown and are relative to the averages of April to June of 1991.

conclusions about future climate come from ensembles of many individual model runs. At the same time, the variation among models helps to quantify the range of uncertainty about future climate.

15.2 Climate Projections

So we have a variety of climate models, and we have some confidence that they can give us reasonably realistic projections of future climate change in response to both anthropogenic and natural factors. Note that I'm using the word *projection* rather than *prediction*. This is partly because climate models advancing decades and centuries into the future can't claim the precision of tomorrow's weather prediction. It's also because future climate depends on factors that are unpredictable, such as volcanic eruptions and human behavior. So model-based studies are more like "what if" stories—for example, What's the climate likely to do if our CO_2 emissions change in such and such a way?— than they are forecasts of what *will* happen.

Doubling of Atmospheric Carbon Dioxide

One benchmark for assessing future climates is a doubling of atmospheric CO_2 from its preindustrial level of 280 ppm to 560 ppm, a condition often abbreviated as $2 \times CO_2$. Given the inertia in the fossil-fueled global economy and the fact that we're already approaching 400 ppm, we're likely to reach 560 ppm of CO_2 sometime in the mid- to late twenty-first century. Only a massive, determined, and almost immediate effort to replace fossil fuels or sequester carbon could stabilize atmospheric CO_2 at a lower level. So $2 \times CO_2$ provides a model input that we can realistically expect to reflect conditions later in the current century.

Figure 15.6 shows responses to CO_2 doubling in both a full, coupled AOGCM and a simpler climate model. The equilibrium doubling, some 3.5°C, is typical of model projections for a world with 560 ppm of atmospheric CO_2. Since Earth warmed by about 0.74°C over the past one hundred years, the particular models of Figure 15.6 suggest an additional warming of somewhat over 2°C if we manage to stabilize CO_2 at twice its preindustrial level.

Figure 15.6 compares just one AOGCM and one simpler model. But our confidence in climate projections comes, as I indicated in the preceding section, from ensembles of different, independent models. So how do other climate models do? Figure 15.10 shows the results of nineteen different runs calculating the same 1 percent per year CO_2 increase used in the first 70 years of Figure 15.6. At 70 years— the time of doubled CO_2—the models show a range of temperature increases from about 1°C to 3°C, with most clustering at just below 2°C. Remember, however, that these are just the transient responses; if the CO_2 increase were to stop at 70 years, then, as Figure 15.6 suggests, we might expect another 1.5°C or so of warming as the climate equilibrates.

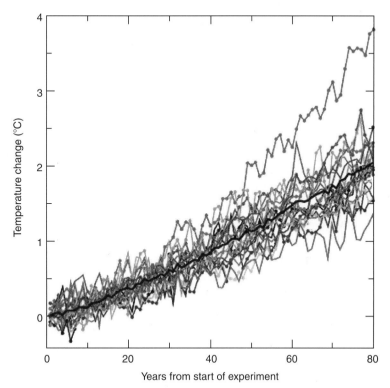

Figure 15.10
Results of nineteen different climate model runs showing response to a 1 percent per year increase in atmospheric CO_2, resulting in a CO_2 doubling at about 70 years. These runs thus compare with the first 70 years of Figure 15.6. Individual model runs are shown in gray, and the thick black curve is the mean of all runs.

What if we don't manage to halt our CO_2 emissions at 560 ppm? Then, as Figure 15.11 shows, we can expect correspondingly higher temperature increases in the current century, with still higher equilibrium values being reached several centuries hence.

The Human Factor

Increases in greenhouse gas concentrations and changes in other forcings depend on human behavior; so, therefore, does future climate change. Will we make a concerted effort to end our dependence on fossil fuels? Will we develop more carbon-free energy technologies, or learn to sequester fossil carbon emissions? Or will it be business as usual? Will population growth continue, or will our numbers peak and then decline? Will global inequities in wealth and the associated resource consumption worsen, or will a higher standard of living spread across the planet? The answers to these and similar questions affect our energy consumption and thus our impact on Earth's climate.

To explore the effects of different human behaviors on climate, the Intergovernmental Panel on Climate Change (IPCC) developed a set of scenarios to represent different possible futures for our species and our planet's climate. The IPCC Special Report on

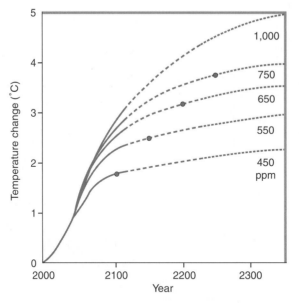

Figure 15.11
Projected global temperature increases from around the year 2000 for CO_2 stabilization at five different concentrations. The black dots indicate when stable CO_2 concentrations are reached. These results are from a simple model calibrated against seven AOGCMs, and the results after 2100 (dashed curves) are increasingly uncertain.

Emissions Scenarios (SRES) includes detailed projections of the radiative forcings from dozens of anthropogenic agents, including greenhouse gases and aerosols. More broadly, the SRES uses two basic classifications to characterize possible human futures. The first is our degree of emphasis on conventional economic growth versus environmental sustainability, efficient use of resources, and social equity. Scenarios in the former category are designated as A scenarios; those in the latter, B scenarios. The second major classification is the degree to which human society becomes more homogeneous as globalization spreads common values, cultural norms, and economic prosperity. Scenarios in which globalization plays a major role are designated 1; those that preserve distinctive regional identities and rely on local solutions to social and environmental problems are designated 2. The result is a two-dimensional grid of four major scenario categories, as shown in Figure 15.12.

A major difference between scenarios 1 and 2 is in population. In the 1 scenarios, globalization spreads education—especially for women—throughout the world. This leads to a rapid decline in the birth rate nearly everywhere, causing the world population to peak in the mid-twenty-first century, then slowly decline. The regionally focused 2 scenarios, in contrast, have a population that's still growing in the year 2100.

The full SRES analysis generates some forty distinct scenarios, each a different "What if?" story about a possible climate future. Here we'll consider only one representative from each broad category, except for A1, which we'll break down into three cases: A1FI is a business-as-usual scenario with continued intensive use of fossil fuels; A1B considers a balanced range of energy sources; and A1T envisions aggressive development of new, nonfossil energy technologies.

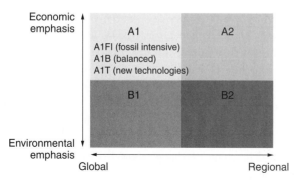

Figure 15.12
Future scenarios from the IPCC's
Special Report on Emissions Scenarios.
The A1 category includes three sub-
scenarios, one a business-as-usual fossil-
intensive economy (A1FI), one using a
more balanced variety of energy sources
(A1B), and one emphasizing new energy
technologies (A1T). Both global
scenarios (A1, B1) have world
population peaking in the mid-twenty-
first century, whereas population growth
continues through 2100 in the more
regional scenarios (A2, B2).

Projected greenhouse emissions vary dramatically across the range of scenarios, as Figure 15.13 shows for CO_2. Here the B1 scenario, with its emphasis on environmental protection, brings a modest rise in CO_2 emissions that is followed by a decline to levels in 2100 that are well below those in 2000. The economic-growth-oriented A1T, with its nonfossil energy technologies, achieves nearly the same emissions pattern. The fossil-intensive A1FI scenario, in contrast, shows substantial emissions growth through 2080, followed by only a modest decline. The A2 scenario—emphasizing regional economic

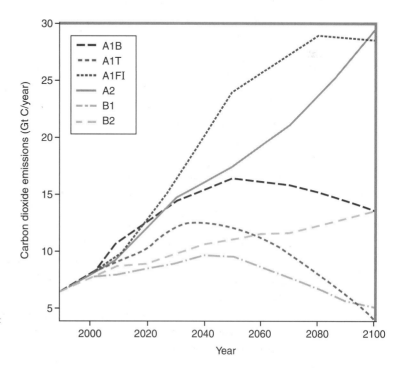

Figure 15.13
Projected CO_2 emissions for the six different SRES scenarios. Emissions are in gigatonnes (billions of tonnes) of carbon per year.

growth without globalization—does even worse: Its CO_2 emissions continue to climb even at the century's end. That's because, despite lower per capita emissions, A2's continued population growth results in greater total emissions. The A1B scenario, perhaps the most realistic because of its balance of fossil and nonfossil energy, peaks in mid-century at over twice our current emissions rate, then declines modestly.

Global temperature change depends on the total accumulated CO_2 and other greenhouse gases, not just the immediate emissions. Furthermore, there's a delay in the temperature response to rising greenhouse gas concentrations. As a result of both factors, projections of global temperature change show less variation across the scenarios than do greenhouse emissions. Figure 15.14 gives projected temperature changes for three of the scenarios we're considering, relative to 1990 values. By 2100 the best estimates range from a high of 4°C for the fossil-intensive A1FI scenario to a low of just under 2°C for the globalized but environmentally friendly B1. Scenario A1B, again perhaps the most realistic, gives an end-of-century rise of nearly 3°C. Taken together, the model runs of Figure 15.14 suggest a likely twenty-first-century temperature rise in the range of 1.1°C to 6.4°C, with 3°C a good "best guess." The uncertainty in these projections is split roughly equally between uncertainties in model calculations and the range of choice in human behavior—again, that's why these projections are "What if?" scenarios rather than forecasts.

Regional Climate Projections

Figure 15.14 shows projected increases for global average temperature over the twenty-first century. But, as we've already discovered when looking at past climates, we can't

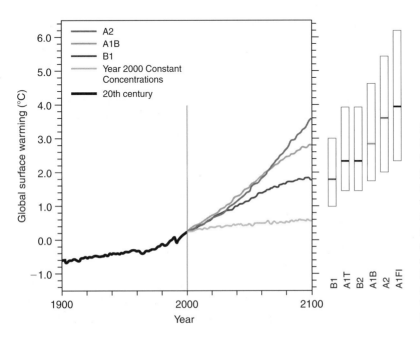

Figure 15.14
Projected global temperature change for the SRES scenarios. The graph shows projected temperature increases for three of the scenarios, as well as the rise that would have resulted if greenhouse gas concentrations had remained constant since the year 2000. The solid lines are averages of many model runs. At right are ranges of likely global temperature in 2100 for all six SRES scenarios considered in Figure 15.13. Here "likely" means with confidence level between 66 and 90 percent.

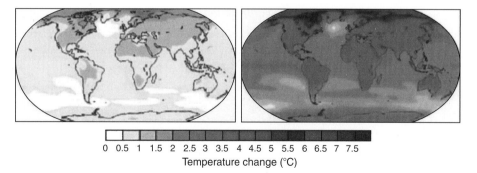

Figure 15.15
Projected temperature increases for the early and late twenty-first century, for the A1B scenario. Left-hand map is for the decade 2020–2029; right-hand map is for 2090–2099. Note that the increases are largest in the Arctic and greater on land than over the oceans.

expect that change to be distributed uniformly across the globe. Although there's more uncertainty in regional details than in global average projections, some regional effects emerge consistently enough that we can have considerable confidence in them. Figure 15.15 maps regional projections of the average temperature change expected for both early and late in the twenty-first century. These particular maps are for the A1B scenario, whose global temperature increase, as shown in Figure 15.14, is about 3°C. Other scenarios show qualitatively similar patterns. For example, all scenarios agree with Figure 15.15 in showing the greatest projected temperature increase in the Arctic. That's entirely consistent with the rapid change observed recently in Arctic climate, as discussed in Chapter 14, and it's a change that occurs for valid scientific reasons, such as ice-albedo feedback. Another obvious pattern visible in Figure 15.15 is that the continents warm more than the oceans. Again, there's a good scientific reason: The oceans' vast heat capacity means it takes a lot of energy to change ocean temperature. The result, though, is that a global average warming of, say, 3°C translates into considerably greater warming over nearly all land areas.

15.3 Consequences of Global Climate Change

Global warming won't just make Earth warmer. It will also bring changes in sea level, altered precipitation patterns, changes in soil moisture content, increases in some extreme weather events, more flooding and more drought, alteration of natural climate cycles such as El Niño, changes in species ranges and the composition of ecosystems, advances of tropical diseases into formerly temperate regions, changes in ocean circulation, melting of Arctic permafrost, and a host of other effects that have an impact on Earth and its inhabitants. Here we'll take a quick look at the more important of these impacts.

Extreme Temperature Events

Obviously, we should expect more hot spells in a warming world. Most of the planet's land area has already seen such increases, and model projections suggest a continued increase in the number of abnormally hot days and in higher maximum temperatures nearly everywhere. At the other end of the temperature spectrum, most land areas can expect fewer exceptionally cold days and longer frost-free growing seasons. Surprisingly, it takes only a modest rise in global average temperatures to increase substantially the frequency of high temperature extremes. Box 15.2 examines the reason.

Box 15.2   Means and Extremes

Many varying quantities, from temperatures to heights of individuals to exam grades, are naturally distributed on a bell-shaped curve (Fig. 15.16). Values of the given quantity cluster around the mean, which, for a symmetric curve, is also the most probable value. The probability that the quantity in question will take a given value decreases with that value's deviation from the mean.

A slight change in the mean value—for example, a degree or so of global warming—can nevertheless cause large changes in the probability of rare extreme values. Figure 15.16 shows why. The probability of temperatures in a given range depends on the relative area under the bell-shaped curve in that range. A slight increase in the mean has little effect on the distribution of the more probable values near the mean, but it substantially raises the curve at the extreme tail end of the distribution. The greater area under the raised curve means that extremely high temperatures, although still rare, increase disproportionately. At the left-hand end of the bell-shaped curve,

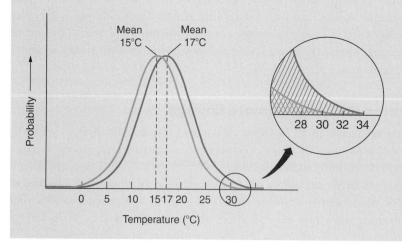

Figure 15.16
Bell-shaped curves representing distributions of temperature for global means of 15°C and 17°C. The two curves don't differ much at the highly probable temperature values near the mean, but the enlargement shows that high-temperature extremes become much more likely with the higher mean.

similar reasoning shows that the likelihood of extremely cold days decreases substantially with a slight upward shift in the mean.

It's difficult to argue that a particular extreme temperature occurrence is the result of a shift in the mean temperature—just as it's difficult to argue that a given case of emphysema, for example, results from air pollution, or that a case of cancer is caused by nuclear radiation. Nevertheless, some events are so far along the tail of the bell curve that they are almost impossibly unlikely absent an increase in the mean. Such is the case with the summer 2003 heat wave in Europe, which killed tens of thousands of people. A statistical analysis suggests that human influence—especially greenhouse gas emissions—at least doubled the probability of this extreme temperature event.

The Hydrologic Cycle and Weather

A warming Earth means a moister atmosphere, as increased thermal energy leads to more evaporation and therefore a more intense hydrologic cycle. So we can generally expect more precipitation with global warming. Again, though, this effect won't be evenly distributed in either space or time. The spatial distribution of projected changes in precipitation is especially wide, with drier areas seeing substantial declines despite the overall trend of an average increase. In general, precipitation and river flows are expected to increase by some 10 to 40 percent at high latitudes and in some parts of the tropics, while decreasing by up to 30 percent over the mid-latitudes and dry tropical regions. In addition, more precipitation is likely to occur in short, intense events. One result of these changes in precipitation patterns will be increased flooding in some areas. Elsewhere—especially in many continental interiors—warmer soils, greater evaporation, and less precipitation will lead to increased drought.

The intense tropical storms known as hurricanes, cyclones, and typhoons get their energy from warm surface water, so it's not surprising that tropical storms are expected to grow more intense, at least some regions. Indeed, as Figure 14.12 indicates, we've seen an increase in hurricane intensity over the past few decades, an increase that correlates well with rising sea-surface temperatures. However, the formation and strength of tropical storms depends on a number of complex factors, and the observed correlation between strength and temperature is actually stronger than theory and models suggest. So we're less certain in attributing hurricane intensity to anthropogenic climate change than we are, say, in making a similar attribution for heat waves. Furthermore, it's not at all clear whether the number of hurricanes should be expected to change significantly.

Sea-Level Rise

Twenty thousand years ago so much water was locked up in continental ice sheets that sea level was some 120 m (nearly 400 feet) below where it is now. As Earth emerged from the ice age, sea level rose at the rapid rate of about 1 cm per year. By six thousand years ago this rate had slowed to about 0.5 mm per year, and by three thousand years ago it was down to 0.1 to 0.2 mm per year. Measurements suggest that, by the twentieth century, sea level was rising ten times faster—at 1 to 2 mm per year.

Measuring sea level isn't easy. The sea surface is rarely smooth, and it's subject to tidal variations. Even weather affects sea level, with the sea rising temporarily beneath regions of low atmospheric pressure. The land itself doesn't stay still either, with some continental edges sinking and others rising—the latter sometimes an ongoing rebound from the time when the weight of massive ice sheets depressed the land surface. Earthquakes and slower tectonic motions also change land elevation. Although we think of water as seeking its own level, across the vastness of the world's oceans there's no single unambiguous value for sea level. Sea-level rise, like other consequences of climate change, isn't distributed evenly around the globe. Nevertheless we have enough information, most of it from a worldwide network of tide-gauging stations, to pin down twentieth-century sea-level rise within a factor of about 2, and it gives that 1 to 2 mm per year figure. More recent satellite-based measurements suggest an acceleration to some 3 mm per year since the early 1990s, although some of this short-term increase may be due to internal variability of the climate system.

You may think that sea-level rise is associated largely with the melting ice I described in Chapter 14, but in fact the single most significant contribution in the past few decades is from the thermal expansion of water as its temperature increases. In effect, the ocean acts like a giant thermometer, its level rising with temperature. However, the thermal expansion rate of water itself depends on temperature, and it takes a long time for warming to penetrate into the ocean depths. As a result, it's not easy to pin down thermal expansion accurately, either in attributing observed sea-level rise or in projecting future sea levels. Nevertheless, it appears that thermal expansion is responsible for about one-third of the twentieth-century rise.

Melting ice makes the second most important contribution, but here, too, things are complicated. First of all, only land-based ice melt is significant. Ice that's already floating in the water makes essentially no change in sea level when it melts, because the greater density of water offsets the volume of ice that wasn't already submerged. (There's a very slight effect due to the fact that ice is freshwater and its density when it melts is different from the surrounding ocean.) Second, melting isn't the only warming-related change in land-based ice cover. Because global warming intensifies the hydrologic cycle, the result can be more snow. In very cold regions, such as the interior of Antarctica, this may lead to an increase in ice accumulation. Over the twentieth century, this effect may have given Antarctica a *negative* contribution to sea-level rise, although here the uncertainty is so great that we can't even be sure of the sign of Antarctica's contribution. In any event, though, it's not enough to offset the positive contribution from the Arctic ice sheets and mountain glaciers. Table 15.1 details these ice-related contributions to sea-level rise.

Table 15.1 Sea-Level Rise from Land Ice

Ice source	Potential sea-level rise equivalent*	Estimated contribution to twentieth-century sea-level rise
Glaciers	0.24 m	3 cm
Ice caps	0.27 m	
Greenland	7.2 m	0.5 mm
Antarctica	61.1 m	−1 cm

*Rise if all the ice melted.

Other factors that contribute to a rise in sea level include melting permafrost, the deposition of sediments by river flows, and, significantly, human alteration of the hydrologic cycle. For example, our depletion of groundwater aquifers moves water from natural underground storage to the oceans. The opposite effect occurs with dams and agricultural irrigation that block the natural flow of water back to the oceans and instead send much of it to the atmosphere via evaporation and transpiration. The result is a lowering of sea level that counters, to some extent, the effect of melting land ice. Although these effects aren't a direct result of climate change, they're another example of human influence on the global environment that happens to coincide with anthropogenic climate change.

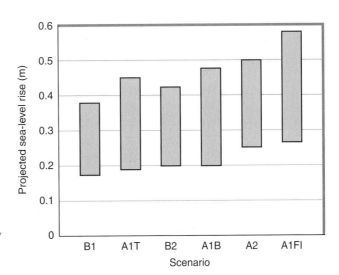

Figure 15.17
Projected sea-level rise under the six SRES scenarios. Each column represents the likely range of sea level rise, from low to high values, for the given scenario. Not included are effects from possible rapid changes in ice dynamics.

Uncertainty about these processes leads to a range of projections for future sea level, mostly from about 20 cm to 60 cm over the twenty-first century. Figure 15.17 shows the projected ranges for the six scenarios we've been considering. In this case the overall uncertainty is considerably greater than the variation across different scenarios, again because of the many imperfectly understood processes that contribute, both positively and negatively, to sea-level rise.

For the A1B scenario, which I've been suggesting may be the most realistic, Figure 15.17 suggests a best estimate for twenty-first-century sea-level rise of around 35 cm (about 14 inches) and in almost any event well under a meter. This is much less than the typical tidal variation and wave action. So why the fuss? Because half the human population lives in coastal areas—some in low-lying river deltas like much of Bangladesh, some already below sea level as in the Netherlands or New Orleans, and some on low-lying islands. A modest rise in sea level will flood some land directly, and in the extreme case of Bangladesh a half-meter rise would submerge more than 10 percent of the country's land area and displace 6 million people. In addition, an overall sea-level rise adds to the highest tides and storm surges, increasing the risk of flooding over much greater areas. Globally, the IPCC projects that by the late twenty-first century many millions of people will be subjected to yearly flooding because of rising sea levels.

A more subtle manifestation of sea-level rise is the intrusion of salt water into the underground water table. Already a problem in such low-lying areas as Bangladesh, Florida, and the Nile delta, salt intrusion leads to reduced agricultural productivity and exacerbates the growing scarcity of freshwater resources. Finally, rising sea level may inundate coastal wetlands, where, as we saw in the discussion on tidal energy in Chapter 8, the interplay of saltwater and freshwater helps to nourish rich ecosystems that act as the nurseries for much of the world's marine life.

So far I've considered only the sea-level rise expected in the twenty-first century, which is significant but hardly overwhelming. For this century you can probably forget apocalyptic visions of New York City, London, Tokyo, Cape Town, Rio de Janeiro, and other great coastal cities going under water. But because it takes a long time for a changing climate to melt land ice and for rising surface temperatures to penetrate the ocean depths, the two major causes of rising sea level—melting ice and thermal expansion—will continue for a very long time. Table 15.1 shows that land ice, in particular, has a huge potential for increasing sea level. We don't expect the many-meter rise implicit in the Greenland and Antarctic values listed in Table 15.1, at least not any time soon. Nevertheless nasty surprises are possible. For example, recent studies suggest that Greenland's ice is melting far more rapidly than anticipated. And land ice needn't melt to raise sea level; the same rise occurs if the ice falls bodily into the sea, as its submerged portion displaces seawater and pushes up the ocean surface. Scientists worry that water from even modest melting can act as a lubricant between the bottommost ice layer and the land, thus allowing large ice sheets to slide into the sea. Of particular concern is the West Antarctic ice sheet, which could raise sea level by about 3 meters, or 10 feet. This

Figure 15.18
The collapse of the West Antarctic ice sheet would raise sea level by some 3 m; as a result, Florida would lose a substantial part of its land area (shaded in dark gray).

amount would be enough to flood many coastal cities, and it would significantly reduce the land area of low-lying regions (Fig. 15.18).

Species Ranges

Every species has a natural range of climatic conditions it can tolerate; this range is broader for some species and narrower for others. As Earth warms, species ranges are expected to shift toward the poles and to higher elevations. A study involving some seventeen hundred species found just this result: On average, species ranges in the Northern Hemisphere shifted northward at an average rate of some 6.1 km per decade. For alpine species, the shift was 6.1 m upward in elevation each decade. You might think this effect would simply redistribute species about the globe, but many climate-induced changes are actually detrimental to species' populations, and even to their long-term survival. For one thing, the warming of the twenty-first century will be more rapid than any natural climate change occurring at least since the end of the last ice age. Some species, especially trees, may not be able to move poleward fast enough to match the changing climate. Thus they may die off at the warmer end of their range, but they won't have time to advance to populate the newly suitable poleward regions. The result could be a decline in some forest ecosystems, including animal species that require particular forest types. Another detrimental effect of climate change follows from the so-called species–area relationship. This long-established principle of ecology says that smaller areas can sustain fewer different species. As global warming moves species ranges upward in both latitude and altitude, the effect is often to shrink the size of regions with particular climatic conditions. The result, according to the species–area relationship, should be fewer species. This climate-induced effect may soon rival the loss of biodiversity resulting from direct human-induced habitat destruction. In addition, outright extinction looms for some species. A study covering some 20 percent of Earth's land area suggests that somewhere between 15 and 37 percent of all species will be on the road to extinction by 2050, depending on just how rapidly global temperatures increase.

Shifts in geographical range aren't the only evidence of climate change to be found in ecological systems and species' behavior. Also changing is the timing of significant events, such as budding of leaves and flowers, mating, nest building, and migration. The same study that gave an average range shift of 6.1 km per decade also found that such significant events are occurring earlier, at an average rate of 2.3 days per decade. (Let me emphasize again that these are statistical results obtained by averaging data from hundreds of species.) Climate change isn't the only factor influencing range and timing in natural systems, and not all species share in the general trends of northward advance and earlier timing. However, the vast majority of the 1,700 species in this study do, leading to a high level of confidence that we're already seeing the effects of climate change on natural systems.

It isn't just birds, frogs, butterflies, and trees that are advancing toward higher latitudes; so are many pathogens and their insect hosts. This movement could make diseases typically associated with the tropics more common in temperate regions. A variety of factors contribute to the spread of disease in our global society, but there's already some indication that illnesses such as malaria, Lyme disease, West Nile virus, and dengue fever are becoming more widespread with increasing temperatures.

Ocean Circulation

Ocean currents play a major role in determining regional climate. Perhaps the best-known example is the Gulf Stream, which carries warm surface water from the Gulf of Mexico toward northern Europe. Its warming influence gives Europe mild weather at the same latitudes that are near-Arctic in North America. One Scottish town situated at the same latitude as Juneau, Alaska, has palm trees lining its main street!

However, water that flows northward must eventually return to keep the oceans in balance. In the case of the Gulf Stream, the return flow begins north of Iceland, between Greenland and Scandinavia. The Gulf Stream water isn't only warm; it also has a higher salt content because evaporation in the warm tropics leaves the salt behind. The water's warmth tends to decrease its density, while the high salt concentration tends to increase it. As the Gulf Stream flows northward and cools, its density therefore increases. Finally, north of Iceland, the Gulf water becomes denser than the water surrounding it and therefore sinks. The cool water then flows southward at a depth of several kilometers. Together, the Gulf Stream and its cold, deep return flow constitute part of a global pattern of **thermohaline circulation**, whose name derives from its two drivers, heat (*thermo*) and salt (*haline*).

As Europe's present climate attests, thermohaline circulation in the North Atlantic represents a significant energy flow—comparable, at high latitudes in the northeastern Atlantic, to the energy delivered by sunlight. A reduction in the North Atlantic thermohaline circulation could therefore cool the climate of northwestern Europe. Ironically, global warming could bring about just such an effect, because increased precipitation and melting ice would dilute the salty North Atlantic surface with freshwater, reducing its density and inhibiting the sinking that drives thermohaline circulation. With the amount of global warming projected for the twenty-first century, this effect is expected to reduce the North Atlantic thermohaline flow by as much as 60 percent. Nevertheless, even this reduction in Gulf Stream warmth won't be enough to keep Europe from warming in response to greenhouse forcing. And thermohaline circulation isn't the only driver of the Gulf Stream; wind-induced currents also contribute. So visions of a deep freeze in Europe or elsewhere seem far-fetched. However, our observational and theoretical knowledge of the thermohaline circulation is incomplete, and this is one area where nonlinearity in the climate system could bring a nasty surprise.

Ocean Acidification

In Chapter 13 I described how land and oceans together absorb slightly more than half of all anthropogenic CO_2 emissions, with the oceans taking a greater share. This is a good thing, since it slows the buildup of the most important greenhouse gas in the atmosphere and thus limits the rate of climate change. But is it a good thing for the oceans? The answer depends on the complex chemistry of oceanic carbon, and recent research suggests the oceans' CO_2 uptake is adversely affecting marine organisms.

When CO_2 dissolves in water, it forms carbonic acid (H_2CO_3), a mild acid that gives carbonated sodas their acidity. Carbonic acid dissociates into hydrogen ions (H^+) and bicarbonate ions (HCO_3^-). The hydrogen ions combine with carbonate (CO_3^-) to form additional bicarbonate. But carbonate is what marine organisms use to build shells and other structures, like the rigid skeleton of coral. So adding CO_2 to the oceans has the net effect of decreasing the amount of carbonate available to marine organisms.

Already the effect of anthropogenic carbon has been to lower the oceans' average pH by 0.1. (Recall from Chapter 6's discussion of acid rain that pH measures the concentration of hydrogen ions in solution, with a decrease of 1 in pH being a tenfold increase in H^+.)

Projections based on IPCC scenarios suggest that the pH could fall another 0.3 to 0.4 by the year 2100. Under such acidic conditions, some species of high-latitude marine plankton may have trouble surviving. Because such organisms are at the base of marine food chains, the populations of larger marine species could suffer as well. Although these problems will be limited at first to polar and high-latitude oceans, they'll spread toward the tropics as the acidification increases.

15.4 Climate Change and Society

Climate change will have broad impacts on human society. Changes projected to occur in the coming century or two won't have the apocalyptic, civilization-destroying character of, say, an all-out nuclear war or a direct hit from a large asteroid. But climate change will be disruptive in many ways to a civilization whose planetary home is already under stress from the demands of billions of human beings. Although the focus of this book is on the scientific and technical issues of energy, environment, and climate, I do want to end this chapter with a brief survey of the major societal impacts expected from twenty-first-century climate change.

We've already seen that the physical impacts of climate change won't be evenly distributed around the globe. Low-lying coastal regions, for example, will suffer more from sea-level rise; continental interiors will see more drought; and high latitudes will warm disproportionately. For that reason alone, impacts will vary from country to country. But social, political, and economic considerations also alter the impact of a changing climate. That's because wealthier countries are better able to adapt to and counter deleterious effects. Wealthy countries can build protective seawalls to ward off the rising oceans; they can engineer massive water projects to ensure continuing supplies of freshwater; they can outbid poorer countries if declines in domestic agriculture require increased food imports; their advanced health care systems can better cope with spreading tropical diseases; and their robust economies can speed recovery from extreme weather events. We're forced to the unfortunate conclusion that climate change will exacerbate the gap between rich and poor nations, as those with the fewest resources will be the most vulnerable. This conclusion should be especially troubling because anthropogenic climate change is almost entirely the result of fossil fuel consumption, deforestation, and other activities carried out by or for the benefit of the developed world.

Not all climate change will be harmful. If global average warming is limited to one or two additional degrees, some regions will experience enhanced agricultural productivity resulting from increased precipitation, more rapid growth associated with higher levels of atmospheric CO_2, and a longer growing season. But beyond 2°C to 3°C of global average warming, essentially all impacts become deleterious, and the chance of surprising nonlinear changes occurring increases significantly. A sobering example is the European summer heat wave of 2003, which reduced plant growth so much that the plants actually became net sources of CO_2, emitting more CO_2 to the atmosphere than they took in through photosynthesis. In this extreme event, any positive effect of CO_2-enhanced plant growth was overwhelmed by the negative effect of heat and drought.

So what are the expected impacts on society? Some, such as increased disease and mortality rates among the elderly and the urban poor, will result directly from increased temperatures—especially from extreme heat waves, whose probability increases disproportionately for the reasons mentioned in Box 15.2. Demand for air conditioning will stress already frail electric power infrastructures, reducing reliability and increasing brownouts, as well as fuel consumption. And tropical diseases such as malaria may spread to higher latitudes. But the most significant impacts on humans will probably come from changes in the hydrologic cycle. More intense precipitation will lead to additional soil erosion and landslides. The former accelerates the loss of agricultural soils while the latter threaten burgeoning hillside populations, both rich and poor. At the opposite extreme, drier summers will decrease crop yields, make scarce freshwater resources even scarcer, and increase the risk of forest fires. These fires also will be more likely to destroy homes, as residential development at the urban fringes spreads into forested areas. In the Arctic, melting permafrost has already caused the collapse of roads, buildings, and pipelines, while the loss of sea ice alters the local ecology, threatening the livelihood of indigenous peoples. Both rising sea level and increased storm intensity will increase damage to coastal regions and may make some entirely uninhabitable. Tourist destinations will shift, precipitating both economic downturns and new opportunities.

No single impact of climate change—barring such surprises as the disintegration of the West Antarctic ice sheet—will in itself be insurmountable. But coming as the world approaches what will probably be its peak population, in a time at once of increasing economic interdependence and growing contrast and hostility between the developed and developing world, climate change will exacerbate economic, political, and humanitarian stresses that we already handle, at best, imperfectly.

For most people, climate change remains low on the list of potential threats to future well-being. But some, such as those in the insurance industry, are seriously worried. Losses from natural disasters have skyrocketed in recent decades. Most of this increased loss results from more people choosing to live in more vulnerable places, but some of it is likely attributable to climate change. Reinsurance companies—those that back up conventional insurance in case of catastrophic losses—are particularly concerned. Indeed, the Swiss company Swiss Re is contributing to the development of carbon-free energy technologies, and along with other European reinsurers is helping to sponsor public education programs on climate change. In the United States, the ski industry has joined with the Natural Resources Defense Council in a campaign entitled "Keep Winter Cool." Even the U.S. military establishment, concerned about the security implications of sudden climate change, has commissioned a study of low-probability but high-impact climate events.

Climate change presents humankind with myriad challenges, but it also presents opportunities. These include opportunities for technological and lifestyle changes that wean us from fossil fuels and thus reduce the greenhouse emissions that drive climate change; opportunities for creative adaptation to a changing planet; and opportunities for wiser, more efficient use of energy, water, and other natural resources. In Chapter 16 I take an upbeat look at how humankind might move forward with energy technologies, practices, and policies that don't have to damage the local environment or upset the global climate.

BIG IDEAS

15.1 Climate models project future climates based on assumptions about greenhouse gas concentrations and other factors. They range from the simplest energy-balance models to large, three-dimensional, time-dependent versions that tax the fastest computers. Models are validated by projecting present-day climate from past conditions, by their successful reproduction of spatial and temporal distributions of climatic conditions, and by comparison with natural experiments such as volcanic eruptions.

15.2 Projections of future climates depend on assumptions about human behavior. Scenarios ranging from business as usual to concentrated efforts at reducing greenhouse emissions suggest a wide range in projected greenhouse gas concentrations by the year 2100. All scenarios show a global temperature rise through the twenty-first century. The greatest rise is expected at higher latitudes, especially in the Arctic.

15.3 Likely impacts of anthropogenic climate change include more heat waves and other high-temperature extremes; increased precipitation, both in amount and intensity; sea-level rise due to thermal expansion and melting land ice; species range shifts and extinctions; changes in growing seasons; and the spread of tropical diseases. Less likely are nonlinear "surprise" events such as abrupt changes in ocean circulation or the large land-based ice caps sliding into the sea. Acidification of the oceans is another effect of increased atmospheric CO_2.

15.4 Climate change will affect human society in a variety of ways. Sea-level rise will threaten low-lying areas; changes in temperature and precipitation will alter agricultural production, positively in some places and negatively in others. Heat waves and the spread of tropical diseases will have an impact on human health. In general, wealthy countries will be better able to adapt to a changing climate than developing countries.

TERMS TO KNOW

atmosphere–ocean general circulation model (p. 428)
equilibrium model (p. 428)
general circulation model (p. 428)
global climate model (p. 428)
grid cell (p. 430)
Moore's law (p. 431)
one-dimensional model (p. 427)

resolution (p. 430)
subgrid parameterization (p. 431)
thermohaline circulation (p. 449)
three-dimensional model (p. 428)
time-dependent model (p. 429)
two-box model (p. 427)
two-dimensional model (p. 427)

GETTING QUANTITATIVE

Global climate model, typical grid cell: ~50 to 200 km

Global warming by 2100, projected: ~1.5°C to 6°C

Sea-level rise by 2100, projected: ~20 to 60 cm

QUESTIONS

1. What was the approximate increase in global average temperature during the twentieth century? What is the range of increases projected for the twenty-first century?

2. Why does halving the size of a climate model's grid cells more than double the computer time required to calculate the equations?

3. Explain what is meant by a coupled climate model.

4. What is Moore's law, and what's its significance for climate modeling?

5. What are subgrid phenomena, and how are they handled in climate models?

6. What are some ways modelers validate their climate models?

7. Figure 15.13 shows a wide range in CO_2 emissions under the various SRES scenarios. Figure 15.14 shows much less variation in projected temperature increases. Why the difference?

8. What factor currently makes the greatest contribution to sea-level rise?

9. Antarctica might have made a negative contribution to sea-level rise during the twentieth century. How is that consistent with global warming?

10. Future climate depends in part on human choices. In that context, explain the two classifications used in IPCC climate projections to describe societal futures.

EXERCISES

1. In Chapter 13 I gave an alternate definition of climate sensitivity as the temperature increase per unit (W/m^2) increase in forcing. Given that a forcing of 3.75 W/m^2 is associated with a doubling of atmospheric CO_2, use Figure 15.6 to estimate both the transient and long-term climate sensitivity of the models used to produce that figure.

2. At the global average temperature of about 15°C, water increases in volume by about 0.02 percent for each degree Celsius of temperature increase. Use this result to make a rough estimate of the contribution to sea-level rise from a 2°C ocean temperature increase, assuming that only the top 500 m initially experiences the warming.

3. The volume of the Greenland ice sheet is some 2.85 million km³. If this ice melts, it will become water occupying about 92 percent of this volume. Given that the ocean covers 71 percent of Earth's surface, estimate the sea-level rise that would result if all of Greenland's ice melted. Compare with Table 15.1.

4. The HadCM3 climate model, developed by the Hadley Centre for Climate Prediction and Research in the United Kingdom, has grid cells whose horizontal extent measures 2.5° in latitude by 3.75° in longitude. (a) How many cells are needed to cover the surface of the globe? (b) What are the actual horizontal dimensions of a grid cell at the equator? (c) At 45° latitude?

5. Verify that the 1 percent per year increase in atmospheric CO_2 used in the model runs shown in Figure 15.6 results in CO_2 doubling in about 70 years.

RESEARCH PROBLEMS

1. In this problem you'll build a two-box climate model based on the energy flows shown in Figures 12.4 and 15.1. The table on the next page gives values of the various energy flows represented by the arrows in Figure 15.1. For example, arrow 1, representing incident solar energy, is taken from Figure 12.4 as 342 W/m². Arrows that involve radiation are described by the familiar expression $e\sigma T^4$ for the energy emitted per square meter from an object at absolute temperature T. For example, arrow 8, the infrared emitted upward from the surface, has the value $e_s\sigma T_s^4$, where the subscript s stands for "surface" and e is the emissivity, the property that describes an object's effectiveness as an emitter and absorber of radiation. Similarly, arrow 4, infrared emitted from the atmosphere to space, is $e_a\sigma T_a^4$. The warm atmosphere emits radiation in both directions, so the downward arrow 9 has the same value as arrow 4 in this simplified model. Arrow 3 is the infrared from the surface (arrow 8) that manages to escape to space. Since e describes both emission and absorption, the atmosphere absorbs a fraction e_a of the arrow 8 energy flow; this is the effect of the greenhouse gases. So the fraction of the surface radiation that gets through the atmosphere is $1 - e_a$; hence the entry for arrow 3 (this point was made earlier in connection with Figure 12.3). For this simplified model, ignore convection and evapotranspiration (arrow 7), as indicated in the table.

    Here's your job: First, consult Figure 12.4 and fill in the remaining quantities in the table. You should be able to read the values right off the figure. Next, set up equations representing energy balance for both the surface and atmosphere. On the left-hand side of each equation should be all the flows *to* the surface or atmosphere; on the right-hand side should be all the flows *away* from the surface or atmosphere. The equals sign between them expresses energy balance. The terms in your equations are the numbers or expressions in the table.

    You're almost ready to solve your equations, but first you need values for the surface and atmospheric emissivities: $e_s = 0.83$ and $e_a = 0.92$. These values aren't necessarily the most accurate, but they've been chosen so this oversimplified model gives reasonable results. The relatively high value of e_a represents the effect of greenhouse gases, which make the atmosphere nearly opaque to infrared. You also need the constant σ, introduced in Chapter 4; it's 5.67×10^{-8} W/m²·K⁴.

You now have two equations in the two unknowns T_s and T_a. Since these appear to the fourth power, it's far easier to solve for T_s^4 and T_a^4, treating each of these quantities as a single unknown. Then take the fourth root (take the square root twice, or raise to the power 0.25). The result is the surface and atmospheric temperatures as predicted by this simple two-box model. Are your answers reasonable? (Don't expect them to be exactly right.) Is the atmospheric temperature lower, as we discussed in Chapter 12?

Arrow	Physical meaning	Value (W/m²)
1	Incident solar energy	342
2	Sunlight reflected from atmosphere	
3	Infrared transmitted through atmosphere to space	$(1 - e_a)e_s\sigma T_s^4$
4	Infrared emitted by atmosphere to space	$e_a\sigma T_s^4$
5	Sunlight absorbed by surface	
6	Sunlight reflected from surface	
7	Convection and evapotranspiration	Ignore
8	Infrared emitted upward from surface	$e_s\sigma T_s^4$
9	Infrared emitted downward from atmosphere	$e_a\sigma T_a^4$

2. Now include the effect of convection and evapotranspiration, which is approximately proportional to the temperature difference between the surface and atmosphere. So use the term $c(T_s - T_a)$ for arrow 7, with $c = 4$ W/m²·K (again, not a particularly accurate value, but chosen for this simplified model). (a) Modify your equations to include this new term. (b) Solve for the two temperatures. Now you have both T and T^4 terms, so this is harder. You'll need to use computer software or a calculator that can solve nonlinear equations. Compare your answer with the result of Research Problem 1 and comment. (c) Calculate the flow represented by arrow 7. Is it roughly comparable to the convection and evapotranspiration arrows in Figure 12.4?

3. Get involved in climate modeling! Join the climate*prediction*.net project by signing up at www.climateprediction.net. In its spare time, your computer will join tens of thousands of others in sharing the effort of running large-scale climate models. You'll be kept up to date on the progress of climate*prediction*'s model experiments, and you'll make a real contribution to climate research.

Chapter 16

ENERGY AND CLIMATE: Breaking the Link

This final chapter asks what we can do to break the link between our energy demand and the deleterious impact our energy consumption has on Earth's environment. In many ways, the same remedies apply both to conventional impacts, such as air pollution and water pollution, and to climate change. But in other ways the truly global problem of climate change presents unique challenges that require new ways of thinking and acting about energy and the environment.

As I explained at the beginning of Chapter 6, there's an important distinction between fossil emissions of traditional pollutants and the greenhouse gas CO_2. Traditional pollutants are undesirable and largely unnecessary by-products of fossil fuel combustion. With enough money and technology we can reduce these pollutant emissions to arbitrarily low levels. For that reason, traditional pollution presents no barriers *in principle* to our continued use of fossil fuels or even to an increase in fossil fuel consumption. I say *in principle* because we have to be willing to pay what it takes to achieve our desired air- and water-quality standards, and to restore environmental quality damaged in fossil fuel extraction and transportation. But we know how to do that if we, as a society, so choose.

Carbon dioxide is different. Along with water, it's a necessary product of fossil fuel consumption. To burn fossil fuels is to make CO_2, so on a certain level we *want* to turn fossil fuels into CO_2! Whereas traditional pollutants occur in relatively small quantities, we produce huge amounts of CO_2 because efficient fossil fuel combustion converts essentially all the fuel carbon into CO_2. For example, the data in Table 7.2 show that a 1-GWe coal-fired power plant produces 1,000 tons of CO_2 per hour, compared with some 75 tons of air pollutants and ash. So climate-changing CO_2 emissions are more than an order of magnitude greater than conventional pollutants. And the impact of CO_2 emissions is truly global, while most pollution has primarily local and regional impacts. For all of these reasons, I want to emphasize the link between energy and climate in this chapter.

16.1 Carbon Emissions: Where We're Going and Where We Need to Be

As I write this chapter, the U.S. National Oceanic and Atmospheric Administration has made headlines with an announcement that last year's atmospheric CO_2 concentration set a new record. Actually, this isn't the least bit surprising, and it doesn't matter what year "last year" was. As Figure 13.8 shows, atmospheric CO_2 has been rising steadily throughout the industrial era of the past few centuries. This means every year sees a new record concentration of atmospheric CO_2. In the early twenty-first century, the annual increase has ranged from about 1.5 ppm to nearly 3 ppm, as you can verify in Research Problem 1. If this trend continues, we're looking at a rise in atmospheric CO_2 of something like 200 ppm by the year 2100, giving a concentration of nearly 600 ppm. However, the rate of increase has itself been increasing—again, try Research Problem 1 to verify this—so we're likely to reach that concentration somewhat sooner.

"Dangerous Anthropogenic Interference"

The ultimate objective of the **United Nations Framework Convention on Climate Change**—an international agreement signed at the 1992 Earth Summit in Rio de Janeiro and whose subsidiary agreements include the Kyoto Protocol—is "stabilization of greenhouse gas concentrations in the atmosphere at a level that would prevent dangerous anthropogenic interference with the climate system." As Chapter 15 suggests, there's no clear threshold at which point our interference with the climate system (read greenhouse emissions) can be considered "dangerous." But most climate scientists believe that we need to hold atmospheric CO_2 to at most a doubling of its preindustrial level of 280 ppm. Such a doubling, to 560 ppm, will likely cause serious and disruptive climate impacts but probably not the most disastrous of the potential upheavals outlined in Chapter 15. However, others would put the level of dangerous anthropogenic influence considerably below the 560-ppm doubling of preindustrial concentrations. And remember that today we're more than 100 ppm above that preindustrial 280 ppm level, and we're closing in fast on 400 ppm.

A more direct measure of "dangerous interference" is the global temperature rise, which many scientists have argued should be limited to 1°C or 2°C additional warming beyond what we've already experienced in the twentieth and early twenty-first centuries. This criterion is based on the amount of sea-level rise that would result from such warming and the significance of coastal land loss; the factor-of-2 uncertainty comes from our imperfect knowledge of all the processes that contribute to sea-level changes. Using a climate sensitivity of 0.75°C per watt per square meter, that 1°C to 2°C "acceptable" temperature rise corresponds to an increase in anthropogenic climate forcing of some 1.3 to 2.6 W/m². The result of Exercise 5 in Chapter 13 indicates that CO_2 doubling—adding 280 ppm to the preindustrial concentration—corresponds to a forcing of about 3.7 W/m², which means our 1.3 to 2.6 W/m² allow for atmospheric CO_2 to increase by about 100 to 200 ppm, to around 490 to 590 ppm. These numbers are rough, and they depend on what one takes for the climate sensitivity and the relationship between CO_2 and forcing, but they confirm that a dangerous anthropogenic influence probably occurs at roughly the 560-ppm doubled CO_2 concentration, and quite possibly at a lower concentration.

How We're Doing

So how are we doing? I've noted that humankind's current rate of CO_2 emissions currently increases atmospheric CO_2 concentration by roughly 2 ppm each year. This rate gives us 50 to 100 years before we reach the maximum "allowable" concentration range of 490 to 590 ppm. However, the rate of CO_2 emissions probably won't remain constant—it could increase or decrease, depending on human decisions. Chapter 15 outlined several scenarios the Intergovernmental Panel on Climate Change (IPCC) has used in projecting future greenhouse emissions and associated temperatures. Figure 15.13, specifically, shows widely varying estimates of future CO_2 emissions, while Figure 15.14 shows less dramatic variations in projected temperatures. It's sobering to note that most of the temperature projections for the year 2100 lie above the 1°C to 2°C rise that we're now considering "acceptable." Those that might meet the 1°C to 2°C standard are based on scenarios involving much more environmentally friendly policies (scenarios B1 and B2) or a massive switch away from fossil-based energy sources (scenario A1T, in which we develop extensive nonfossil energy technologies). Coupled with global carbon-cycle models, the emissions projections of Figure 15.13 lead to projections of the atmospheric carbon concentrations shown in Figure 16.1. Again the results are sobering: Only two of the scenarios remain near or below a doubling of preindustrial CO_2 through the twenty-first century, and none of the other scenarios even begin to stabilize CO_2 concentrations

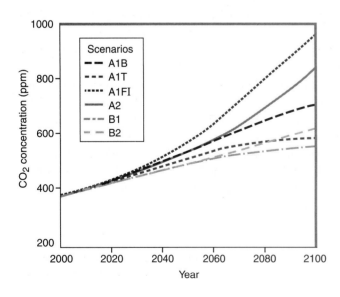

Figure 16.1
Projections of twenty-first-century atmospheric CO_2 concentration, based on the IPCC SRES scenarios introduced in Chapter 15 and a carbon-cycle model.

within the present century. Even more sobering is the fact that so far we're doing very little to move away from the business-as-usual fossil-intensive A1FI scenario. The longer we wait, the more difficult it will be to stabilize atmospheric CO_2 at a "safe" level.

We as a species need to take steps to minimize anthropogenic climate change. I can think of four approaches we might follow. We could compensate for our climate-changing emissions with global-scale engineering projects that alter Earth's energy balance or the carbon cycle. Examples include putting giant mirrors in orbit to reflect sunlight and lower the solar energy input, or fertilizing the open oceans with iron to encourage the growth of phytoplankton that take up CO_2. Although such schemes have their proponents, I'm not going to say any more about them because I think we need to know a whole lot more about Earth and its complex natural systems before we start tinkering with the planet in such global ways. Dismissing global engineering leaves us three options, all of which require changes in our energy behavior. The first of these options is to continue producing CO_2, but to treat it as just another pollutant and try to keep it out of the atmosphere. Our second option is a switch to energy sources that entail less CO_2 production. Finally, we could simply use less energy—either by changing our lifestyles to require less or by using energy more efficiently. The remainder of this chapter explores these three options.

16.2 Carbon Capture and Sequestration

Capturing and removing CO_2 from the exhaust streams of fossil fuel combustion facilities is a daunting task, given that CO_2 is a necessary product of fossil fuel burning. For the mobile fossil-fueled engines that power today's cars, trucks, and airplanes, it's for all practical purposes impossible. But CO_2 capture might be feasible for large stationary power plants and industrial boilers. The IPCC estimates that worldwide today there are some 8,000 individual sources producing annually at least 100,000 tons of CO_2 each, and that these might be candidates for CO_2 capture. Together, these large sources account for nearly 14 Gt of CO_2 emissions annually, which is equivalent to 3.7 Gt of carbon or about half of the 7 Gt of carbon emissions from fossil fuel combustion for the carbon cycle as depicted in Figure 13.10. The larger of these sources produce CO_2 at prodigious rates; for example, I alluded earlier to the 1,000 tons per hour of CO_2 from a 1-GWe coal-fired power plant. This CO_2 is released in gaseous form at approximately atmospheric pressure, so it occupies a huge volume. (Exercise 1 shows the volume of that hourly 1,000 tons to be about half a million cubic meters.) So there's simply an enormous mass and volume of CO_2 to be captured from the exhaust stream, and once it's captured it has to be sequestered in permanent storage where it can't get into the atmosphere.

Capturing Carbon Dioxide

A conventional fossil-fueled power plant or boiler combusts its fuel in the presence of air, which is roughly 20 percent oxygen and 80 percent nitrogen. Most of the oxygen combines with the fuel to make CO_2 and H_2O, but very little of the nitrogen does (although it's the main source of NO_x pollution). So the exhaust stream is largely harmless atmospheric nitrogen, along with water vapor and the CO_2 that we'd like to capture. Typically, this mix—called **flue gas**—is only about 15 percent CO_2. Separating the CO_2 from the diffuse flue gas isn't easy.

One approach is to react the flue gas with chemicals that absorb CO_2, and then heat the chemicals to release concentrated CO_2 for subsequent processing and sequestration. That second step is necessary to make the absorber chemicals available for reuse. The vast quantities of CO_2 produced in fossil fuel combustion essentially rule out one-time chemical removal.

Another approach is to remove carbon before combustion. This is especially appropriate with the efficient combined-cycle power plants I described in Chapter 5. Many combined-cycle plants burn natural gas, which already has a lower carbon content than coal or oil. But some plants use coal, not directly but by first gasifying the coal through a reaction with limited oxygen to produce a mix of hydrogen and carbon monoxide—a product called *syngas* (short for synthetic gas). The syngas then burns in the gas turbine of the combined-cycle power-generating system. By adding more oxygen in the gasification phase, it's possible to produce a mix of mostly CO_2 and hydrogen, rather than carbon monoxide and hydrogen. Capturing the CO_2 at this precombustion stage is relatively easy, leaving a combustible gas that's mostly hydrogen.

A final proposed approach to CO_2 capture is to use pure oxygen in place of air to support fossil fuel combustion. The flue gas is then mostly CO_2 and water, making it easier to separate the CO_2. But extracting oxygen from air is expensive and consumes energy, and combustion with pure oxygen occurs at temperatures too high for most common materials to withstand.

I emphasize that these CO_2-capture options are practical only for large stationary sources of CO_2. There's simply no hope of removing CO_2 from mobile sources. There's also little hope of removing CO_2 from residential and commercial heating systems or from smaller industrial facilities. Yet, as I mentioned in the previous section, the large sources account for only about half of all global CO_2 emissions. Figure 16.2 quantifies this point for U.S. emissions.

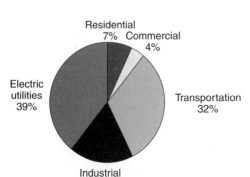

Figure 16.2

Sources of CO_2 emissions for the United States in the early twenty-first century. Capture of CO_2 is practical for electric utilities and for only some industrial facilities; together, these account for about half of all CO_2 emissions.

Sequestration

Once we've captured CO_2 we have to sequester it away where it can't escape to the atmosphere. But where? Many deep sedimentary rock formations contain dense brine—extremely salty water—and injecting CO_2 at high pressure would replace the brine with CO_2. At a depth of 1 km, the pressure is some one hundred times atmospheric pressure, and under these conditions CO_2 is a dense fluid that combines the properties of a gas and a liquid. If impermeable layers cap the porous rock formation, then injected CO_2 is likely to remain in place.

Another place to sequester CO_2 is in the very wells that produce fossil petroleum. Carbon dioxide is actually a commercial commodity for the oil industry, because its physical and chemical properties help free viscous oil that clings to rock in oil deposits. Oil workers routinely pump pressurized CO_2 into old wells to extract all the oil possible. As the price of oil rises, so does the economic advantage of this procedure. A recent pilot project at a Canadian oil field sequestered 5 million tons of CO_2 from a North Dakota coal-gasification plant and, in the process, increased the oil field's output by 10,000 barrels per day (see Exercises 3 and 4 to get a feel for these numbers). Oil-field sequestration isn't ideal, however, because a typical oil field is riddled with thousands of oil-well penetrations, each offering a potential escape route for sequestered CO_2. Even abandoned, cement-filled wells are vulnerable, because CO_2 reacts with underground water to make carbonic acid (H_2CO_3), a mild acid that, over time, can corrode cement.

Yet another place to put CO_2 is the deep ocean. Figure 13.10 shows that Earth's oceans already store nearly 40,000 Gt of carbon—far more than any reservoir other than the rocks that make up the planet itself. You can show in Exercise 5 that it would take more than a century for all the carbon from the world's power plants to increase oceanic carbon by a mere 1 percent. But most oceanic carbon isn't in the form of CO_2. Can we be sure that excess carbon in the form of CO_2 will remain sequestered in the ocean? What effects might it have on ocean chemistry and marine ecosystems? Is it even legal, under the UN Convention on the Law of the Sea, to dump such a waste product into international waters? And what's the cost—in dollars and energy—of compressing CO_2 to such high pressure that it can be forced into the deep ocean? All these questions make ocean sequestration a tentative proposition. On the other hand, we know that dumping CO_2 into the atmosphere is having a deleterious effect on Earth's climate. Furthermore, much of the CO_2 that goes into the atmosphere ends up in the oceans, so we're already increasing oceanic CO_2, but in the more vulnerable, life-supporting surface layers. So might we be better off short circuiting the process with direct CO_2 injection into the deep ocean?

Effectiveness and Safety of Carbon Sequestration

Carbon sequestration is viable only if we can be sure that the sequestered CO_2 can't find its way back into the atmosphere. Even relatively slow leakage would defeat the purpose and lead to climate-changing increases in atmospheric CO_2. Sudden, catastrophic leakage is outright dangerous, as heavier-than-air CO_2 collects at ground level and causes asphyxiation. A natural disaster of this sort killed thousands of people in Cameroon in

1986 when volcanic CO_2 at the bottom of Lake Nyos suddenly burst to the surface. Accidents like this one are unlikely with properly designed and managed CO_2 sequestration, but slow leaks are a very real concern.

Carbon dioxide sequestration is probably a viable way to reduce greenhouse emissions if we can be sure that sequestration facilities will retain at least 90 to 95 percent of the stored CO_2 for the first century, and 65 to 95 percent for 500 years. Those numbers appear feasible: Studies cited in the IPCC's report *Carbon Dioxide Capture and Storage* suggest that well-engineered geologic sequestration facilities are very likely to retain 99 percent or more of their stored CO_2 for at least a century and are likely to retain that much for a millennium or more. But designing a well-engineered sequestration facility is not easy. Because fluid CO_2 at high pressures is buoyant in brine, it rises until it encounters an impermeable rock layer. So it's incumbent on geologists to find truly impermeable rock under which CO_2 can be stored—layers with no hidden cracks or faults, no uncharted oil wells or other penetrations, and no upturned edges around which the CO_2 could escape.

Prospects for Carbon Capture and Sequestration

Technically, carbon capture and sequestration (CCS) seems a feasible strategy for helping to mitigate anthropogenic climate change. However, there are economic and policy considerations as well. Although the successful implementation of CCS might prolong the fossil fuel era and delay the need for disruptive changes in our energy infrastructure, this development would be both good and bad. In the short run, it might put off the need to make difficult energy choices. But resource limitations dim the long-term prospect for fossil fuels, so even with CCS we need to start diversifying our mix of energy sources.

Carbon capture and sequestration can help, but it's far from a complete solution to anthropogenic climate change. First, as I've pointed out, only about half of today's sources of anthropogenic CO_2 are candidates for CCS. Even though the implementation of CCS doesn't mean a wholesale change in our energy sources, construction of CCS facilities is a massive enterprise and not one that's likely to be done quickly. Therefore, the IPCC estimates that only 20 to 40 percent of fossil carbon emissions might be amenable to CCS by the year 2050. Everything you know from Chapters 12 to 15 suggests that we can't wait that long for major reductions in CO_2 emissions.

Finally, there are energy and economic costs to CCS. The CO_2 capture process is energy intensive, and it takes even more energy to pressurize CO_2 for injection into sequestration facilities. For a modern, conventional coal-fired power plant, adding CCS would result in 24 to 40 percent greater fuel consumption for a given electric power production. This increases emissions of pollutants and solid waste, and it increases the cost of electric power by some 40 to 85 percent for a conventional coal plant. Although CCS can capture as much as 95 percent of the CO_2, the increased fuel consumption somewhat offsets this impressive reduction in emissions. Figure 16.3 illustrates the balance between CCS emissions reductions and increased fuel demand. It's important to note that this

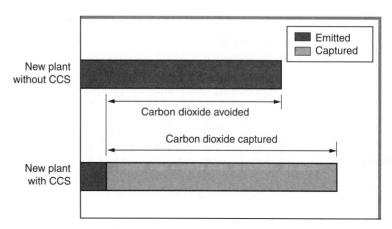

Figure 16.3
A new power plant with carbon capture and sequestration technology consumes more fuel per kilowatt-hour of electrical energy produced. This additional fuel consumption causes increased CO_2 emissions, but because the carbon captured and sequestered isn't emitted, there is a net reduction of CO_2 emissions by the amount labeled as "carbon dioxide avoided."

discussion applies to new power plants, built from the ground up with integrated CCS facilities. Retrofitting CCS to existing power plants is a more difficult and costly prospect. On the other hand, economic cost might drop and overall fuel supply could increase if the sequestered carbon were pumped into oil fields to enhance oil production.

Carbon capture and sequestration is an interesting approach, one that might let us continue our fossil-fueled business as usual a bit longer, at least for large, stationary CO_2 sources. But it's no panacea, and CCS alone won't stop anthropogenic climate change. Like many other technological fixes, CCS hasn't yet been proven effective and economical on a large scale. CCS may well play a role in our efforts to mitigate anthropogenic climate change, but it isn't the whole answer or even a major part of it. So there's no getting around it: We must move away from fossil fuels.

16.3 Alternative Energy Sources

Chapters 7 to 11 introduced alternative energy sources that don't involve fossil fuels. Some, such as solar energy and its derivatives, wind and hydropower, have the advantage of being renewable—they come to Earth in a steady stream and can't run out. Another solar derivative, biomass, is renewable if it's harvested sustainably. Other sources, such as nuclear energy, aren't strictly renewable but promise nearly unlimited fuel supplies in their more advanced manifestations, namely breeder reactors and, especially, nuclear fusion. Geothermal and tidal energy can probably make only modest contributions to our global energy supply but could be significant in localized regions. All of these alternatives have the advantage that their emissions of CO_2 and other greenhouse gases are far lower than those of fossil fuels.

Far lower—but not zero. In the chapters on the various energy sources, I discussed the environmental impacts of each, including greenhouse emissions. One conclusion was clear: No energy source is completely benign. But some are better than others, and how much better depends on which measures you apply, as well as on factors that are harder to quantify, such as the risk of catastrophic nuclear accidents or dam failures. So how do the climate impacts of the different energy sources compare?

It's easy to measure greenhouse emissions from fossil energy, since nearly all of it comes from the combustion process. Fossil fuel is a commercial commodity, so we know pretty accurately how much of it we're burning. But greenhouse emissions from other energy sources are harder to quantify, because they don't involve the ongoing production of energy as much as they do processes such as manufacturing energy facilities, producing cement, mining nuclear fuels, growing biomass, and transporting materials. Chapter 7 mentioned a study of greenhouse emissions associated with nuclear power; Chapter 10 gave the surprising result that building hydroelectric dams in tropical regions may cause methane emissions with greater climate impact than comparable fossil-fueled facilities. A comprehensive study for the International Atomic Energy Agency examined nearly all the alternative energy sources and gave estimated high and low ranges for greenhouse emissions; Figure 16.4 summarizes the results. In most cases, the high estimate includes existing technologies, whereas the low estimate refers to technologies expected to be available by the year 2020. For hydropower, the high category is for tropical dams and the low category is for run-of-the river power plants that require far less concrete and don't entail large, stagnant reservoirs. For other sources, the high and low estimates are simply ranges of quantities that are difficult to determine accurately.

Much of the CO_2 emissions implied in Figure 16.4 comes from fossil fuels consumed in the manufacture and installation of energy facilities. That's why solar energy ranks fairly high in CO_2 emissions, because present-day semiconductor fabrication is very energy intensive (although this will change as techniques for manufacturing thin-film photovoltaic cells become widespread). Carbon dioxide emissions for some alternative energy sources also result from cement production; that's the case for nuclear power plants, hydro dams, and the massive concrete foundations of large-scale wind turbines. Once nonfossil energy sources are deployed on a large scale, however, the energy used in manufacturing

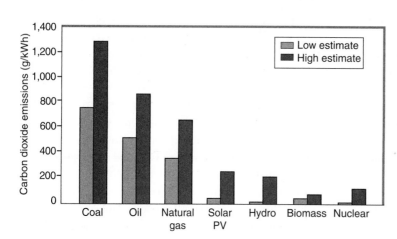

Figure 16.4
High and low estimates of CO_2 emissions from various energy sources. Emissions include those from combustion (from fossil fuels only) as well as all other aspects of power-plant construction, transportation, fuel and materials mining, among others. Values are in grams of CO_2 per kilowatt-hour of electricity generated. The high estimates for solar and hydro may be misleading: For solar, the estimate is based on outdated manufacturing techniques, and for hydro, it is based on a typical hydro dam in tropical Brazil.

such facilities can itself shift away from fossil fuels, lowering the associated carbon emissions.

Although alternative-energy advocates may be disappointed to find that their favorite alternative isn't necessarily "pure" from a climate-change standpoint, it's nevertheless clear from Figure 16.4 that all the alternatives have far less climate impact than fossil energy. So why not switch to one or more energy alternatives and be done with the climate-change problem? That may well be what's needed, but it's not necessarily an easy switch. Of the energy alternatives considered in Chapters 7 to 11, only hydro and nuclear power currently make significant contributions to the global energy supply. Both are best suited to producing electrical energy, so they're not good matches for powering transportation vehicles. Hydropower potential in the developed world is largely tapped already, and nuclear power continues to face an uncertain future. Geothermal and tidal energy, though useful in some local situations, just aren't sufficient to make a major contribution. Direct solar, wind, and biomass have the potential to supply far more energy than we use. Although their contribution is growing rapidly, it will be some time before they can achieve their full potential. Direct solar and wind, too, aren't generally adaptable for transportation. And although wind and some biomass fuels can now compete with conventional energy sources, solar photovoltaics remain economically out of the running for all but remote applications.

Substituting nonfossil sources in transportation is particularly challenging. Biofuels can help, but it will be decades, if ever, before we have enough biofuel capacity to supply the world's burgeoning vehicle fleet. Hydrogen has promise for powering transportation, but remember that there's no hydrogen energy resource here on Earth. So unless we make hydrogen with some nonfossil means—for example, using nuclear, hydro, wind, or photovoltaic-generated electricity—then hydrogen doesn't help much with climate change (although fossil-based hydrogen production may be more amenable to CCS, as I described in Section 16.2). Nuclear fusion could offer virtually unlimited energy for all our needs, including hydrogen for transportation, but fusion is decades or more in the future.

A look at Figure 16.4 suggests a more modest approach to carbon reduction within the fossil-fueled economy—switching from coal and oil to natural gas. You've seen how the chemistry of these different fuels gives roughly a factor-of-2 spread in carbon emissions per unit of energy produced. And natural gas has featured more prominently in the world's fuel mix since 1980 (Fig. 16.5), although the data in Figure 16.5 for the years after 2000 show that this situation may be changing. As Chapter 5's discussion of fossil fuel resources suggests, the long-term prospects for oil and natural gas aren't good, and as resources decline, these fuels increasingly will be subject to price spikes and other market uncertainties. A judicious switch to natural gas, where possible, is a useful short-term strategy for reducing the climate impact of fossil fuels, but it's not an answer for the long term. At best, natural gas can act as a bridge to a future economy that's not based heavily on fossil fuels.

So which energy source is the long-term solution? Over the next half-century or so, there simply isn't one answer. Maybe someday we'll live in a society powered entirely by

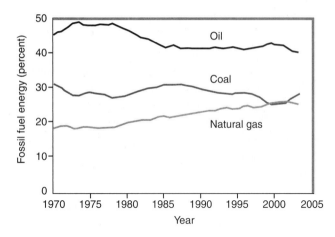

Figure 16.5
Percentage of fossil fuel energy attributable to each of the three fossil fuels from 1970 to the early twenty-first century. Natural gas's share increased until about 2000 but has been relatively stagnant since then.

direct solar energy or nuclear fusion—the two alternatives whose availability far exceeds humanity's energy demand. We ought to be working toward that day, but in the meantime we need to adopt an energy strategy that can reduce carbon emissions over the next few decades. That strategy will surely mean using the many different alternative energy sources available to us. But alternative energy can't be the whole solution to anthropogenic climate change. Carbon capture from fossil fuels might play a modest role, too, but there's still one much more important and easier approach to consider.

16.4 Using Less Energy

Phrases such as *using less energy* and *energy conservation* conjure up images of people clad in bulky sweaters shivering in dimly lit rooms or crammed into impossibly tiny vehicles. But that's not what I mean by using less energy. I mean something a lot smarter—namely, using energy more efficiently so that we get the same benefits and living standards we're used to, but with a lot lower energy consumption.

Using less energy is the simplest, most straightforward, and often the most economical way to reduce climate-changing CO_2 emissions, not to mention many other deleterious environmental impacts. And the "resource" here is huge, thanks to the inefficiencies inherent in most of today's energy technologies. Even the most advanced technologies now in commercial use have energy efficiencies far below their theoretical maxima, and well below the efficiencies of the best prototype models. Figure 16.6 suggests the energy efficiency improvements available with two common technologies, namely vehicles and refrigerators.

Unlike some of the "greenest" alternative energies, which have yet to demonstrate that they're economically or technologically feasible on a large scale, energy efficiency is a proven way to reduce carbon emissions. We've done it before, and we continue—sometimes modestly, sometimes dramatically—to improve efficiency in some energy-use sectors. In Chapter 2 I introduced the notion of energy intensity, the ratio of a country's

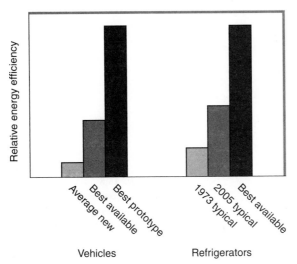

Figure 16.6
Potential efficiency improvements for light-duty vehicles and household refrigerators. The new and best commercially available vehicles are for model year 2006; the new-vehicle average is 21 miles per gallon and the best is the Honda Insight hybrid at 66 miles per gallon. The best prototype is a Volkswagen built in 2002 that goes 282 miles on a gallon of diesel fuel. The best available refrigerator is a Sunfrost R-19, a medium-size vertical refrigerator designed for solar-powered homes and other applications demanding maximum efficiency. Both categories show that large energy savings would result if existing units were replaced with the best commercially available models.

energy consumption to its gross domestic product (GDP). Figure 2.8 indicates almost steady reductions in U.S. energy intensity throughout the twentieth century, with a drop of nearly 50 percent since as recently as 1975. Many industrialized countries show similar trends. A reduction in energy intensity is like discovering an equivalent amount of oil, or natural gas, or other fuel—usually at a price far below what it would have cost to purchase the actual fuel.

It's particularly instructive to consider the times when we consciously tried to reduce energy consumption under pressure from market forces, shortages, or government incentives. Figure 16.7 focuses on the period since 1965 in the United States, revealing some

Figure 16.7
U.S. energy intensity (energy consumption per dollar of gross domestic product) since 1965. The actual intensity dropped by half between 1970 and 2004, but extrapolation of the steep drop during the late 1970s and early 1980s (dashed curve) would have resulted in a greater decline, which could have displaced two-thirds of our current oil imports.

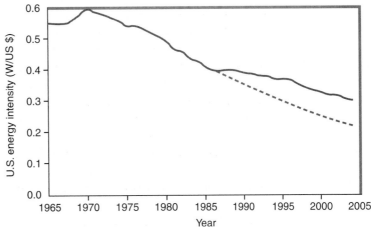

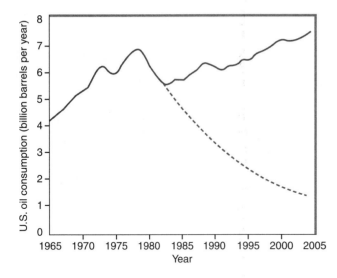

Figure 16.8
U.S. oil consumption since 1965. The dashed curve extrapolates the steep decline in the late 1970s and early 1980s; by 2004 the difference would have been enough to displace all of our imports and make the United States a net oil exporter.

15 years of rapid decline in consumption as the energy shortages of the early 1970s drove Americans to smaller cars, better home insulation, more efficient appliances and heating systems, and even solar energy. I've fit an exponential curve to this data, suggesting that U.S. energy intensity was then decreasing annually at about 3 percent, and I've extrapolated that curve to show where we would be today had the trend continued. Although it may not look huge, the difference between the actual and extrapolated declines would be the energy equivalent of some 3 billion barrels of oil per year, as you can verify in Exercise 6. This amount is two-thirds of our current oil imports. The decline in energy use is even more dramatic if we look at oil alone; there, Figure 16.8 shows that our consumption declined more than 6 percent per year in the early 1980s. Had that trend continued, our oil consumption today would be only about 18 percent of what it actually is, and the United States could be a net oil exporter!

You might argue that we surely couldn't have continued those downward trends indefinitely, and you'd probably be right. But it wasn't technological barriers that slowed Figure 16.7's trend toward greater energy efficiency and reversed Figure 16.8's decline in oil consumption—it was a drop in the price of oil. And it wasn't technological failures that derailed the nascent solar-energy industry of the late 1970s and early 1980s—it was a conscious political decision to eliminate tax credits and research funding designed to grow the solar alternative. Had that not happened, we might now have a robust solar industry whose contribution to the U.S. energy mix could greatly exceed its pathetic one-eighteenth of 1 percent share in the "other" category in Figure 2.5. On a more positive note, California's energy policies over the past several decades has resulted in that state's substantial deviation from the national rate of growth in per capita electricity consumption, as shown in Figure 16.9.

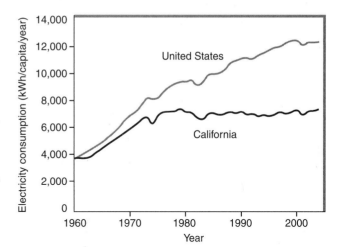

Figure 16.9
Per capita electrical energy consumption in California compared to the rest of the United States. The two were closely correlated in the 1960s, but subsequent policies encouraging energy efficiency leveled California's per capita electricity consumption while the rest of the country's consumption continued to grow. Note that these curves show per capita consumption and don't account for actual consumption growth associated with population increases.

A look at specific energy uses shows how we can achieve substantial reductions in energy intensity or, equivalently, increases in efficiency. Refrigerators, for example, are among the greatest energy consumers in a typical home, after heating and cooling systems and hot-water heaters. Their energy intensity has fallen by nearly 80 percent since the 1970s. Figure 16.10 shows this trend and makes clear that the implementation of state and national standards has been a significant factor; you can explore the associated energy savings in Exercise 7. Airplanes are another example. Cost pressures have driven airlines to scrap older models for newer, more fuel-efficient ones; as a result, the U.S. airline fleet today uses only about one-third of the energy it did in 1970 per passenger per mile flown. Similarly, the most energy-efficient cars available have been getting better

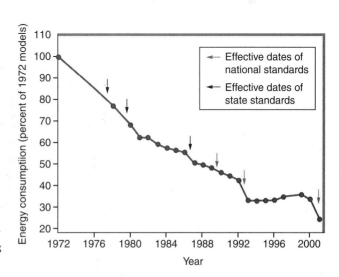

Figure 16.10
Energy consumption of household refrigerators in the United States has dropped dramatically in recent decades. Arrows mark the introduction of state and federal efficiency standards, which have played a significant role in decreasing energy consumption.

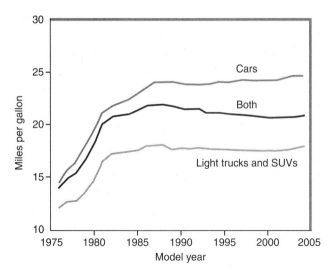

Figure 16.11
Fuel efficiency of the U.S. car and light-truck fleet over time. Although car and truck fuel efficiency has increased very slightly in recent years, the the trend away from smaller cars to light trucks and SUVs has lowered the combined average since the mid-1980s.

over time, although here the picture is more mixed as we—especially we in the United States—have chosen to drive larger cars and the so-called light trucks that include most SUVs (Fig. 16.11).

One caution here: Energy intensity is not a direct indicator of energy consumption. We can make refrigerators more efficient, but if we become so affluent that we put two or three refrigerators in each home, then we defeat those gains. We can make cars twice as efficient, but if we drive them twice as far then we've gained nothing. We can decrease our per capita energy consumption, but if there are more of us then our total consumption still increases. For the period shown in Figure 16.7, in fact, U.S. per capita energy consumption did rise, although only by a modest 0.6 percent. The per capita GDP, in contrast, rose 78 percent, which is why our energy intensity declined significantly. Our overall energy consumption nevertheless increased substantially because the population increased, but it would have increased a lot more if our energy efficiency hadn't improved.

Compounding Efficiency Gains

Energy efficiency has a way of compounding. Take home heating, for example: A homeowner achieves modest energy savings by installing insulating window shades and by caulking to eliminate air infiltration. This reduces the amount of natural gas or oil burned to heat the house, but it's still burned in the same furnace. Building a new home presents greater opportunities for savings—a lot more insulation goes into the structure, windows are upgraded to double-glazed argon-filled units with substantial R values, and air infiltration is minimized. So the new house can use a smaller, less-expensive furnace—one that not only burns less fuel but that also took less material and energy to manufacture. The energy-efficient house maintains a more even temperature, so it needs less piping or ductwork to deliver heat throughout the house, which means more savings in costs and manufacturing energy. Insulate the house to an even greater extent, and you may be able to eliminate the heating system altogether, relying entirely on solar gain, the metabolism of the occupants (remember, that's about 100 W each), and the heat generated by appliances, lighting, computers, and other energy-consuming devices. Down, down, down go the home's energy consumption, the amount of energy and materials that go into making it, and the costs. You can make similar arguments for virtually every other energy-consuming product, from cars to airplanes to appliances to entire factories.

Economics of Energy Efficiency

Imagine that you're the owner of an electric utility faced with a 10 percent increase in power demand because of population growth, the trend toward larger homes, the proliferation of electrical gadgets, and the increasing "leakage" of power from supposedly "smart" electronic devices that consume energy even when they're turned off (see Exercise 8). So what do you do? You could crank up your generating facilities to maximum output, if they aren't already there. You could buy power from other utilities, but this solution is subject to the vagaries of the market, because your neighboring utilities, too, face increasing demand. You could ration your power through browouts of individual towns or neighborhoods on a rotating basis—a draconian approach that's common in developing countries and a rare last resort in the developed world. You could build another power plant, at huge capital cost and with a long lead time before your new generators come online. Or you could pay your customers to use less electricity by providing them with compact fluorescent lamps that use only 20 percent as much energy as incandescents; giving them extra insulation for their water heaters, reducing so-called standby losses from the hot water stored in the tank; providing rebates for purchasing newer, more efficient appliances; or subsidizing individual wind and solar electric installations. If these energy-saving measures cost less than the fuel it would take to generate more electricity, or if they obviate the need to build that expensive new power plant, then you've saved money. You've also prevented more carbon emissions, or radioactive waste, or the disruption of a river ecosystem with dams.

The idea that reducing demand for a product can actually save money for its producer is a radical one; after all, it's unusual to find a manufacturer who doesn't want to increase sales. But it's an idea whose time has come, and one that benefits all parties involved: the power company, the consumer, and the environment.

The same arguments apply to other sectors of the energy industry. Take oil, for example: We've cut our energy intensity for oil in half since 1975, meaning that we now get twice as much GDP out of each barrel of oil. We can probably cut it in half again using currently available energy technologies. The cost of applying those technologies would be about $12 per barrel of oil saved, which is a huge bargain. As I write these words, the cost of oil stands at $95 per barrel. If you run a company that is looking to invest in a new piece of equipment, you'd happily pay the extra $12 for the more efficient version: This expenditure would save you $95 in energy costs, so you'd pocket the $83 difference.

What we're talking about here are *negawatts* and *negabarrels*—watts of power we don't need to generate and barrels of oil we don't need to burn. We can ascribe a cost to them just as we do for the "real thing." This cost is associated with buying more-efficient energy technologies rather than more energy. If the cost of greater efficiency is less than the cost to produce real energy, then it's more economical to buy the *negenergy*. And, because of the compounding effect described in the preceding section, moving from modest efficiency gains to super-efficient designs may actually lower the cost of the negenergy.

Box 16.1 Negawatts and Hypercars: The Rocky Mountain Institute

The term *negawatt* is the brainchild of Amory Lovins, cofounder of the Rocky Mountain Institute (RMI) in Snowmass, Colorado. Lovins champions the view that dramatic gains in energy efficiency are both technologically possible and economically preferable to business as usual. Lovins and and his colleagues at RMI make their case in numerous well-documented publications, such as *Winning the Oil Endgame: Innovation for Profits, Jobs, and Security*. Such efficiency gains would not only buy us reduced greenhouse emissions and less conventional pollution, but also security from the economic, political, and even military upheavals that could accompany the final decades of a fossil-fueled economy running on empty.

RMI's physical facilities carry the philosophy of energy efficiency to its extreme. The 4,000-square-foot headquarters building is nestled into a south-facing slope high in Colorado's Rocky Mountains (Fig. 16.12), and it doesn't have a conventional heat source. Solar gain through advanced windows and the glazing of a built-in greenhouse provides all the energy needed. The glazing boasts R values as high as 8, far greater than that of conventional windows. Reflective coatings let visible light through, but they block outgoing infrared to reduce losses even further. The building retains its solar-supplied energy with walls insulated to $R = 40$ and its roof to $R = 80$, and some 500 tons of concrete and soil provide a thermal mass that maintains an even temperature despite fluctuations in sunlight. The RMI building cost only about 1 percent more than a conventional structure, and its energy savings paid for that excess in a mere three years. What was state of the art when RMI was constructed in the 1980s is now widely available commercial technology. Lovins cites contemporary examples of RMI's philosophy applied to large office buildings that use 80 to 90 percent less energy than conventional structures while taking less time and money to build.

Amory Lovins is also chair of Hypercar, Inc., an RMI spin-off aimed at commercializing the vehicular equivalent of its super-efficient building. Hypercar envisions lightweight carbon-composite materials reducing vehicle weight, engine power requirements, and fuel consumption while increasing safety. Early Hypercar designs would use gasoline-electric hybrid power systems, and could achieve 90 miles per

Figure 16.12
The headquarters building of the Rocky Mountain Institute has no internal heat source, despite its location in the cold climate of the Colorado Rockies. Technology that was cutting-edge when the building was constructed is now widely available.

gallon. More advanced designs would use hydrogen fuel cells and could get the equivalent of 200 miles per gallon. These vehicles would have the range of conventional gasoline-powered cars and would meet all safety and crash-test standards.

Are RMI's visions pie-in-the-sky dreams? Or are they grounded in a solid reality of physics, technology, and economics? RMI's emphasis on working with the market economy suggests a mix of reality and radical but perhaps achievable vision. RMI has already helped major corporations redesign their energy facilities, saving money in the process. Time will tell whether RMI's super-efficient designs become the standard. If they do, we'll have gone a long way toward breaking the link between energy and climate.

Energy and Living Standards

Let me emphasize again that increased energy efficiency is a very different proposition than simply reducing our energy consumption. The former is a smart approach that buys us the same energy benefits at a lower cost; the latter requires that we give up some of the comforts provided by our energy consumption. But might we do a little of each? Could we have a comfortable, technologically advanced, economically prosperous, and culturally rich society that offers its citizens satisfying lives with, say, half the energy consumption we're now used to? Figure 16.13 suggests that, for North Americans, the answer is yes. We would hardly consider the countries of Europe to be technologically, economically, or culturally backward, yet Figure 16.13 shows that Europeans use only

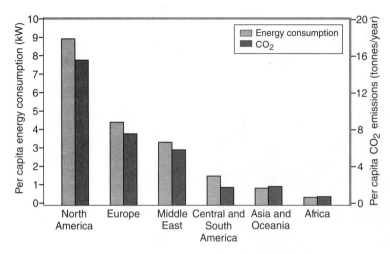

Figure 16.13

Per capita energy consumption (light gray) and CO_2 emissions (dark gray) for different world regions.

about half as much energy per capita as North Americans do and that their greenhouse emissions are comparably lower. There are, of course, some legitimate reasons why North Americans consume more energy. Some of these I discussed in Chapter 2, including geography and North America's role as a world food supplier. But others are simply a matter of societal choice. European countries have chosen high gasoline taxes to help curb consumption (recall Fig. 5.26). Even many smaller European cities have vibrant public transportation, and reliable, convenient rail systems link European communities. Zoning regulations encourage city-, town-, and village-based development over suburban and exurban sprawl. As a result of such tax, public-transportation, and community-development policies, Europeans have fewer cars per capita and their cars are smaller and more fuel efficient (see Research Problem 5).

In Chapter 2 you did deep knee bends to experience physically your body's typical 100-W power output, and I made a point of tying energy-consumption figures to that 100-W human equivalent. Figure 2.3, for example, shows that it would take 100 energy servants working around the clock to satisfy the average American's energy demand. You've now seen how greater energy efficiency, coupled with sensible public policies that discourage profligate energy consumption, might make that figure considerably lower. How much lower? And how does lower per capita energy use translate into future global energy consumption and greenhouse emissions?

Projections of future energy consumption range widely. Analyses based on the IPCC's Special Report on Emissions Scenarios (SRES) discussed in Chapter 15 show a fivefold range in projections of total world energy consumption in the year 2100. Energy expert Vaclav Smil, in his book *Energy at the Crossroads*, explains how simple assumptions bracket this huge range in energy futures. A conservative projection, in Smil's view, would be doubled energy efficiency over the coming century (conservative because we did much better during the twentieth century) and a world population of 7 billion averaging today's per capita energy consumption. This scenario, he argues, leads to a global energy-

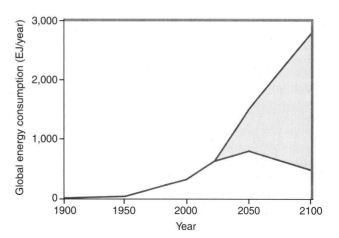

Figure 16.14

Projections of future world energy use show a fivefold uncertainty (shaded area), depending on assumptions made about population growth, energy efficiency gains, and per capita energy consumption.

consumption rate in 2100 that's just a little over what it is today, and it provides an energy-based living standard for everyone that's equivalent to what Italy enjoys today. That clearly requires some reduction in per capita consumption among the more profligate energy users, particularly the United States. But the potential for energy savings described earlier in this chapter suggests that it would not be at all difficult to halve the number of energy servants working for each of us, and that with a concerted effort we could go a lot lower.

On the other hand, if per capita energy consumption continues the twentieth century's nearly fourfold increase, and if population is still climbing in 2100, then the world could be using some six times as much energy at the end of the century as it does today. Everything you know about available energy resources and their environmental and climatic impacts says this scenario is one to be avoided, if not actually impossible.

Figure 16.14 shows the projected spread in possible global energy futures based on the SRES scenarios. Where we end up is up to us, and the outcome depends on choices we make now.

16.5 Strategy for a Sustainable Future

I started this chapter by asking what we need to do to stabilize atmospheric CO_2 at levels that will prevent "dangerous anthropogenic interference" with Earth's climate. The quick answer is that we need to reduce our CO_2 emissions, most of which come from burning fossil fuels. But how? We've now seen three distinct approaches: to keep burning fossil fuels but capture and sequester the CO_2, to substitute alternative energy sources for fossil fuels, and to reduce overall energy consumption.

Any realistic strategy must get us to a more climate-friendly energy future without massive disruption of the world economy or reliance on theoretically wonderful but technologically unproven energy alternatives. However much we might want to live in a world powered by direct solar energy alone, this isn't going to happen overnight. Getting to that or any other sustainable energy future will entail a gradual transition using any and all available strategies for reducing our climate impact as quickly as is realistically possible.

The Wedge Approach

There's no single magic bullet that will solve the world's energy, environmental, and climate problems overnight. Recognizing that, Princeton University energy experts Stephen Pacala and Robert Socolow have proposed a set of interchangeable "wedges," each of which eliminates some of the projected increase in future greenhouse emissions. Their idea is to break a seemingly overwhelming problem into manageable pieces, and to provide enough such pieces to allow some choice in how we solve the problem.

Pacala and Socolow begin with a 50-year extrapolation of historical carbon emissions, which projects an approximate doubling of anthropogenic emissions over that period. Their goal is to prevent that doubling by holding global emissions to their current levels for the next 50 years. As Figure 16.15 shows, the discrepancy between projected growth and the flat-path goal is a triangular region in the emissions-versus-time graph. The beauty of this idea is that it doesn't require a major upheaval right at first, because the emissions we're trying to eliminate don't yet exist. That is, the difference between the growing emissions projection and the flat emissions goal starts out small and very manageable. But over 50 years the flat path results in far less carbon in the atmosphere. Pacala and Socolow call the missing emissions region the *stabilization triangle*.

Global anthropogenic CO_2 emissions from fossil fuel combustion are currently about 7 Gt of carbon per year, so the projected 50-year doubling would add another 7 Gt per year. Socolow and Pacala propose to prevent this additional 7 Gt per year through seven "wedges," each of which would eliminate 1 Gt per year of carbon emissions 50 years

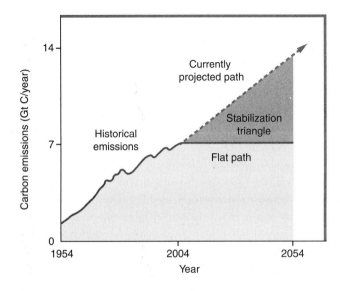

Figure 16.15
The stabilization triangle represents emissions to be avoided over 50 years if we're to stabilize atmospheric CO_2 at less than a doubling from its preindustrial concentration. The shaded areas represent total CO_2 emissions (in gigatonnes of carbon).

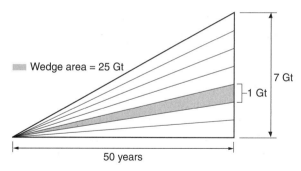

Figure 16.16
Stabilization can be achieved with seven wedges, each of which grows to the point where it eliminates 1 Gt of carbon emissions per year 50 years from now. Over the period of 50 years, each wedge keeps 25 Gt of carbon out of the atmosphere. The wedges represent individual energy strategies, each developed to achieve the same emissions reductions.

from now (Fig. 16.16). But again, the wedges' initial job is more modest; in the first year of the wedge strategy, each wedge has to displace a relatively small amount of carbon emissions. But over 50 years, each wedge will have kept some 25 Gt of carbon out of the atmosphere. Exercise 9 is a quantitative exploration of this idea.

So what are the wedges? Pacala and Socolow identify more than the seven required wedges, giving us a range of choices. A sampling of their proposed wedges includes energy efficiency; renewable energy sources; nuclear energy; substituting natural gas for coal; carbon capture and sequestration; biofuels and carbon-free hydrogen for transportation; afforestation (planting trees to remove atmospheric CO_2); and better management of methane from agriculture, landfills, and natural gas operations. They quantify each wedge, showing the magnitude of the effort involved after 50 years. If we choose to include a CCS wedge, for example, we would need to have CCS installed at eight hundred 1-GW coal-fired power plants by the end of the 50-year period. If we choose a wind-power wedge, we'd have to build 2 million 1-MW wind turbines over the 50 years to replace coal power. A solar photovoltaic wedge would mean 2,000 GW of peak photovoltaic power in 50 years, requiring about 2 million hectares of land—about the area of New Jersey. A nuclear wedge would have us doubling the amount of nuclear-generated electricity and displacing an equivalent amount of coal-generated electricity. Two billion cars running on biofuels and getting 60 miles per gallon would provide another wedge, and would require one-sixth of the world's cropland for biofuel production. Or we could run those 2 billion vehicles on hydrogen produced from fossil fuels with CCS or from 4 million wind turbines. Stopping tropical deforestation and rehabilitating 300 million hectares of tropical forest would give us a forest-based wedge. This is just a sampling of the list of well-documented choices, each capable of keeping 25 Gt of carbon out of the atmosphere over the next 50 years.

So would Pacala and Socolow's wedges solve our energy-climate problem? And do they really make the problem manageable? On the second point, the wedge advocates concede that their approach breaks a "heroic challenge" into what's still a "set of monumental tasks." But we as a species already have the tools needed for each of these tasks, and there are enough possible wedges that we can pick and choose among them. Each wedge represents the same interchangeable unit, namely 25 Gt of eliminated carbon, which makes it straightforward to compare their relative costs, risks, and environmental impacts.

The wedge approach stabilizes carbon *emissions*, but does it achieve the UN goal of "stabilization of greenhouse gas *concentrations* in the atmosphere at a level that would prevent dangerous anthropogenic interference with the climate system" (my italics)? Not in itself: Pacala and Socolow argue that after 50 years of stable carbon *emissions*, we'll then need to begin reducing emissions if we're to stabilize atmospheric carbon *concentration* at somewhat below a doubling of the preindustrial level. Furthermore, they suggest that we really need to start now: If we don't implement something like the wedge approach right away, then we'll face a much more difficult task to stabilize atmospheric CO_2 even at *triple* its preindustrial level.

Parting Words

Here, at the end of what has been a mostly scientific textbook, I'll wrap up with some personal thoughts. I've stressed from the beginning the huge magnitude of humankind's energy consumption, and I've tried throughout to get you thinking quantitatively about energy and its environmental impacts. You can now talk fluently of gigawatts, exajoules, and quads with a feel for those energy units' colossal size. You can also quantify conventional pollutants, and you know how gigatonnes of anthropogenic carbon emissions alter the natural carbon cycle. You can scrutinize an alternative energy proposal and decide whether it's quantitatively and technologically realistic. In the end, though, all of these energy numbers come down to one quantitative fact: The magnitude of humanity's energy enterprise has become large enough to have a global impact on our planet. Reducing that impact and mitigating its effects will require people who, like you, can undertake a serious quantitative analysis of the problem and its potential solutions. But it also will take visionaries with new technological, economic, and social ideas that can reconcile energy-consuming humanity with a healthy planet. And it will need leaders, activists, and policymakers with the optimism, courage, and confidence to guide us through what, by any measure, will be some challenging decades. I hope you'll play a part.

Chapter 16 Chapter Review

BIG IDEAS

16.1 The goal of the **United Nations Framework Convention on Climate Change** is "stabilization of greenhouse gas concentrations in the atmosphere at a level that would prevent dangerous anthropogenic interference with the climate system." This goal probably entails limiting atmospheric CO_2 to, at most, a doubling of its preindustrial value, and limiting any additional global temperature rise to under 2°C. There are four possible paths to that goal, but this book dismisses one—global engineering projects—as dangerous tampering with a system we don't yet understand well enough.

16.2 **Carbon capture and sequestration** captures the carbon from fossil fuels, either before or after combustion, and sequesters it underground or in the deep ocean. Its large-scale technological or economic viability has not yet been demonstrated, but if proven it would allow continued use of fossil fuels in stationary sources without significant climate impact.

16.3 The alternative energy sources described in Chapters 7 to 11 offer the prospect of greatly reduced—although not zero—carbon emissions.

16.4 Using energy more efficiently results in less energy use and lower greenhouse emissions without reducing the benefits that energy provides.

16.5 Achieving a sustainable future with a stable climate will require using many of the different approaches available for reducing greenhouse emissions.

TERMS TO KNOW

carbon capture and sequestration (p. 459)
flue gas (p. 460)
United Nations Framework Convention on Climate Change (p. 457)

GETTING QUANTITATIVE

Estimated maximum CO_2 concentration to avoid "dangerous anthropogenic interference with the climate system": double the preindustrial level, or 560 ppm

CCS requirement for typical coal-fired power plant: 1,000 tons of CO_2 every hour

Estimated range of carbon emissions from different energy sources:
high estimate = 755 g/kWh (coal); low estimate = 4 g/kWh (hydro)

QUESTIONS

1. Why isn't carbon capture a straightforward problem in pollution control—as is, for example, the removal of sulfur emissions from flue gases?

2. The oceans already hold nearly 40,000 Gt of carbon. So why do we worry about the effects of adding a little more in the form of CO_2, as we might do if we decided to capture CO_2 and sequester it in the oceans?

3. Why aren't energy alternatives such as solar photovoltaics and nuclear power totally free of greenhouse emissions?

4. Why might the "wedge" approach be easier to implement than some other strategies for stabilizing atmospheric CO_2?

5. What would *you* consider to be "dangerous anthropogenic interference with the climate system"?

EXERCISES

1. (a) Estimate the volume occupied by the 1,000 tons of CO_2 produced each hour in a 1-GWe coal-fired power plant, assuming it's at approximately normal atmospheric temperature and pressure. Under these conditions, 1 mole of any gas occupies 22.4 liters. (b) Use your answer to determine the volume occupied by the power plant's yearly CO_2 production, expressed in cubic kilometers.

2. If you drive a car getting 22 miles per gallon 11,000 miles per year, how much CO_2 do you produce in a year? See Exercise 1 in Chapter 6.

3. Section 16.2 mentions a pilot project that's sequestered 5 million tons of CO_2 in a Canadian oil field. How much time would it take a 1-GWe coal-fired power plant to produce that much CO_2?

4. The pilot sequestration project mentioned in Section 16.2 increased the oil field's output by 10,000 barrels per day. How much oil-fired electric power could that excess generate, assuming 40 percent efficiency?

5. How long could we inject CO_2 from one thousand 1-GWe power plants into the deep ocean before we increased the stored carbon (now at 39,000 Gt) by 1 percent? (That 39,000 Gt is *carbon*, and the emissions from a 1-GWe power plant is given in this chapter in tons of *carbon dioxide*.)

6. Assuming a 100-EJ annual U.S. energy consumption, and assuming no change in GDP, show that the difference between actual and extrapolated energy intensities at the end of the curve in Figure 16.7 is the energy equivalent of about 3 billion barrels of oil annually, and compare this quantity with our annual oil imports.

7. Figures 16.6 and 16.10 reflect dramatic gains in the efficiency of U.S. household refrigerators since the 1970s. In 1973 the average refrigerator used 1,800 kWh of electricity per year; in 2005 the average was 500 kWh. (a) Estimate the total yearly energy savings in the United States today that is associated with this drop in refrigerator energy consumption, and estimate how many large electric power plants this improvement in efficiency displaced.

8. Today's "smart" electronic devices and appliances typically consume about 5 W continuously, even when they're off. Suppose the typical household has ten such devices. Estimate the total of this standby power consumption in the United States. How many large power plants are at work just to supply this standby power?

9. Estimate the total carbon emissions associated with each of the three shaded regions in Figure 16.15—that is, the historical emissions from 1954 to 2054, the emissions from 2004 to 2054 under the flat-path scenario, and the emissions associated with the stabilization triangle.

10. A 20-W compact fluorescent lamp replaces a 100-W incandescent bulb. How much energy does it save over its 10,000-hour lifetime? At 10¢ per kilowatt-hour, how much money does it save?

RESEARCH PROBLEMS

1. Consult the Mauna Loa CO_2 data used for Figure 13.7, and make a table or graph of the annual *change* in atmospheric CO_2 concentration since 1960.

2. Has California continued to hold its per capita electrical energy consumption steady beyond the time shown in Figure 16.9? Find the data to continue the graph as far as you can.

3. What has happened to the U.S. vehicle fleet's average fuel efficiency since the last data point shown in Figure. 16.11? Continue this graph as far as you can.

4. Find the appropriate data and make a graph comparable to Figure 16.8, but for world oil consumption. Is there any period of rapid decline? If so, fit an exponential curve and determine the annual percentage decrease in consumption.

5. Find data and prepare a graph showing the number of motor vehicles per one thousand people in about ten countries. Include the United States, at least one country in Western Europe, and at least one developing country.

APPENDIX

Properties of Materials

| Table A1 | **Air and Water** | | | | | |

MATERIAL	Density (kg/m^3)	Specific heat heat (kJ/kg·K)	Melting point (K)	Boiling point (K)	Heat of fusion (kJ/kg)	Heat of vaporization (kJ/kg)
AIR	1.2	1.01	O$_2$: 54.8 N$_2$: 63.3	O$_2$: 90.2 N$_2$: 77.4		
WATER	Liquid: 1,000 Ice: 917	Liquid: 4.184 Ice: 2.05	273 (0°C)	373 (100°C)	334	2,257

Table A2	**Thermal Conductivities of Selected Materials**	
MATERIAL	**Thermal conductivity (W/m·K)**	**Thermal conductivity (Btu·inch/hour·ft²·°F)**
AIR	0.026	0.18
ALUMINUM	237	1,644
CONCRETE (TYPICAL)	1	7
FIBERGLASS	0.042	0.29
GLASS (TYPICAL)	0.8	5.5
ROCK (GRANITE)	3.37	23.4
STEEL	46	319
STYROFOAM (EXTRUDED POLY-STYRENE FOAM)	0.029	0.2
WATER	0.61	4.2
WOOD (PINE)	0.11	0.78
URETHANE FOAM	0.019	0.13

Table A3 *R* Values (ft²·°F·h/Btu) of Some Common Building Materials

AIR LAYER
Adjacent to inside wall	0.68
Adjacent to outside wall, no wind	0.17

CONCRETE, 8-inch	1.1

FIBERGLASS
3.5-inch	12
(fits "4-inch" lumber)	
5.5-inch	19
(fits "6-inch" lumber)	

GLASS, ⅛-inch single pane	0.023

GYPSUM BOARD (Sheetrock), ½-inch	0.45

POLYSTYRENE FOAM, 1-inch	5

URETHANE FOAM, 1-inch	7.5

WINDOW (*R* values include adjacent air layer)
Single-glazed wood	0.9
Standard double-glazed wood	2.0
Argon-filled double-glazed with low-E coating	2.9
Argon-filled triple-glazed with low-E coating	5.5

WOOD
½-inch cedar	0.68
¾-inch oak	0.74
½-inch plywood	0.63
¾-inch white pine	0.96

Table A4 Specific Heats (J/kg/K) of Some Common Materials

ALUMINUM	900
CONCRETE	880
GLASS	753
STEEL	502
STONE (granite)	840
WATER	
Liquid	4,184
Ice	2,050
WOOD	1,400

Table A5   Global Warming Potentials

GAS	Atmospheric lifetime (years)	Global warming potential relative to CO_2		
		Time frame: 20 years	100 years	500 years
CARBON DIOXIDE (CO_2)	~1,000*	1	1	1
METHANE (CH_4)	11	67	23	6.9
NITROUS OXIDE (N_2O)	114	291	298	153
CFC-11 (CCl_3F)	45	6,700	4,750	1,620
HCFC-22 ($CHClF_2$)	12	5,200	1,800	550
HFC-23 (CHF_3)	270	12,000	14,800	12,200

*The lifetime of CO_2 is ambiguous; see Section 13.5.

Table A6   Some Important Radioactive Isotopes

ISOTOPE	Half-life (approximate)
CARBON-14	5,730 years
IODINE-131	8 days
POTASSIUM-40	1.25 billion years
PLUTONIUM-239	24,000 years
RADON-222	3.8 days
STRONTIUM-90	29 years
TRITIUM (hydrogen-3)	12 years
URANIUM-235	704 million years
URANIUM-238	4.5 billion years

Periodic Table of the Elements

Key
- ■ Alkaline earth metals
- ■ Alkali metals
- □ Transition metals
- □ Rare earths
- ■ Nonmetals
- ■ Halogens
- □ Noble gases
- □ Other metals

Legend for each cell: Atomic number (of protons) · Symbol for element · Name · Atomic mass. Group designation shown at top of each column.

Period	IA 1	IIA 2	IIIB 3	IVB 4	VB 5	VIB 6	VIIB 7	VIIIB 8	VIIIB 9	VIIIB 10	IB 11	IIB 12	IIIA 13	IVA 14	VA 15	VIA 16	VIIA 17	VIIIA 18
1	1 H Hydrogen 1.00794																	2 He Helium 4.002602
2	3 Li Lithium 6.941	4 Be Beryllium 9.012182											5 B Boron 10.811	6 C Carbon 12.0107	7 N Nitrogen 14.0067	8 O Oxygen 15.9994	9 F Fluorine 18.99840	10 Ne Neon 20.1797
3	11 Na Sodium 22.98977	12 Mg Magnesium 24.3050											13 Al Aluminum 26.98154	14 Si Silicon 28.0855	15 P Phosphorus 30.97376	16 S Sulfur 32.065	17 Cl Chlorine 35.453	18 Ar Argon 39.948
4	19 K Potassium 39.0983	20 Ca Calcium 40.078	21 Sc Scandium 44.955910	22 Ti Titanium 47.867	23 V Vanadium 50.9415	24 Cr Chromium 51.9961	25 Mn Manganese 54.938049	26 Fe Iron 55.845	27 Co Cobalt 58.93320	28 Ni Nickel 58.6934	29 Cu Copper 63.546	30 Zn Zinc 65.409	31 Ga Gallium 69.723	32 Ge Germanium 72.64	33 As Arsenic 74.92160	34 Se Selenium 78.96	35 Br Bromine 79.904	36 Kr Krypton 83.798
5	37 Rb Rubidium 85.4678	38 Sr Strontium 87.62	39 Y Yttrium 88.90585	40 Zr Zirconium 91.224	41 Nb Niobium 92.90638	42 Mo Molybdenum 95.94	43 Tc Technetium 98.9072	44 Ru Ruthenium 101.07	45 Rh Rhodium 102.90550	46 Pd Palladium 106.42	47 Ag Silver 107.8682	48 Cd Cadmium 112.411	49 In Indium 114.818	50 Sn Tin 118.710	51 Sb Antimony 121.760	52 Te Tellurium 127.60	53 I Iodine 126.90447	54 Xe Xenon 131.293
6	55 Cs Cesium 132.90545	56 Ba Barium 137.327	57 La Lanthanum 138.9055	72 Hf Hafnium 178.49	73 Ta Tantalum 180.9479	74 W Tungsten 183.84	75 Re Rhenium 186.207	76 Os Osmium 190.23	77 Ir Iridium 192.217	78 Pt Platinum 195.078	79 Au Gold 196.96654	80 Hg Mercury 200.59	81 Tl Thallium 204.3833	82 Pb Lead 207.2	83 Bi Bismuth 208.98037	84 Po Polonium 208.9824	85 At Astatine 209.9871	86 Rn Radon 222.0176
7	87 Fr Francium 223.01197	88 Ra Radium 226.0277	89 Ac Actinium 227.0277	104 Rf Rutherfordium 261.1089	105 Db Dubnium 262.1144	106 Sg Seaborgium 263.118	107 Bh Bohrium 262.12	108 Hs Hassium 265.1306	109 Mt Meitnerium (268)	110 Ds Darmstadtium (271)	111 Uuu Unununium (272)	112 Uub Ununbium (285)		114 Uuq Ununquadium (289)		116 Uuh Ununhexium (292)		118 ---- 0

Lanthanides

58 Ce Cerium 140.116	59 Pr Praseodymium 140.90765	60 Nd Neodymium 144.24	61 Pm Promethium 144.9127	62 Sm Samarium 150.36	63 Eu Europium 151.964	64 Gd Gadolinium 157.25	65 Tb Terbium 158.92534	66 Dy Dysprosium 162.50	67 Ho Holmium 164.93032	68 Er Erbium 167.26	69 Tm Thulium 168.93421	70 Yb Ytterbium 173.04	71 Lu Lutetium 174.967

Actinides

90 Th Thorium 232.0381	91 Pa Protactinium 231.0359	92 U Uranium 238.0289	93 Np Neptunium 237.0482	94 Pu Plutonium 244.0642	95 Am Americium 243.0614	96 Cm Curium 247.07003	97 Bk Berkelium 247.0703	98 Cf Californium 251.0796	99 Es Einsteinium 252.083	100 Fm Fermium 257.0951	101 Md Mendelevium 258.0984	102 No Nobelium 259.1011	103 Lr Lawrencium 262.110

GLOSSARY

absolute zero the lowest possible temperature, corresponding to a system's state of minimum energy

absorber plate the black-coated element in a solar collector that absorbs solar radiation

acid mine drainage seepage of acidic water from coal mines into surface streams

acid rain precipitation of higher than normal acidity, resulting from the reaction of sulfur dioxide pollution with water

active solar heating a solar heating system that uses active devices such as pumps or fans to move heated air or liquid

aerobic respiration respiration that uses oxygen and produces carbon dioxide

albedo a number between 0 and 1 representing the fraction of incident sunlight that a planet reflects back to space

alpha radiation radiation consisting of high-energy alpha particles (helium nuclei, $^4_2$He)

amorphous silicon silicon that lacks a crystal structure; it is inexpensive to produce, but solar photovoltaic cells made from amorphous silicon are less efficient than those made from crystalline silicon

anaerobic respiration respiration in the absence of oxygen, often producing methane

anthropogenic of human origin

atmosphere-ocean general circulation model (AOGCM) a climate model that couples together separate modules for computing atmospheric and oceanic conditions

atmospheric lifetime the time that a molecule or particle remains in the atmosphere before being removed by physical or chemical processes

atomic nucleus the inner core of the atom, consisting of protons and neutrons and containing nearly all the atomic mass

atomic number (Z) the number of protons in a nucleus; determines the chemical element

background radiation the level of radiation from both natural and anthropogenic sources to which a population is typically exposed

baghouse a chamber filled with fabric bags used for filtering particulates from an exhaust stream

barrel of oil equivalent (BOE) an energy unit equal to 6.12 GJ; the energy contained in one 42-gallon barrel of oil

baseload the average minimum electrical energy use in a given region; baseload power plants such as large nuclear and coal-fired facilities generally run continually to supply this load

battery a device that uses chemical reactions to separate electric charge and thus supply electrical energy

beadwall a window with two transparent surfaces that fills with beads of insulating material to minimize heat loss when sunlight isn't available

becquerel (Bq) the SI unit of radioactivity, equal to 1 decay per second

beta radiation radiation consisting of high-energy electrons (e^-) or positrons (e^+)

binary system a geothermal power system that uses geothermal fluids to heat a working fluid, which in turn boils to vapor and drives a turbine

binding energy the energy needed to disassemble a nucleus; equivalently, the energy released when the nucleus forms from individual nucleons; usually expressed as binding energy per nucleon

biodiesel a biofuel made from plant oils that can be used in diesel engines

biofuel a fuel made from biomass; examples include ethanol, biodiesel, and methane

biological pump the process whereby dead organisms and organic waste sink into the deep ocean thus removing carbon from the surface waters

biomass matter derived from living things and useful as fuel; dry biomass has an energy content of typically 15 to 20 MJ/kg

boiling-water reactor (BWR) a light-water reactor in which water boils within the reactor vessel and the resulting steam drives a turbine

borehole a hole drilled into the ground to produce a profile of temperature with depth, which is useful for assessing climate history

breeder reactor a reactor designed to breed plutonium-239 from uranium-238, thus producing more fissile fuel than it consumes

British thermal unit (Btu) an energy unit equal to 1,054 joules; the energy needed to raise 1 pound of water by 1°F

calorie (cal) an energy unit equal to 4.184 joules; the energy needed to raise the temperature of 1 gram of water by 1°C

carbon capture and sequestration (CCS) the process of capturing carbon dioxide and sequestering it underground or in the deep ocean to prevent fossil carbon emissions from adding to atmospheric greenhouse gases

carbon cycle the system of flows and reservoirs whereby carbon moves through the Earth-ocean-atmosphere-biosphere system

carbon neutral a system that produces zero net carbon emissions

carbon sink any system that removes carbon from the atmosphere

Carnot engine a heat engine that provides the theoretical maximum efficiency

catalytic converter a device that uses a catalyst to accelerate chemical reactions that render pollutants less harmful; for example, catalytic converters in cars convert nitrogen oxides into nitrogen and oxygen; carbon monoxide into carbon dioxide; and unburned hydrocarbons into carbon dioxide and water

chain reaction a nuclear fission reaction wherein neutrons released in one fission go on to cause additional fissions

Clean Air Act a law, first enacted in 1963 and subsequently amended substantially, that has been instrumental in improving air quality in the United States

Clean Air Act Amendments Amendments to the Clean Air Act that have generally strengthened the legislation

climate the average conditions prevailing in Earth's atmosphere

climate forcing an upset of Earth's energy balance, generally considered to occur at the tropopause; generally leads to a temperature change

climate sensitivity a measure of the response of the climate system to a change in forcing; expressed either as a temperature change for a doubling of atmospheric carbon dioxide or as a temperature change per watt per square meter of forcing

coal a black, solid fossil fuel high in carbon

coefficient of performance in a heat pump, the ratio of heat delivered to high-quality energy required to run the pump

cogeneration the process whereby fuel energy is used to produce both electricity and heat for industrial use or space heating, thus putting to use the waste heat required by the second law of thermodynamics

combined-cycle power plant a power plant using gas turbines whose waste heat then drives a conventional steam cycle, resulting in high overall efficiency

compound parabolic concentrator a solar concentrator that boosts light intensity without focusing the light to a point or line

compression-ignition engine an intermittent-combustion engine in which fuel ignites as the fuel-air mixture in the cylinder is heated by compression to ignition temperature, as in diesel engines

concentration ratio the ratio by which a solar concentrator boosts the light intensity

condenser in a steam power plant, the device in which steam is recondensed to water

conduction heat flow by physical contact, when energy is transferred by collisions among molecules

confinement time the length of time that a fusion reactor can contain hot plasma

continuous-combustion engine a heat engine in which combustion takes place continuously, as in a power plant, in contrast to the intermittent combustion in gasoline and diesel engines

control rods neutron-absorbing rods that move in and out of a reactor to control the chain reaction

convection heat transfer by the bulk motion of a fluid

coolant a material used to carry heat away from the core of a nuclear reactor or other heat engine

cooling tower a structure used in power plants to cool the water that flows through the plant's condenser; extracts waste heat and discharges it to the atmosphere

cornucopian one who believes that human ingenuity will alleviate the problem of finite resources through new discoveries and technologies

crankshaft a rotating shaft in an engine that converts the back-and-forth motion of pistons into rotary motion

criteria pollutant one of six substances—carbon monoxide, lead, nitrogen dioxide, ozone, sulfur oxides, and two classes of particulates—for which the Clean Air Act sets maximum concentrations in ambient air

critical mass the amount of nuclear material needed for a self-sustaining chain reaction

crude oil oil as it comes from the ground, before refining or other processing

curie an older unit of radioactivity, equal to 37 billion decays per second, and approximately the radioactivity of 1 gram of radium-226

curve of binding energy a graph of binding energy per nucleon versus mass number; the curve peaks near mass number 56 (iron), indicating that nuclear energy release is possible by fusion of nuclei lighter than this mass and by fission of heavier nuclei

cyclone a device that swirls exhaust gases in a high-speed vortex to extract particulate pollution

diffuse insolation solar energy that has been scattered in the atmosphere before reaching Earth's surface

direct insolation solar energy coming directly from the Sun

dry cooling tower structure designed to transfer waste heat from a power plant's cooling water to the atmosphere; dry towers keep the water in a closed system where it can't evaporate

electric charge a fundamental property of matter that comes in two kinds (positive and negative) and that gives rise to electric fields

electric current a flow of electric charge

electric field the influence of electric charge on the space around it, giving rise to electric forces on other charges; also arises from a changing magnetic field

elastic potential energy energy stored in the altered configuration of elastic materials such as rubber bands or springs

electrolysis the use of electric current to dissociate water into hydrogen and oxygen

electromagnetic force a force associated with electric charge (electric force) or moving electric charge (magnetic force)

electromagnetic induction a fundamental phenomenon whereby a changing magnetic field produces an electric field; the basis of electric generators and essential for the existence of electromagnetic waves

electromagnetic radiation a flow of electromagnetic energy in the form of electromagnetic waves

electromagnetic wave a wave of self-regenerating electric and magnetic fields that propagates through empty space, carrying electromagnetic energy; light is one example of an electromagnetic wave

electrostatic precipitator a device that uses a strong electric field to extract particulates from an exhaust stream

emissivity (e) a number between 0 and 1 that gives the efficiency with which a material radiates electromagnetic waves; equal to the efficiency of absorption at a given wavelength

end use the final use to which energy is being put

energy balance a state in which a system maintains a constant temperature because its energy input balances its energy loss

energy carrier a substance, such as hydrogen (H_2), which contains energy but isn't available naturally as an energy resource

energy flow a movement of energy from one place to another, as in the flows of solar and geothermal energy that bring energy to Earth's surface

energy intensity a measure of energy consumption per unit of gross domestic product

energy payback time the time it takes a system to produce as much energy as was required to construct it

enrichment the process of enhancing the fraction of fissile uranium-235 in nuclear fuel to levels above its natural concentration of 0.7 percent

entropy a measure of disorder; systems with higher entropy have lower energy quality

equilibrium model a climate model that determines the equilibrium climate after all changes have stopped occurring

ethanol ethyl alcohol, used as a biofuel and generally produced by fermentation of plant matter

evapotranspiration the two processes of evaporation and plant transpiration, both of which emit water vapor into the atmosphere

external-combustion engine a heat engine in which combustion takes place outside the cylinders or other chambers holding the engine's working fluid; an example is a steam power plant

feedback effect an additional effect that follows as the result of some change in a system; a feedback is positive if it tends to enhance the original change and negative if it tends to diminish it

filter a simple device for removing particulates via filtration

first law of thermodynamics a statement of energy conservation; the change in a system's internal energy is the sum of the work done on the system and the heat that flows into it

fissile isotope an isotope that fissions when struck by low-energy neutrons; uranium-235 and plutonium-239 are the most important fissile isotopes

fission product a substance produced as a result of nuclear fission; typical fission products span a range of middleweight nuclei and are highly radioactive

fissionable isotope an isotope that fissions when struck by neutrons; the required neutron energy is high except in the subcategory of fissile isotopes

flat-plate collector a solar collector with a flat absorber plate

flue gas the mix of gaseous and particulate effluent that emerges from the combustion chamber of a fossil- or biomass-fueled power plant or industrial boiler

flue gas desulfurization a chemical process for removing sulfur compounds from flue gas before it is emitted to the atmosphere

fly ash the ash produced in coal-burning power plants

focus the point or line on which a solar concentrator focuses sunlight

force a push or a pull

forced convection convection driven by a mechanical device such as a fan or pump, or by wind

fossil fuel a hydrocarbon fuel formed over tens to hundreds of millions of years from once-living matter, containing stored solar energy originally captured by photosynthetic plants; coal, oil, and natural gas are the common fossil fuels

frictional force a force that acts between two surfaces to oppose their relative motion

fuel a substance that stores energy

fuel cell a device that uses a steady stream of fuel to produce electricity directly, without combustion; hydrogen fuel cells, which combine hydrogen and oxygen are particularly promising and produce only water as a by-product

gamma radiation radiation consisting of high-energy photons

gas-electric hybrid a vehicle propulsion system that uses both a gasoline engine and an electric motor; the batteries powering the electric motor are charged directly by the gasoline engine or by the vehicle's motion

gas turbine a continuous-combustion rotary engine operating at high temperature; used in advanced power plants and jet aircraft engines

general circulation model (GCM) a large, complex model that includes details of atmospheric circulation

geopressured system a geothermal system in which energy is stored in pressurized water

geothermal energy energy from Earth's interior

geothermal gradient the rate at which temperature increases with depth below Earth's surface

global climate model (GCM) a large, complex climate model capable of projecting future climate with detailed spatial resolution; often used interchangeably with general circulation model

global warming potential (GWP) a measure of a substance's effectiveness in causing global warming, measured relative to that of carbon dioxide; depends on the time frame involved

gravitational force the force that every piece of matter exerts on every other piece

gravitational potential energy potential energy associated with objects that have been lifted against gravity

gray (Gy) a unit of radiation dose, equal to 1 joule of radiation energy per kilogram of absorbing material

greenhouse effect the warming of a planet's surface due to the absorption of outgoing infrared radiation by atmospheric greenhouse gases

greenhouse gas an atmospheric gas that absorbs infrared radiation; water vapor and carbon dioxide are the most important greenhouse gases

greenhouse term a term in the energy-balance equation at the top of the atmosphere that represents the greenhouse effect; shows that the greenhouse effect depends on the upper atmosphere being cooler than the surface; the greenhouse term is $-e_a\sigma(T_s^4 - T_a^4)$

grid cell the fundamental unit of atmosphere, ocean, or surface in which a climate model computes a single value for each quantity

gross domestic product (GDP) a measure of the total economic value of a country's goods and services

gross primary productivity see *primary productivity*

ground-source heat pump a heat pump that takes its energy from underground, where the temperature is roughly constant year round

half-life the time it takes half the nuclei of a given radioactive substance to decay

head the vertical distance through which water drops to drive a hydropower turbine

heat energy that is flowing as a result of a temperature difference

heat capacity a measure of the energy required to raise a particular object's temperature; in SI, its units are J/K or J/°C

heat engine a device that extracts thermal energy from a hot source and delivers mechanical energy; the second law of thermodynamics states that this process cannot be 100 percent efficient

heat exchanger a device that transfers energy from one fluid to another

heat of fusion the energy per unit mass required to change a material from the solid to the liquid state

heat of transformation the energy per unit mass required to effect a phase transition

heat of vaporization the energy per unit mass required to change a material from a liquid to a gas

heat pump a refrigerator operated so as to transfer heat from exterior air or the ground into a building for purposes of heating; can also operate in reverse to provide summer cooling

heavy water (D₂O) water made with deuterium (D) instead of ordinary hydrogen (H) and used as a moderator and coolant in some nuclear reactor designs, especially Canadian reactors

heliostat a Sun-tracking mirror

hole the absence of an electron in an interatomic bond in a semiconductor; acts like a positive charge that can move through the material

horsepower (hp) a unit of power equal to 746 watts

Hubbert's peak the point at which production of a resource peaks, after which there may be a discrepancy between falling production and rising demand

hydride a chemical compound of hydrogen and another light element; possibly useful for storing hydrogen fuel in hydrogen-powered vehicles

hydrologic cycle the natural cycle whereby water evaporates from oceans and lakes, forms clouds, precipitates, and runs back into surface waters or underground aquifers

hydropower power produced by running or falling water

hydrothermal system a geothermal resource in which water and steam circulate freely

ice age a period of cool climate marked by the advance of polar ice sheets

ice-albedo feedback a positive feedback associated with global warming in which reduced ice cover decreases Earth's reflectivity and thus results in more solar energy being absorbed, enhancing the original warming

ignition temperature the temperature at which a plasma undergoes nuclear fusion; the lowest ignition temperature, for deuterium-tritium fusion, is still 50 million kelvins

indirect gain a passive solar design in which solar radiation warms a thermal mass and the energy is then circulated through the building by natural convection

inertial confinement a fusion scheme for confining plasma at high density for such short times that the inertia of the plasma keeps its particles from leaving the fusion site; usually induced with high-power laser or particle beams

insolation the intensity of sunlight at a given location and time, measured in watts per square meter

interglacial the warm, relatively short period between ice ages

Intergovernmental Panel on Climate Change (IPCC) an international body established by the World Meteorological Organization and the United Nations Environment Program; issues major reports on climate change approximately every six years

intermittent-combustion engine refers to a heat engine in which combustion takes place intermittently, as in the spark-initiated ignition of a gasoline engine

internal-combustion engine a heat engine in which combustion takes place within a closed chamber, such as the cylinders of a gasoline engine

internal energy energy associated with random thermal motion of molecules; also called *thermal energy*

inversion an atmospheric condition in which temperature increases with altitude, resulting in a mass of stable air that traps pollutants

inverter a device that converts direct current, such as that produced by photovoltaic cells, into standard alternating current

irreversible process a thermodynamic process that converts some mechanical energy into random thermal energy, resulting in a loss of energy quality

isotope a particular nuclear species determined by both its atomic number (Z) and mass number (N); different isotopes of the same element have the same atomic number but different mass numbers

joule (J) the SI unit of energy, equal to 1 watt-second or 1 newton-meter

Kelvin scale the SI temperature scale; a temperature difference of 1 K is the same as a difference of 1°C, but the zero in the Kelvin scale is absolute zero (−273 K)

kerogen a waxy substance formed from organic material that is an intermediate stage in the formation of petroleum

kinetic energy energy of motion

lapse rate the rate at which temperature changes with altitude, normally a decrease of about 6.5°C per kilometer

latent heat the energy stored in a material as a result of its having changed from a lower- to a higher-energy phase

Lawson criterion a criterion for successful fusion; the product of plasma density and confinement time; must exceed 10^{20} s/m^3 for deuterium-tritium fusion

light-water reactor (LWR) a nuclear reactor using ordinary water as coolant and moderator; the most common reactor design, especially in the United States

linear response a response that's directly proportional to its cause

magnetic confinement a fusion scheme that uses magnetic fields to confine plasma at relatively low density and for relatively long times

magnetic field the influence of moving electric charge on the space around it, giving rise to magnetic forces on other moving charges; also arises from a changing electric field

mass number the total number of nucleons (neutrons plus protons) in a nucleus; a rough measure of the nuclear mass

mesosphere the atmospheric layer above the stratosphere, characterized by a temperature that decreases with altitude

methane clathrate, methane hydrate a compound formed underwater at great depth, containing methane (CH_4; natural gas) trapped in an ice-like structure

microwave sounding unit (MSU) a satellite-based instrument that measures microwave radiation from atmospheric oxygen; the data can be used to calculate temperature at different levels in the atmosphere

moderator a material used in a nuclear reactor to slow down neutrons, making them more likely to cause fission

Moore's law a prediction that computer power would double approximately every 18 months; has been an accurate forecast for the past 40 years

mountaintop removal a coal-mining technique in which entire mountaintops are removed to get at coal; the residue is often dumped into adjacent valleys

multiplication factor the factor by which the number of fissioning nuclei is multiplied in each successive generation of fission events; should be exactly 1 for a controlled reaction as in a nuclear reactor; a multiplication factor much greater than 1 leads to an explosive chain reaction, as in a fission bomb

National Ambient Air Quality Standards (NAAQS) the maximum allowed concentrations of the criteria pollutants under the Clean Air Act

natural convection convection that occurs naturally when a fluid is heated and the warm fluid then rises, carrying energy upward

natural gas a gaseous fossil fuel, consisting mostly of methane (CH_4)

natural greenhouse effect the greenhouse effect caused by naturally occurring greenhouse gases, especially water vapor and carbon dioxide; for Earth, this effect increases the surface temperature by some 33°C

negative feedback a feedback effect that tends to diminish the original change

net primary productivity see *primary productivity*

neutron a neutral elementary particle that, along with the proton, makes up the atomic nucleus

neutron-induced fission fission that results when a neutron strikes a nucleus

newton (N) the SI unit of force, equal to 1 kg·m/s^2

nonlinear response a response that's not in direct proportion to its cause; in extreme cases, a nonlinear response may be an abrupt jump in some quantity in response to a very small change in another quantity

N-type semiconductor a semiconductor that's been doped with impurities so that the dominant carriers of electric current are negative electrons

nuclear difference the factor of about 10 million difference between the energy released in nuclear reactions compared with chemical reactions

nuclear fission the process of splitting a heavy nucleus to produce two lighter ones and release energy

nuclear force the strong, short-range force that binds neutrons and protons in the atomic nucleus

nuclear fuel cycle the entire sequence of steps from mining of nuclear fuel through disposal of nuclear waste

nuclear fusion the process of joining two light nuclei to produce a heavier nucleus and release energy

nucleon a constituent of the atomic nucleus—a neutron or proton

ocean thermal energy conversion (OTEC) a technological scheme using the temperature difference between warm, surface ocean water and cool, deep water to drive a heat engine

oil shale rock containing kerogen that can be extracted and processed into oil

one-dimensional model a climate model in which conditions vary in only one direction, often height

ozone the oxygen compound O_3; a greenhouse gas; in the stratosphere, ozone is an absorber of solar ultraviolet radiation, but it is a highly reactive, toxic pollutant near ground level

ozone layer the region of the stratosphere containing ozone that absorbs solar ultraviolet radiation

parabolic dish a dish-shaped solar concentrator with a parabolic shape

parabolic reflector a solar concentrator with a parabolic shape that provides the optimum focusing of light

parabolic trough concentrator a trough-shaped parabolic concentrator that focuses light to a line

parallel hybrid a hybrid vehicle in which a gasoline engine provides most of the motive power, assisted occasionally by the electric motor

particulate pollution fine particles of solid material suspended in the air

passive solar heating solar heating technology that does not make use of active devices such as pumps and fans to move energy around

peaking power smaller power facilities that can be turned on and off as needed to provide extra power during times of peak demand; hydropower is an excellent form of peaking power

peat a crumbly, brown material formed from decaying vegetation; a first step in the formation of coal, peat itself is burned as a fuel in some parts of the world

petroleum strictly speaking, a liquid or gaseous fossil fuel, but often used as a synonym for oil

pH scale the scale used to designate acidity, with 7 being neutral and lower values being acidic

photochemical reaction a reaction driven by the energy of sunlight; important in forming smog from other air pollutants

photochemical smog a mix of harmful chemicals formed by the action of sunlight on air pollutants

photolysis a process using sunlight to break water into hydrogen and oxygen

photon a particle-like bundle of electromagnetic wave energy

photosynthesis the process whereby green plants take in carbon dioxide and solar energy to make carbohydrates that serve as energy sources for life

photovoltaic cell a semiconductor device that converts light energy directly into electricity

photovoltaic module an assemblage of photovoltaic cells

Planck's constant a fundamental constant of nature that quantifies the essential "graininess" of nature in quantum physics

plankton tiny marine organisms at the base of the marine food chain; plankton are important precursors of petroleum

plasma a gas that's so hot it's ionized, so its particles are positively charged ions and negative electrons

PN junction a junction between *P*-type and *N*-type semiconductors; such a junction contains a strong electric field that accelerates charges in a photovoltaic cell; a *PN* junction conducts electricity in only one direction

pollution traditionally, an unwanted by-product of combustion or other human processes that is toxic to people or damaging to the environment

polycrystalline silicon silicon that consists of many distinct crystals; less expensive to make than single-crystal silicon

positive feedback a feedback effect that tends to enhance the original change

potential energy stored energy, associated with work done against a force

power coefficient the fraction of the wind's power that a turbine actually extracts; the theoretical maximum is 59 percent

power curve a plot of power output versus wind speed for a wind turbine

power tower a solar thermal system in which a field of Sun-tracking mirrors concentrates light on an absorber high in a central tower

power (*P*) the rate of energy production or consumption, measured in watts

pressure vessel the heavy steel chamber that contains the nuclear fuel in a light-water reactor

pressurized-water reactor (PWR) a light-water reactor in which water in the reactor vessel is under such high pressure that it doesn't boil, but instead transfers its heat to a secondary loop in which water boils and drives a turbine

primary energy the energy input to a system such as a power plant, including any that ends up as waste

primary productivity the rate at which plants capture solar energy through photosynthesis; gross primary productivity is the total amount, whereas net primary productivity—typically about half of gross—measures what's left over after plants use energy themselves

primary standards air-quality standards aimed at protecting human health

proton-exchange membrane fuel cell (PEMFC) a fuel cell using a special membrane that permits only protons to pass through

proton a positively charged elementary particle that, along with the neutron, makes up the atomic nucleus

proxy a quantity that "stands in" for another, as when tree rings or isotope ratios serve as proxies for climatic conditions

P-type semiconductor a semiconductor that's been doped with impurities so that the dominant carriers of electric current are positive holes

pumped storage a scheme for storing energy by pumping water from a lower source to a higher reservoir; running the water downhill to turn turbines then recovers the stored energy as electricity

quad (Q) 1 quadrillion (10^{15}) Btu

rad an older unit of radiation dose, equal to 0.01 Gy

radiation energy flow by electromagnetic waves

radiative forcing forcing due to an upset in the balance of radiation energy flows

radioactive decay the process whereby a radioactive nucleus emits one or more high-energy particles, thus turning into a different nuclear species (or, with gamma decay, a lower-energy state of the same nucleus)

radiosonde a balloon-borne suite of instruments for measuring conditions at different altitudes in the atmosphere and transmitting the results via radio

rechargeable battery a battery capable of storing electrical energy via chemical reactions when current is run "backward" through the battery

refine the process of chemically altering crude oil to form a variety of products, including heavy oils, aviation fuels, and gasoline

refrigerator a device that transfers heat from cooler objects to warmer surroundings; the second law of thermodynamics says this can't happen without an input of high-quality energy

regenerative braking braking in a hybrid vehicle by using an electric generator to slow the wheels; the generator charges the battery, thus recovering some of the energy of the vehicle's motion

rem an older unit of absorbed radiation dose, equal to 0.01 Sv

repair mechanism a cellular mechanism that repairs radiation damage to DNA

reprocessing processing of spent nuclear fuel to extract different substances, usually plutonium-239

reserve/production ratio a time obtained by dividing the reserves of a fuel by the current rate of production; provides a rough estimate of how long the fuel will last, assuming the production rate stays approximately constant

reserve the amount of a fuel that is known or reasonably certain to exist

reservoir a repository where materials reside temporarily or permanently, as in the atmospheric carbon reservoir or the carbon reservoir associated with biomass

resolution a measure of the fineness in which a climate model divides the physical domain it is modeling; smaller grid cells give higher resolution

resource the total amount of a fuel in the ground, whether or not it's been discovered or even estimated to exist

run-of-the-river a hydropower installation that uses river flow directly, rather than backing up the river behind a high dam; may still involve a low dam

R value a measure of the insulating value of a material; in English units, an R value of 1 corresponds to an energy loss rate of 1 Btu/hour through every square foot for every degree Fahrenheit temperature difference across the material

saturation the complete absorption of all infrared radiation by greenhouse gases in certain wavelength bands, meaning that further increases in greenhouse gas concentrations don't result in more absorption in those bands; makes greenhouse gas forcing a nonlinear function of concentration

scrubbing removal of chemical pollutants, especially sulfur compounds, from flue gas by reacting it with another substance

second law of thermodynamics the statement that systems naturally evolve toward states of lower organization and therefore lower energy quality; applied to heat engines, the second law says it's impossible to extract thermal energy from a hot source and convert it to mechanical energy with 100 percent efficiency

secondary standards air-quality standards aimed at protecting the general welfare, including property and environmental quality

selective surface a solar absorber surface whose properties are wavelength dependent, typically absorbing in the visible but not in the infrared, and therefore minimizing infrared radiation loss

semiconductor a material that conducts electricity by virtue of either negative electrons or positive holes, and whose electrical properties can be manipulated precisely; at the heart of all modern electronics, including photovoltaic cells

sensible heat thermal energy associated with an object's being at a given temperature; as opposed to latent heat

series hybrid a hybrid vehicle in which the gasoline engine is used solely to generate electricity that runs the electric motor; not in widespread use

sievert (Sv) a unit of radiation dose, similar to the gray but adjusted for the biological impact of different types of radiation

single-axis concentrator a concentrator that tracks the Sun in only one direction and focuses light to a line

solar chimney a system that uses the flow of solar-heated air rising up a huge chimney-like structure to power turbines

solar collector a device consisting of one or more glass covers and an absorber plate, used in active solar heating systems

solar constant the rate at which solar energy crosses a square meter of area oriented perpendicular to the Sun's rays, at Earth's distance from the Sun; equal to approximately 1,368 W/m^2

solar lighting use of sunlight for illumination inside buildings

solar oven a cooking device that works by concentrating sunlight

solar panel a photovoltaic module

solar pond a pond that functions as a solar collector; typically uses layers of salty water to inhibit convective energy loss

solar still a device that uses sunlight to distill water, making freshwater from saltwater

solar thermal power system a system in which sunlight is concentrated to boil a fluid and drive a turbine or other engine

spark-ignition engine an intermittent-combustion engine in which a spark initiates combustion, as in gasoline engines

specific heat a measure of the energy per unit mass required to raise a material's temperature; in SI, its units are J/kg·K or J/kg·°C

stagnation temperature the maximum temperature attainable in a solar collector under given conditions; equal to the temperature that would be reached under stagnation conditions—that is, if the heat-transfer fluid stops flowing

steam generator the device in a pressurized-water reactor in which heat from the primary coolant water transfers energy to boil water in the secondary loop

steam reforming a reaction of high-temperature steam with natural gas to produce hydrogen and carbon monoxide; the resulting mixture can be burned as a fuel

Stefan-Boltzmann constant The constant in the Stefan-Boltzmann law, given by $\sigma = 5.67 \times 10^{-8}$ W/m²·K⁴

Stefan-Boltzmann law the statement that the power radiated from an object with surface area A and emissivity e at temperature T is given by $P = e\sigma A T^4$

stratification in a solar hot-water tank, the desirable situation in which hotter water is at the top of the tank

stratosphere the second level of the atmosphere, extending from the tropopause to some 50 km

strip mining extracting coal by stripping off overlying surface layers

subgrid parameterization a procedure for dealing with phenomena that occur on size scales smaller than the grid size of a climate model

sulfate aerosol sulfur-based particulates, particularly from coal burning; these are highly reflective and thus have a cooling effect on climate

tar sands sand containing a heavy tar from which oil can be extracted

temperature anomalies deviations in temperature from established values or averages

temperature (T) a measure of the average thermal energy of the molecules in a substance

thermal conductivity (k) a measure of a material's ability to conduct heat

thermal energy energy associated with random thermal motion of molecules; also called *internal energy*

thermal mass a large, massive structure whose heat capacity helps store solar energy and even out temperature fluctuations in a building

thermal pollution waste heat, typically from a power plant, discharged to the environment and possibly disrupting the local ecology

thermal power plant a power plant that extracts energy from a hot source, usually a fossil-fueled boiler or nuclear fission reactor

thermal splitting the use of high temperatures to dissociate water into hydrogen and oxygen

thermodynamic efficiency the theoretical maximum efficiency for a heat engine, given by $e = 1 - \dfrac{T_c}{T_h}$, where T_h and T_c are the maximum and minimum temperatures, respectively

thermohaline circulation ocean circulation driven by variations in temperature and salinity

thermosiphon a solar hot-water system in which heat-transfer fluid circulates by natural convection

thermosphere the atmospheric layer above the mesosphere, characterized by hot, diffuse gas whose temperature increases with altitude

three-dimensional model a climate model that includes variations in all three spatial dimensions

tidal energy energy originating in the motions of Earth and Moon, resulting in ocean tides

time-dependent model a model that computes climatic conditions as a function of time

tokamak a toroidal plasma device that appears to be the most promising candidate for a nuclear fusion reactor

ton, tonne, metric ton (t) units of mass; 1 ton is 1 English ton, or 2,000 pounds; 1 tonne (or metric ton) is 1,000 kg or 2,200 pounds, roughly the same as 1 ton

tonne oil equivalent (TOE) an energy unit equal to 41.9 GJ; the energy contained in 1 metric ton of oil

total primary energy supply (TPES) the total primary energy supplied to a country or the world

transpiration the process whereby plants take in water from their roots and release it to the atmosphere

transuranic an element heavier than uranium; not found in nature but formed by neutron capture in fission reactors

Trombe wall an indirect-gain solar heating scheme in which a massive wall serves as both absorber and thermal mass

tropopause the boundary between the troposphere and stratosphere

troposphere the lowest level of the atmosphere, extending 8 to 18 km above the surface

turbine a fanlike arrangement of blades driven by a fluid flow; includes steam turbines in power plants and wind turbines

two-axis concentrator a concentrator, such as a parabolic dish, that tracks the Sun in two directions and focuses light to a point

two-box model a simple climate model that treats Earth's surface and atmosphere as separate systems, with each characterized by a single value for relevant physical quantities

two-dimensional model a climate model in which conditions vary in two directions, usually height and latitude

United Nations Framework Convention on Climate Change (UNFCCC): the framework, originally forged at the Rio Earth Summit in 1992, that establishes international mechanisms for dealing with climate change; the Kyoto Protocol is a subsidiary agreement to the UNFCCC

urban heat island effect the apparent warming observed over time at a fixed location due not to global climate change but to increasing urbanization at that location

vapor-dominated system a geothermal system saturated with steam at high pressure; the most valuable type of geothermal resource

water vapor feedback a positive feedback effect associated with global warming; warming results in more evaporation, which increases atmospheric water vapor, and since water vapor is a greenhouse gas, this enhances the original warming

watt (W) the SI unit of power, equal to 1 J/s

weight the force that gravity exerts on an object; near Earth's surface, weight is the product of mass and the strength of gravity (mg)

well-mixed greenhouse gas a greenhouse gas that remains in the atmosphere long enough to reach an approximately uniform global concentration; these include carbon dioxide, methane, and most other common greenhouse gases

wet cooling tower structure designed to transfer waste heat from a power plant's cooling water to the atmosphere; wet towers do so by evaporation from water exposed directly to the air

work a measure of the mechanical energy supplied to an object; equal to the product of applied force times the distance moved in the direction of the force

zero-dimensional energy-balance model an energy-balance climate model that treats Earth as a single point and yields only a single global average temperature

SUGGESTED READINGS

Books/Publications

Annual Review of Environment and Resources. Vol. 31. Palo Alto, CA: Annual Reviews, 2006. Each volume of this ongoing series features papers on energy and environmental issues by leading scholars in these fields; emphasis varies with volume. Note: Volumes 1 to 15 (1976–1990) were published as the *Annual Review of Energy,* and volumes 16 to 27 (1991–2002) were published as the *Annual Review of Energy and the Environment.* Relevant to the entire book.

Bent, Robert, Lloyd Orr, and Randall Baker, eds. *Energy: Science, Policy, and the Pursuit of Sustainability.* Washington, DC: Island Press, 2002. In seven chapters, individual experts look at technological and policy aspects of energy and its relationship to the environment. Relevant to Chapters 5 to 11 and 16.

Bodansky, David. *Nuclear Energy: Principles, Practices, and Prospects.* 2nd ed. New York: American Institute of Physics Press, 2004. A detailed scholarly account of all aspects of contemporary nuclear energy applications. Relevant to Chapter 7.

Boeker, Egbert, and Rienk van Grondelle. *Environmental Physics.* Chichester, UK: John Wiley & Sons, 1995. Scientific analysis of energy sources and technologies, climate change, pollution, and environmental analysis. Not for the mathematically challenged! Relevant to the entire book, for those who want quantitative depth.

Brower, Michael. *Cool Energy: Renewable Solutions to Environmental Problems.* Rev. ed. Cambridge, MA: MIT Press, 1992. A survey of renewable energy in light of the climate challenge, written by a research director for the Union of Concerned Scientists. Relevant to Chapters 8 to 11 and 16.

Brown, Robert C. *Biorenewable Resources: Engineering New Products from Agriculture.* Ames, IA: Iowa State Press, 2003. An overview of biofuels and related agricultural products. Readable, but assumes some math and chemistry. Relevant to Chapter 10.

Deffeyes, Kenneth S. *Beyond Oil: The View from Hubbert's Peak.* New York: Hill and Wang, 2006. Deffeyes, an oil geologist who worked with M. King Hubbert, is a firm believer that global peak oil production is upon us. Relevant to Chapter 5.

Garwin, Richard L., and Georges Charpak. *Megawatts and Megatons: The Future of Nuclear Power and Nuclear Weapons.* Chicago: University of Chicago Press, 2002. Long-time nuclear science and policy expert Garwin teams up with Nobel laureate Charpak in this up-to-date account of nuclear power and nuclear weapons. The authors make a reasoned argument for their stand in support of nuclear power while urging nuclear disarmament. Relevant to Chapter 7.

Gipe, Paul. *Wind Energy Comes of Age.* New York: John Wiley & Sons, 1995. Now rather dated, this book remains a good history of wind development through the early 1990s. Relevant to Chapter 10.

Gipe, Paul. *Wind Power: Renewable Energy for Home, Farm, and Business.* White River Junction, VT: Chelsea Green Publishing, 2004. A good survey of wind technology and a practical guide to those who are thinking seriously about installing small- and medium-scale wind systems, written by a tireless wind energy enthusiast. Relevant to Chapter 10.

Glantz, Michael. *Climate Affairs: A Primer.* Washington, DC: Island Press, 2003. A discussion of climate change and its impacts written for a general audience by a social scientist who is the former head of the Environmental and Societal Impacts Group at the National Center for Atmospheric Research. Complements Chapters 12 to 15 from a social science perspective.

Harvey, L. D. Danny. *Global Warming: The Hard Science.* Harlow, UK: Prentice Hall/Pearson Education, 2000. An often-quantitative survey of the science behind global climate change. Relevant to Chapters 12 to 15.

Hoffmann, Peter. *Tomorrow's Energy: Hydrogen, Fuel Cells, and the Prospects for a Cleaner Planet.* Cambridge, MA: MIT Press, 2001. A clear and readable account of hydrogen energy that is especially strong on the history of hydrogen technology. Relevant to Chapter 11.

Houghton, John. *Global Warming: The Complete Briefing.* 3rd ed. Cambridge, UK: Cambridge University Press, 2004. A thorough, authoritative, yet eminently readable account of global climate change. Sir John Houghton was co-chair of the science working group for the Intergovernmental Panel for Climate Change's Third Assessment Report. Relevant to Chapters 12 to 15.

Intergovernmental Panel on Climate Change. *Climate Change 2007: The Physical Science Basis.* IPCC 4. Cambridge, UK: Cambridge University Press, 2007. This is the science portion of the IPCC's fourth assessment report—a comprehensive survey that synthesizes a vast array of theory, data, and modeling results to describe the state of climate science. It also is available at www.ipcc.ch. Relevant to Chapters 12 to 15.

International Energy Agency. *Biofuels for Transport: An International Perspective.* Paris: Chirat, 2004. A well-organized wealth of technical information and statistics on biofuels. Relevant to Chapter 10.

Lovins, Amory B., E. K. Datta, O-E Bustnes, J. G. Koomey, and N. J. Glasgow. *Winning the Oil Endgame: Innovation for Profits, Jobs, and Security.* Snowmass, CO: Rocky Mountain Institute, 2005. This elaborately documented report shows how humankind might end its dependence on climate-changing fossil fuels. It also is available in full at www.oilendgame.com. Relevant to Chapters 6 to 11 and 16.

Manwell, J. F., J. G. McGowan, and A. L. Rogers. *Wind Energy Explained: Theory, Design, and Application.* Chichester, UK: Wiley, 2002. A thorough and mathematically detailed account of wind turbine engineering, this book also covers economics and environmental impacts. Relevant to Chapter 10.

National Research Council, and National Academy of Engineering. *The Hydrogen Economy: Opportunities, Costs, Barriers, and R&D Needs.* Washington, DC: National Academies Press, 2004. A thoughtful, serious study of the prospects for hydrogen energy produced by a diverse group ranging from industry executives to environmentalists, this publication contains a good mix of technology, economics, and policy. Unfortunately, there's no index. Relevant to Chapter 11.

Pahl, Greg. *Biodiesel: Growing a New Energy Economy.* White River Junction, VT: Chelsea Green Publishing, 2005. A look at the technology, policies, and worldwide spread of biodiesel fuel. Pahl is a biodiesel enthusiast, but he presents a realistic analysis of the challenges that biodiesel faces. Relevant to Chapter 10.

Pepper, Ian, Charles Gerba, and Mark Brusseau, eds. *Pollution Science.* San Diego: Academic Press, 1996. This textbook gives a thorough survey of pollution science at about the same level of *Energy, Environment, and Climate.* Especially relevant to Chapter 6.

Romm, Joseph J. *The Hype about Hydrogen: Fact and Fiction in the Race to Save the Planet.* Washington, DC: Island Press, 2004. A realistic assessment of hydrogen's prospects (it's far off) by a former assistant secretary in the U.S. Department of Energy's Office of Energy Efficiency and Renewable Energy. Romm sees fossil fuel shortages and climate change as looming events, and he warns that a hydrogen economy won't develop soon enough to avoid them. Relevant to Chapters 11 and 16.

Ruddiman, William. *Earth's Climate: Past and Future.* New York: W. H. Freeman, 2001. A beautifully illustrated textbook that is especially strong on paleoclimate. Relevant to Chapters 12 to 15.

Smil, Vaclav. *Energy: A Beginner's Guide*. Oxford, UK Oneworld Publications, 2006. A quick guide to energy that's particularly strong on energy history. Relevant to Chapters 1, 2, 6 to 11, and 16.

Smil, Vaclav. *Energy at the Crossroads: Global Perspectives and Uncertainty*. Cambridge, MA: MIT Press, 2003. A thorough, scholarly, but readable account of energy history, alternatives, and energy futures by an independent thinker. Relevant to Chapters 1, 2, 6 to 11, and 16.

Sørensen, Bent. *Renewable Energy: Its Physics, Engineering, Use, Environmental Impacts, Economy and Planning Aspects.* 2nd ed. London: Academic Press, 2000. A thorough, quantitative, and highly detailed account of the basic physics and engineering of renewable energy sources and technologies. Relevant to Chapters 7 to 11.

U.S. Department of Energy. *Solar Energy: Complete Guide to Solar Power and Photovoltaics, Practical Information on Heating, Lighting, and Concentrating, Energy Department Research.* This set of two CD-ROMs contains practical tips on solar energy, as well as a survey of U.S. Department of Energy research on solar technologies. Relevant to Chapter 9.

Weart, Spencer R. *The Discovery of Global Warming.* Cambridge, MA: Harvard University Press, 2004. An authoritative history of the science that has led to our present-day understanding of climate change. Relevant to Chapters 12 to 15.

Wolfson, Richard. *Nuclear Choices: A Citizen's Guide to Nuclear Technology*. Rev. ed. Cambridge, MA: MIT Press, 1993. Your author describes nuclear technology and policy issues associated with both nuclear power and nuclear weapons. Although somewhat dated, especially on weapons, it's useful for a simple explanation of nuclear technologies. Relevant to Chapter 7.

World Meteorological Organization. *Climate into the 21st Century.* Cambridge, UK: Cambridge University Press, 2003. A quick introduction to climate issues in a profusely illustrated coffee-table–style book. Relevant to Chapters 12 to 15.

Film/Video

Kohn, Walter, and Alan Heeger. *Power of the Sun.* DVD. Santa Barbara, CA: University of California at Santa Barbara, 2005. Nobel laureates Kohn and Heeger explore the promise of photovoltaic technology. The DVD includes a 22-minute film on the physics of photovoltaic devices and a 57-minute feature on photovoltaics and their applications.

Wolfson, Richard. *Earth's Changing Climate.* DVD. Chantilly, VA: The Teaching Company, 2007. In this twelve-lecture video course (also on audio), your author explores the science of climate and the role of humankind in the ongoing changes in Earth's climate. It's less mathematical than this book, with more emphasis on climate and less on energy.

Authoritative Web Sites

http://cdiac.esd.ornl.gov The Carbon Dioxide Information Analysis Center at Oak Ridge National Laboratory makes available data on concentrations of CO_2 and other greenhouse gases, as well as historical temperature records from a host of sources such as ice cores.

www.cru.uea.ac.uk The Climatic Research Unit of the University of East Anglia in England maintains one of the most comprehensive and accessible datasets on global temperature.

www.eia.doe.gov/emeu/aer/ The U.S. Department of Energy's Energy Information Administration posts the *Annual Energy Review* at this site; a new version appears each summer. The site also contains a wealth of tables on all aspects of energy production and use, mostly in the United States but also with a section on international energy.

www.eia.doe.gov The portal to the U.S. Department of Energy's Energy Information Administration gives access to a wealth of energy statistics including but not limited to those in the *Annual Energy Review*. The link under "Electricity" and then "Existing generating capacity" turns up, under the heading "More Tables . . . ," links entitled "Existing electric generating units in the United States" for several different years. These links bring up spreadsheets that are used in a number of research problems throughout this book.

www.epa.gov The home page of the U.S. Environmental Protection Agency. The "Clean Air Act" and "Climate Change" links are especially relevant; follow the latter link to find the latest inventory of greenhouse gas emissions.

www.iea.org The home page of the International Energy Agency offers links to a wide range of energy information. The "Statistics" link lets you extract desired data for any country or region, and the "Key Statistics" link therein gets you to a downloadable version of the IEA's *Key World Energy Statistics*, published yearly.

www.ipcc.ch The home page of the Intergovernmental Panel on Climate Change has links to the IPCC's comprehensive reports.

www.ncdc.noaa.gov/oa/ncdc.html The National Climatic Data Center, operated by the National Oceanographic and Atmospheric Administration, has a host of climate data ranging from individual weather stations to national and global averages.

www.noaa.gov The U.S. National Oceanic and Atmospheric Administration posts frequent updates and studies on climate and weather at this site.

www.nrel.gov The U.S. Department of Energy's National Renewable Energy Laboratory describes its many research programs on this site.

http://nsidc.org The National Snow and Ice Data Center, located at the University of Colorado's Cooperative Institute for Research in Environmental Sciences, is an affiliate of the National Oceanographic and Atmospheric Administration. It maintains data archives on snow and ice coverage as measured from satellites and ground surveys.

www.pewclimate.org Here the Pew Center on Climate Change posts its many reports on anthropogenic climate change and strategies for dealing with it. The Pew Center is especially strong on getting the business community involved with climate issues.

www.realclimate.org This site contains running commentaries by reputable climate scientists on climate change, including refutations of claims by "climate skeptics."

www.undp.org The home page of the United Nations Development Programme, which publishes the annual *Human Development Report* and makes it available on this site. The section on Human Development Data includes the Human Development Index, an alternative to gross national product.

ANSWERS TO ODD-NUMBERED EXERCISES

Chapter 1

1. 1.7×10^{17} W

3. In 1985: 1965 = 68 million; 1985 = 83 million; 2000 = 73 million

5. 10^{13} W

Chapter 2

1. $80

3. About 40%

5. 625 W

7. $3 per year per watt. This is the inverse of the energy intensity for the countries whose points lie near the line.

Chapter 3

1. 97 W

3. 3.3 TW

5. (a) 32 kW; (b) 4.8 kW

7. Reduction of 4 million barrels per day, or about one third of our imports

9. 26 minutes

11. 82 kW or 110 hp

Chapter 4

3. 25 gallons

5. (a) 94 kJ; (b) 105 s; (c) 156 s

7. (a) 170 PJ; (b) 18 days

9. Savings increase to $440 for the month

11. 22 mm^2

13. $e_{\text{winter}} = 0.53$, $P_{\text{winter}} = 1.3$ GWe; $e_{\text{summer}} = 0.48$, $P_{\text{summer}} = 1.2$ GWe

Chapter 5

1. 8 CO_2, 9 H_2O, 12.5 O_2

3. 79%

11. (b) 24 years; (c) 30 years; (d) 39 years

13. 18 years

Chapter 6

1. 8.5 kg, or about 19 pounds (A nice round number is 20 pounds per gallon of gasoline burned.)

3. (a) 10^{13} kg; (b) roughly 60 million tonnes

5. 1.57 kg

7. (a) 5.6 billion gallons; (b) ninety-one ships with the capacity of the Exxon *Valdez*

9. 49,000 tonnes

11. Tier 1: 0.96 kg; Tier 2: 0.12 kg

13. Increases by a factor of $\sqrt{10}$, or about 3.2

15. (a) 1.26 GW; (b) 13°C

Chapter 7

1. Cu-65

3. 0.08%

5. Using 29 MJ/kg for the energy content of coal, the three trainloads carry 8.7×10^{14} J.

7. $^{234}_{90}\text{Th} \rightarrow {}^{234}_{91}\text{Pa} + {}^{0}_{-1}e + \bar{\nu}$, $^{234}_{91}\text{Pa} \rightarrow {}^{234}_{92}\text{U} + {}^{0}_{-1}e + \bar{\nu}$, $^{234}_{92}\text{U} \rightarrow {}^{230}_{90}\text{Th} + {}^{4}_{2}\text{He}$, $^{230}_{90}\text{Th} \rightarrow {}^{226}_{88}\text{Ra} + {}^{4}_{2}\text{He}$

9. (a) about 108,000; (b) natural: about 88,000; artificial: about 20,000 (nearly all of which is from medical procedures)

Chapter 8

1. 43 TW
3. 67 mW/m^2
5. (a) 29%; (b) 76%
7. (a) 6.66; (b) 2.25 kW
9. 18%

Chapter 9

1. 1,361 to 1,372 W/m^2
3. About 40 minutes
5. 8.8 W/m^2/°C
7. 16 m^2
9. 76 minutes
11. (a) 24%; (b) maximum possible efficiency is 62%
13. 1%
15. (a) 450 MW; (b) 2,360 hectares, about 32% of the Cheviot mine

Chapter 10

1. Gravitational potential energy mgh = 30 kJ, about 1% of the latent energy
3. 2.9 GW. In fact, the Niagara power plants generate up to 4.2 GW at higher-than-average flows.
5. 29 m
7. 0.39
9. 0.42
13. (a) 250 billion kWh, about 7% of the total; (b) 29 GW, enough to displace 29 large power plants
15. 6.7%

Chapter 11

3. 15 TJ, capable of powering civilization for about 1 second
5. The fuel-cell car is responsible for twice as much CO_2 emission (0.96 kg CO_2 per mile, versus 0.5 kg CO_2 per mile for gasoline).
7. 2.8 pJ, in agreement with the reaction in Equation 11.5
9. 10^{20} particles per cubic meter

Chapter 12

1. 279 K
3. (a) 328 K; (b) 232 K
5. 190 K
7. It would increase by 1.8 K.
9. 700 W/m^2

Chapter 13

3. 1.35 K
5. (a) 1.7 W/m^2, slightly higher than the value 1.63 W/m^2 shown for CO_2 in the well-mixed greenhouse gas bar in Figure 13.4; (b) 3.7 W/m^2
9. About 20 tonnes, also roughly 20 English tons

Chapter 14

There are no exercises for this chapter.

Chapter 15

1. Transient: 0.64 °C per watt per square meter; long-term: 0.93 °C per watt per square meter
3. 7.2 m, the same as Table 15.1

   Research Problem 1. Arrow 2: 77 W/m^2; arrow 5: 168 W/m^2; arrow 6: 30 W/m^2. Solution: T_s = 293 K (20°C); T_a = 251 K (–22°C)

   Research Problem 2. (b) T_s = 274 K (1°C), lower than in Research Problem 1 since less energy is radiated from the surface because convection and evapotranspiration carry away some energy; (c) arrow 7 flow is 85 W/m^2, somewhat less than Figure 12.4's total of 102 W/m^2

Chapter 16

1. (a) 500,000 m^3; (b) 4.5 km^3
3. About 7 months
5. 163 years
7. (a) 130 billion kWh per year of electrical energy, or 15 GWe—the equivalent of fifteen large (1 GWe) power plants
9. 1954–2004: ~200 Gt; 2004–2054 under the "flat path": 350 Gt; stabilization triangle: 175 Gt

CREDITS AND DATA SOURCES

Abbreviations:

AER: Energy Information Administration, *Annual Energy Review* (Washington, DC: U.S. Department of Energy)

CDIAC *Trends:* *Trends: A Compendium of Data on Global Change,* Carbon Dioxide Information Analysis Center, Oak Ridge National Laboratory, United States Department of Energy (DOE)

EIA: United States Department of Energy, Energy Information Administration

EPA: United States Environmental Protection Agency

IAEA: International Atomic Energy Agency

IEA: Energy Information Administration, *International Energy Annual* (Washington, DC: U.S. Department of Energy)

IPCC 3: Intergovernmental Panel on Climate Change, *Climate Change 2001: The Scientific Basis* (Cambridge, UK: Cambridge University Press, 2001)

IPCC 4: Intergovernmental Panel on Climate Change, *Climate Change 2007: The Physical Science Basis* (Cambridge, UK: Cambridge University Press, 2007)

NASA: National Aeronautics and Space Administration

Figures
Chapter 1

Fig. 1.3 © John Sibbick.

Fig. 1.5 Adapted from L. D. Danny Harvey, *Global Warming: The Hard Science* (Harlow, England: Prentice Hall/Pearson Education, 2000), Fig. 1.1a; based on L. A. Frakes, *Climates Throughout Geologic Time* (New York: Elsevier, 1979).

Fig. 1.7 Adapted from Hoffman and Schrag, "Snowball Earth," *Scientific American* (Jan 2000), 72, stage 2 picture.

Fig. 1.8 Adapted and updated from Richard Wolfson, *Nuclear Choices: A Citizen's Guide to Nuclear Technology* (Cambridge, MA: MIT Press, 1993), 245; which in turn is adapted from Robert H. Romer, *Energy – An Introduction to Physics* (W. H. Freeman, 1976); Reprinted with permission.

Fig. 1.9 Data from United States Census Bureau, www.census.gov/ipc/www/worldpop.html.

Chapter 2

Fig. 2.1 Tad Merrick.

Fig. 2.3 Energy data from Vaclav Smil, "Energy in the Twentieth Century: Resources, Conversions, Costs, Uses, and Consequences," Table 1, p. 24 in *Annual Review of Energy and the Environment 2000*, vol. 25 (Palo Alto, CA Annual Reviews, 2000) and IEA (2003), Table E1; Population data from United Nations, www.un.org/esa/population/publications/sixbillion/sixbilpart1.pdf.

Fig. 2.4 Data from AER (2004), Table 2.1a (2004 data).

Fig. 2.5 Data from AER (2004), Table 1.3.

Fig. 2.6 Data from IEA (2003), Table 1.8.

Fig. 2.7 Data from International Energy Agency, *Key World Energy Statistics 2004* (Paris: International Energy Agency, 2004), 48ff.

Fig. 2.8 U.S. energy 1900–1945 from AER (2005), Table E1; 1950–2004 from AER (2005), Table 1.1. U.S. real gross domestic product (in 2000 dollars) from Economic History Services, http://eh.net/hmit/gdp/. World energy and gross world product (in 1990 dollars) from Vaclav Smil, "Energy in the Twentieth Century: Resources, Conversions, Costs, Uses, and Consequences," *Annual Review of Energy and the Environment 2000*, vol. 25 (Palo Alto, CA: Annual Reviews, 2000), Table 1, p. 24, adjusted to 2000 dollars using conversion factor 1.23 obtained from Economic History Services calculator at http://eh.net/hmit/compare/ using GDP deflator.

Fig. 2.9 Energy data from International Energy Agency, *Key World Energy Statistics 2004* (Paris: International Energy Agency, 2004), p. 48ff. Life expectancy data from *Human Development Report 2005* (New York: UN Development Programme, 2005), Table 10, p. 250ff, http://hdr.undp.org/reports/global/2005/.

Chapter 3

Fig. 3.3 Data from AER (2004), Tables 1.1 and 8.1.

Fig. 3.5b,c Peter Bowater / Photo Researchers, Inc.

Fig. 3.6   Data from AER 2004, Table 8.2a.
Fig. 3.9   G. R. "Dick" Roberts, © Natural Sciences Image Library.

Chapter 4

Fig. 4.5   Courtesy Dow Building Solutions.
Fig. 4.7b   Courtesy Manuel G. Velarde, Instituto Pluridisciplinar, Universidad Complutense, Madrid, Spain.
Fig. 4.14   © Comet, Zürich.
Fig. 4.15   Courtesy Michael Moser, Assistant Director of Facilities Services, Middlebury College.
Fig. 4.17   From G. M. Whitesides et al., Science 315: 796–798 (9 February 2007). Reprinted with permission from AAAS.

Chapter 5

Fig. 5.1   Ted Clutter / Photo Researchers, Inc.
Fig. 5.2   Loosely adapted from web-based animation associated with Stephen Marshak's *Essentials of Geology*, (W. W. Norton, 2006).
Fig. 5.4   Data from AER (2004), Table 5.11 (2004 data).
Fig. 5.5   Pamela A. Miller, Northern Alaska Environmental Center.
Fig. 5.6   Data from Worldwatch Institute, *Vital Signs 2006–2007* (New York: W. W. Norton, 2006), p. 33.
Fig. 5.7   Data from Natural Gas Supply Association, www.naturalgas.org/environment/naturalgas.asp.
Fig. 5.8   © Colin Garratt; Milepost $92^1/_2$/Corbis.
Fig. 5.9   (top right): © G. Bowater/Corbis.
Fig. 5.9   (bottom right): © Lowell Georgia/Corbis.
Fig. 5.9   (top left): G. R. "Dick" Roberts, © Natural Sciences Image Library.
Fig. 5.9   (bottom left): G. R. "Dick" Roberts, © Natural Sciences Image Library.
Fig. 5.9   (middle right): G. R. "Dick" Roberts, © Natural Sciences Image Library.
Fig. 5.10   © Earl Richardson.
Fig. 5.11   Courtesy Chevron Corporation.
Fig. 5.12   Adapted from Marc Ross, "Fuel Efficiency and the Physics of Automobiles" (revised paper, Physics Department, University of Michigan), pp. 16–17, http://sitemaker.umich.edu/mhross/files/fueleff_physicsautossanders.pdf.
Fig. 5.13a Leonard Lessin / Photo Researchers, Inc.
Fig 5.13b   Courtesy Toyota Motor Sales, USA.
Fig. 5.14   Data from U.S. Department of Energy and EPA at www.fueleconomy.gov (2006 models).
Fig. 5.15   Rolls-Royce plc.
Fig. 5.17   Data from AER (2005), Table 11.13 (2002 data).
Fig. 5.18   Data from AER (2005), Table 11.4 (2004 data).
Fig. 5.19   Data from AER (2005), Table 5.1 (2004 data).
Fig. 5.21   G. R. "Dick" Roberts, © Natural Sciences Image Library.

Fig. 5.22a Sarah Leen / National Geographic Image Collection.

Fig. 5.22b Sarah Leen / National Geographic Image Collection.

Fig. 5.25 Historical data from AER (2005), Table 11.5, except 1950 and 1955 from Worldwatch Institute, *Vital Signs 2006–2007* (New York: W. W. Norton, 2006), p. 33. Other curves calculated using methodology described in Robert L. Hirsch, Roger Bezdek, and Robert Wendling, "Peaking of World Oil Production: Impacts, Mitigation, and Risk Management" (online publication, U.S. Department of Energy National Energy Technology Laboratory, February 2005), www.netl.doe.gov/publications/others/pdf/Oil_Peaking_NETL.pdf.

Fig. 5.26 Data from Gerhard P. Metschies, *International Fuel Prices 2005: 172 Countries*, 4th ed. (Eschborn, Germany: Deutsche Gesellschaft für Technische Zusammenarbeit, 2005), www.gtz.de/de/dokumente/en_International_Fuel_Prices_2005.pdf; gasoline prices (US¢ per liter) from individual area maps; taxes from Table 12.5, pp. 104–105 (2004 data).

Chapter 6

Fig. 6.1 Adapted with permission from images supplied by Volkswagen of America, Inc.

Fig. 6.2 Data from "Power Plant Emissions: Particulate Matter-Related Health Damages and the Benefits of Alternative Emission Reduction Scenarios" (Boston: Clean Air Task Force, June 2004), Table 6-1, www.catf.us/publications/reports/Power_Plant_Emissions.pdf. Tall bars are all cases attributable to power plants ("no EGU" column); short bars subtract the Jeffords reductions (last column).

Fig. 6.3 Archive Holdings Inc./Getty Images.

Fig. 6.4b Courtesy Rees-Memphis, Inc.

Fig. 6.5b Photo courtesy FLSmidth & Co.

Fig. 6.6 National Atmospheric Deposition Program (NRSP-3), 2006. NADP Program Office, Illinois State Water Survey, 2204 Griffith Drive, Champaign, IL 61820 (2004 data), http://nadp.sws.uiuc.edu/isopleths/maps2004/phfield.pdf.

Fig. 6.7 Data from EPA, www.epa.gov/airtrends/2005/pdfs/CONational.pdf (2003 data).

Fig. 6.8 Data from EPA, www.epa.gov/airtrends/2005/pdfs/NOXNational.pdf (2003 data).

Fig. 6.10a Peggy & Yoram Kahana / Peter Arnold, Inc.

Fig. 6.10b Peggy & Yoram Kahana / Peter Arnold, Inc.

Fig. 6.11 Data from California Air Resources Board, www.arb.ca.gov/aqmis2/paqdselect.php. Data is for all day Thursday, January 5, 2006, at North Main Street location in Los Angeles County, with downloads for specific pollutants.

Fig. 6.12 The Image Bank/Getty (Creative) Images

Fig. 6.13 EPA, *Mercury Maps: A Quantitative Spatial Link Between Air Deposition and Fish Tissue* (EPA-823-R-01-009), September 2001, www.epa.gov/waterscience/models/maps/report.pdf.

Fig. 6.14 Photo C Vivian Stockman / www.ohvec.org. Flyover courtesy SouthWings.

Fig. 6.15 AP Images/ Glenwood Springs Post Independent, Mike Brinson.

Fig. 6.16 © Corbis.

Fig. 6.17 Sarah Leen / National Geographic Image Collection.

Fig. 6.18a ESA/Getty Images.

Fig. 6.18b   © Reuters/Corbis.

Fig. 6.19   Data from International Tanker Owners Pollution Federation, www.itopf.com/stats.html.

Fig. 6.20   Data from EPA Office of Mobile Sources, *Emission Facts: The History of Reducing Tailpipe Emissions* (EPA420-F-99-017), May 1999, www.epa.gov/oms/consumer/f99017.pdf.

Fig. 6.21   Data from EPA, www.epa.gov/airtrends/2005/econ-emissions.html.

Fig. 6.22   Map generated on 1/8/06 from EPA, www.epa.gov/air/data/nonat.html?us~usa~ United%20States, for eight-hour ozone; nonattainment counties shown as of April 2005.

Chapter 7

Fig. 7.1   National data from IAEA, *Energy, Electricity, and Nuclear Power Estimates for the Period Up to 2030,* July 2003 ed. (IAEA-RDS-1/24), 2004. State data from Nuclear Energy Institute, www.nei.org/documents/State_by_State_Electricity_Fuel_Shares_2003.pdf.

Fig. 7.2   Data from IAEA, *Energy, Electricity, and Nuclear Power Estimates for the Period Up to 2030,* July 2003 ed. (IAEA-RDS-1/24), 2004, Table 1.

Fig 7.6   Courtesy Entergy Vermont Yankee.

Fig 7.9   Courtesy of General Electric Nuclear Energy.

Fig 7.11   Courtesy of Pennsylvania Power & Light Corporation.

Fig. 7.14   Fig. 7.14: National Council on Radiation Protection and Measurements, *Ionizing Radiation Exposure of the Population of the United States* (Report No. 93).

Fig. 7.16   From "The Disposal of Radioactive Wastes from Fission Reactors," by Bernard L. Cohen. Copyright © by Scientific American, Inc. All rights reserved; Reprinted with permission.

Fig. 7.17   Courtesy of Entergy Vermont Yankee.

Fig. 7.18   Adapted from Richard Garwin and Georges Charpak, *Megawatts and Megatons: The Future of Nuclear Power and Nuclear Weapons* (Chicago: University of Chicago Press, 2001), 150. *We have made every effort to contact the copyright holder to obtain permission to reprint this selection. If you have information that would help us, please write to W. W. Norton & Company, Inc., 500 Fifth Avenue, New York, NY 10110, Attn: Permissions Department.*

Fig. 7.19   Data from EIA, *International Energy Outlook 2006* (DOE/EIA-0484[2006]), Figure 10 data.

Chapter 8

Fig. 8.1   Adapted from Wendell A. Duffield and John H. Sass, "Geothermal Energy: Clean Power from the Earth's Heat," *U.S. Geological Survey Circular* 1249 (2003).

Fig. 8.2   Adapted from United States Department of Energy, Office of Energy Efficiency and Renewable Energy, "U.S. Geothermal Resource Map," www.eere.energy.gov/geothermal/geomap.html.

Fig. 8.3   © Bo Zaunders/Corbis.

Fig. 8.4   Courtesy Geo-Heat Center, Oregon Institute of Technology.

Fig. 8.5   Data from Energy & Geoscience Institute, University of Utah, "Geothermal Energy" brochure (2001), www.egi.utah.edu/geothermal/GeothermalBrochure.pdf.

Fig. 8.7   Courtesy of United States Department of Energy.

Fig. 8.8   Adapted from California Energy Commission, "Santa Rosa Geysers Recharge Project: GEO-98-001, Final Report," (500-02-078V1), October 2002.

Fig. 8.10   Geothermal Resources Council.

Fig. 8.14   Courtesy Hopewell Rocks, Bay of Fundy, New Brunswick.

Fig. 8.15   Adapted from Bent Sørensen, *Renewable Energy*, 2nd ed. (San Diego: Academic Press, 2000), 280.

Fig. 8.16   © Photofusion Picture Library / Alamy.

Fig. 8.17   Courtesy Ocean Power Delivery, Ltd.

Chapter 9

Fig. 9.2   Courtesy of Bill Marshall.

Fig. 9.4   Theoretical curves adapted from John A. Duffie and William A. Beckman, *Solar Energy Thermal Processes* (New York: Wiley, 1974), 44; Reprinted with permission. City values from NASA, "Surface Meteorology and Solar Energy Data Set for Renewable Energy Use," Release 3 (October 2000), www.apricus-solar.com/html/solar_collector_insolation.htm.

Fig. 9.5   Adapted from National Renewable Energy Laboratory.

Fig. 9.6   Data from National Renewable Energy Laboratory, *Solar Radiation Data Manual for Flat-Plate and Concentrating Collectors* (April 1994), http://rredc.nrel.gov/solar/pubs/redbook/.

Fig. 9.7   Data from Peter Foukal, *Solar Astrophysics* (New York: Wiley, 1990), 68, Table 3.1.

Fig. 9.11   Adapted from Duffie & Beckman, *Solar Energy Thermal Processes* (New York: Wiley, 1974), 170. Reprinted with permission.

Fig. 9.13   Tad Merrick.

Fig. 9.16a   Courtesy Australian National University.

Fig. 9.16b   Courtesy Stirling Energy Systems.

Fig. 9.17   Sandia National Laboratories.

Fig. 9.19   Hank Morgan / Photo Researchers, Inc.

Fig. 9.22   Tommaso Guicciardini / Photo Researchers, Inc.

Fig. 9.23   Photograph courtesy of BP Solar.

Fig. 9.24   www.solarhouse.com.

Fig. 9.25   Courtesy Tucson Electric Power.

Fig. 9.26   Courtesy of International Energy Agency Photovoltaic Power Systems Programme.

Fig. 9.28   1980–1990 data from Richard Wolfson, *Nuclear Choices: A Citizen's Guide to Nuclear Technology* (Cambridge, MA: MIT Press, 1993), Fig. 11.19; 1990–2002 data from AER (2003), Table 10.5; 2002–2005 data from www.solarbuzz.com/moduleprices.htm. Data values have been scaled to agree with the U.S. Department of Energy values at the 1990 and 2002 overlap years.

Fig. 9.29   Fossil energy data from United States Department of Energy (DOE), Fossil Energy Fiscal Year 2006 Budget, www.fossil.energy.gov/aboutus/budget/06/FY2006_Budget_.html. Renewable energy data from U.S. Department of Energy, Energy Efficiency and Renewable Energy Fiscal Year 2006 Budget, www1.eere.energy.gov/ba/pba/budget_06.html.

Chapter 10

Fig. 10.1   Data from AER (2004), Tables 1.2 and 10.1.

Fig. 10.2   Data from IEA (2003), Table 6.3.

Fig. 10.3   Data from United Nations Educational, Scientific and Cultural Organization (UNESCO), www.unesco.org/science/waterday2000/Cycle.htm.

Fig. 10.4   Data from World Energy Council, *Survey of Energy Resources* (1999 data), www.worldenergy.org/wec-geis/publications/reports/ser/hydro/hydro.asp.

Fig. 10.6   Photo Jonas N. Jordan / United States Army Corps of Engineers.

Fig. 10.7a   Bill Johnson / United States Army Corps of Engineers.

Fig. 10.7b   © Corbis.

Fig. 10.9   Courtesy of FirstLight Power Resources.

Fig. 10.12   National Renewable Energy Laboratory, *Wind Energy Resource Atlas of the United States,* Map 2.1, accessed through http://rredc.nrel.gov/wind/pubs/atlas/maps.html.

Fig. 10.13a   © tomroster.com.

Fig. 10.13b   © Paul Gipe.

Fig. 10.16a   © William Manning Photography / Alamy.

Fig. 10.16b   © Bjorn Svensson / Alamy.

Fig. 10.17   Data from Wind Service Holland, *Wind Energy Statistics Worldwide,* http://home.wxs.nl/~windsh/stats.html.

Fig. 10.18   Data from Global Wind Energy Council brochure, p. 4, www.gwec.net/fileadmin/documents/Publications/GWEC_brochure_2006.pdf.

Fig. 10.19   Data from Michael Pidwirny, *Fundamentals of Physical Geography,* 2nd ed. (online textbook; 2006), Table 9l-1, www.physicalgeography.net/fundamentals/9l.html. Cornfield data from Jim Johnston and Matt Bowman, "Performance of Conventional, Dwarf, Grazing, and Silage Corn at New Liskeard," New Liskeard Agricultural Research Station, Ontario, Canada.

Fig. 10.20   Courtesy of Burlington Electric Company.

Fig. 10.21   Adapted from International Energy Agency, *Biofuels for Transport: An International Perspective* (Paris: Chirat, 2004), Figure 1.1.

Fig. 10.22   Ibid., Figure 1.2.

Fig. 10.23   Courtesy Luis A. Vega, Ph.D.

Chapter 11

Fig. 11.1   Courtesy of NASA.

Fig. 11.4   Data from U.S. Department of Energy, Energy Efficiency and Renewable Energy, "Hydrogen Storage," *2004 Annual Hydrogen Program Review,* p. 8, www.eere.energy.gov/hydrogenandfuelcells/pdfs/review04/st_1_miliken.pdf. Gasoline datum calculated using 36 kWh/gallon from Table 3.3.

Fig. 11.5   Courtesy of Honda Motor Company, Inc.

Fig. 11.6   Adapted from *The Hydrogen Economy: Opportunities, Costs, Barriers, and R&D Needs* (Washington, DC: National Academies Press, 2004), 67.

Fig. 11.9 Photo courtesy of the United States Department of Energy (DOE).

Fig. 11.10a Photo courtesy of Princeton Plasma Physics Laboratory (PPPL).

Fig. 11.10b Published with permission of ITER.

Fig. 11.12 Adapted from United States Department of Energy (DOE), Office of Science, "Fusion Energy Sciences" brochure, April 2004.

Chapter 12

Fig. 12.4 *Climate Change 2007: The Physical Science Basis*, Intergovermental Panel on Climate Change, p. 96. Reprinted with permission.

Fig. 12.5 Courtesy of Lunar and Planetary Institute, NASA.

Chapter 13

Fig. 13.4 Data from Intergovernmental Panel on Climate Change, (IPCC) "Summary for Policymakers," Fourth Assessment Report (February 2007), Fig. SFP-2.

Fig. 13.5 Carbon dioxide data from G. Marland, T. A. Boden, and R. J. Andres, "Global, Regional, and National CO_2 Emissions," in CDIAC *Trends* (2006), http://cdiac.esd.ornl.gov/trends/emis/em_cont.htm (2002 data). Land-use data from R. A. Houghton and J. L. Hackler, "Carbon Flux to the Atmosphere from Land-Use Changes," in CDIAC *Trends* (2002), http://cdiac.esd.ornl.gov/trends/landuse/houghton/houghton.html (2000 data).

Fig. 13.6 1750–1992 methane data from D. M. Etheridge, L. P. Steele, R. J. Francey, and R. L. Langenfelds, "Historical CH_4 Records Since About 1000 A.D. from Ice Core Data," in CDIAC *Trends* (2002), http://cdiac.esd.ornl.gov/ftp/trends/atm_meth/EthCH498B.txt; 1992–2001 data from L. P. Steele, P. B. Krummel, and R. L. Langenfelds, "Atmospheric CH_4 Concentrations from Sites in the CSIRO Atmospheric Research GASLAB Air Sampling Network" (October 2002 version), in CDIAC *Trends* (2002), http://cdiac.esd.ornl.gov/trends/atm_meth/csiro/csiro-mloch4.html (the 1992 datum is an average of both datasets). 1978–1995 HFC-23 data from D. E. Oram, W. T. Sturges, S. A. Penkett, et al., "Atmospheric Fluoroform (CHF_3, HFC-23) at Cape Grim, Tasmania," in CDIAC *Trends* (2000), http://cdiac.esd.ornl.gov/ftp/trends/otheratg/oram/oramdata.txt; 1998 data from Atsushi Kurosawa, "Climate Target: Multigas and CO_2 Sequestration" (presentation, Workshop on GHG Stabilization Scenarios, Tsukuba, Japan, January 22–23, 2004), http://www-iam.nies.go.jp/aim/AIM_workshop/GHG/KUROSAWA.pdf (1960 and 1970 HFC-23 data are estimated; 1996–1997 data are interpolated).

Fig. 13.7 Main graph data from A. Neftel, H. Friedli, E. Moor, et al., "Historical CO_2 Record from the Siple Station Ice Core," in CDIAC *Trends* (1994), http://cdiac.ornl.gov/ftp/trends/co2/siple2.013. Mauna Loa inset data from C. D. Keeling and T. P. Whorf, "Atmospheric CO_2 Records from Sites in the SIO Air Sampling Network," in CDIAC *Trends* (2005), http://cdiac.ornl.gov/ftp/trends/co2/maunaloa.co2 (2005 data estimated).

Fig. 13.8a 1010–1975 data from D. M. Etheridge, L. P. Steele, R. L. Langenfelds, et al., "Historical CO_2 Record from the Law Dome DE08, DE08-2, and DSS Ice Cores," in CDIAC *Trends*

(1998), http://cdiac.ornl.gov/ftp/trends/co2/lawdome.combined.dat; 1976–2004 data from C. D. Keeling and T. P. Whorf, "Atmospheric CO_2 Records from Sites in the SIO Air Sampling Network," in CDIAC *Trends* (2005), http://cdiac.ornl.gov/ftp/trends/co2/maunaloa.co2.

Fig. 13.8b   Vostok ice core data from J. M. Barnola, D. Raynaud, C. Lorius, and N. I. Barkov, "Historical CO_2 Record from the Vostok Ice Core," in CDIAC *Trends* (2003), http://cdiac.ornl.gov/ftp/trends/co2/vostok.icecore.co2.

Fig. 13.9   NASA, Goddard Institute for Space Studies.

Fig. 13.10   Data (rounded to integers) assembled from IPCC 3 and IPCC 4; John Houghton, *Global Warming,* 3rd ed. (Cambridge, UK: Cambridge University Press, 2004); L. D. Danny Harvey, *Global Warming: The Hard Science* (Harlow, England: Prentice-Hall/Pearson Education, 2000); G. Marland, T. A. Boden, and R. J. Andres, "Global, Regional, and National CO_2 Emissions," in CDIAC *Trends* (2006), http://cdiac.esd.ornl.gov/trends/emis/em_cont.htm; R. A. Houghton and J. L. Hackler, "Carbon Flux to the Atmosphere from Land-Use Changes," in CDIAC *Trends* (2002), http://cdiac.esd.ornl.gov/trends/landuse/houghton/houghton.html; private communication from Dr. Peter Vitousek, Stanford University; Woods Hole Oceanographic Institute, www.whrc.org/carbon/. Data have been adjusted for consistency.

Fig. 13.11   Data from G. Marland, T. A. Boden, and R. J. Andres, "Global, Regional, and National CO_2 Emissions," in CDIAC *Trends* (2006), http://cdiac.esd.ornl.gov/trends/emis/em_cont.htm.

Fig. 13.12   Plotted using carbon dioxide pulse decay equation from IPCC 4, Table TS-2, footnote (a).

Chapter 14

Fig. 14.1   Data from University of East Anglia, Climatic Research Unit, Global (GL) data from HadCRUT3v dataset, www.cru.uea.ac.uk/cru/data/temperature/; error bars estimated from IPCC 3, Fig. 2.1.

Fig. 14.2   Based on map generated with NASA Goddard Institute for Space Studies Surface Temperature Analysis, http://data.giss.nasa.gov/gistemp/maps/, using the following parameters: GISS 2001 for land, Had/Reyn_v2 for oceans, map type: anomalies, time interval: 2005–2005, base period: 1961–1990, smoothing radius: 1,200 km.

Fig. 14.3   From James Hansen, Larissa Nazarenko, Reto Ruedy, et al., "Earth's Energy Imbalance: Confirmation and Implications," *Science* 308 (June 3, 2005): 1433; Reprinted with permission from AAAS.

Fig. 14.4   Data for top two plots from Remote Sensing Systems, MSU and AMSU Data, version 3.0 (February 2007), www.ssmi.com/msu/msu_data_description.html#msu_amsu_time_series. Bottom plot is a portion of Fig. 14.1.

Fig. 14.5   Robert A. Rohde / Global Warming Art, http://en.wikipedia.org/wiki/Image:1000_Year_Temperature_Comparison.png.

Fig. 14.6   Data from J. R. Petit, D. Raynaud, and C. Lorius, "Historical Isotopic Temperature Record from the Vostok Ice Core," in CDIAC *Trends* (2000), http://cdiac.ornl.gov/trends/temp/vostok/jouz_tem.htm.

Fig. 14.7   United States Geological Survey, http://pubs.usgs.gov/gip/continents/map.jpg.

Fig. 14.8   Carbon dioxide data from J. M. Barnola, D. Raynaud, C. Lorius, and N. I. Barkov, "Historical CO_2 Record from the Vostok Ice Core," in CDIAC *Trends* (2003), http://cdiac.ornl.gov/ftp/trends/co2/vostok.icecore.co2. Temperature data from J. R. Petit, D. Raynaud, and C. Lorius, "Historical Isotopic Temperature Record from the Vostok Ice Core," in CDIAC *Trends* (2000), http://cdiac.ornl.gov/trends/temp/vostok/jouz_tem.htm.

Fig. 14.9   Data from R. B. Alley, "GISP2 Ice Core Temperature and Accumulation Data," International Geosphere-Biosphere Progamme (IGBP) PAGES/World Data Center for Paleoclimatology, Data Contribution Series #2004-013, NOAA/NGDC Paleoclimatology Program, Boulder, CO (2004).

Fig. 14.10   United States Geological Survey.

Fig. 14.11a NASA, Goddard Institute for Space Studies, www.nasa.gov/centers/Goddard/news/topstory/2005/arcticice_decline.html.

Fig. 14.11b Data from User Services, National Snow and Ice Data Center, Boulder, CO.

Fig. 14.12   Data from Kerry Emanuel, Massachusetts Institute of Technology, Cambridge, MA, http://wind.mit.edu/~emanuel/Papers_data_graphics.htm.

Fig. 14.13   Adapted from Gerald A. Meehl, Warren M. Washington, Caspar M. Ammann, et al., "Combinations of Natural and Anthropogenic Forcings in Twentieth-Century Climate," *Journal of Climate* 17, no. 19 (2004): 3723. © American Meteorological Society (AMS). Reprinted with permission.

Chapter 15

Fig. 15.2   Adapted from William F. Ruddiman, *Earth's Climate: Past and Future* (New York: W. H. Freeman, 2001), p. 74.

Fig. 15.3   Adapted from William F. Ruddiman, *Earth's Climate: Past and Future* (New York: W. H. Freeman, 2001), p. 74.

Fig. 15.5   Data from Chris Carothers, Intel Corporation, www.cs.rpi.edu/~chrisc/COURSES/CSCI-4250/ SPRING-2004/slides/cpu.pdf.

Fig. 15.6   Adapted from IPCC 3, Fig. 9.1.

Fig. 15.7   Adapted from IPCC 3, Fig. 8.8a.

Fig. 15.8   Adapted from IPCC 3, Fig. 8.3.

Fig. 15.9   Intergovernmental Panel on Climate Change, *Climate Change 1995: The Science of Climate Change* (Cambridge, UK: Cambridge University Press, 1996), p. 258.

Fig. 15.10   Adapted from IPCC 3, Fig. 9.3a.

Fig. 15.11   Adapted from IPCC 3, Fig. 9.16.

Fig. 15.13   Adapted from IPCC 3, Fig. TS-17.

Fig. 15.14   Adapted from IPCC 4, Fig. SPM-5.

Fig. 15.15   Adapted from IPCC 4, Fig. SPM-6.

Fig. 15.17   Data from IPCC 4, Table SPM-3.

Fig. 15.18   Courtesy of J. Weiss, Environmental Studies Laboratory, Department of Geosciences, The University of Arizona. Reprinted with permission.

Chapter 16

Fig. 16.1   Adapted from IPCC 3, Fig. 3.12b.

Fig. 16.2   Data from AER (2005), Table 12.2 (2003 data).

Fig. 16.3   Intergovernmental Panel on Climate Change, "Summary for Policymakers," *Carbon Dioxide Capture and Storage* (Geneva: IPCC, 2005), Fig. SPM.2.

Fig. 16.4   All data except nuclear high from Joseph V. Spadaro, Lucille Langlois, and Bruce Hamilton, "Greenhouse Gas Emissions of Electricity Generation Chains: Assessing the Difference," *IAEA Bulletin* 42, no. 2 (2000): 21; nuclear high is based on an assessment of several studies, including Uwe R. Fritsche, "Comparing Greenhouse-Gas Emissions and Abatement Costs of Nuclear and Alternative Energy Options from a Life-Cycle Perspective" (lecture, CNIC Conference on Nuclear Energy and Greenhouse-Gas Emissions, Tokyo, November 1997), p. 5, www.oeko.de/service/gemis/files/info/nuke_co2_en.pdf.

Fig. 16.5   Data from Worldwatch Institute, *Vital Signs 2005* (New York: Norton, 2005), 31.

Fig. 16.6   Data for cars from EPA, Honda Motor Company, Volkswagen of America; data for refrigerators from EPA.

Fig. 16.7   Energy data from AER (2005), Table 1.1. Real gross domestic product (in 2000 dollars) from Economic History Services, http://eh.net/hmit/gdp/.

Fig. 16.8   Data from AER (2004), Table 5.1.

Fig. 16.9   Data from Pat McAuliffe, California Energy Commission (4/26/06).

Fig. 16.10   Adapted from Arthur Rosenfeld, Pat McAuliffe, and John Wilson, "Energy Efficiency and Climate Change," *Encyclopedia of Energy*, vol. 2 (Boston: Elsevier Press, 2004), p. 373.

Fig. 16.11   Office of Transportation and Air Quality, *Light-Duty Automotive Technology and Fuel Economy Trends: 1975 through 2005*, EPA420-S-05-001 (Washington, DC: EPA, July 2005), p. iii.

Fig. 16.12   Courtesy of The Rocky Mountain Institute.

Fig. 16.13   Data from IEA (2003), Tables E.1 and H.1 (2003 data).

Fig. 16.14   Vaclav Smil, *Energy at the Crossroads: Global Perspectives and Uncertainties* (Cambridge, MA: MIT Press, 2003), Fig. 3.27, p. 171.

Fig. 16.15   Adapted from Robert Socolow, "Stabilization Wedges: Mitigation Tools for the Next Half-Century" (lecture, Avoiding Dangerous Climate Change: A Scientific Symposium on Stabilisation of Greenhouse Gases, Exeter, UK, February 3, 2005); based on S. Pacala and R. Socolow, "Stabilization Wedges: Solving the Climate Problem for the Next 50 Years with Current Technologies," *Science* 305, no. 5686 (August 13, 2004): 968–972.

Fig. 16.16   Ibid.

INDEX

Page numbers in italics refer to figures and illustrations.